Metallic Nanoparticles for Health and the Environment

Metallic Nanoparticles for Health and the Environment covers different routes of synthesis for metallic nanoparticles and their process variables. Both the functions and roles of these particles as a drug delivery system and diagnostic agent and other potential theranostic purposes against metabolic disorders, photocatalysis applications, as well as wastewater treatments, are discussed. The book compares the different properties of bulk metallic forms and their nanoparticulated forms. It discusses the mechanisms and impacts of different process variables in different synthesis routes, as well as emerging trends in clinics and so forth.

Features:

- Covers different routes of synthesis to create metallic nanoparticles (MNPs) of different characteristics with reference to bulk forms of metals.
- Describes formulation parameters that have a significant effect on these MNPs including dimensions, morphology, mechanism, surface properties, and other characteristics.
- Discusses different roles and performances of MNPs in photothermal therapy, metabolic disorders, mechanisms in bacterial, fungal, and viral infections, and inflammatory pathways.
- Reviews the potential and emerging roles of different MNPs with site target delivery applications and genetic manipulation purposes.
- Examines the advantages and challenges of these MNPs against remediation of pollutants and toxicants, owing to their superior surface catalytic activities.

This book is aimed at researchers and professionals in nanomaterials, pharmaceuticals, and drug delivery.

Advances in Bionanotechnology

Series Editors: Ravindra Pratap Singh, *Department of Biotechnology, Indira Gandhi National Tribal University, Anuppur, Madhya Pradesh, India*, **Jay Singh,** *Department of Chemistry, Institute of Science, Banaras Hindu University, Varanasi, Uttar Pradesh, India* **and Charles Oluwaseun Adetunji,** *Department of Microbiology, Edo State University Uzairue, Iyamho, Edo State, Nigeria*

Bionanotechnology is a multi-disciplinary field that shows immense applicability in different domains, namely chemistry, physics, material sciences, biomedical, agriculture, environment, robotics, aeronautics, energy, electronics and so forth. This book series will explore the enormous utility of bionanotechnology for biomedical, agricultural, environmental, food technology, space industry, and many other fields. It aims to highlight all the spheres of bionanotechnological applications and its safety and regulations for using biogenic nanomaterials that are a key focus of the researchers globally.

Bionanotechnology Towards Sustainable Management of Environmental Pollution
Edited by Naveen Dwivedi and Shubha Dwivedi

Natural Products and Nano-formulations in Cancer Chemoprevention
Edited by Shiv Kumar Dubey

Bionanotechnology towards Green Energy
Innovative and Sustainable Approach
Edited by Shubha Dwivedi and Naveen Dwivedi

Biotic Stress Management of Crop Plants using Nanomaterials
Edited by Krishna Kant Mishra and Santosh Kumar

Metallic Nanoparticles for Health and the Environment
Edited by Md Sabir Alam, Md Noushad Javed and Jamilur R. Ansari

For more information about this series, please visit: www.routledge.com/Advances-in-Bionanotechnology/book-series/CRCBIONAN

Metallic Nanoparticles for Health and the Environment

Edited by
Md Sabir Alam,
Md Noushad Javed, and
Jamilur R. Ansari

CRC Press is an imprint of the
Taylor & Francis Group, an **informa** business

Designed cover image: © Shutterstock

First edition published 2024
by CRC Press
6000 Broken Sound Parkway NW, Suite 300, Boca Raton, FL 33487–2742

and by CRC Press
4 Park Square, Milton Park, Abingdon, Oxon, OX14 4RN

CRC Press is an imprint of Taylor & Francis Group, LLC

ISBN: 978-1-032-32912-3 (hbk)
ISBN: 978-1-032-32913-0 (pbk)
ISBN: 978-1-003-31731-9 (ebk)

DOI: 10.1201/9781003317319

Typeset in Times
by Apex CoVantage, LLC

Contents

Preface

In Chapter 1, we discuss Metallic (Inorganic) Nanoparticles: Classification, Synthesis, Mechanism, and Scope. Nanoscience has become a discipline that focuses on the synthesis of metal nanoparticles with sizes ranging from 1 to 100 nanometers, along with particle shape and scale modification, employing numerous synthesis processes. In Chapter 2, we discuss recent trends in the Biomedical Application of Metallic Nanoparticles. These unique characteristics and their similarity in size with biomolecules make them suitable candidates for various biomedical applications that include biosensing, imaging, therapy, and drug delivery. Noble MNPs have been used in the diagnosis and treatment of human diseases, including anticancer, diagnostic assays, drug delivery, bioimaging, biosensing, photoablation, hyperthermia, and gene delivery, due to their distinctive characteristics and exceptional ability. In Chapter 3, we discuss Application of Metallic Nanoparticles in Targeted Delivery and Genetic Manipulations. Recently, scientists have been focusing on hybrid nanoparticles, either by coating them or by combing them with other dosage carriers to achieve targeted delivery of the drug to the desired site. MNP could work in this direction in combination with other drug delivery options by using smart polymers for targeted drug delivery to the tumor site specifically. In Chapter 4, we discuss Emerging Trends in Anti-infectious Metallic Nanoparticles. The top-down method, on the other hand, is a different approach used to obtain nanostructures. In this principle, nanolevels are reached by different methods by leaving the bulk material. The methods used for this can vary from physical etching methods to laser etching methods. In Chapter 5, we discuss Anti-inflammatory Metallic NPs: Type, Role, and Mechanisms. The inflammatory response is mediated by immune cells (mast cells, neutrophils, dendritic cells, and macrophages) and non-invasive cells like endothelium and fibroblasts. In both acute and chronic inflammation, systemic TNF generation by macrophages activates the primary microglia cells, which are the core components of the innate immune system. In Chapter 6, we discuss Trends in Theranostic Applications of Metallic Nanoparticles. It has been noted that conventional therapeutic techniques (such as chemotherapy, radiation, and surgeries) could have adverse effects. Due to inadequate therapy outcomes, interest has shifted to introducing nanotechnology in cancer management. However, correct diagnosis is necessary for the effective and appropriate treatment of cancer, since each type of tumor involves a unique therapeutic regimen. The ability of metallic nanoparticles may be enhanced by external stimuli including light, heat, ultrasonic radiation, and magnetic fields to target biological systems by modifying their redox potential and producing reactive oxygen species (ROS), which further sensitize target tissues. In Chapter 7, we discuss In Vivo and In Vitro Toxicity Study of Metallic Nanoparticles. By monitoring the effects of chemical substances in living animal beings (in vivo) or in animal and human cell lines, toxicity that affects humans is explored (in vitro). In standardized experimental designs using various model systems, toxico-genomics research gathers gene expression profiles and histopathological evaluation data for hundreds of medicines and contaminants. These data are a priceless resource for studying how drugs affect biological systems across the entire genome. In Chapter 8, we discuss Metallic Nanoparticles for Skins and Photothermal Therapy. Generally, nanoparticles utilized in biotechnology have particle sizes ranging from 10 to 500 nm, rarely reaching 700 nm. Cell surfaces and inside cells can be evaluated and assigned various aspects of the cell because of the tiny size of these particles. Nanoparticles like "gold and silver nanoparticles, magnetic nanoparticles (iron oxide), nano-shells, and nano-cages" have been employed and modified over the course of the year to allow their usage as a therapeutic agent and diagnostic agent for various disease-targeting systems like anticancer, antibacterial, neurological applications. In Chapter 9, we discuss Advancement Toward the Brain-Targeted Metallic Nanoparticles in Neurological Disorders. Acute brain injury and chronic neurodegenerative diseases are two categories of CNS circumstances that damage the structure and functionalities of the brain's neurons (NDs). Numerous therapeutic drug moieties have been discovered to combat neurological disorders, but their effectiveness is restricted due to the existence of various barriers in the CNS among the BBB is the main impediment in drug transport to the brain. This contextual concern investigates NP implementations for brain illnesses as well as how a wide collection of NPs are employed in clinical settings. In Chapter 10, we discuss Catalyst Metallic Nanoparticles: Types, Mechanism, and Trends. Nanoparticles have recently been utilized in the practical domains of nanofabrication, nano-biosensors, and optoelectronic devices. Nanocatalysis is an essential branch of nano-science, where nanoparticles (NPs) are used as catalysts. These features of metallic NPs abruptly affect their catalytic properties and behaviors. In order to overcome the stability issue, metallic NPs are encapsulated in nanoshells or nanopores. Metallic nanoparticle catalysts act as a bridge builder between homogeneous and heterogeneous catalysts. For ensuing industrial applications and commercialization based on metallic nanoparticles as prominent catalysts, there are still some barricades that need to be knocked down, such as inaccessible synthesis methods, toxicity, designing strategies, and maintenance of the catalytic property under extreme situations. In Chapter 11, we discuss Metallic Nanoparticles for Remediation of Toxicants and Wastewater Treatment. Wastewater is stemmed from numerous sources such as hospitals, industrial and agricultural activities, and residential areas. Facing these threats, various classical methods of wastewater treatment are developed, including precipitation, chlorination, ionization, ion exchange, electrochemical treatments, membrane filtration and reverse osmosis, evaporation, and floatation. Metallic nanoparticles, carbonaceous nanomaterials, zeolites, and dendrimers are the most commonly used nanomaterials in wastewater treatment. In Chapter 12, we discuss Toxicology and Regulatory Challenges of Metallic Nanoparticles. MNPs with small sizes can easily penetrate the bodies to accommodate and enter inside cells and tissues of living organisms and human beings. The toxicity levels and properties of the bulk materials from which the MNPs are made are known. Still, the toxicity level of these materials is changed at the nanoscale appearance, and it becomes difficult to determine whether the dosage and size of the MNP will have a toxicity impact. However, it is impossible to generalize that "the smaller, the more toxic" because some MNPs showed that the toxicity is reduced when the particle size is decreased. In Chapter 13, we discussed Toxicity of Metallic Nanoparticles: Assessment and Impacts. Many of the MNPs can be used in imaging modalities because they have distinct physicochemical characteristics. Despite the abundance of remarkable physicochemical features, there have been indications of toxicity also associated with MNPs. It is necessary to determine the dangers associated with metal NPs before using them in applications involving biological systems. As a result, metal NP surfaces must be designed in such a way that their toxicity is minimized while their potentialities are maximized. This chapter covers the physicochemical properties of MNPs, diverse production methods (chemical, biological and physical), factors impacting properties and biochemical behavior of NPs, and uses of NPs in treatment, imaging, tissue engineering, and drug administration. The toxicity concerns connected with the usage of these NPs are also addressed.

About the Editors

Dr. Md Sabir Alam, Associate Professor at the Department of Pharmaceutics, SGT College of Pharmacy, SGT University Gurgaon, Haryana, India

Dr. Md Sabir Alam obtained an M. Pharm in pharmaceutics from RGUHS Bangalore and a Ph.D. degree in nano-pharmaceutics. His main area of research is pharmaceutical nanotechnology and its biomedical application, with specialization in drug delivery for various targeted diseases, including psoriasis, neurological disorders, cancer targeting, antimicrobial, and anti-malarial activity. He currently works as an associate professor in the Department of Pharmaceutics, SGT College of Pharmacy, SGT University Gurgaon, Haryana, India. He has 8 years of teaching and research experience and has authored more than 20 research and review articles in international peer-review journals, 1 book, 20 book chapters, and 4 patents. His areas of interest include synthesis, characterization application of various nanocarriers for drug delivery/delivery for treating neurological disease, cancer, and malarial disease. Dr. Sabir Alam also focuses on revealing the role and impact of process variables in synthesizing various nanoparticles via process, optimization approaches using QbD and DoE. In *Bentham Science Journal* he was recently appointed as Executive Guest Editor for Pharmaceutical Nanotechnology (PNT) and has recently submitted book titled *Emerging Role of Metallic Nanocarriers for Health and Environment* (2022). His book proposal was accepted by CRC Press (Taylor and Francis Group) having (Impact Factor-3.2). He is also a reviewer of *IET Nanotechnology Journal* (Impact Factor-1.84); as advisory board member of *Heliyon Pharmaceutical Sciences, Pharmacology and Toxicology* (Cell Press), and editorial board member of *ILIVER Journal* (Elsevier). He is consistently among the world's top 10,000 scientists as per AD Scientific index 2023 version 1 for the last three years.

Dr. Md Noushad Javed, Department of Mechanical Engineering, University of Texas, Rio-Grande Valley, Edinburg, Texas TX 78539, USA.

Dr. Md Noushad Javed is associated with the Department of Mechanical Engineering, University of Texas Rio-Grande Valley, Edinburg, Texas TX 78539, USA. He is a senior researcher in the field of nanomaterials and an alumnus of the prestigious SPER of Jamia Hamdard (India). He has authored many patents, publications, and chapters on the role of different nanomaterials for pharmacological interventions in neurodegenerative disorders, epilepsy, cancer, and hypoxic disorders. His research also focuses on revealing the role and impact of process variables in the process of synthesizing various types of nanoparticles via process, optimization approaches using QbD, DoE, and artificial intelligence. Apart from his profession, he has served as an editor of many reputed scientific journals and published many high-impact papers and book chapters. He is currently associated with many book projects.

Dr. Jamilur R. Ansari, Associate Professor of Physics, Dronacharya College of Engineering, Gurugram, India

Dr. Jamilur R. Ansari is currently working as an Associate Professor of Physics at Dronacharya College of Engineering, Khentawas, Gurugram, India. He obtained his Ph.D. degree in the field of Physics/Nanotechnology from Guru Gobind Singh Indraprastha University, Delhi, India, M.Phil. in Nanotechnology, M.Sc. (Physics) from Jamia Millia Islamia, New Delhi, and MBA from Rajasthan Technical University, Rajasthan. He is having 17 years of teaching and research experience with 35 publications in SCI journals. Dr. Jamilur R. Ansari has been acknowledged as the outstanding contributor for reviewing in 2020 by IOP Journal. He has presented many talks in International Conference in India and abroad, beside he was an Invited Speaker in an International Conference. Four students are pursuing their Ph.D. under his supervision and they have published a few papers in SCI referred International Journals; beside he has also guided 11 students in their M.Sc. Project. Two of his students are working at NPL and they have published a few papers and are planning to join Germany University for their doctorate program. He is a material scientist and his research focuses on MXenes, perovskites, metal-semiconductor, 2D materials, QDs, self-assembly which involves photo-catalysis, solar cells, Artificial Intelligence, sensing, packaging and biomedical applications. He is a reviewer in many SCI index journals.

Contributors

Md Sabir Alam
SGT College of Pharmacy, SGT University, Gurgaon-Badli Road Chandu, Budhera, Gurugram
Haryana, India

Md Naushad Alam
Department of Physics, Lalit Narayan Mithila University, Darbhanga
Bihar, India

Azize Alayli
Department of Nursing, Faculty of Health Sciences
Sakarya University of Applied Sciences
Sakarya, Turkey

Md Meraj Anjum
Department of Pharmaceutical Sciences, Babasaheb Bhimrao Ambedkar University
Lucknow, Uttar Pradesh, India

Jamilur R. Ansari
Department of Applied Science & Humanities, Dronacharya College of Engineering, Khentawas, Farrukh Nagar, Gurugram
Haryana, India

Tarique Mahmood Ansari
Faculty of Pharmacy, Integral University, Lucknow
Uttar Pradesh, India

Arti
Department of Physics, Lalit Narayan Mithila University, Darbhanga
Bihar, India

Reena Badhwar
SGT College of Pharmacy, SGT University, Gurgaon-Badli Road Chandu, Budhera, Gurugram
Haryana, India

Vijay Bhalla
SGT College of Pharmacy, SGT University, Gurgaon-Badli Road Chandu, Budhera, Gurugram
Haryana, India

Charu Bharti
School of Pharmacy, Bharat Institute of Technology, Partapur bypass Road Meerut
UP, India

Shailendra Bhatt
Department of Pharmacy, School of Medical and Allied Sciences, G.D. Goenka University
Gurugram, India

Anindya Bose
School of Pharmaceutical Sciences (SPS), Siksha O Anusandhan University, Kalinganagar, Bhubaneswar
Odisha, India

Tulasi C D S L N
Centre for Biotechnology, UPGCST, Jawaharlal Nehru Technological University
Hyderabad, India

CH. Shilpa Chakra
Centre for Nanoscience & Technology, UPGCST, Jawaharlal Nehru Technological University
Hyderabad, India

Pranabesh Chakraborty
Bengal School of Technology
(A College of Pharmacy), Sugandha, Delhi Road, Hooghly
West Bengal, India

Soumalya Chakraborty
National Institute of Pharmaceutical Education and Research (NIPER), Sector 67, S.A.S. Nagar, Mohali
Punjab, India

Kalyani Chepuri
Centre for Biotechnology, UPGCST, Jawaharlal Nehru Technological University
Hyderabad, India

Sanjoy Kumar Das
Institute of Pharmacy, Netaji Subhas Chandra Bose Road, Hospital Para, Jalpaiguri
West Bengal, India

Laxmi Devi
Faculty of Pharmacy, Integral University, Lucknow
Uttar Pradesh, India

Manikantha Dunna
Centre for Biotechnology, UPGCST, Jawaharlal Nehru Technological University
Hyderabad, India

Hamza Elfil
Desalination and Natural Water Valorization Laboratory (LaDVEN), Water Researches and Technologies Center (CERTE)
Soliman, Tunisia

Jannatul Fardous
Comilla University
Cumilla, Bangladesh

Arun Garg
NIMS Institute of Pharmacy, NIMS University, Shobha Nagar, Jaipur
Rajasthan, India

Mansi Garg
SGT College of Pharmacy, SGT University, Gurgaon-Badli Road Chandu, Budhera, Gurugram
Haryana, India

Rupesh K. Gautam
Department of Pharmacology, Indore Institute of Pharmacy, IIST Campus, Rau, Indore
Madhya Pradesh, India

Shushay Hagos Gebre
Department of Chemistry, College of Natural and Computational Science, Jigjiga University
Jijiga, Ethiopia

Berhane Gebremedhin Gebrezgabher
Department of Biology, College of Natural and Computational Science, Jigjiga University
Jijiga, Ethiopia
School of Pharmaceutical Sciences, MVN University, Aurangabad, Palwal
Haryana, India

Tinku Gupta
Department of Pharmacognosy, SRM Modinagar College of Pharmacy, SRMIST, SRM University, Delhi NCR Campus, Modinagar
Ghaziabad, (Uttar Pradesh), India

Parvez Hassan
Institute of Biological Sciences, University of Rajshahi
Rajshahi, Bangladesh

Md Noushad Javed
Department of Mechanical Engineering, University of Texas Rio-Grande Valley
Edinburg, Texas, USA

Dr. Vikas Jogpal
Department of Pharmacology, School of Medical and Allied Sciences, GD Goenka University
Gurugram, Haryana, India

Renu Kadian
Ram Gopal College of Pharmacy, Sultanpur, Gurugram
Haryana, India

A. K. M. Shafiul Kadir
Institute of Biological Sciences, University of Rajshahi
Rajshahi, Bangladesh

Renu Kadyan
Research Associate, Department of Pharmaceutics, School of Pharmaceutical Education & Research, Jamia Hamdard
New Delhi, India

Biswakanth Kar
School of Pharmaceutical Sciences (SPS), Siksha O Anusandhan University, Kalinganagar, Bhubaneswar
Odisha, India

Ashish Kumar
Government Medical College, Jalaun, Orai
Uttar Pradesh, India

Girish Kumar
School of Pharmaceutical Sciences, MVN University, Aurangabad, Palwal
Haryana, India

Sainu Gopika
School of Pharmaceutical Sciences, MVN University, Aurangabad, Palwal
Haryana, India

Manish Kumar
M.M. College of Pharmacy, Maharishi Markandeshwar (Deemed to be University), Mullana, Ambala, Haryana, India

Mukesh Kumar
Department of Physics, Faculty of Science, Shree Guru Gobind Singh Tricentenary University Gurgaon
India

Puja Kumari
SGT College of Pharmacy, SGT University, Budhera
Gurugram, Haryana, India

Poonam Kushwaha
Faculty of Pharmacy, Integral University, Lucknow
Uttar Pradesh, India

Abdelmoneim Mars
Desalination and Natural Water Valorization Laboratory (LaDVEN), Water Researches and Technologies Center (CERTE)
Soliman, Tunisia

Navneet Mehan
M.M. College of Pharmacy, Maharishi Markandeshwar (Deemed to be University), Mullana, Ambala
Haryana, India

Alma Mejri
Desalination and Natural Water Valorization Laboratory (LaDVEN), Water Researches and Technologies Center (CERTE)
Soliman, Tunisia

Emna Melliti
Desalination and Natural Water Valorization Laboratory (LaDVEN), Water Researches and Technologies Center (CERTE)
Soliman, Tunisia

Shivani Munagala
Centre for Biotechnology, UPGCST, Jawaharlal Nehru Technological University
Hyderabad, India

Syed Muzammil Munawar
Department of Chemistry & Biochemistry, C. Abdul Hakeem College (Autonomous), Melvisharam, Ranipet District
Tamil Nadu, India

Hayrunnisa Nadaroglu
Department of Nano-Science and Nano-Engineering, Institute of Science and Technology
Ataturk University, Turkey

Namita
Department of Physics, Lalit Narayan Mithila University, Darbhanga
Bihar, India

Gautam Pal
Institute of Pharmacy, Block A2, Kalyani
West Bengal, India

Jayamanti Pandit
Research Associate, Department of Pharmaceutics, School of Pharmaceutical Education & Research, Jamia Hamdard
New Delhi, India

Rabiya
Department of Pharmaceutics, SD College of Pharmacy and Vocational Studies, Muzaffarnagar
Uttar Pradesh, India

Tanzilur Rahman
North South University
Dhaka, Bangladesh

Dhandayuthabani Rajendiran
Department of Chemistry & Biochemistry, C. Abdul Hakeem College (Autonomous), Melvisharam
Ranipet District, Tamil Nadu, India.

Sudipta Roy
Bengal School of Technology (A College of Pharmacy), Sugandha, Hooghly
West Bengal, India

Anjali Sharma
Freelancer, Pharmacovigilance Expert
India

Ashwani Sharma
School of Pharmaceutical Sciences, MVN University, NH-2 Delhi-Agra Highway, NCR, Aurangabad
Palwal, Haryana, India

Khaleel Basha Sabjan
Department of Chemistry & Biochemistry, C. Abdul Hakeem College (Autonomous), Melvisharam, Ranipet District
Tamil Nadu, India.

Dr. Mohit Sanduja
Department of Pharmaceutical Chemistry, School of Medical and Allied Sciences, GD Goenka University, Gurugram
Haryana, India

Mohammad Hossain Shariare
North South University
Dhaka, Bangladesh

Swami Prasad Sinha
Department of Physics, Faculty of Physical Sciences, PDM University, Bahadurgarh
Haryana, India

Tarun Virmani
School of Pharmaceutical Sciences, MVN University, Aurangabad
Palwal, Haryana, India

Reshu Virmani
School of Pharmaceutical Sciences, MVN University, Aurangabad, Palwal
Haryana, India

Aafrin Waziri
University School of Biotechnology, Guru Gobind Singh Indraprastha University, Dwarka
Delhi, India

Farheen Waziri
Department of Zoology, Gautam Buddha Mahila College, Gaya
Bihar, India

Madhu Yadav
Department of Physics, Faculty of Science, Shree Guru Gobind Singh Tricentenary University
Gurgaon, India

Tejpal Yadav
NIMS Institute of Pharmacy, NIMS University, Sobha Nagar
Jaipur, Rajasthan, India

1 Metallic (Inorganic) Nanoparticles

Classification, Synthesis, Mechanism, and Scope

Laxmi Devi, Tarique Mahmood Ansari, Md Sabir Alam, Ashish Kumar, and Poonam Kushwaha

1.1 INTRODUCTION

Nanoscience has become a discipline that focuses on the synthesis of metal nanoparticles with sizes ranging from 1 to 100 nanometers, along with particle shape and scale modification, employing numerous synthesis processes. Nanomaterials have currently employed in biological sciences, physics, organic and inorganic science, pharmaceuticals, and nanoscience [1, 2]. Nanostructured particle size reduction exhibits distinct or advanced attributes, including particle size distribution and geometry, which arise from changes in nanoparticles [3]. The word 'nanoparticle' was derived from the Greek expression 'nano,' which means 'small or tiny'; it also indicates a particle with a diameter of 10^{-9}m, 1 billionth of a meter, or 1 nm, which is used as a prefix [4, 5]. Nanoparticles exhibit combined solute and different component-stage properties. Nanoparticles have a surface-to-volume ratio of 35–45% higher than a larger component or molecule. One external property of a nanoparticle's specific surface area relates to its significant value and also to its unique intrinsic qualities, such as surface tension reactivity [6]. As an outcome, nanoparticles' distinctive features are important for their versatile properties. There is a rising prevalence of their use in fields including industry, medicine, and nutrition [7].

In the field of nanoparticles, a new terminology has developed in recent decades: metallic nanoparticles or nanostructured materials. Metal complexes including gold, silver, and platinum are being used to manufacture metallic nanoparticles, which already have therapeutic benefits [8]. Researchers are continuing to investigate metallic nanoparticles, nanostructures, and nanomaterial production due to their specific attributes, which can be exploited in catalytic reactions [9], polymer matrix processes [10], medical detection and therapy [11], and monitoring devices [12–14].

Several physicochemical and biological methods are frequently used in the synthesis and safety of nanoparticles, including electrochemical changes, chemical reduction, and photochemical reduction [15, 16]. Approaches including the kinetic model of metal ions interaction with reducing and stabilizing agents, the adsorption process

DOI: 10.1201/9781003317319-1

of stabilizing agents with nanoparticles, and different experimental processes have a significant influence on the morphological characters (structure and size), stability, and physical and chemical properties of metallic nanoparticles [17].

Some metal particles included in beauty products, cleansers, care products, shampoos, conditioners, medications, and pharmaceuticals affect users. Gold is commonly used in Ayurvedic formulations and remedies in India and China [8]. Nanoparticles are employed in a variety of diagnostics and therapies [18]. Silver nanoparticles, for example, are being used in pharmaceutical applications [19], such as nanoscience and nanotechnology [20] and novel drug delivery systems [8, 21]. Silver's antibacterial and anti-inflammatory activities are significant. Its attributes are also employed to promote wound dressings, pharmaceutical dose regimens, and implantable device coatings to hasten wound healing.

Several metal nanoparticles, including platinum nanoparticles, have already been investigated for their safety attributes and have also been successfully used in biomedical applications as a single or combined effect with several metallic nanostructures in either pure state or metallic doped state. Metallic nanoparticles have been widely employed in biomedicine and associated industries globally [22, 23]. In this chapter, we'll explore how metallic nanoparticles are classified, how they're synthesized (chemical and green nanotechnology methods), how they perform, how they're delivered, and what the future will bring for nanoparticles.

1.1.1 Metallic Nanoparticles

Metallic nanoparticles can constitute of nanometric structures which consists of solid metals (e.g., gold, platinum, silver, titanium, zinc, cerium, iron, and thallium) or related composites (e.g., oxides, hydroxides, sulfides, phosphates, fluorides, and chlorides). Several types of components can be employed to deliver small nanoparticles or aggregates; these can be classified as metallic, semiconductor, ionic, unique gas, or molecular based on their composition. If a nanoparticle contains only one type of atom, it is called homogeneous; otherwise, it is called heterogeneous [24, 25]. They might be neutral or charged (anions or cations).

Metallic nanoparticles have a variety of diagnostic and therapeutic applications [26]. Inorganic NPs, including quantum dots [27], gold NPs [28], and magnetic NPs [29], have therapeutic efficacy as drug delivery carriers or biosensing agents [30, 31]. Due to various physicochemical and biological attributes, including high stability, high reactivity, photothermal, antibacterial activity, electrical conductivity, chemical properties, biocompatibility, and photonic aspects, metallic NPs are beneficial clinical transporters. As a basis, photo-acoustic treatment with photonic NPs including gold NPs is used to target a variety of tumor cells [32, 33]. Qds, gold nanoparticles, and magnetic iron NPs have all been used during cell imaging [34, 35]. According to their diagnostic and therapeutic applications, a variety of metallic nanoparticles are used in theranostics [36, 37]. Manufactured nanoparticles can be formed by changing their surface, modifying their diameter and form, and adding certain ligands. These changes allow nanomaterials to be used to treat patients and other diseases.

1.1.2 Classification of Metallic Nanoparticles (Inorganic Particles)

Non-carbon nanostructures are classified as metallic NPs. Metal-based nanostructures account for the majority of metallic nanoparticles, as shown in Figure 1.1.

1.1.2.1 Metal-Based Nanostructures

Metal-based nanostructures are nanoparticles synthesized using metals to nanosized dimensions, possibly destructively or constructively. Almost all metal nanoparticles can be synthesized [38]. Aluminum (Al), cadmium (Cd), cobalt (Co), copper (Cu), gold (Au), iron (Fe), lead (Pb), silver (Ag), and zinc (Zn) are some of the most frequently employed metals for nanoparticle production. Nanostructures have a size and distribution that vary from 10 to 100 nm, as well as morphology such as enhanced surface area to volume, permeability, surface charge and surface charge density, amorphous and crystalline structures, spherical and spherical shapes, color, reaction, and exposure to environmental aspects such as air, moisture, heat, and natural daylight [39], as shown in Figure 1.2.

1.1.2.2 Metallic Oxide-Based Nanostructures

Metallic oxide-based nanostructures are being established to improve on the properties of their metal-based counterparts. Nanocomposites (Fe) decay rapidly to iron oxide (Fe_2O_3) in the atmospheric oxygen at ambient temperature, increasing their

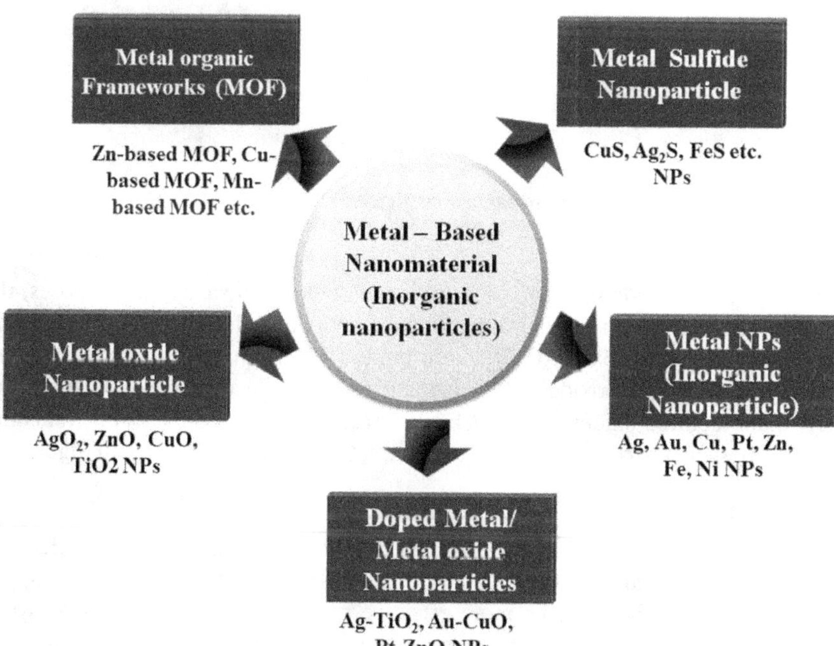

FIGURE 1.1 Different types of metal-based nanoparticles.

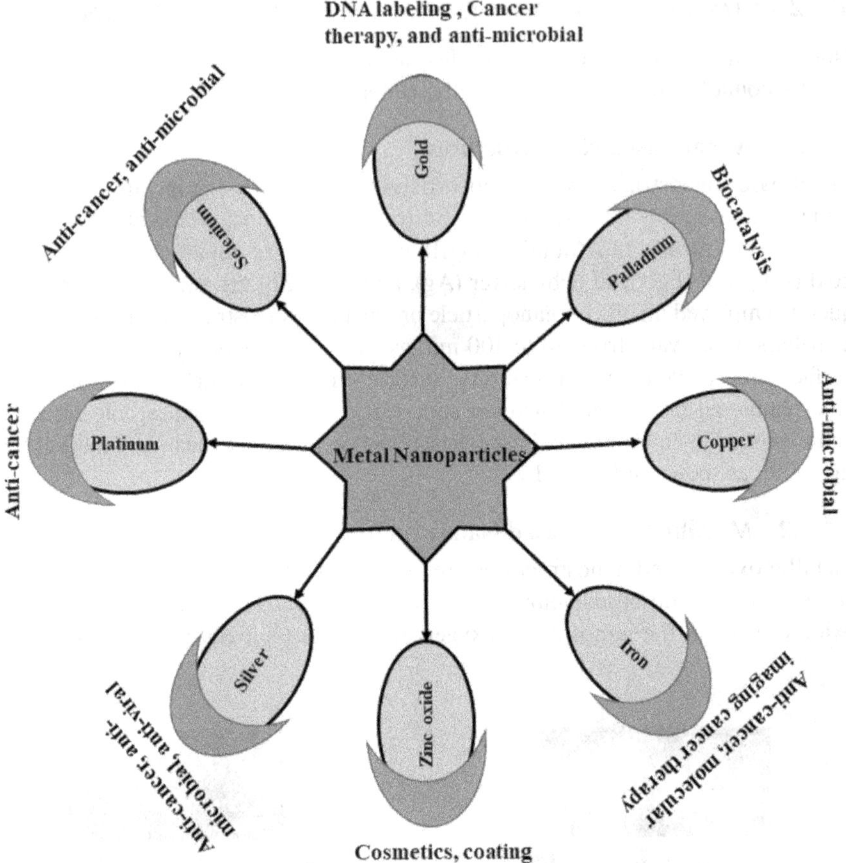

FIGURE 1.2 Different types of metal nanoparticles.

sensitivity. Metal oxide nanoparticles are synthesized for a variety of purposes, the most important of which are their improved reactivity and efficiency [40]. Aluminum oxide (Al_2O_3), cerium oxide (CeO_2), iron oxide (Fe_2O_3), magnetite (Fe_3O_4), silicon dioxide (SiO_2), titanium oxide (TiO_2), and zinc oxide (ZnO) are some of the most routinely manufactured oxides (ZnO). When contrasted to their equivalent metal counterparts, these nanostructures have better characteristics [39], as shown in Figure 1.3.

1.1.2.3 Doped Metal/Metal Oxide-Based Nanoparticles

Scientists across the globe are focusing on producing non-toxic nanoparticles that are chemically safe and environmentally acceptable. Metal oxide biomedical applications have been shown to be more effective and have substantial benefits when metal oxides NPs are employed. Antimicrobial activities may be enhanced by ZnO nanoparticles doped with Sb or Mg. Due to their minimal degradation limitations, doped metallic NPs have already been explored for utilization in a variety of medicinal applications [5].

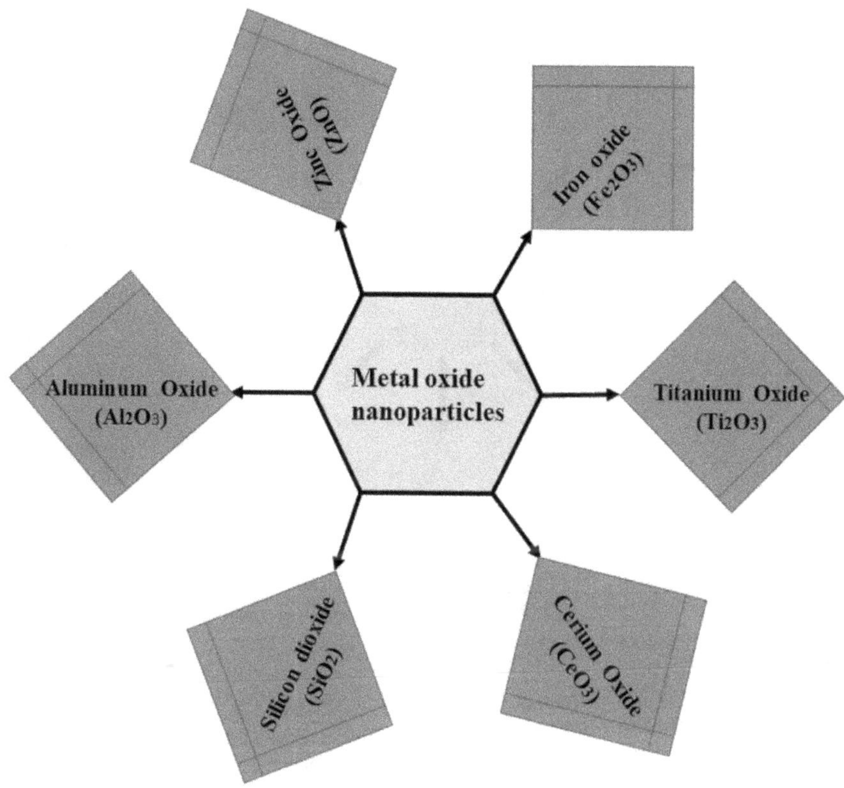

FIGURE 1.3 Different types of metal oxide-based NPs.

1.1.2.4 Metal Sulfide-Based Nanoparticles

Due to the general presence of specific features in nanosized regions, metal-based semiconductors are recognized as a commonly used semiconductor resource with a variety of medicinal uses. Its semiconductor has a high fluorescence, a small optical conductivity, and outstanding magnetic, thermal, mechanical, and systemic properties [41]. Biosensors, drug delivery, biolabeling, bioimaging, and detection have already been described as applications for quantum dots, nanoparticles, and metallic-stabilized organic–inorganic hybrid nanostructures. The structures of CdSe, PbS, CdSe-CdTe, CdSeZnTe, CdTe-CdSe, and other metal-based chalcogenides are very complex [42]. Metallic sulfides containing chalcogenide sulfur and strongly linked to a destructive metal have recently attracted a lot of interest in the medical community [43]. AgS, CuS, FeS, and ZnS are the most prevalent metallic sulfides, and they are the most precious resources for biomedical applications, as shown in Figure 1.4.

1.1.2.5 Metallic Organic Frameworks

Metal organic frameworks (MOFs) are a type of crystalline nanostructured material that is still pretty recent. MOFs are synthesized by the identity of metallic ions that

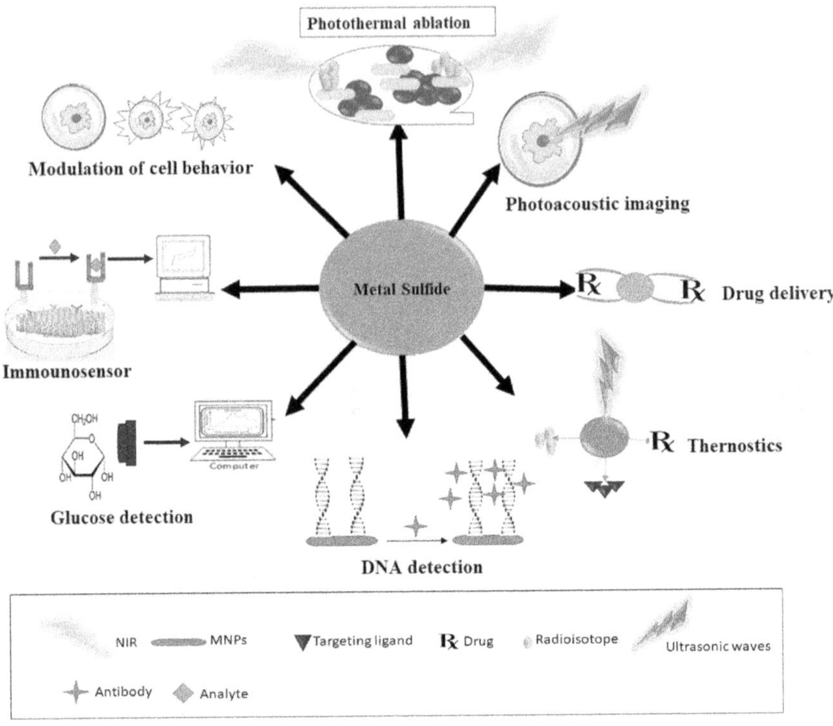

FIGURE 1.4 Potential medical applications of metal sulfide-based nanoparticles.

perform as connection components for biological substances, including as intermediates in metal centers [44]. One of the most significant major innovations in nanoscale technologies involves high porosity coordination, often known as mixed organic–inorganic coordination systems. Strong porosity and surface area, which vary from 1,000 to 10,000 m^2/g, make MOFs more prevalent than normal adsorbents [45]. Due to their excellent drug-loading potential, irregular fabrication, optimum biodegradability, and biocompatibility, use of MOFs as drug delivery systems has been widely researched in recent years [44]. Because of their ability to vary particle sizes and moieties, MOFs are deemed an ideal product for the delivery of drugs, as shown in Figure 1.5.

1.1.3 Importance of Metallic Nanoparticles in Biology

Metal ions of extensive variety can be found in all living systems and are involved in a broad variety of biological activities. Metal-based particles have been used in numerous biomedical fields since the beginning of time, as explained previously. Modern toxicity investigations, on the other hand, reveal that substances in medicine have a lot of disadvantages (toxicity, low bioavailability, non-specificity, and so on)

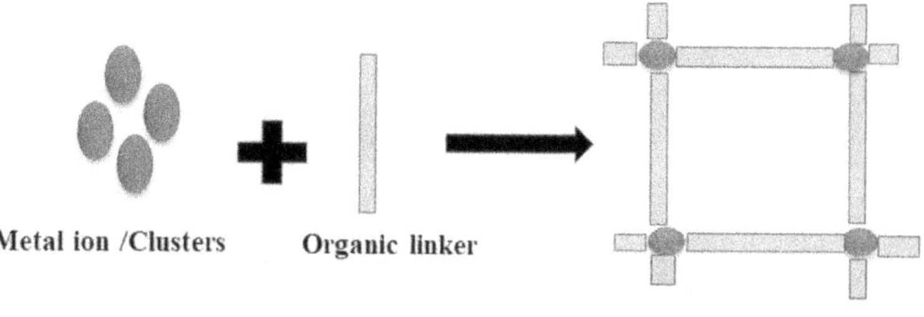

FIGURE 1.5 Metallic organic framework based MNPs.

[46]. As a consequence, researchers have concentrated their efforts on creating nanostructures from bulk materials and metals. Bulk materials' physicochemical properties are altered when reduced to the nanoscale, rendering them valuable in a number of biomedical applications [47–49]. In therapeutic diagnostics, metallic nanoparticles are commonly utilized as polymeric vehicles (medicines, enzymes, antigens, nucleotides, oligonucleotides, etc.) [50]. Metallic nanoparticles, in contrast to transporting biomolecules, have already been successfully used for diagnostic and therapeutic therapies for a number of chronic diseases (cancer, cardiovascular disease, diabetes, retinal disorders, neurodegenerative disorders, microbial infections, etc.). Because there are so many studies on metal nanoparticles in the literature, this study will elaborate on their current innovations in therapies [51, 52].

1.1.4 Methods for Preparation of Inorganic Nanoparticles

1.1.4.1 Physical Methods

Inorganic nanoparticles have been synthesized using a variety of processes, which can be divided into two categories: bottom-up methods and top-down approaches [53, 54]. The key difference between the two techniques is in the production of the nanoscale starting material. In top-down approaches, the bulk material is utilized as the developing medium, and the nanoparticle size is reduced to nanoscale using various physical, chemical, and mechanical processes, while in bottom-up methods, atoms or molecules are applied as the starting materials, as shown in Figure 1.6 [55, 56].

1.1.4.2 Top-Down Methods

The synthesis is used to transform bulk material into nanoscale nanoparticles. The fabrication of nanoparticles is based on physically and chemically reducing the size of the starting material [57]. Mechanical milling, laser therapy, and laser ablation are just a few of the approaches that have been used. Although top-down approaches are easy to explain, they are unsuitable for forming irregularly shaped or very small nanoparticles. Surface modification and the physicochemical properties of nanostructures are two significant drawbacks of this approach [58].

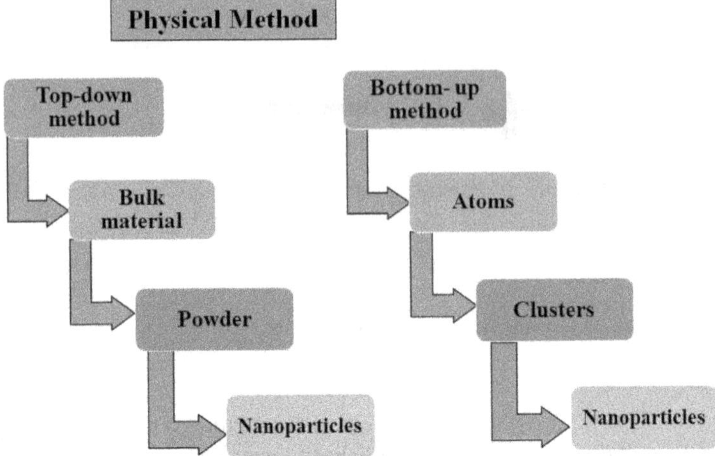

FIGURE 1.6 Synthesis process for physical method.

1.1.4.3 Bottom-Up Methods

This approach to the synthesis of nanomaterials is premised on synthesizing nanoparticles from smaller molecules, such as atoms, molecules, or smaller particles. The nanomaterial component parts of nanoparticles are produced first, and the nanoparticles are then integrated to form a finished product [21].

1.1.4.4 Chemical Reduction Method

In this method, ionic salt is usually reduced in an appropriate medium in the presence of surfactant using various reducing agents in a chemical reduction process [59]. For synthesizing metallic nanoparticles, a reducing agent, including sodium borohydride, is added to the aqueous medium. Trisodium citrate (TSC) or sodium lauryl sulfate are used to cap metallic nanoparticles (SLS). A stabilizing agent is sometimes used in association with such a reducing agent. The stability of metallic nanoparticles inside a dispersion is determined using spectrophotometer measurements [60]. Silver nanoparticles are synthesized with sodium borohydrate ($NaBH_4$), glucose, ethylene glycol, ethanol, sodium citrate, and hydrazine hydrate, among other reducing agents [61], as shown in Figure 1.7.

Advantages

The easiest approach toward synthesizing metallic nanoparticles [62].

Disadvantages

There are several drawbacks to reducing agents, including toxicity, high costs, limited reducing ability, and contaminants [63].

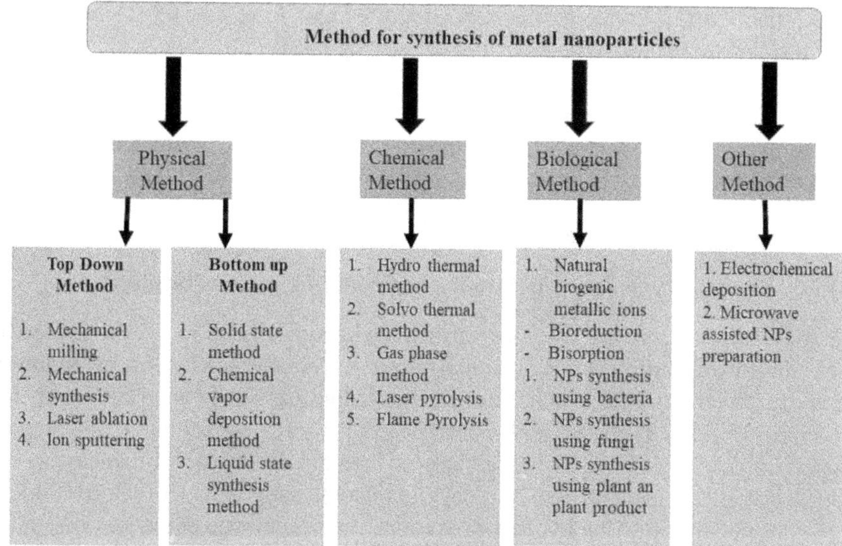

FIGURE 1.7 Various methods for the synthesis of metal nanoparticles.

Applications

Synthesis of copper oxide nanoparticles using potassium borohydrate as a reducing agent [63].

1.1.4.5 Biological Method/Biomimetic Method/Green Synthesis Method

An emerging trend in nanotechnology is the formation of nanoparticles employing green synthesis methods. These strategies have emerged to address issues with traditional procedures such as reactive complexities, high costs, and safety concerns. Green chemistry methods integrate new methodologies into the synthesis of chemicals and their potential uses in order to reduce harm to human health and the environment [64].

In the research method for nanoparticle formation, various microbes and their enzymes, as well as plant material such as isolates and crude extracts, are employed. Such methods have numerous benefits over traditional physical and chemical approaches, such as cost-effectiveness, environmental friendliness, and the possibility of significantly increasing large-scale synthesis. Furthermore, during green synthesis, excessive pressure, energy, temperature, or toxic substances are not utilized [65].

1.1.4.6 There Are Two Types of Natural Biogenic Metallic Nanoparticle Production

1.1.4.6.1 Bioreduction

Microbes as well as their enzymes use reducing agents to bring metal ions into biologically stable forms.

The stable and inert inorganic nanostructures that form can be safely isolated from toxic materials [66, 67].

1.1.4.6.2 Biosorption

It is an approach to nanoparticle synthesis in which metallic cations in the water phase are permitted to interact with microorganism cell walls, producing stabilized nanoparticles [68].

1.1.5 Pharmacological Mechanism for Metallic Nanoparticles

Metallic nanoparticles are used for various applications, including food, preservatives, cosmetics, and medicine. The toxicological characteristics of metal nanoparticles are affected by their concentration and exposure time. Metal nanoparticles induce cytotoxicity at higher dosages and degrees of contact, leading to a variety of diseases in people [69]. Furthermore, the predicted biodistribution of metal nanoparticles to specific organs or tissues, along with access paths, impact their usefulness in biomedical applications. In order to examine the whole toxic potential of metallic nanoparticles in animal studies, involving biosafety, long-term toxicity, pharmacokinetic profiles, and metabolic long-term fate [70].

In this context, researchers have already progressed to use metabolic pathway pharmacokinetic models to study metallic nanoparticles permeability, distribution, metabolism, and excretion, in accordance with the OECD's approval and regulations for the assessment of nanoparticle risks in preclinical models [71, 72]. When nanoparticles are administered orally, for instance, they may be eliminated in the GI tract, absorbed in the body system, or excreted in the feces [73]. Intraperitoneally (IP), intramuscularly, intradermally, or subcutaneously administered nanoparticles, on either side, are incorporated into lymph nodes [74, 75]. Numerous studies have shown that nanoparticles are transported throughout the body via blood circulation after intake, depending on the delivery route (oral, IP, intravenous (IV), inhalation, etc.). Therefore, in this case, the nanostructures may remain on the outside surface of the organs or invade their inside cells. The bioavailability of nanostructures in a specific tissue may be altered by the systemic circulation. Nanostructure formation is revealed to be very significant inside the reticuloendothelial (RES) system of a number of patients, specifically in the liver and spleen.

In these situations, experts have changed their surfaces using hydrophilic PEG molecules to prevent nanoparticle entrapment by RES [76]. The development of nanoparticles in the lungs and GI tract is reported to be higher after pulmonary and oral delivery of nanoparticles, respectively.

The number of research, on the other hand, reported that nanoparticle accumulation in the brain, heart, and kidney was very low to moderate. Nanostructures must be digested in the biological body during uptake and biodispersion, resulting in changes in their physical and chemical characteristics. The way nanostructures are metabolized in the human body may be influenced by their structure. Many metallic nanoparticles are stable enough to be absorbed in the human system while being biodegradable and biocompatible. Current research has reported the biocompatibility of several metallic nanoparticles, such as ZnO [77–79], gold [80, 81], silver [82], iron oxide [83, 84], europium [85], and MgO nanoparticles [86]. As researchers discover that metallic nanoparticles are processed in the liver by the kidneys, all of those nanoparticles could be metabolized and eliminated by urine or feces.

Choi et al. [87] researched the renal route of CdSe/ZnS quantum dots and indicated that small nanoparticles can be eliminated quickly by urine. Several inorganic nanostructures, such as copper, iron, titanium, and europium, were also extracted from feces and urine in other studies [88–90].

1.1.6 Routes of Administrations

In most cases, the absorption mechanism of MNPs by biological systems determines the type and level of possible harm in diverse tissues and organs. More material, mostly on different routes of administration and doses used during experimental animal models, may be found further along, as shown in Figure 1.8.

1.1.6.1 Skin Absorption

This route of administration includes both active and passive NP delivery into the skin (e.g., inhalation following nebulization in a chamber or directly inside the water for aquatic animals) [91]. The zebra fish is a physiologically recognizable animal study organism commonly used to deliver MNP toxicity in vivo. Zebra fish have a number of advantages when compared to other vertebrate animal models. Housing and administrative expenditures are substantially cheaper, making it a good model for preliminary toxicological and/or environmental exposure research. The zebra fish safety evaluation process can simply be applied to people and other animals.

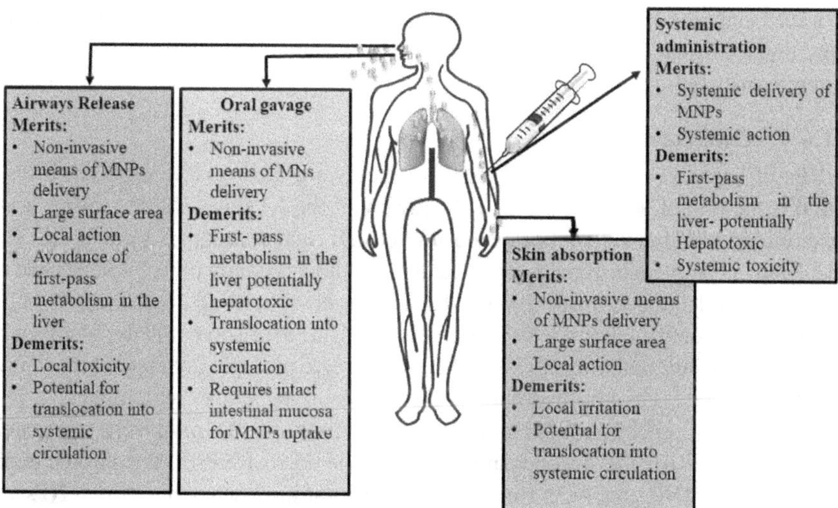

FIGURE 1.8 Various routes of administration for MNPs.

1.1.6.2 Airways Release

In order to accomplish a detailed safety assessment that analyzes the adsorption, distribution, metabolism, and excretion (ADME) of MNPs in vivo in rats, the in vivo cytotoxicity of AgNPs and AuNPs has been examined in models of sub-chronic inhalation. A study is vital for both global corporations and to evaluate the impact of MNPs on humans, given their prevalence. The dose that had no adverse toxic effect in the pre-clinical study and the smallest dosage for an adverse effect (LOAEL) are additional crucial factors (NOAEL). Slightly different dosages were tested in these investigations (e.g., AuNPs in different groups of male and female rats with 6-hrs inhalation treatment consecutively for three months). Weekly, the animals' body-weight was recorded, and their airways, skin, and mucosa (oral, nasal, and vaginal) were examined daily. Biochemistry and hematologic markers, as well as nephro-toxicity and lung function, were all examined prior to the death. Multi-organ 14 histopathology analysis was included in the post-mortem investigation (including broncho-alveolar lavage). Atomic Absorption Spectroscopy (AAS) [92] was used to determine AuNP accumulation in tissue.

1.1.6.3 Systemic Administration

The most popular methods for MNP systemic administration are IV and intra-peritoneal (IP) injections. MNPs can be delivered systemically through these two pathways, allowing the researchers to monitor their organ distribution across time. Fluorescence imaging using confocal microscopy is used to evaluate the formation of NPs in the organs of injected animals. Small silica-coated magnetic (50 nm) NPs expressing rhodamine B isothiocyanate were attracted to a brain through the blood-brain barrier [93], but AuNPs with a size of 100–250 nm were found in a variety of organs and tissues (e.g., liver, spleen, kidney, or lungs) [94]. These animals are subjected to biochemical and hematological assays to detect the cytotoxicity of NPs in tissues and blood cells. Assessment of the potential toxicity, chromosomal aberrations, and low toxicity for germinal cell viability research is a relevant technique with systemic distribution of NPs [95, 96].

1.1.6.4 Oral Gavage

CuNPs of various sizes were gavaged into mice via a naso-gastric probe to investigate this method of administration. The efficacy of the NPs was studied in this acute toxicity scenario after only a brief exposure to high concentrations. Once the LD50 is calculated, a complete histopathological and hematological study must be performed on the animal model to determine the ADME features in vivo (lethal doses for 50% of the population). Systemic cytotoxicity must be assessed due to the potential use of NPs with meals or orally ingested drugs [97–99]. Furthermore, to properly understand the in vivo toxicity of MNPs, oral administration of NPs via drinking water contaminants or passive uptake from other administration routes (e.g., airways or skin) must be adequately evaluated [100, 101]. The animals are given a single dose of NPs in the chronic exposure model for MNP toxicity evaluation; hence, escalating doses of the several NPs and the comparison (vehicle) are provided to the animals to determine the effective dosage.

Chen et al. examined the effect of two diameters of CuNPs (25 nm and 17 nm) on soluble ionic Cu delivered by a gavage probe in a new study. The diameter of the NPs is important in producing toxicity in this study; larger NPs have a lower surface area with the animal and are thus possibly more harmful than smaller NPs or ionic Cu. One important outcome of this study is that determining the differences and specificity of the cytotoxicity generated by MNPs necessitates investigation into different methods of delivering NPs. Inside the stomach, for instance, the combination of NPs and H^+ rate is much higher in the creation of HCO_3^-, which causes an elevation in pH, leading to renal injury and electrolytic metabolic changes [98]. One noteworthy issue is the difference in toxicity testing findings between males and females after oral ingestion of NPs. H^+ production in the stomach is more common in men than in women and increases the risk of renal disease [97]. All of these reasons highlight the importance of developing an approved in vivo hazardous testing method. Furthermore, further specific descriptions of the biology of the chosen animals are needed in order to accurately assess the influence of systemic delivery on MNP-induced cytotoxicity [98, 102].

1.1.7 Present and Future Scope of Metallic Nanoparticles

Along with their critical significance in nanoparticle and other biological applications, inorganic nanoparticles appear to have a dominant position in the 21st century [103–109]. The nanoparticle can be made using a number of synthetic approaches and then used in a variety of nanomedicine and biomedical research [110–116]. However, in order to cut costs, these nanoparticles must be manufactured on a commercial scale. The natural resources used to make these nanoparticles should be long-lasting, low-cost, environmentally friendly, and free of hazardous chemicals. In order to conduct future research, monodispersed nanoparticles must be created. However, the method for synthesizing nanoparticles has yet to be fully described. As a result, future studies should focus on the mechanisms that affect the form and size of nanoparticles. Another key problem is increasing the usage of nanoparticles in medicinal applications like cancer, antimicrobial and neurological activity while lowering toxicity [117–128]. With the potential progress in nanoscience, new ways are being proposed to solve the hurdles by employing noble metal nanoparticles. The impact on biological health parameters must be evaluated before widespread use.

Nanotechnology has the potential to revolutionize biomedicine. It will need to be investigated in a much more targeted manner in the future. Since they have new capabilities just at molecular and supramolecular levels are employed as therapeutic drugs in diagnostics or other medical approaches. Wide varieties of nanostructures and associated chemicals are currently generating a lot of attention in the healthcare system, medical, and scientific domains. As a result, in order to protect human health, safety standards must take precedence. Important safety profile analyses are required if nanostructured materials are employed in health care. The use of different nanostructured materials for different applications is the focus of scientific research.

In the future, noble metal nanoparticle preparation will need to be scaled up from the laboratory to commercial medicinal scale. Metallic nanoparticles have been produced and are being tested in a variety of directions. It's primarily utilized to treat malignant cells at the moment. Metal nanoparticles have shown promise as a novel therapeutic therapy for the treatment of cancer in the future. Is it feasible to determine if noble metal nanoparticles are biocompatible or harmful? It's really critical to explore nanoparticle interaction at the lab scale and change nanoparticles for higher biocompatibility with malignant cells in order to avoid injury to healthy cells. Because the majority of research is currently done in laboratories, modern research should concentrate on commercial applications of nanotechnology. Exploration on a commercial scale has the potential to transform human life.

1.1.8 Conclusions

This publication emphasizes the importance of nanoparticles in biological and medicinal research. Due to the extensive diagnostic procedures and a variety of other biomedical applications, metal-based nanostructures have inspired huge interest. Owing to their unique physicochemical features, metal-based nanoparticles can regulate antiproliferative effects on a variety of organs, body tissues, and subcellular, cellular, and protein levels. Furthermore, even though they appear harmless on a wide scale, many metallic nanoparticles (Ag, Au, Zn, and Cu) can be harmful when their diameter is decreased. In antimicrobial research, metal–metal oxide or pure metal nanoparticles have been employed as a worthy alternative. Antimicrobial nanoparticles can be used to treat infections as well as develop identity pharmaceutical products. Metal oxide or metal oxide/metal-doped composites could be used to make concise, reduced antibacterial agents. It will be used in place of antibiotics. Antibacterial properties of inorganic nanoparticles and similar combinations have been demonstrated and are used in medicine administration and a variety of targeted therapies. Moreover, due to their harmful behavior at large concentrations, metal oxide nanoparticles have such serious limitations. Ion doping, functionalization, and attached polymeric metal oxide nanoparticles have all been used to reduce identity. Ultimately, in the future, low-toxicity metal, metal oxide, or composite nanoparticles could be employed to treat a variety of severe bacterial infections and tumor cells.

CONFLICT OF INTEREST

The authors declare that there is no conflict of interest, financial, or otherwise.

ACKNOWLEDGMENTS

The authors are also thankful to Dr. Jamilur R. Ansari, Dronacharya College of Engineering, Gurgaon, India, for the needful editing in the Chapter. Our sincere thanks to Aafrin Wazari for editing the manuscript. The authors gratefully acknowledge the contributions of the collaborators and co-workers mentioned in the cited reference. We want to express our sincere thanks to Prof. S.W. Akhtar, Honorable Chancellor, Integral University, Dasauli, Lucknow, India, for providing an excellent research atmosphere. LD, TMA, and PK acknowledge Integral University; Ashish Kumar thankfully acknowledges the *Government Medical College*. MSA gratefully acknowledges the Vice Chancellor and Director of Pharmacy at SGT University for their kind cooperation and support.

REFERENCES

[1] F.J. Heiligtag, M. Niederberger, The fascinating world of nanoparticle research, Mater. Today 16 (7) (2013) 262–271.

[2] M. De, P.S. Ghosh, V.M. Rotello, Applications of nanoparticles in biology, Adv. Mater. 20 (22) (2008) 4225–4241.

[3] V.D. Willems, Roadmap report on nanoparticles, W&W Espana SL, Barcelona (2005), p. 157.

[4] H. You, S. Yang, B. Ding, H. Yang, Synthesis of colloidal metal and metal alloy nanoparticles for electrochemical energy applications, Chem. Soc. Rev. 42 (7) (2013) 2880–2904.

[5] A.A. Yaqoob, H. Ahmad, T. Parveen, A. Ahmad, M. Oves, I.M.I. Ismail, et al., Recent advances in metal decorated nanomaterials and their various biological applications: a review. Front. Chem. 8 (2020) 341.

[6] M. Auffan, J. Rose, M.R. Wiesner, J.Y. Bottero, Chemical stability of metallic nanoparticles: a parameter controlling their potential cellular toxicity in vitro, Environ. Pollut. 157 (4) (2009) 1127–1133.

[7] R. Chandra, A.K. Chawla, P. Ayyub, Optical and structural properties of sputter-deposited nanocrystalline Cu_2O films: effect of sputtering gas, J. Nanosci. Nanotechnol. 6 (4) (2006) 1119–1123.

[8] R. Bhattacharya, P. Mukherjee, Biological properties of "naked" metal nanoparticles, Adv. Drug Deliv. Rev. 60 (11) (2008) 1289–1306.

[9] R. Narayanan, M.A. El-Sayed, Shape-dependent catalytic activity of platinum nanoparticles in colloidal solution, Nano Lett. 4 (7) (2004) 1343–1348.

[10] D. Moura, M.T. Souza, L. Liverani, G. Rella, G.M. Luz, J.F. Mano, et al., Development of a bioactive glass-polymer composite for wound healing applications, Mater. Sci. Eng. C 76 (2017) 224–232.

[11] K. Banerjee, S. Das, P. Choudhury, S. Ghosh, R. Baral, S.K. Choudhuri, A novel approach of synthesizing and evaluating the anticancer potential of silver oxide nanoparticles in vitro, Chemotherapy 62 (5) (2017) 279–289.

[12] P. Gomez-Romero, Hybrid organic–inorganic materials—in search of synergic activity, Adv. Mater. 13 (3) (2001) 163–174.

[13] S.F. Shaikh, R.S. Mane, B.K. Min, Y.J. Hwang, O.S. Joo, D-sorbitol-induced phase control of TiO_2 nanoparticles and its application for dye-sensitized solar cells, Sci. Rep. 6 (2016) 2010–2013.

[14] D.H. Gracias, J. Tien, T.L. Breen, C. Hsu, G.M. Whitesides, Forming electrical networks in three dimensions by self-assembly, Sci. 289 (5482) (2000) 1170–1172.

[15] W. Chen, W. Cai, L. Zhang, G. Wang, L. Zhang, Sonochemical processes and formation of gold nanoparticles within pores of mesoporous silica, J. Colloid Interface Sci. 238 (2) (2001) 291–295.

[16] R. Geethalakshmi, D.V. Sarada, Gold and silver nanoparticles from *Trianthema decandra*: synthesis, characterization, and antimicrobial properties, Int. J. Nanomed. 7 (2012) 5375–5384.

[17] M. Vijayakumar, K. Priya, F.T. Nancy, A. Noorlidah, A.B. Ahmed, Biosynthesis, characterization and anti-bacterial effect of plant-mediated silver nanoparticles using *Artemisia nilagirica*, Ind. Crops Prod. 41 (2013) 235–240.

[18] D.R. Bhumkar, H.M. Joshi, M. Sastry, V.B. Pokharkar, Chitosan reduced gold nanoparticles as novel carriers for transmucosal delivery of insulin, Pharm. Res. 24 (8) (2007) 1415–1426.

[19] P. Puvanakrishnan, J. Park, D. Chatterjee, S. Krishnan, J.W. Tunnell, In vivo tumor targeting of gold nanoparticles: effect of particle type and dosing strategy, Int. J. Nanomed. 7 (2012) 1251.

[20] D. Sykora, V. Kasicka, I. Miksik, P. Rezanka, K. Zaruba, P. Matejka, et al., Application of gold nanoparticles in separation sciences, J. Sep. Sci. 33 (3) (2010) 372–387.

[21] P.G. Jamkhande, N.W. Ghule, A.H. Bamer, M.G. Kalaskar, Metal nanoparticles synthesis: an overview on methods of preparation, advantages and disadvantages, and applications, J. Drug Deliv. Sci. Technol. (2019) 101174.

[22] I.M. Chung, I. Park, K. Seung-Hyun, M. Thiruvengadam, G. Rajakumar, Plantmediated synthesis of silver nanoparticles: their characteristic properties and therapeutic applications, Nanoscale. Res. Lett. 11 (1) (2016) 40.

[23] P.V. Asharani, G. Low Kah Mun, M.P. Hande, S. Valiyaveettil, Cytotoxicity and genotoxicity of silver nanoparticles in human cells, ACS Nano 3 (2) (2008).

[24] S. Lanone, J. Boczkowski, Biomedical applications and potential health risks of nanomaterials: molecular mechanisms, Curr. Mol. Med. 6 (6) (2006) 651–663.

[25] V. Wagner, A. Dullaart, A.K. Bock, A. Zweck, The emerging nanomedicine landscape, Nat. Biotechnol. 24 (10) (2006) 1211–1217.

[26] L.E. Euliss, J.A. DuPont, S. Gratton, J. DeSimone, Imparting size, shape, and composition control of materials for nanomedicine, Chem. Soc. Rev. 35 (11) (2006) 1095–1104.

[27] O.C. Farokhzad, R. Langer, Nanomedicine: developing smarter therapeutic and diagnostic modalities, Adv. Drug Deliv. Rev. 58 (14) (2006) 1456–1459.

[28] H. Mertens, J.S. Biteen, H.A. Atwater, A. Polman, Polarization-selective plasmon-enhanced silicon quantum-dot luminescence, Nano Lett. 6 (11) (2006) 2622–2625.

[29] E.C. Dreaden, L.A. Austin, M.A. Mackey, M.A. El-Sayed, Size matters: gold nanoparticles in targeted cancer drug delivery, Ther. Deliv. 3 (4) (2012) 457–478.

[30] L.B. Thomsen, M.S. Thomsen, T. Moos, Targeted drug delivery to the brain using magnetic nanoparticles, Ther. Deliv. 6 (10) (2015) 1145–1155.

[31] H.S. Sharma, S.F. Ali, W. Dong, Z.R. Tian, R. Patnaik, S. Patnaik, et al., Drug delivery to the spinal cord tagged with nanowire enhances neuroprotective efficacy and functional recovery following trauma to the rat spinal cord, Ann. N. Y. Acad. Sci. 1122 (2007) 197–218.

[32] D. Williams, Quantum dots in medical technology, Med. Device Technol. 17 (4) (2006) 8–9.

[33] J.Y. Fang, Nano- or submicron-sized liposomes as carriers for drug delivery, Chang Gung Med. J. 29 (4) (2006) 358–362.

[34] S. Jain, D.G. Hirst, J.M. O'Sullivan, Gold nanoparticles as novel agents for cancer therapy, Br. J. Radiol. 85 (1010) (2012) 101–113.

[35] Y. Wang, R. Hu, G. Lin, I. Roy, K.T. Yong, Functionalized quantum dots for biosensing and bioimaging and concerns on toxicity, ACS Appl. Mater. Interfaces 5 (8) (2013) 2786–2799.

[36] R. Qiao, Q. Jia, S. Huwel, R. Xia, T. Liu, F. Gao, et al., Receptor-mediated delivery of magnetic nanoparticles across the blood-brain barrier, ACS Nano 6 (4) (2012) 3304–3310.

[37] F. Chen, E.B. Ehlerding, W. Cai, Theranostic nanoparticles, J. Nucl. Med. 55 (12) (2014) 1919–1922.

[38] M. Salavati-Niasari, F. Davar, N. Mir, Synthesis and characterization of metallic copper nanoparticles via thermal decomposition, Polyhedron 27 (2008) 3514–3518.

[39] S. Anu Mary Ealia, M.P. Saravanakumar, A review on the classification, characterisation, synthesis of nanoparticles and their application, IOP Conf. Ser. Mater. Sci. Eng. 263 (2017) 032019.

[40] C.Y. Tai, C. Tai, M. Chang, H. Liu, Synthesis of magnesium hydroxide and oxide nanoparticles using a spinning disk reactor, Ind. Eng. Chem. Res. (2007) 5536–5541.

[41] M.P. Vena, M. Jobbágy, S.A. Bilmes, Microorganism mediated biosynthesis of metal chalcogenides; a powerful tool to transform toxic effluents into functional nanomaterials, Sci. Total Environ. 565 (2016) 804–810.

[42] Z. Li, S.L. Wong, Functionalization of 2D transition metal dichalcogenides for biomedical applications, Mater. Sci. Eng. C 70 (2017) 1095–1106.
[43] S.A. Dahoumane, E.K. Wujcik, C. Jeffryes, Noble metal, oxide and chalcogenide-based nanomaterials from scalable phototrophic culture systems, Enzyme Microb. Technol. 95 (2016) 13–27.
[44] S. Keskin, S. Kizilel, Biomedical applications of metal organic frameworks, Ind. Eng. Chem. Res. 50 (2011) 1799–1812.
[45] A. Arenas-Vivo, G. Amariei, S. Aguado, R. Rosal, P. Horcajada, An Ag-loaded photoactive nano-metal organic framework as a promising biofilm treatment, Acta Biomater. 97 (2019) 490–500.
[46] A. El-Ansary, S. Al-Daihan, On the toxicity of therapeutically used nanoparticles: an overview, J. Toxicol. (2009) 754810.
[47] M. Gielen, E.R.T. Tiekink, Metallotherapeutic drugs and metal-based diagnostic agents, John Wiley & Sons Ltd., Chichester (2005).
[48] K.K. Jain, Applications of nanobiotechnology in clinical diagnostics, Clin. Chem. 53 (2007) 2002–2009.
[49] Y. Li, Z. Tang, P.N. Prasad, M.R. Knecht, M.T. Swihart, Peptide-mediated synthesis of gold nanoparticles: effects of peptide sequence and nature of binding on physicochemical properties, Nanoscale 6 (2014) 3165–3172.
[50] N. Chanda, V. Kattumuri, R. Shukla, A. Zambre, K. Katti, A. Upendran, et al., Bombesin functionalized gold nanoparticles show in vitro and in vivo cancer receptor specificity, Proc. Natl. Acad. Sci. U.S.A. 107 (2010) 8760–8765.
[51] S. Andreescu, M. Ornatska, J.S. Erlichman, A. Estevez, J.C. Leiter, Fine particles in medicine and pharmacy (ed. E. Matijevic), Springer, Boston (2011), pp. 57–100.
[52] E. Boisselier, D. Astruc, Gold nanoparticles in nanomedicine: preparations, imaging, diagnostics, therapies and toxicity, Chem. Soc. Rev. 38 (2009) 1759–1782.
[53] N.L. Pacioni, C.D. Borsarelli, V. Rey, A.V. Veglia, Synthetic routes for the preparation of silver nanoparticles, silver nanoparticle applications, Springer International Publishing, Cham (2015), pp. 13–46.
[54] S. Ahmed, M. Ahmad, B.L. Swami, S. Ikram, A review on plants extract mediated synthesis of silver nanoparticles for antimicrobial applications: a green expertise, J. Adv. Res. 7 (1) (2016) 17–28.
[55] N.L. Pacioni, C.D. Borsarelli, V. Rey, A.V. Veglia, Synthetic routes for the preparation of silver nanoparticles, silver nanoparticle applications, Springer International Publishing, Cham (2015), pp. 13–46.
[56] N. Rajput, Methods of preparation of nanoparticles-a review, Int. J. Adv. Eng. Technol. 7 (6) (2015) 1806.
[57] M.A. Meyers, A. Mishra, D.J. Benson, Mechanical properties of nanocrystalline materials, Prog. Mater. Sci. 51 (4) (2006) 427–556.
[58] M.N. Nadagouda, T.F. Speth, R.S. Varma, Microwave-assisted green synthesis of silver nanostructures, Acc. Chem. Res. 44 (7) (2011) 469–478.
[59] M.G. Guzman, J. Dille, S. Godet, Synthesis of silver nanoparticles by chemical reduction method and their antibacterial activity, Int. J. Chem. Biomol. Eng. 2 (3) (2009) 104–111.
[60] D.P. Chattopadhyay, B.H. Patel, Nano metal particles: synthesis, characterization and application to textiles, in: W. Ahmed, N. Ali (eds.), Manufacturing nanostructures, One Central Press (OCP), Manchester (2014), pp. 184–215.
[61] S.M. Landage, A.I. Wasif, P. Dhuppe, Synthesis of nano silver using chemical reduction methods, Int. J. Adv. Res. Eng. Appl. Sci. 2278 (2014) 6252.
[62] M.G. Guzman, J. Dille, S. Godet, Synthesis of silver nanoparticles by chemical reduction method and their antibacterial activity, Int. J. Chem. Biomol. Eng. 2 (3) (2009) 104–111.

[63] Q.L. Zhang, Z.M. Yang, B.J. Ding, X.Z. Lan, Y.J. Guo, Preparation of copper nanoparticles by chemical reduction method using potassium borohydride, Trans. Nonferrous Metals Soc. China 20 (2010) S240–S244.

[64] R.R. Bhosale, A.S. Kulkarni, S.S. Gilda, N.H. Aloorkar, R.A. Osmani, B.R. Harkare, Innovative eco-friendly approaches for green synthesis of silver nanoparticles, Int. J. Pharm. Sci. Nanotech. 7 (2014) 2328–2337.

[65] R.L. Karnani, A. Chowdhary, Biosynthesis of silver nanoparticle by eco-friendly method, Indian J. Sci. 1 (1) (2013) 25–31.

[66] N. Pantidos, L.E. Horsfall, Biological synthesis of metallic nanoparticles by bacteria, fungi and plants, J. Nanomed. Nanotechnol. 5 (5) (2014) 1.

[67] K. Deplanche, I. Caldelari, I.P. Mikheenko, F. Sargent, L.E. Macaskie, Involvement of hydrogenases in the formation of highly catalytic Pd (0) nanoparticles by bioreduction of Pd (II) using Escherichia coli mutant strains, Microbiol. 156 (9) (2010) 2630–2640.

[68] P. Yong, N.A. Rowson, J.P. Farr, I.R. Harris, L.E. Macaskie, Bioreduction and biocrystallization of palladium by Desulfovibrio desulfuricans NCIMB 8307, Biotechnol. Bioeng. 80 (4) (2002) 369–379.

[69] G. Oberdorster, E. Oberdorster, J. Oberdorster, Nanotoxicology: an emerging discipline evolving from studies of ultrafine particles, Environ. Health Perspect. 113 (2005) 823.

[70] S.T. Holgate, Exposure, uptake, distribution and toxicity of nanomaterials in humans, J. Biomed. Nanotechnol. 6 (2010) 1–19.

[71] S.C. Sahu, D.A. Casciano, Nanotoxicity: from in vivo and in vitro models to health risks, Wiley-VCH Verlag GmbH, Weinheim (2009).

[72] A. Seaton, L. Tran, R. Aitken, K. Donaldson, Nanoparticles, human health hazard and regulation, J. R. Soc. Interface 7 (2010) S119–S129.

[73] P. Li, J. Li, C.Z. Wu, Q.S. Wu, J. Li, Synergistic antibacterial effects of beta-lactam antibiotic combined with silver nanoparticles, Nanotechnol. 16 (2005) 1912–1917.

[74] Y. Yun, Y.W. Cho, K. Park, Nanoparticles for oral delivery: targeted nanoparticles with peptidic ligands for oral protein delivery, Adv. Drug Delivery Rev. 65 (2013) 822–832.

[75] M. Li, K.T. Al-Jamal, K. Kostarelos, J. Reineke, Physiologically based pharmacokinetic modeling of nanoparticles, ACS Nano 4 (2010) 6303–6317.

[76] J. Kim, L. Cao, D. Shvartsman, E.A. Silva, D.J. Mooney, Targeted delivery of nanoparticles to ischemic muscle for imaging and therapeutic angiogenesis, Nano Lett. 11 (2011) 694–700.

[77] Y. Zhang, T.R. Nayak, H. Hong, W. Cai, Biomedical applications of zinc oxide nanomaterials, Curr. Mol. Med. 13 (2013) 1633–1645.

[78] J.H. Park, L. Gu, G. von Maltzahn, E. Ruoslahti, S.N. Bhatia, M.J. Sailor, Biodegradable luminescent porous silicon nanoparticles for in vivo applications, Nat. Mater. 8 (2009) 331–336.

[79] J. Zhou, N.S. Xu, Z.L. Wang, Dissolving behavior and stability of ZnO wires in biofluids: a study on biodegradability and biocompatibility of ZnO nanostructures, Adv. Mater. 18 (2006) 2432–2435.

[80] P. Huang, J. Lin, W. Li, P. Rong, Z. Wang, S. Wang, et al., Biodegradable gold nanovesicles with an ultrastrong plasmonic coupling effect for photoacoustic imaging and photothermal therapy, Angew. Chem. Int. Ed. 52 (2013) 13958–13964.

[81] L. Devi, R. Gupta, S.K. Jain, S. Singh, P. Kesharwani, Synthesis, characterization and in vitro assessment of colloidal gold nanoparticles of Gemcitabine with natural polysaccharides for treatment of breast cancer, J. Drug Deliv. Sci. Technol. (2020) 101565.

[82] P. Cheviron, F. Gouanve, E. Espuche, Green synthesis of colloid silver nanoparticles and resulting biodegradable starch/silver nanocomposites, Carbohydr. Polym. 108 (2014) 291–298.

[83] L. Lartigue, D. Alloyeau, J. Kolosnjaj-Tabi, Y. Javed, P. Guardia, A. Riedinger, et al., Biodegradation of iron oxide nanocubes: high-resolution in situ monitoring, ACS Nano 7 (2013) 3939–3952.

[84] H.W. Chen, J. Burnett, F.X. Zhang, J.M. Zhang, H. Paholak, D.X. Sun, Highly crystallized iron oxide nanoparticles as effective and biodegradable mediators for photothermal cancer therapy, J. Mater. Chem. B 2 (2014) 757–765.
[85] V.S. Bollu, S.K. Nethi, R.K. Dasari, S.S. Rao, S. Misra, C.R. Patra, Evaluation of in vivo cytogenetic toxicity of europium hydroxide nanorods (EHNs) in male and female Swiss albino mice, Nanotoxicol. 10 (2015) 413–425.
[86] D.R. Di, Z.Z. He, Z.Q. Sun, J. Liu, A new nano-cryosurgical modality for tumor treatment using biodegradable MgO nanoparticles, Nanomed. Nanotechnol. Biol. Med. 8 (2012) 1233–1241.
[87] H.S. Choi, W. Liu, P. Misra, E. Tanaka, J.P. Zimmer, B.I. Ipe, et al., Renal clearance of quantum dots, Nat. Biotechnol. 25 (2007) 1165–1170.
[88] J. Liu, M. Yu, X. Ning, C. Zhou, S. Yang, J. Zheng, PEGylation and zwitterionization: pros and cons in the renal clearance and tumor targeting of near-IR-emitting gold nanoparticles, Angew. Chem. Int. Ed. 52 (2013) 12572–12576.
[89] C. Zhou, M. Long, Y. Qin, X. Sun, J. Zheng, Luminescent gold nanoparticles with efficient renal clearance, Angew. Chem. Int. Ed. 50 (2011) 3168–3172.
[90] H.S. Choi, W. Liu, P. Misra, E. Tanaka, J.P. Zimmer, B.I. Ipe, et al., Renal clearance of quantum dots, Nat. Biotechnol. 25 (2007) 1165–1170.
[91] R.J. Griffitt, J. Luo, J. Gao, J.C. Bonzongo, D.S. Barber, Effects of particle composition and species on toxicity of metallic nanomaterials in aquatic organisms, Environ. Toxicol. Chem. 27 (2008) 1972.
[92] Y.L. Hu, W. Qi, F. Han, J.Z. Shao, J.O. Gao, Toxicity evaluation of biodegradable chitosan nanoparticles using a zebrafish embryo model, Int. J. Nanomed. 6 (2011) 3351.
[93] J.H. Sung, J.H. Ji, J.D. Park, M.Y. Song, K.S. Song, H.R. Ryu, et al., Subchronic inhalation toxicity of gold nanoparticles, Particle Fibre Toxicol. 8 (2011) 16.
[94] J.S. Kim, T.J. Yoon, K.N. Yu, B.G. Kim, S.J. Park, H.W. Kim, et al., Toxicity and tissue distribution of magnetic nanoparticles in mice, Toxicol. Sci. 89 (2006) 338.
[95] W.H. De Jong, W.I. Hagens, P. Krystek, M.C. Burger, A.J. Sips, R.E. Geertsma, Particle size-dependent organ distribution of gold nanoparticles after intravenous administration, Biomater. 29 (2008) 1912.
[96] H. Xie, M.M. Mason, J.P. Wise, Genotoxicity of metal nanoparticles, Rev. Environ. Health 26 (2011) 251.
[97] G.M. Hodges, E.A. Carr, R.A. Hazzard, C. O'Reilly, K.E. Carr, A commentary on morphological and quantitative aspects of microparticle translocation across the gastrointestinal mucosa, J. Drug Target. 3 (1995) 57.
[98] Z. Chen, H. Meng, G. Xing, C. Chen, Y. Zhao, G. Jia, et al., Acute toxicological effects of copper nanoparticles in vivo, Toxicol. Lett. 163 (2006) 109.
[99] T. Xia, M. Kovochich, M. Liong, L. Madler, B. Gilbert, H. Shi, et al., Comparison of the mechanism of toxicity of zinc oxide and cerium oxide nanoparticles based on dissolution and oxidative stress properties, ACS Nano 2 (2008) 2121–2134.
[100] A. Elsaesser, C.V. Howard, Toxicology of nanoparticles, Adv. Drug Delivery Rev. 64 (2012) 129–137.
[101] L. Yildirimer, N.T.K. Thanh, M. Loizidou, A. M. Seifalian, Toxicological considerations of clinically applicable nanoparticles, Nano Today 6 (2011) 585–607.
[102] L. Luque-Garcia, R. Sanchez-Díaz, I. Lopez-Heras, C. Camara, P. Martin, Bioanalytical strategies for in-vitro and in-vivo evaluation of the toxicity induced by metallic nanoparticles, TrAC, Trends Anal. Chem. 43 (2013) 254–268.
[103] M.S. Hasnain, M.N. Javed, M.S. Alam, P. Rishishwar, S. Rishishwar, S. Ali, et al., Purple heart plant leaves extract-mediated silver nanoparticle synthesis: optimization by Box-Behnken design, Mater. Sci. Eng. C 99 (2019) 1105–1114.
[104] M.S. Alam, M.N. Javed, F.H. Pottoo, A. Waziri, F.A. Almalki, M.S. Hasnain, et al., QbD approached comparison of reaction mechanism in microwave synthesized gold nanoparticles and their superior catalytic role against hazardous nirto-dye, Appl. Organomet. Chem. 33 (9) (2019) e5071.

[105] J. Pandit, M.S. Alam, J.R. Ansari, M. Singhal, N. Gupta, A. Waziri, et al., Multifaced applications of nanoparticles in biological science, in: Nanomaterials in the battle against pathogens and disease vectors, CRC Press, Boca Raton (2022), pp. 17–50.

[106] S. Mishra, S. Sharma, M.N. Javed, F.H. Pottoo, M.A. Barkat, M.S. Alam, et al., Bioinspired nanocomposites: applications in disease diagnosis and treatment, Pharm. Nanotechnol. 7 (3) (2019 June 1) 206–219.

[107] M.S. Alam, M.F. Naseh, J.R. Ansari, A. Waziri, M.N. Javed, A. Ahmadi, et al., Synthesis approaches for higher yields of nanoparticles, in: Nanomaterials in the battle against pathogens and disease vectors, CRC Press, Boca Raton (2022), pp. 51–82.

[108] M.S. Alam, A. Garg, F.H. Pottoo, M.K. Saifullah, A.I. Tareq, O. Manzoor, et al., Gum ghatti mediated, one pot green synthesis of optimized gold nanoparticles: investigation of process-variables impact using Box-Behnken based statistical design, Int. J. Biol. Macromol. 104 (2017) 758–767.

[109] M.N. Javed, F.H. Pottoo, M.S. Alam, Metallic nanoparticle alone and/or in combination as novel agent for the treatment of uncontrolled electric conductance related disorders and/or seizure, epilepsy & convulsions, Pat. Acq. 10 (2016 October) 40.

[110] F.H. Pottoo, S. Sharma, M.N. Javed, M.A. Barkat, M.S. Alam, M.J. Naim, et al., Lipid-based nanoformulations in the treatment of neurological disorders, Drug Metab. Rev. 52 (1) (2020) 185–204.

[111] M.N. Javed, M.S. Alam, A. Waziri, F.H. Pottoo, A.K. Yadav, M.S. Hasnain, et al., QbD applications for the development of nanopharmaceutical products, in: Pharmaceutical quality by design, Academic Press, Cambridge (2019), pp. 229–253.

[112] M.N. Javed, E.S. Dahiya, A.M. Ibrahim, M. Alam, F.A. Khan, F.H. Pottoo, Recent advancement in clinical application of nanotechnological approached targeted delivery of herbal drugs, in: Nanophytomedicine, Springer, Singapore (2020), pp. 151–172.

[113] M. Aslam, M. Javed, H.H. Deeb, M.K. Nicola, M. Mirza, M. Alam, et al., Lipid nanocarriers for neurotherapeutics: introduction, challenges, blood-brain barrier, and promises of delivery approaches, CNS Neurol. Disord. Drug Targets (Formerly Curr. Drug Targets-CNS & Neurol. Disord.) 21 (10) (2022).

[114] M.N. Javed, F.H. Pottoo, A. Shamim, M.S. Hasnain, M.S. Alam, Design of experiments for the development of nanoparticles, nanomaterials, and nanocomposites, in: Design of experiments for pharmaceutical product development, Springer, Singapore (2021), pp. 151–169.

[115] S. Singhal, M. Gupta, M.S. Alam, M.N. Javed, J.R. Ansari, Carbon allotropes-based nanodevices: graphene in biomedical applications, in: Nanotechnology, CRC Press, Boca Raton (2022), pp. 241–269.

[116] R. Kumar, G. Dhamija, J.R. Ansari, M.N. Javed, M.S. Alam, C-dot nanoparticulated devices for biomedical applications, in: Nanotechnology, CRC Press, Boca Raton (2022), pp. 271–299.

[117] F.H. Pottoo, N. Tabassum, M. Javed, S. Nigar, R. Rasheed, A. Khan, et al., The synergistic effect of raloxifene, fluoxetine, and bromocriptine protects against pilocarpine-induced status epilepticus and temporal lobe epilepsy, Mol. Neurobiol. 56 (2) (2019) 1233–1247.

[118] F.H. Pottoo, S. Sharma, M.N. Javed, M.A. Barkat, M.S. Alam, M.J. Naim, et al., Lipid-based nanoformulations in the treatment of neurological disorders, Drug Metab. Rev. 52 (1) (2020) 185–204.

[120] F.H. Pottoo, M. Javed, M. Barkat, M. Alam, J.A. Nowshehri, D.M. Alshayban, M.A. Ansari, Estrogen and serotonin: complexity of interactions and implications for epileptic seizures and epileptogenesis, Curr. Neuropharmacol. 17 (3) (2019) 214–231.

[121] F.H. Pottoo, N. Tabassum, M.N. Javed, S. Nigar, S. Sharma, M.A. Barkat, et al., Raloxifene potentiates the effect of fluoxetine against maximal electroshock induced seizures in mice, Eur. J. Pharm. Sci. 146 (2020) 105261.

[122] M. Aslam, M. Javed, H.H. Deeb, M.K. Nicola, M. Mirza, M. Alam, et al., Lipid nanocarriers for neurotherapeutics: introduction, challenges, blood-brain barrier, and promises of delivery approaches, CNS Neurol. Disord. Drug Targets (Formerly Curr. Drug Targets-CNS & Neurol. Disord.) 21 (10) (2022).

[123] A. Waziri, C. Bharti, M. Aslam, P. Jamil, M. Mirza, M.N. Javed, U. Pottoo, et al., Probiotics for the chemoprotective role against the toxic effect of cancer chemotherapy, Anticancer Agents Med. Chem. (Formerly Curr. Med. Chem. Anti-Cancer Agents) 22(4) (2022) 654–667.

[124] M.N. Javed, M.H. Akhter, M. Taleuzzaman, M. Faiyazudin, M.S. Alam, Cationic nanoparticles for treatment of neurological diseases, in: Fundamentals of bionanomaterials, Elsevier, Amsterdam (2022), pp. 273–292.

[125] A.M. Ibrahim, L. Chauhan, A. Bhardwaj, A. Sharma, F. Fayaz, B. Kumar, et al., Brain-derived neurotropic factor in neurodegenerative disorders, Biomed. 10 (5) (2022) 1143.

[126] N. Kumari, N., Daram, N., M.S. Alam, A.K. Verma, Rationalizing the use of polyphenol nano-formulations in the therapy of neurodegenerative diseases, CNS Neurol. Disord. Drug Targets (Formerly Curr. Drug Targets-CNS & Neurol. Disord.) 21 (10) (2022) 966–976.

[127] M.F. Naseh, J.R. Ansari, M.S. Alam, M.N. Javed, Sustainable nanotorus for biosensing and therapeutical applications, in: Handbook of green and sustainable nanotechnology: fundamentals, developments and applications, Springer International Publishing, Cham (2022), pp. 1–21.

[128] S. Raj, R. Manchanda, M. Bhandari, M. Alam, Review on natural bioactive products as radioprotective therapeutics: present and past perspective, Curr. Pharm. Biotechnol. 23 (14) (2022).

2 Recent Trends in the Biomedical Application of Metallic Nanoparticles

Parvez Hassan, Mohammad Hossain Shariare, Jannatul Fardous, Tanzilur Rahman, Swami Prasad Sinha, Farheen Waziri, and A. K. M. Shafiul Kadir

2.1 INTRODUCTION

Nanotechnology is the study of matter at the atomic level and its use to create novel materials and apparatuses. Nanotechnology has recently emerged as a promising area. Nanoparticles (NP) that range from 10 nm to 1000 nm [1] in size are the building blocks of nanotechnology. Such materials can be synthesized and manipulated in various forms, which result in increased interest in using them in various fields, including environmental, medical, energy, catalytic, and pharmaceutical [2]. Their optical properties, reactivity, toughness, and many others are heavily influenced by their size, shape, and structure [3]. Their physicochemical properties include quantum effects, larger surface area, and self-assembly [4]. These unique characteristics and their similarity in size with biomolecules make them suitable candidates for various biomedical applications that include biosensing, imaging, therapy, and drug delivery [5–7].

The history of colloidal solutions dates back as far as the Roman era. They were then frequently used to stain glass for decorating items [8]. People used silver coins for the preservation of food too. Indian people used "bhasma", an ancient Indian nanomedicine in Ayurveda-based medication, for treating patients against anemia, fever, asthma, sleeplessness, digestion, cosmetic application, and so on [9]. In 1850, Michael Faraday prepared colloidal gold by naming it "divided metals" and observed that the properties of the so-called "divided metals" solutions differed from those of bulk gold [10]. This can be marked as the modern scientific evaluation of colloidal gold or NPs. Later in 1890, Nobel Prize–winning German microbiologist Robert Koch demonstrated that gold-containing compounds inhibited the growth of bacteria. This antibacterial particularity can be considered the first indication for the application of NPs in foods or food packaging, which was later extended to other sectors such as medicine, water waste treatment, and biosensing [11].

Depending on their size, morphology, and physicochemical properties, NPs can be categorized into metallic nanoparticles (MNPs), ceramic nanoparticles,

 DOI: 10.1201/9781003317319-2

semiconductor nanoparticles, ceramic nanoparticles, and so on [3]. Metal nanoparticles (MNPs) can be monometallic NPs like gold (Au), silver (Ag), copper (Cu), and platinum (Pt), metal oxides like copper oxide (CuO), silica (SiO_2), magnesium oxide (MgO), and zinc oxide (ZnO), or alloys of two or more metals like iron-cobalt (Fe-Co) and iron platinum [12, 13].

MNPs are obtained using various synthesis techniques. This involves physical, chemical, biological, or a combination of them. Chemical reduction is one of the most frequently used methods of synthesis because it generates uniform-sized particles. This method is also scalable, relatively straightforward, and affordable. The green synthesis approach has recently gained popularity too due to its non-toxic nature, resilience, and environmental friendliness [7, 14–19]. Various plants, animals, and microorganisms have been used as promising sources for the green synthesis of MNPs. Recent progress in cost-effective and robust synthesis methods has contributed to the formation of MNPs with adjustable thermal, optical, and physiochemical properties and functionalities [20].

Noble MNPs have been used in the diagnosis and treatment of human diseases, including cancer, diagnostic assays, drug delivery, bioimaging, biosensing, photoablation, hyperthermia, and gene delivery, due to their distinctive characteristics and exceptional ability. Owing to their small (nano) size, NPs can interact with molecules at the surface level and in vivo in cells making them useful in the diagnosis as well as therapeutic processes. Specific functionalization of MNPs with peptides, antibodies, RNA, DNA, and other functional groups provides them with excellent properties for use in contrast imaging, cellular tracking, and image-guided interventions. AuNPs have been commonly used for the identification of protein interactions and the detection of DNA from biological samples for quite some time [5, 6, 21–24]. The role of AgNP in cell imaging, cancer therapy, and genetic delivery is quite significant [25]. The antimicrobial properties of Ag or Cu NPs [26], Ag_2O [27], and ZnO [28] and their improved activities against *E.coli* have been demonstrated numerous times. FeO [29] and MnO_2 NPs [30] functionalized with proteins are used for in vivo bioimaging of tumor cells, biosensing, chemotherapy, and other treatment purposes [31–36]. This chapter provides an overview of MNPs and their synthesis process, the potential platform offered by NPs in various biological-related applications, the challenges associated with metallic NPs, and their future perspective [37–39].

2.2 SYNTHESIS OF METALLIC NANOPARTICLES

2.2.1 Chemical Synthesis of Metallic Nanoparticles

Nanoparticles with given sizes and characteristics are required in different biomedical applications. Chemical synthesis of nanoparticles is a very challenging process because the synthesis technique determines the size, shape, and surface properties of nanoparticles [40]. A range of chemical synthesis processes are available, which include co-precipitation, thermal decomposition, hydrodermal synthesis, sol-gel, microemulsion, and so on.

2.2.1.1 Co-precipitation

The most popular technique for creating MNPs is co-precipitation. Because it uses benign materials, it is frequently employed in biomedical applications [41]. The term "co-precipitation" is used because in this method, a precipitate carries out a lot of substances that are normally soluble in that solution. During this technique, two aqueous salt solutions are mixed and a base is added under inert conditions at room temperature. An inert atmosphere is employed in the co-precipitation technique using nitrogen. Inert atmosphere maintenance is critical because magnetic particles can be oxidized into maghemite when they come in contact with air [42]. The MNP size can be controlled to 5–40 nm using the co-precipitation synthesis method. By adjusting the pH and ionic strength of the medium, it is possible to control the MNPs' size, shape, and magnetic properties [43].

2.2.1.2 Microemulsion

Microemulsion method is used for MNP synthesis, where two immiscible liquids are dispersed to form a stable emulsion system. Water in oil (W/O) microemulsion system isa widely used method to synthesize uniform-size MNPs, which are generally stabilized using a surfactant [44]. In this system, the water pools remain spherical in shape, and the surfactant molecules surround those water pools. Water pool walls surround the growing MNPs and help reduce the particle size through collision and aggregation [42]. The precipitate is developed using a solvent like ethyl alcohol or acetone, and extraction of the precipitate is done by filtration or centrifugation [45]. The size of the MNPs can be controlled between 10 and 25 nm using this synthesis method, and particles remain monodispersed and spherical [46]. The scale-up procedure of the microemulsion technique is critical due to the remaining surfactant, and this is a drawback of this method. However, this technique needs a massive volume of solvent, and the quantity of MNPs produced is less [45].

2.2.1.3 Hydrothermal Synthesis

Hydrothermal technique is also a popular method of synthesizing crystalline MNPs. This process involves maintaining a metal linoleate, a liquid phase of ethanol and linoleic acid, and a solution of water and ethanol under high temperature and pressure [47]. The processis performed at 220°C, pressure is higher than 107 Pa, and the overall reaction time is about 72 h. The MNPs synthesized by this method are uniform in size and shape, and the size is controllable from a few nanometers to hundreds [48]. Nanoparticles synthesized by this method have better crystallinity and are within the range of 200–800 nm [49].

2.2.1.4 Sol-Gel

This method of synthesis involves the hydroxylation and condensation of a colloidal nanoparticle solution, which is then dried by removing the solvent to produce MNPs. The reaction is performed at room temperature using water as a solvent [48]. The starting material could be a metal alkoxide dissolved in an appropriate solvent, usually ethyl alcohol. The addition of a small volume of water hydrolyzes the material in order to form a polymeric material, which can be improved by making the solution slightly acidic. This method can generate particle sizes of the MNPs in the range of 15 to 50 nm [50]. Previous research studies suggest that the use of a basic catalyst results in a colloidal gel and the use of an acid catalyst results in polymeric particles, where heat is critical to getting a crystalline final product. The volume and phase of the solvent are important in this method for the magnetic ability of the synthesized material [51].

2.2.1.5 Thermal Decomposition

To synthesize particles with a narrow size distribution, thermal decomposition can be an excellent approach that provides MNPs with a narrow size distribution and a better particle mean diameter [52]. In this technique, heat is needed to break chemical bonds within the compound undergoing decomposition, and therefore the reaction is endothermal. Thermal decomposition is accomplished by two different policies, specifically "heating-up" and "hot-injection". Heat is applied continuously in a pre-mixed solution of formulation compounds during the heating-up process, while surfactants and solvent are added at a fixed temperature, and then the MNPs start clustering and growing [53]. In the hot-injection approach, reagents are injected into a hot surface-active agent solution, followed by a controlled growth stage, resulting in rapid and uniform nucleation. In both processes, organic solvents and surfactants are present while heating a non-magnetic formulation ingredient [54].

2.3 GREEN SYNTHESIS OF METALLIC NANOPARTICLES

The green synthesis technique enables microbes and plants to make nanoparticles that are biologically safe, cost-effective, and environmentally benign. Physical, chemical, and biological mechanisms can be used to generate nanoparticles of different shapes and sizes. However, using severe corrosion inhibitors to utilize physical and chemical processes results in high energy requirements, low yield, high expense, and environmental impact. Different chemical factors, such as solvent, temperature, pressure, and pH, affect green synthesis approaches based on biologically based precursors. The biological routes use microorganisms (bacteria, fungi, yeast, algae, and so on) or plants, and using microorganisms is riskier due to pathogenicity concerns; it also necessitates the care of huge cultures [55]. Green metal nanoparticles are harmless in origin, with no deleterious byproducts formed during the production process. This methodology also produces a larger output of nanoparticles than other standard procedures. Numerous physiochemical production processes demand high radiation doses, extremely toxic reductants, and stabilizing agents—all of which can be harmful to both people and marine life. On the other hand, a single or multiple environmentally friendly bio-reduction processes that start with relatively little energy are known as "green synthesis" of MNPs [56].

2.3.1 Plant-Based Synthesis of Metallic Nanoparticles

Because there are so many plants being studied for green synthesis, it is not possible to include them all. Plants with pharmacological or therapeutic significance that have been investigated for green synthesis have been prioritized above randomly chosen ones. For a better understanding of the plants' capacity to reduce metallic ions and to support the capping nature of MNPs produced by green synthesis, the mechanism of MNP production has been thoroughly investigated. The binding of biomolecules to the surfaces of MNPs during synthesis is assumed to be caused by green synthesis. This phenomenon is known as "capping" in the field of green synthesis [57].

Azadirachta indica and *Camellia sinensis* are two plant species that have been thoroughly investigated for their ability to produce plant-mediated MNPs. By interacting with Au^{3+} cations, *C. sinensis* (tea) biomolecules were extracted and purified, demonstrating their participation in the environmentally friendly creation of gold nanoparticles. The green manufacture of platinum nanoparticles utilizes pure tea polyphenol, which is commercially accessible [58]. *Azadirachta indica*'s reducing and capping activity in the environmentally friendly creation of gold and silver nanoparticles was also confirmed by the extraction and purification of the tetranortriterpenoid azadirachtin. *Asparagus racemosus* was used in the photosynthesis of palladium nanoparticles, which is an outstanding attempt to reduce the cost of green synthesis. Changes in reaction temperature mostly caused changes in the morphology of the created MNPs [59]. Common phytochemical elements like phenols, alkaloids, and terpenoids have been attributed to the green synthesis of many MNPs [60]. Because their purest preparations are readily available on the market, phenolic substances like epicatechin and quercetin-glucuronide are essential in the green synthesis of MNPs (like Fe_2O_3). It has been suggested that the cyclic peptides curcacycline A and curcacycline B discovered in the latex of *Jatropha curcas L.* can decrease metal cations [61]. *Alfa sprouts* have been proposed for in situ manufacturing of silver nanoparticles [62].

Cycascircinalis leaf extract, *Pinus thunbergii* pine cone extract, aqueous extract of *Thujaorientalis* leaf, *Torreya nucifera* leaf extract, *Pterisbiaurita* leaf extract, flavonoids in *Azollamicrophylla*, fern *Nephrolepis exaltata* leaflet extracts, thallus of *Anthoceros*, *Riccia* and *Fissidens minutes, Oscillatoria willei* cyanobacteria, aqueous extract of *Hypneamusciformis* leaf are responsible for the green synthesis of silver nanoparticles [63]. *Azollapinnata* aqueous extract and cell extracts/cultures from other algae, such as *Sesbania drummondii, Rhizoclonium fontinale, Lyngbya majuscule, Spirulina subsalsa, Spirulina platensis, Chlorella vulgaris, Tetraselmis kochinensis*, and *Fucus vesiculosus*, are used to create gold nanoparticles. The *Adiantum philippense* L. For the creation of gold and silver nanoparticles, extracts can be used [64]. Copper nanoparticles are synthesized at room temperature using *Ginkgo biloba* Linn leaf extract [65].

2.3.2 Animal-Based Synthesis of Metallic Nanoparticles

For the green synthesis of MNPs, animal hosts were evaluated to see their cellular response to the administered nanomaterials for the prospective outcome.

Silk protein, for example, fibroin, produced by the various species of spiders, silkworms, and tree ants, is responsible for synthesizing fibroin-titanium dioxide nanocomposite and nano-hydroxyapatite (which is a form of calcium NPs) [66]. Another silk protein, for example, sericin, can be implemented to synthesize NPs [67]. Marine organisms, for example, sponges and starfish, can synthesize biological silica NPs with enzymatic activity. Earthworms, for example, *Eisenia andrei* extracts, are used to synthesize nano-red gold in the presence of trichlorogold;hydrochloride. Marine worms, for example, Polychaeta, are used for the green synthesis of silver NPs [68]. Chitosan found in chitins, derived from the shells of crustaceans, was capable of synthesizing titanium dioxide-chitosan nanocomposites [69].

2.3.3 Microbiome-Based Synthesis of Metallic Nanoparticles

Toxic wastes generated during the chemical synthesis of MNPs have paved the way for the green synthesis of MNPs using microbiomes. Bacteria, viruses, fungi, and yeast are the microbiomes that produce abundant, cost-effective, and eco-sustainable nanoparticles [70]. The characteristics of MNPs manufactured by green synthesis employing microbiomes are pretty similar to those of MNPs created by chemical synthesis [71]. The synthesis of MNPs in microbiomes occurs either intracellularly or extracellularly. It varies among different microorganisms. As a prelude to the existence of iron oxide, the intracellular formation of iron oxide (Fe_3O_4) crystals occurs in *Aquaspirillum magnetotacticum* during the crystal blooming process. When it comes to metal binding, the bacterial cell wall is highly essential. The intracellular metal deposition was seen after chemical treatment of *Bacillus subtilis* peptidoglycan, which was twenty to forty thousand times higher than the extracellular quantity [72]. During the culture of *Klebsiella pneumonia* in the culture medium containing cadmium ions, it was seen to form cadmium sulfide on the surface of the bacterial cells mentioned earlier. Silver NPs were seen to form between the cell wall and the plasma membrane of *Pseudomonas stutzeri*. Sulfate-reducing bacteria, *Desulfovibrio desulfuricans*, can produce palladium nanoparticles in the presence of hydrogen. Gold (Au), silver (Ag), and titanium (Ti) were seen to form intracellularly in *Lactobacillus* sp [73].

Gold (Au) and silver (Ag) nanoparticles are synthesized extracellularly when the NADH-enzyme *of Fusarium oxysporum* fungi reacts with aqueous tetrachloroaurate and silver nitrate ions. Also, in the presence of aqueous potassium fluorozirconate, this fungus can synthesize zirconia. Eventually, platinum, silica, and titanium nanoparticles are produced when hexachloroplatinic acid, aqueous silica, and titanium anions are present. *Verticillium luteoalbum* can synthesize gold (Au) NPs. *Aspergillus flavus* and *Coriolus versicolor* were able to synthesize silver (Ag) NPs [74]. Bacteriophages, for example, *tobacco mosaic virus* (*TMV*) and *M13 bacteriophage*, have shown the ability to synthesize various MNPs, namely silicon dioxide, cadmium sulfide, zinc sulfide, lead sulfide, and iron oxide [75].

2.4 BIOMEDICAL APPLICATIONS OF METALLIC NANOPARTICLES

2.4.1 Applications of Metallic Nanoparticles in Medical Diagnosis and Molecular Biology

Because of their inherent physicochemical features, MNPs have found numerous biomedical applications. As particle size decreases, the high surface area of particles produces specific properties that differ significantly from those of a macro-sized structure. There are numerous processes involved in the synthesis of MNPs, which fall under the top-down or bottom-up categories. The top-down approach uses physical, chemical, or mechanical processes to break down bulk materials into nano-sized particles, whereas the underside approach combines individual atoms and molecules. The most common technique for synthesizing metallic nanoparticles is thermal decomposition in the bottom-up approach, which is flexible, easier, less expensive, and generates homogeneous particles. The biological approach to synthesizing nanoparticles has recently gained popularity due to its non-toxic nature, low cost, sustainability, and environmental friendliness [76].

2.4.2 Application of Metallic Nanoparticles in Bioimaging

Despite the fact that X-ray technology was utilized in diagnostic imaging, a few non-intrusive approaches were developed and effectively applied in a number of medical research fields, including drug development, therapeutic agents, diagnostic analysis, and cellular biology research. Clinical imaging research is carried out in various fields of study and significantly advances medical science. Modifications have been made to MRI, electro-optic imaging, ultrasonic scans, positron emission tomography, and other molecular bioimaging structures [77]. Noble metals, such as gold and silver, nanoparticles can play a vital role in instantaneous ignition and monitoring since living tissues emit photons at near-infrared wavelengths. Common noble MNPs were used in clinical research and treatments due to their potent absorption of near radiation. These nanoparticles can therefore be viewed as highly effective diverse agents [76].

2.4.3 Application of Metallic Nanoparticles in Magnetic Resonance Imaging

Because of its better spatial resolution, noninvasiveness, and nondestructive nature, magnetic resonance imaging (MRI) is by far the most widely used medical imaging modality in clinical medicine. Nanotechnology has transformed the capabilities of the MRI imaging modality. In the field of medical imaging, nanoparticles (NPs) continue to garner attention due to their potential for specific clinical applications found in experimental studies [78]. The combined effect of multimodal imaging and the need for an appropriate response will result in cutting-edge innovations that will exploit the benefits of MNPs. However, molecular imaging responsiveness is clearly lower than that of computed tomography and optical imaging. MRI contrast agents, such as super-paramagnetic iron oxide MNP, are emerging as some of the best potential probes for quality and highly at the cellular or perhaps even molecular and cellular levels [76].

2.4.4 Application of Metallic Nanoparticles as Contrast Agents in X-ray Computed Tomography (CT)

X-ray computed tomography is widely used in both medicine and the study of biological and chemical samples. Some medical applications performed in clinics necessitate special procedures in which contrast agents may be required to improve the visual. Iodine-based contrast agents have been found to be environmentally hazardous, and some patients have iodine prejudice due to thyroid-related diseases [78]. Scientists from various fields are working to develop contrast agents that can be used instead of iodine-based contrast agents. Because materials have high X-ray absorption and specific gravity, MNPs have the potential to be used as CT contrast agents. Metal nanoparticles are thought to be toxic to living organisms. Trying to cover nanoparticles with organic molecules frequently reduces toxicity while improving bioactivity [77].

2.4.5 Application of Metallic Nanoparticles in Tissue Construction

Recently, tissue engineering has paid a lot of attention to nanoparticles. Functional replacements for different kinds of wounded tissue are made using engineering and physiological principles. For usage in developing or regenerating tissues, nanoparticles with low toxicity, adaptable properties, diverse agent activities, active targeting capability, and exact behavioral control are a good choice [79]. Au nanoparticles may be employed in stem cell treatment, which is used in tissue switching due to the formation of malignant cells. BGC nanoparticles, which are made of bioactive glass ceramic, are another promising substance frequently employed in repair and regeneration [80].

2.4.6 Application of Metallic Nanoparticles in Protein Detection

Proteins are a crucial component of the body's language, industrial equipment, and framework, and recognizing their functions is critical for further advancement in human health. In immunocytochemistry, gold nanoparticles are commonly used to identify protein-protein interactions. Surface-enhanced Raman scattering spectroscopy is a well-known method for detecting and identifying single dye molecules. Combining both methodologies in a single nanoparticle investigation can significantly improve protein probe multicarrier functionality [81].

2.5 APPLICATIONS OF METALLIC NANOPARTICLES IN DRUG DELIVERY

Metal nanoparticles show great promise for the accurate treatment and detection of a variety of disorders in the medical profession [82]. Due to their distinctive physicochemical properties, such as their high electrical properties, high mechanical and thermal stability, large surface area, high optical and magnetic properties, and particularly their biological application of metal- and metal oxide supported nanomaterials in drug delivery, these materials have drawn more attention [77]. Increased bioavailability, drug stability, targeted action, required half-life of the drug carrier in circulation; and so on make MNPs a useful carrier in drug delivery system (DDS) [83]. Due to their enhanced permeability and retention (EPR) impact, MNPs [77].

2.5.1 Metallic Nanoparticles in Cancer Drug Delivery

The objective of anticancer therapy is to ensure maximum bioavailability of anticancer drugs with site-specific action, thereby reducing the harmful effects on healthy tissues [84]. MNPs have a large surface-area-to-volume ratio, which allows a surface for chemical modification, resulting in enhanced cellular uptake of anticancer drugs along with their improved bioavailability [85]. Different types of metal nanoparticles, including gold (Au), palladium (Pd), silver (Ag), zinc (Zn), titanium (Ti), copper (Cu), and so on, are extensively used as effective carriers for chemotherapeutic agents now [83].

Gold nanoparticles (AuNPs) are one of the most investigated MNPs in the field of cancer drug delivery. Their low toxicity, easy surface modification prosperity, increased bioavailability, and immunogenicity make them suitable DDS with targeted action against cancer cells. AuNPs were found effective against several human cancer cell lines, namely liver cancer, colon cancer, breast cancer, lung cancer, and so on. AuNPs increase cancer cell apoptosis and inhibit their proliferation by means of enhanced reactive oxygen species (ROS) production, increased caspase expression, and reduced antiapoptic protein expression [86]. Different types of anticancer drugs like methotrexate, oxaliplatin, paclitaxel, cyclophosphamide, doxorubicin, and so on showed higher bioavailability and targeted drug delivery after conjugation with gold nanoparticles [87].

Silver nanoparticles (AgNPs) are another widely studied MNP as a vehicle for anticancer drugs. Like the AuNPs, AgNPs also have antitumor activity against prostate cancer, lung cancer, liver cancer, ovarian cancer, osteosarcoma, and so on [88]. AgNPs exert cytotoxicity against cancer cells by altering membrane fluidity, inducing early apoptosis, and increasing ROS production. AgNPs release Ag^+ cations in mitochondria and nucleus, resulting in DNA fragmentation and cancer cell death. Recent studies revealed the possibility of successful delivery of chemotherapeutic agents, for example, paclitaxel, cisplatin, methotrexate, and doxorubicin, from AgNPs to the tumor site with enhanced antitumor activity [89].

One of the possible prospects in the field of nanomedicine, notably for the treatment, diagnosis, and imaging of cancer, is iron oxide nanoparticles (IONPs). Because of their high targeting capabilities under external magnetic fields and surface tenability, drug-loaded IONPs have become widely used in the treatment of cancer. IONPs are able to move through the blood capillaries to the target site in the presence of an applied magnetic field, leading to increased therapeutic efficacy along with minimal damage to surrounding healthy cells [90]. Additionally, the magnetic characteristics of IONPs convert radiant energy into heat energy, or ROS, after the application of an external magnetic field, helping to lessen the side effects of chemotherapy. Similar to AuNPs and AgNPs, IPNPs are a prospective carrier for many chemotherapeutic drugs (such as doxorubicin, docetaxel, and curcumin), either by themselves or in conjunction with other nanomaterials, for the treatment of various malignancies [86].

Besides, other metal nanoparticles are also used in cancer therapy as DDS. For example, zinc oxide (ZnO) nanoparticles have anticancer activities against several human cancer cell lines by forming micronuclei in the tumor cells [91]. Likewise, cuprous and copper oxide nanoparticles trigger apoptosis, ROS generation, stimulate autophagy, and inhibit metastasis when used in tumor therapy. Titanium dioxide (TiO_2), cerium oxide (CeO) nanoparticles also have intrinsic cytotoxic activity against cancer cells caused by radiation and oxidative stress. Palladium nanoparticles (PdNPs) have strong antitumor efficacy against human cervical cancer cell lines as well as other cancer cells owing to their high porosity and pH-sensitive drug release properties. Additionally, doxorubicin-loaded platinum nanoparticles (Pt NPs) show a significant reduction in melanoma tumor growth by increasing the tumor suppressor protein p53 and diminishing tumor proliferation markers [92].

2.5.2 Metallic Nanoparticles in Antimicrobial Drug Delivery

Because of their enormous surface area and nanoscale, AgNPs have drawn increased attention for their antibacterial action against a variety of pathogens, including viruses, bacteria, and fungi. AgNPs are being used as antiviral medicines more frequently these days as a result of their effectiveness in inhibiting various hepatitis, coronavirus, influenza, herpes, recombinant respiratory syncytial virus, and human immunodeficiency virus strains. In order to inactivate viral biomolecules including sulfhydryl, amino, carboxyl, phosphate, and imidazole groups and prevent viral treatment resistance, AgNPs interact with these molecules. Similarly, AgNPs also showed significant antibacterial efficacy where Ag^+ cations damaged the bacterial membrane and caused cellular leakage followed by bacterial death [93]. Nanoparticle attachment to the cell membrane is facilitated by the electrostatic attraction between the positively charged Ag ions and the negatively charged bacterial cell membrane. However, AgNPs have a stronger bactericidal effect against gram-negative bacteria compared to gram-positive bacteria because of their high membrane thickness. Beside this, silver nanoparticles are extensively used as antifungal agents with maximum therapeutic efficacy against several pathogenic fungi [94].

Gold nanoparticles have the ideal combination of physical, chemical, optical, and electrical properties, which makes them a promising medication carrier for antimicrobial medicines thanks to advancements in biomedical nanotechnologies [83]. Owing to the large surface-to-volume ratio of AuNPs, they can conjugate with antibiotics by expressing abundant functional ligands on their surface, followed by increased multivalent interactions with target bacteria. AuNPs are widely used in multi-drug-resistant bacteria and are effective against both gram-positive and gram-negative bacteria [95]. Copper nanoparticles' antibacterial properties are currently receiving more attention, despite the fact that they are less effective than AgNPs and ZnO NPs. *Bacillus subtilis* and *Bacillus anthracis* have more carboxyl and amine groups on their cell surfaces, which supports the increased affinity of Cu ions for the cell surface. Cu NPs damage membranes, induce oxidative stress through increased ROS generation, and cause microbial mortality [96]. These nanoparticles are also active against various bacterial pathogens like *Escherichia coli, Streptococcus aureus,* and *Klebsiella pneumonia*. However, use of Cu NPs is limited due to their poor stability under atmospheric conditions and hazardous effects on the environment [97].

Because ZnO nanoparticles have a negligible impact on healthy cells, they have been demonstrated to be safe for use as antibacterial agents. These MNPs have potent activity both against temperature-resistant spores and gram-positive and gram-negative bacteria. Cells are damaged by oxidation when ROS are produced as a result of H_2O_2 formed on the surface of ZnO NPs [98]. ZnO NPs significantly inhibit the growth of fungi by the same mechanism as bacteria. In addition to the previously described metal nanoparticles, MgO nanoparticles, and TiO_2 nanoparticles also have antimicrobial activities against a broad spectrum of bacteria. Both of these MNPs exert prominent antifungal activity against some fungi [99].

2.5.3 Metallic Nanoparticles in Insulin and Anti-inflammatory Agents' Delivery

Very recently, some studies have shown the efficacy of gold nanoparticles as a carrier for insulin in the treatment of type I diabetes. These AuNPs are able to release the insulin immediately into the hypoglycemic state, followed by maintaining the normoglycemic level for up to 48 hours in type I diabetes [100]. On the other hand, rheumatoid arthritis was successfully treated with Au/Fe/Au plasmonic nanoparticles loaded with methotrexate (MTX) and combined with arginylglycylaspartic acid (RGD). While iron NPs transport the medicine to swollen joints in the presence of an external magnetic field, AuNPs release the drug in the inflammatory area when exposed to near-infrared (NIR) radiation. Use of these MNPs ensures superior efficacy of anti-inflammatory agents in arthritis, even at a lower dose [101].

2.5.4 Metallic Nanoparticles in Vaccine and Gene Delivery

Tumor/cancer vaccines are efficient tools for cancer prevention and treatment in addition to chemotherapy drugs. MNPs are effective adjuvants for nano-vaccines because they can trigger an immune response to neoplastic transformation via a transcellular and/or intracellular signaling cascade [102]. Tumor antigens, for example, tumor lysate, nucleic acid, and short peptides, are administered as cancer vaccines to induce strong immunity. MNPs like iron oxide NPs and zinc oxide NPs deliver these antigens effectively with enhanced cellular uptake by antigen-presenting cells, resulting in a greater immune response against tumor cells. Likewise, the high surface charge of MNPs raises the possibility of using them as a suitable carrier for antisense nucleotides, that is, DNA and RNA, which inhibit specific gene expression in tumor cells [103].

2.6 CYTOTOXICITIES OF METALLIC NANOPARTICLES

2.6.1 Main Consideration

A major problem of nanotechnology is the release of metal nanoparticles into the environment and into humans. The growing commercial use of nanoparticles will almost probably lead to increased exposure, which could have negative effects on human health. This suspicion is described in terms of the large subject of nanotoxicology. Numerous nanomaterials can be ingested by humans through their gastrointestinal tracts, absorbed through their skin, or inhaled by them. The skin is among the most significant physiological barriers in people's and animals' systems, with many layers. It is used to prevent disease or the detrimental consequences of any material. When nanoparticles are consumed, they are reported to increase more in the liver, which is a highly metabolic organ [104]. The condition caused glutathione depletion, damaged mitochondrial possibilities, and a boost in oxidative stress. Similarly, the respiratory system is an innovative target for ingested nanoparticles. At the moment, many metal nanoparticles have been shown to cause peroxidation and other harmful effects on the pulmonary system at the moment [105].

2.6.2 Cytotoxicity of Metallic Nanoparticles Manufactured by Green Synthesis

Nanomaterials, especially those produced utilizing environmentally benign methods, have considerable potential for use in the medical industry as controlled DDSs and anticancer agents. On the other side, there is still concern about the biosecurity of MNPs utilized as anticancer drugs. It was found that the size of nanoparticles had an adverse effect on their cytotoxicity. The majority of the MNPs that plants produced were circular or almost spherical, with a mean average lethal dose of 120 g/mL. Less potent nanoparticle designs existed. The majority of plant-mediated MNP synthesis methods were discovered to be cytotoxic, but some were not [106]. Ag-NPs were tested for cytotoxicity on the MOLT-4 cell line using the MTT assay. These nanoparticles showed impressive antibacterial activity against microbes at low concentrations. According to the cytotoxicity assays, green produced Ag-NPs had an IC50 of 0.011 m as opposed to 1.8 for cisplatin, a more potent monoclonal antibody for the MOLT-4 cell line [107].

2.6.3 Cytotoxicity of Metallic Nanoparticle Manufactured by Chemical Synthesis

MNPs, particularly gold nanoparticles (AuNPs), have wide applicability in biomedicine. A significant problem is their cytotoxicity, which is highly dependent on a variety of factors, including nanoparticle morphology. Even though MNPs have had an effect on cell membrane stability, their physical form as well as their physiochemical interrelations with cell membranes may affect cell survivability. Differences in the appearance of gold nanoparticles might very well indicate specific nanoparticle morphologies with high cytotoxicity. Gold nanospheres and nanorods were found to be more toxic than gold nanostructures such as stars, flowers, and prisms. This is due to their tiny dimensions and the agglomeration procedure. Because of their antibacterial activities, silver nanoparticles (Ag-np) are progressively being used in wound dressings, catheters, and a variety of household items [108]. Human glioblastoma cells (IMR-90) and common human respiratory fibroblast cells (IMR-90) were used to investigate the toxicity of starch-coated silver nanoparticles (U251). To evaluate toxicity, changes in cellular shape, viability, metabolic processes, and oxidative stress were taken into account. AgNPs decreased the amount of ATP in the cell, harming the mitochondria, and increased the production of ROS over the course of a day [109].

2.7 CURRENT CHALLENGES ASSOCIATED WITH METALLIC NANOPARTICLES

MNPs have been increasingly applied in the field of medicine currently owing to their increased bioavailability, EPR effect, and ability to deliver several therapeutic agents and biologics preferentially to the target site. Nevertheless, despite multiple difficulties and obstacles at various phases of development, only a very small number of MNP-based medications have been licensed for clinical use. In order to produce consistent nanoparticles with the appropriate physicochemical properties, biological behaviors, and pharmacological profiles, rigorous design and engineering, thorough analytical methodologies, and reproducible scale-up and production processes are needed [84].

2.7.1 Challenges of Targeted Action

Successful delivery of drugs from MNPs to the disease site is a prerequisite for therapeutic action. Nanoparticles have to aid in the passage of drug molecules through different biological barriers; for example, oral NPs should be stable in gastric acid and penetrate through the intestinal epithelium to attain systemic availability. Similarly, after intravenous injection, nanoparticles need to pass through several biological barriers to reach the target site. In the case of cancer therapy with MNPs, drug delivery is often hampered because of high cancer cell density and dense tumor stroma [110]. High interstitial fluid pressure in tumors, which acts as a barrier and lessens transvascular drug delivery across the tumor, is brought about by impaired lymphatic drainage. Additionally, in order to provide greater efficiency and ideal internalization, the affinity and efficacy of nanocarrier ligands must be tuned [111].

2.7.2 Size, Shape, and Surface Chemistry of Metallic Nanoparticles

The biological behavior of metal NPs largely depends on their size, shape, and surface chemistry of NPs. In the case of MNP-based nanomedicine, both particle size and their distribution have a significant impact on their distribution, pharmacokinetics, and safety [84]. Particle size should be controlled carefully since small size (<20 nm) NPs are subjected to rapid clearance while larger NPs (>200 nm) cause decreased cellular uptake. A crucial criterion that assesses the heterogeneity of particles with regard to size, shape, or mass is the size distribution of nanoparticles expressed as polydispersity (PD). The targeting properties, drug release rate, biocompatibility, toxicity, and in vivo behaviors of MNPs with the same average size but varying PD can be significantly altered, in addition to their stability. The encapsulating medicine has been delivered to the target location more effectively by spherical nanoparticles than by rod-shaped, helical-shaped, and cuboidal-shaped nanoparticles [112]. In another context, surface properties, namely charge, hydrophobicity, functional groups, and so on, are critical parameters while designing a metal nanoparticle. These surface properties are determinants of nanoparticle stability, their behavior, and their interaction with proteins and cells after administration [113]. High surface affinity of AuNPs readily adsorbs proteins, resulting in modification of surface properties and hence required surface modification for desired therapeutic action. Similarly, AgNPs tend to agglomerate in culture media, cytoplasm, and cell nuclei because of their surface charge and may exert toxic effects. Dissolution rate of MNPs is also affected by surface charge along with the surrounding media, thereby requiring critical observation during nanoparticle production [93].

2.7.3 Challenges with MNPs' Analysis and Characterization

Due to the complexity of metal nanoparticles, advanced testing techniques are needed to completely describe the created NPs' physical, chemical, and biological properties. Despite the fact that a number of cutting-edge testing techniques are being developed for MNP analysis, they frequently fall short in their ability to distinguish between active formulation and inactive or less active formulation [84].

2.7.4 Pharmacology and Safety Challenges

Evaluation of pharmacokinetic properties of nanoparticles is done by measuring the plasma drug concentration as one of the standard approaches. However, it is challenging for MNPs due to their nanosize range and may often lead to inconsistent pharmacokinetic properties. Therefore, the determination of bioequivalency and therapeutic activity of nanoparticles has been more challenging until today. Additionally, there is no sufficient data on MNP toxicity, particularly invivo and for prolonged administration [114].

2.7.5 Regulatory Challenges

Although various nanoparticles are already approved by FDA and European Medicines Agency (EMA) for cancer therapy, no guideline for drug products containing soft matter has been approved by any regulatory authorities. As a result, NPs evaluation became solely dependent on individuals risk and benefit assessments, which are time consuming and require high-level expertise, leading to regulatory delays. At this moment, it is vital to develop advanced and multifunctional tools along with standardized guidelines for nanoparticles in order to ensure the safe use of MNPs [84].

2.8 FUTURE PERSPECTIVES OF METALLIC NANOPARTICLES

MNPs are essential in the twenty-first century because of how widely they are used in biology and medicine. MNPs have demonstrated a valuable and dominant function in the treatment of cancer by improving drug targeting and delivery. The MNPs have evolved over time to serve as an imaging diagnostic tool for cancer cells. In the biomedical field, MNPs have been proven to be safe, efficient, and affordable in comparison to currently used treatments. Despite having many benefits, MNPs still have certain issues when used on people. One of the biggest difficulties when creating nanoparticles is getting MNPs to get through biological barriers. These biological limitations may significantly reduce pharmaceutical businesses' ability to invest in metallic nanomedicines. In order for MNPs to overcome biological obstacles, it is crucial to understand disease pathophysiology and disease heterogeneity among different people. To provide better drug targeting to the illness site through the simple passage of MNPs through the membrane, changing the physicochemical features of MNPs should also be a major issue. Often, optimistic MNPs are unsuccessful in clinical trials since the correlation between patient biology and MNP behavior is poorly understood. Therefore, a comprehensive evaluation of preclinical data for safety, efficacy, biodistribution, and pharmacokinetics in an appropriate animal model is a must to overcome this challenge.

On the other hand, it is considered that most of the MNPs work through the EPR effect when used as carriers for chemotherapeutic agents. However, drug accumulation by EPR was reported only in few tumors owing to the heterogeneous nature of tumors and inter-patient and intra-patient variability. Therefore, a particular size of nanoparticles for all approaches should be avaoided while designing MNPs for cancer therapy. As discussed in Section 2.2, MNPs are synthesized either by chemical synthesis or by green synthesis, and the synthesis process is multifaceted and difficult. Complexity of synthesis processes as well as the complex structure of MNPs limit their production on a large scale because of their high manufacturing costs. Modification of the nanoparticle synthesis process should be taken into consideration for the successful implementation of MNPs in the biomedical field in an economical way. The use of natural resources during green synthesis helps to some extent to solve this issue, but the materials used must also be cheap, eco-friendly, and devoid of hazardous chemicals. Additionally, MNPs have the potential to cause clinical toxicities, necessitating a previous evaluation of their general safety for human usage. While this is going on, several metal nanoparticle kinds and their composites have recently gained popularity in the medical, biological, and health care industries, necessitating thorough toxicological and safety profile investigations. More ways to lessen the self-toxicity of MNPs include ion doping, functionalization, and conjugated polymer metal oxide nanoparticles. Finally, thanks to technological advancements, MNPs are currently undergoing advanced testing for multidirectional applications. Metal nanoparticles are mostly employed today to treat malignant cells. The time has come to look into MNPs uses as a DDS for several therapeutic drug classes. Additionally, MNPs' antibacterial properties could be employed therapeutically, to create medical devices with auto-disinfection capabilities, and there is a strong likelihood that a low-cost alternative to conventional antibiotics will be discovered. Therefore, resolving the issues with MNPs in the near future will undoubtedly aid in improving the clinical translation of MNPs for their biological application in the treatment of cancer as well as other diseases.

2.9 CONCLUSIONS

This chapter explores MNPs, their techniques of manufacture, and numerous biological applications. Through a thorough analysis of the literature and the clinical use of nano-based systems, it presents an overview of recent developments in nanomedicine, notably in DDSs. The effectiveness of these systems in enhancing the conventional drug-based system and their function in selective diagnosis are also covered.

In the field of nanomedicine, future perspectives, opportunities, and problems have also been noted. Over the past two decades, numerous innovative, economical, and reliable synthesis techniques have been created. The enhanced physicochemical characteristics of the resultant hybrid MNPs led to new potential applications in disease diagnosis and treatment. Recent research trends indicate that the enormous potential of various MNPs has been extensively explored through their application in the biomedical domain. Surface modifications and specific functionalization of MNPs have also been keys to identifying their multifaceted applications. Emerging roles of MNPs in healthcare may include controlling antagonistic effects on tissues and organs, applying targeted drug delivery or therapy to specific sites through nanocarriers, performing both disease diagnosis and therapy through a single agent, and so on. MNP-based low-cost antimicrobial agents can be a future alternative to traditional antibiotics too. It won't be far stretched if one expects that in the future many chronic and highly infectious diseases will be cured through the application of noble MNPs.

ACKNOWLEDGEMENT

The authors are also thankful to Dr. Jamilur R. Ansari, Dronacharya College of Engineering, Gurgaon, India, for the needful editing in the chapter. PH and AKMSK is thankful to their organization for the support and cooperation. SPS gratefully thanks PDM University for their support. FW gratefully acknowledge the support from the Principal, GBM College, Gaya in writing this chapter.

REFERENCES

[1] K. McNamara, S.A.M. Tofail, Nanoparticles in biomedical applications, Advances in Physics: X. 2 (2017) 54–88. https://doi.org/10.1080/23746149.2016.1254570.

[2] P. Biswas, C.-Y. Wu, Nanoparticles and the environment, Journal of the Air & Waste Management Association. 55 (2005) 708–746. https://doi.org/10.1080/10473289.2005.10464656.

[3] I. Khan, K. Saeed, I. Khan, Nanoparticles: Properties, applications and toxicities, Arabian Journal of Chemistry. 12 (2019) 908–931. https://doi.org/10.1016/j.arabjc.2017.05.011.

[4] N. Kumar, S. Kumbhat, Unique properties, in: Essentials in Nanoscience and Nanotechnology, Wiley, Hoboken, NJ (2016): pp. 326–360.

[5] M.N. Javed, E.S. Dahiya, A.M. Ibrahim, M.S. Alam, F.A. Khan, F.H. Pottoo, Recent advancement in clinical application of nanotechnological approached targeted delivery of herbal drugs, in: S. Beg, M.A. Barkat, F.J. Ahmad (Eds.), Nanophytomedicine, Springer International Publishing, Singapore (2020): pp. 151–172. https://doi.org/10.1007/978-981-15-4909-0_9.

[6] M.S. Alam, M.N. Javed, F.H. Pottoo, A. Waziri, F.A. Almalki, M.S. Hasnain, A. Garg, M.K. Saifullah, QbD approached comparison of reaction mechanism in microwave synthesized gold nanoparticles and their superior catalytic role against hazardous nirto-dye, Applied Organometallic Chemistry. (2019). https://doi.org/10.1002/aoc.5071.

[7] M.N. Javed, F.H. Pottoo, A. Shamim, M.S. Hasnain, M.S. Alam, Design of experiments for the development of nanoparticles, nanomaterials, and nanocomposites, in: S. Beg (Ed.), Design of Experiments for Pharmaceutical Product Development, Springer International Publishing, Singapore (2021): pp. 151–169. https://doi.org/10.1007/978-981-33-4351-1_9.

[8] D.A. Giljohann, D.S. Seferos, W.L. Daniel, M.D. Massich, P.C. Patel, C.A. Mirkin, Gold nanoparticles for biology and medicine, Angewandte Chemie International Edition. 49 (2010) 3280–3294. https://doi.org/10.1002/anie.200904359.

[9] D. Pal, C. Sahu, A. Haldar, Bhasma: The ancient Indian nanomedicine, Journal of Advanced Pharmaceutical Technology & Research. 5 (2014) 4. https://doi.org/10.4103/2231-4040.126980.

[10] P.P. Edwards, J.M. Thomas, Fein verteiltes gold—Faradays Beitragzu den heutigen Nanowissenschaften, Angewandte Chemie. 119 (2007) 5576–5582. https://doi.org/10.1002/ange.200700428.

[11] V. Mody, R. Siwale, A. Singh, H. Mody, Introduction to metallic nanoparticles, Journal of Pharmacy and Bioallied Sciences. 2 (2010) 282. https://doi.org/10.4103/0975-7406.72127.

[12] J.R. Ansari, N. Singh, S. Anwar, S. Mohapatra, A. Datta, Silver nanoparticles decorated two dimensional MoS_2 nanosheets for enhanced photocatalytic activity, Colloids and Surfaces A: Physicochemical and Engineering Aspects. 635 (2022) 128102. https://doi.org/10.1016/j.colsurfa.2021.128102.

[13] J.R. Ansari, N. Singh, S. Mohapatra, R. Ahmad, N.R. Saha, D. Chattopadhyay, M. Mukherjee, A. Datta, Enhanced near infrared luminescence in Ag@Ag2S core-shell nanoparticles, Applied Surface Science. 463 (2019) 573–580. https://doi.org/10.1016/j.apsusc.2018.08.244.

[14] A.M. Ibrahim, L. Chauhan, A. Bhardwaj, A. Sharma, F. Fayaz, B. Kumar, M. Alhashmi, N. AlHajri, M.S. Alam, F.H. Pottoo, Brain-derived neurotropic factor in neurodegenerative disorders, Biomedicines. 10 (2022) 1143. https://doi.org/10.3390/biomedicines10051143.

[15] F.H. Pottoo, M.N. Javed, M.A. Barkat, M.S. Alam, J.A. Nowshehri, D.M. Alshayban, M.A. Ansari, Estrogen and serotonin: Complexity of interactions and implications for epileptic seizures and epileptogenesis, Current Neuropharmacology. 17 (2019) 214–231. https://doi.org/10.2174/1570159X16666180628164432.

[16] M. Aslam, M.N. Javed, H.H. Deeb, M.K. Nicola, M.A. Mirza, M.S. Alam, M.H. Akhtar, A. Waziri, Lipid nanocarriers for neurotherapeutics: Introduction, challenges, blood-brain barrier, and promises of delivery approaches, CNS & Neurological Disorders, Drug Targets. 21 (2022) 952–965. https://doi.org/10.2174/18715273206662107061 04240.

[17] F.H. Pottoo, S. Sharma, M.N. Javed, M.A. Barkat, Harshita, M.S. Alam, M.J. Naim, O. Alam, M.A. Ansari, G.E. Barreto, G.M. Ashraf, Lipid-based nanoformulations in the treatment of neurological disorders, Drug Metabolism Reviews. 52 (2020) 185–204. https://doi.org/10.1080/03602532.2020.1726942.

[18] A. Waziri, C. Bharti, M. Aslam, P. Jamil, M.A. Mirza, M.N. Javed, U. Pottoo, A. Ahmadi, M.S. Alam, Probiotics for the chemoprotective role against the toxic effect of cancer chemotherapy, Anti-Cancer Agents in Medicinal Chemistry. 22 (2022) 654–667. https://doi.org/10.2174/1871520621666210514000615.

[19] M.S. Hasnain, M.N. Javed, M.S. Alam, P. Rishishwar, S. Rishishwar, S. Ali, A.K. Nayak, S. Beg, Purple heart plant leaves extract-mediated silver nanoparticle synthesis: Optimization by Box-Behnken design, Materials Science and Engineering: C. 99 (2019) 1105–1114. https://doi.org/10.1016/j.msec.2019.02.061.

[20] G. Habibullah, J. Viktorova, T. Ruml, Current strategies for noble metal nanoparticle synthesis, Nanoscale Research Letters. 16 (2021) 47. https://doi.org/10.1186/s11671-021-03480-8.

[21] F.H. Pottoo, N. Tabassum, M.N. Javed, S. Nigar, S. Sharma, M.A. Barkat, Harshita, M.S. Alam, M.A. Ansari, G.E. Barreto, G.M. Ashraf, Raloxifene potentiates the effect of fluoxetine against maximal electroshock induced seizures in mice, European Journal of Pharmaceutical Sciences. 146 (2020) 105261. https://doi.org/10.1016/j.ejps.2020.105261.

[22] N. Kumari, N. Daram, M.S. Alam, A.K. Verma, Rationalizing the use of polyphenol nano-formulations in the therapy of neurodegenerative diseases, CNS & Neurological Disorders, Drug Targets. 21 (2022) 966–976. https://doi.org/10.2174/1871527321666220512153854.

[23] S. Raj, R. Manchanda, M. Bhandari, M.S. Alam, Review on natural bioactive products as radioprotective therapeutics: Present and past perspective, Current Pharmaceutical Biotechnology. 23 (2022) 1721–1738. https://doi.org/10.2174/1389201023666220110104645.

[24] F.H. Pottoo, N. Tabassum, M.N. Javed, S. Nigar, R. Rasheed, A. Khan, M.A. Barkat, M.S. Alam, A. Maqbool, M.A. Ansari, G.E. Barreto, G.M. Ashraf, The synergistic effect of raloxifene, fluoxetine, and bromocriptine protects against pilocarpine-induced status epilepticus and temporal lobe epilepsy, Molecular Neurobiology. 56 (2019) 1233–1247. https://doi.org/10.1007/s12035-018-1121-x.

[25] C.L. Keat, A. Aziz, A.M. Eid, N.A. Elmarzugi, Biosynthesis of nanoparticles and silver nanoparticles, Bioresources and Bioprocessing. 2 (2015) 47. https://doi.org/10.1186/s40643-015-0076-2.

[26] O. Fellahi, R.K. Sarma, M.R. Das, R. Saikia, L. Marcon, Y. Coffinier, T. Hadjersi, M. Maamache, R. Boukherroub, The antimicrobial effect of silicon nanowires decorated with silver and copper nanoparticles, Nanotechnology. 24 (2013) 495101. https://doi.org/10.1088/0957-4484/24/49/495101.

[27] P. Salas, N. Odzak, Y. Echegoyen, R. Kägi, M.C. Sancho, E. Navarro, The role of size and protein shells in the toxicity to algal photosynthesis induced by ionic silver delivered from silver nanoparticles, Science of the Total Environment. 692 (2019) 233–239. https://doi.org/10.1016/j.scitotenv.2019.07.237.

[28] R.V.S.P. Sanka, B. Krishnakumar, P. Shyam, R. Sravendra, Metal oxide based nanomaterials and their polymer nanocomposites, in: K. Niranjan (Ed.), Nanomaterials and Polymer Nanocomposites: Raw Materials to Applications, Elsevier, Amsterdam; Cambridge, MA (2019): pp. 123–144.

[29] Q. Gao, W. Xie, Y. Wang, D. Wang, Z. Guo, F. Gao, L. Zhao, Q. Cai, A theranostic nanocomposite system based on radial mesoporous silica hybridized with $Fe_3\,O_4$ nanoparticles for targeted magnetic field responsive chemotherapy of breast cancer, RSC Advances. 8 (2018) 4321–4328. https://doi.org/10.1039/C7RA12446E.

[30] L. Wang, C. Hu, L. Shao, The antimicrobial activity of nanoparticles: Present situation and prospects for the future, International Journal of Nanomedicine. 12 (2017) 1227–1249. https://doi.org/10.2147/IJN.S121956.

[31] M. Kumar, M. Na, J.R. Ansari, Analysis of the heating ability by varying the size of Fe_3O_4 magnetic nanoparticles for hyperthermia, Nanoscience and Technology: An International Journal. (2022). https://doi.org/10.1615/NanoSciTechnolIntJ.2022040075.

[32] R. Kumar, G. Dhamija, J.R. Ansari, M.N. Javed, M.S. Alam, C-Dot nanoparticulated devices for biomedical applications, in: Nanotechnology, 1st ed., CRC Press, Boca Raton (2022): pp. 271–299. https://doi.org/10.1201/9781003220350-15.

[33] S. Singhal, M. Gupta, M.S. Alam, M.N. Javed, J.R. Ansari, Carbon allotropes-based nanodevices, in: Nanotechnology, 1st ed., CRC Press, Boca Raton (2022): pp. 241–269. https://doi.org/10.1201/9781003220350-14.

[34] M.S. Alam, M.F. Naseh, J.R. Ansari, A. Waziri, M.N. Javed, A. Ahmadi, M.K. Saifullah, A. Garg, Synthesis approaches for higher yields of nanoparticles, in: Nanomaterials in the Battle Against Pathogens and Disease Vectors, 1st ed., CRC Press, Boca Raton (2022): pp. 51–82. https://doi.org/10.1201/9781003126256-3.

[35] M.F. Naseh, J.R. Ansari, M.S. Alam, M.N. Javed, Sustainable nanotorus for biosensing and therapeutical applications, in: U. Shanker, C.M. Hussain, M. Rani (Eds.), Handbook of Green and Sustainable Nanotechnology, Springer International Publishing, Cham (2022): pp. 1–21. https://doi.org/10.1007/978-3-030-69023-6_47-1.

[36] C.A. Sunilbhai, M.S. Alam, K.K. Sadasivuni, J.R. Ansari, SPR assisted diabetes detection, in: K.K. Sadasivuni, J.-J. Cabibihan, A.K. Al-Ali, R.A. Malik (Eds.), Advanced Bioscience and Biosystems for Detection and Management of Diabetes, Springer International Publishing, Cham (2022): pp. 91–131. https://doi.org/10.1007/978-3-030-99728-1_6.
[37] S. Mishra, S. Sharma, M.N. Javed, F.H. Pottoo, M.A. Barkat, Harshita, M.S. Alam, M. Amir, M. Sarafroz, Bioinspired nanocomposites: Applications in disease diagnosis and treatment, Pharmaceutical Nanotechnology. 7 (2019) 206–219. https://doi.org/10.2174/2211738507666190425121509.
[38] M. Kumar, J.R. Ansari, M.T. Beig, Studies on superparamagnetic behaviour of Ni100-xCux alloy films deposited by DC magnetron sputtering, Materials Research Innovations. (2021) 1–6. https://doi.org/10.1080/14328917.2021.1987069.
[39] J.R. Ansari, N. Singh, R. Ahmad, D. Chattopadhyay, A. Datta, Controlling self-assembly of ultra-small silver nanoparticles: Surface enhancement of Raman and fluorescent spectra, Optical Materials. 94 (2019) 138–147. https://doi.org/10.1016/j.optmat.2019.05.023.
[40] T.K. Indira, P.K. Lakshmi, Magnetic nanoparticles—a review, International Journal of Pharmaceutical Sciences and Nanotechnology. 3 (2010).
[41] Y.S. Kang, S. Risbud, J.F. Rabolt, P. Stroeve, Synthesis and characterization of nanometer-size Fe_3O_4 and γ-Fe_2O_3 particles, Chemistry of Materials. 8 (1996) 2209–2211. https://doi.org/10.1021/cm960157j.
[42] C. Liu, B. Zou, A.J. Rondinone, Z.J. Zhang, Reverse micelle synthesis and characterization of superparamagnetic $MnFe_2O_4$ spinel ferrite nanocrystallites, The Journal of Physical Chemistry B. 104 (2000) 1141–1145. https://doi.org/10.1021/jp993552g.
[43] P. Biehl, M. Von der Lühe, S. Dutz, F.H. Schacher, Synthesis, characterization, and applications of magnetic nanoparticles featuring polyzwitterionic coatings, Polymers. 10 (2018). https://doi.org/10.3390/polym10010091.
[44] Y. Wang, C.-X. Yang, X.-P. Yan, Hydrothermal and biomineralization synthesis of a dual-modal nanoprobe for targeted near-infrared persistent luminescence and magnetic resonance imaging, Nanoscale. 9 (2017) 9049–9055. https://doi.org/10.1039/C7NR02038D.
[45] S.A.M.K. Ansari, E. Ficiarà, F.A. Ruffinatti, I. Stura, M. Argenziano, O. Abollino, R. Cavalli, C. Guiot, F. D'Agata, Magnetic iron oxide nanoparticles: Synthesis, characterization and functionalization for biomedical applications in the central nervous system, Materials (Basel, Switzerland). 12 (2019) 465. https://doi.org/10.3390/ma12030465.
[46] A. Lassoued, M.S. Lassoued, B. Dkhil, S. Ammar, A. Gadri, Synthesis, photoluminescence and magnetic properties of iron oxide (α-Fe_2O_3) nanoparticles through precipitation or hydrothermal methods, Physica E Low-Dimensional Systems and Nanostructures. 101 (2018) 212–219. https://doi.org/10.1016/j.physe.2018.04.009.
[47] K. Raja, M. Mary Jaculine, M. Jose, S. Verma, A.A.M. Prince, K. Ilangovan, K. Sethusankar, S. Jerome Das, Sol–gel synthesis and characterization of α-Fe_2O_3 nanoparticles, Superlattices and Microstructures. 86 (2015) 306–312. https://doi.org/10.1016/j.spmi.2015.07.044.
[48] S.F. Hasany, I. Ahmed, J. Rajan, A. Rehman, Systematic review of the preparation techniques of iron oxide magnetic nanoparticles, Nanoscience & Nanotechnology. 2 (2012) 148–158. https://doi.org/10.5923/j.nn.20120206.01.
[49] L.E. Mathevula, L.L. Noto, B.M. Mothudi, M. Chithambo, M.S. Dhlamini, Structural and optical properties of sol-gel derived α-Fe_2O_3 nanoparticles, Journal of Luminescence. 192 (2017) 879–887. https://doi.org/10.1016/j.jlumin.2017.07.055.
[50] S. Belaïd, D. Stanicki, L. Vander Elst, R.N. Muller, S. Laurent, Influence of experimental parameters on iron oxide nanoparticle properties synthesized by thermal decomposition: Size and nuclear magnetic resonance studies, Nanotechnology. 29 (2018) 165603. https://doi.org/10.1088/1361-6528/aaae59.
[51] F.B. Effenberger, R.A. Couto, P.K. Kiyohara, G. Machado, S.H. Masunaga, R.F. Jardim, L.M. Rossi, Economically attractive route for the preparation of high quality magnetic

nanoparticles by the thermal decomposition of iron (III) acetylacetonate, Nanotechnology. 28 (2017) 115603. https://doi.org/10.1088/1361-6528/aa5ab0.

[52] A. Mohammadi, M. Barikani, M. Barmar, Effect of surface modification of Fe_3O_4 nanoparticles on thermal and mechanical properties of magnetic polyurethane elastomer nanocomposites, Journal of Materials Science. 48 (2013) 7493–7502.

[53] S.-N. Sun, C. Wei, Z.-Z. Zhu, Y.-L. Hou, S.S. Venkatraman, Z.-C. Xu, Magnetic iron oxide nanoparticles: Synthesis and surface coating techniques for biomedical applications, Chinese Physics B. 23 (2014) 37503. https://doi.org/10.1088/1674-1056/23/3/037503.

[54] M.L. Immordino, F. Dosio, L. Cattel, Stealth liposomes: Review of the basic science, rationale, and clinical applications, existing and potential, International Journal of Nanomedicine. 1 (2006) 297–315.

[55] G. Pal, P. Rai, A. Pandey, Green synthesis of nanoparticles: A greener approach for a cleaner future, in: Green Synthesis, Characterization and Applications of Nanoparticles, Elsevier, Amsterdam (2019).

[56] P. Kurhade, S. Kodape, R. Choudhury, Overview on green synthesis of metallic nanoparticles, Chemical Papers. 75 (2021). https://doi.org/10.1007/s11696-021-01693-w.

[57] J. Singh, T. Dutta, K.-H. Kim, M. Rawat, P. Samddar, P. Kumar, "Green" synthesis of metals and their oxide nanoparticles: Applications for environmental remediation, Journal of Nanobiotechnology. 16 (2018) 84. https://doi.org/10.1186/s12951-018-0408-4.

[58] S.K. Nune, N. Chanda, R. Shukla, K. Katti, R.R. Kulkarni, S. Thilakavathi, S. Mekapothula, R. Kannan, K.V.Katti, Green nanotechnology from tea: Phytochemicals in tea as building blocks for production of biocompatible gold nanoparticles, Journal of Materials Chemistry. 19 (2009) 2912–2920. https://doi.org/10.1039/b822015h.

[59] D. Zhang, X. Ma, Y. Gu, H. Huang, G. Zhang, Green synthesis of metallic nanoparticles and their potential applications to treat cancer, Frontiers in Chemistry. 8 (2020). https://doi.org/10.3389/fchem.2020.00799.

[60] K. Shameli, M. Bin Ahmad, A. Zamanian, P. Sangpour, P. Shabanzadeh, Y. Abdollahi, M. Zargar, Green biosynthesis of silver nanoparticles using Curcuma longa tuber powder, International Journal of Nanomedicine. 7 (2012) 5603–5610. https://doi.org/10.2147/IJN.S36786.

[61] S. Joglekar, K. Kodam, M. Dhaygude, M. Hudlikar, Novel route for rapid biosynthesis of lead nanoparticles using aqueous extract of Jatropha curcas L. latex, Materials Letters. 65 (2011) 3170–3172. https://doi.org/10.1016/j.matlet.2011.06.075.

[62] J.L. Gardea-Torresdey, E. Gomez, J.R. Peralta-Videa, J.G. Parsons, H. Troiani, M. Jose-Yacaman, Alfalfa sprouts: A natural source for the synthesis of silver nanoparticles, Langmuir. 19 (2003) 1357–1361. https://doi.org/10.1021/la020835i.

[63] R.K. Das, S.K. Brar, Plant mediated green synthesis: Modified approaches, Nanoscale. 5 (2013) 10155–10162. https://doi.org/10.1039/C3NR02548A.

[64] H.M. El-Rafie, M.H. El-Rafie, M.K. Zahran, Green synthesis of silver nanoparticles using polysaccharides extracted from marine macro algae, Carbohydrate Polymers. 96 (2013) 403–410. https://doi.org/10.1016/j.carbpol.2013.03.071.

[65] S.K. St. Angelo, E.L. Hartz, Ginkgo as a green reducing agent for gold nanoparticles and nanoplatelets, International Journal of Green Nanotechnology. 4 (2012) 111–116. https://doi.org/10.1080/19430892.2012.678706.

[66] A. Singh, S. Hede, M. Sastry, Spider silk as an active scaffold in the assembly of gold nanoparticles and application of the gold–silk bioconjugate in vapor sensing, Small (Weinheim an Der Bergstrasse, Germany). 3 (2007) 466–473. https://doi.org/10.1002/smll.200600413.

[67] P. Aramwit, N. Bang, J. Ratanavaraporn, S. Ekgasit, Green synthesis of silk sericin-capped silver nanoparticles and their potent anti-bacterial activity, Nanoscale Research Letters. 9 (2014) 79. https://doi.org/10.1186/1556-276X-9-79.

[68] G. Romano, M. Almeida, A.V. Coelho, A. Cutignano, L.G. Gonçalves, E. Hansen, D. Khnykin, T. Mass, A. Ramšak, M.S. Rocha, T.H. Silva, Biomaterials and bioactive natural products from marine invertebrates: From basic research to innovative applications, Marine Drugs. 20 (2022).

[69] K. Tokarek, J.L. Hueso, P. Kuśtrowski, G. Stochel, A. Kyzioł, Green synthesis of chitosan-stabilized copper nanoparticles, European Journal of Inorganic Chemistry. 2013 (2013) 4940–4947. https://doi.org/10.1002/ejic.201300594.

[70] X. Li, H. Xu, Z.-S. Chen, G. Chen, Biosynthesis of nanoparticles by microorganisms and their applications, Journal of Nanomaterials. 2011 (2011) 270974. https://doi.org/10.1155/2011/270974.

[71] N.I. Hulkoti, T.C. Taranath, Biosynthesis of nanoparticles using microbes—a review, Colloids and Surfaces B: Biointerfaces. 121 (2014) 474–483. https://doi.org/10.1016/j.colsurfb.2014.05.027.

[72] K.B. Narayanan, N. Sakthivel, Biological synthesis of metal nanoparticles by microbes, Advances in Colloid and Interface Science. 156 (2010) 1–13. https://doi.org/10.1016/j.cis.2010.02.001.

[73] S. Iravani, Bacteria in nanoparticle synthesis: Current status and future prospects, International Scholarly Research Notices. 2014 (2014) 359316.

[74] G.S. Dhillon, S.K. Brar, S. Kaur, M. Verma, Green approach for nanoparticle biosynthesis by fungi: Current trends and applications, Critical Reviews in Biotechnology. 32 (2012) 49–73. https://doi.org/10.3109/07388551.2010.550568.

[75] C. Mao, C.E. Flynn, A. Hayhurst, R. Sweeney, J. Qi, G. Georgiou, B. Iverson, A.M. Belcher, Viral assembly of oriented quantum dot nanowires, Proceedings of the National Academy of Sciences. 100 (2003) 6946–6951. https://doi.org/10.1073/pnas.0832310100.

[76] B. Klębowski, J. Depciuch, M. Parlińska-Wojtan, J. Baran, Applications of noble metal-based nanoparticles in medicine, International Journal of Molecular Sciences. 19 (2018) 4031. https://doi.org/10.3390/ijms19124031.

[77] A.A. Yaqoob, H. Ahmad, T. Parveen, A. Ahmad, M. Oves, I.M.I. Ismail, H.A. Qari, K. Umar, M.N. Mohamad Ibrahim, Recent advances in metal decorated nanomaterials and their various biological applications: A review, Frontiers in Chemistry. 8 (2020). https://doi.org/10.3389/fchem.2020.00341.

[78] A. Yadollahpour, H.M. Asl, S. Rashidi, Applications of nanoparticles in magnetic, Asian Journal of Pharmaceutics. 2017 (2017) 10–11.

[79] C.-Y. Tsai, A.-L. Shiau, S.-Y. Chen, Y.-H. Chen, P.-C. Cheng, M.-Y. Chang, D.-H. Chen, C.-H. Chou, C.-R. Wang, C.-L. Wu, Amelioration of collagen-induced arthritis in rats by nanogold, Arthritis & Rheumatism. 56 (2007) 544–554. https://doi.org/10.1002/art.22401.

[80] O.V.Salata, Applications of nanoparticles in biology and medicine, Journal of Nanobiotechnology. 2 (2004) 3. https://doi.org/10.1186/1477-3155-2-3.

[81] S.S. Agasti, S. Rana, M.-H. Park, C.K. Kim, C.-C. You, V.M. Rotello, Nanoparticles for detection and diagnosis, Advanced Drug Delivery Reviews. 62 (2010) 316–328. https://doi.org/10.1016/j.addr.2009.11.004.

[82] S.D. Anderson, V.V.Gwenin, C.D. Gwenin, Magnetic functionalized nanoparticles for biomedical, drug delivery and imaging applications, Nanoscale Research Letters. 14 (2019) 188. https://doi.org/10.1186/s11671-019-3019-6.

[83] V. Chandrakala, V. Aruna, G. Angajala, Review on metal nanoparticles as nanocarriers: Current challenges and perspectives in drug delivery systems, Emergent Materials. (2022). https://doi.org/10.1007/s42247-021-00335-x.

[84] N. Desai, M. Momin, T. Khan, S. Gharat, R.S. Ningthoujam, A. Omri, Metallic nanoparticles as drug delivery system for the treatment of cancer, Expert Opinion on Drug Delivery. 18 (2021) 1261–1290. https://doi.org/10.1080/17425247.2021.1912008.

[85] Y. Yao, Y. Zhou, L. Liu, Y. Xu, Q. Chen, Y. Wang, S. Wu, Y. Deng, J. Zhang, A. Shao, Nanoparticle-based drug delivery in cancer therapy and its role in overcoming drug

resistance, Frontiers in Molecular Biosciences. 7 (2020). https://doi.org/10.3389/fmolb.2020.00193.

[86] D.N. Păduraru, D. Ion, A.-G. Niculescu, F. Muşat, O. Andronic, A.M. Grumezescu, A. Bolocan, Recent developments in metallic nanomaterials for cancer therapy, diagnosing and imaging applications, Pharmaceutics. 14 (2022). https://doi.org/10.3390/pharmaceutics14020435.

[87] M. Yafout, A. Ousaid, Y. Khayati, I.S. El Otmani, Gold nanoparticles as a drug delivery system for standard chemotherapeutics: A new lead for targeted pharmacological cancer treatments, Scientific African. 11 (2021) e00685. https://doi.org/10.1016/j.sciaf.2020.e00685.

[88] D. Kovács, N. Igaz, M.K. Gopisetty, M. Kiricsi, Cancer therapy by silver nanoparticles: Fiction or reality? International Journal of Molecular Sciences. 23 (2022). https://doi.org/10.3390/ijms23020839.

[89] R.R. Miranda, I. Sampaio, V. Zucolotto, Exploring silver nanoparticles for cancer therapy and diagnosis, Colloids and Surfaces B, Biointerfaces. 210 (2022) 112254. https://doi.org/10.1016/j.colsurfb.2021.112254.

[90] F. Soetaert, P. Korangath, D. Serantes, S. Fiering, R. Ivkov, Cancer therapy with iron oxide nanoparticles: Agents of thermal and immune therapies, Advanced Drug Delivery Reviews. 163–164 (2020) 65–83. https://doi.org/10.1016/j.addr.2020.06.025.

[91] N. Wiesmann, W. Tremel, J. Brieger, Zinc oxide nanoparticles for therapeutic purposes in cancer medicine, Journal of Materials Chemistry B. 8 (2020) 4973–4989. https://doi.org/10.1039/D0TB00739K.

[92] S. Mukherjee, R. Kotcherlakota, S. Haque, D. Bhattacharya, J.M. Kumar, S. Chakravarty, C.R. Patra, Improved delivery of doxorubicin using rationally designed PEGylated platinum nanoparticles for the treatment of melanoma, Materials Science & Engineering C, Materials for Biological Applications. 108 (2020) 110375. https://doi.org/10.1016/j.msec.2019.110375.

[93] Y.-G. Yuan, S. Zhang, J.-Y. Hwang, I.-K. Kong, Silver nanoparticles potentiates cytotoxicity and apoptotic potential of camptothecin in human cervical cancer cells, Oxidative Medicine and Cellular Longevity. 2018 (2018) 6121328. https://doi.org/10.1155/2018/6121328.

[94] T. Chatterjee, B.K. Chatterjee, D. Majumdar, P. Chakrabarti, Antibacterial effect of silver nanoparticles and the modeling of bacterial growth kinetics using a modified Gompertz model, Biochimica et Biophysica Acta. 1850 (2015) 299–306. https://doi.org/10.1016/j.bbagen.2014.10.022.

[95] X. Li, S.M. Robinson, A. Gupta, K. Saha, Z. Jiang, D.F. Moyano, A. Sahar, M.A. Riley, V.M. Rotello, Functional gold nanoparticles as potent antimicrobial agents against multi-drug-resistant bacteria, ACS Nano. 8 (2014) 10682–10686. https://doi.org/10.1021/nn5042625.

[96] N. Beyth, Y. Houri-Haddad, A. Domb, W. Khan, R. Hazan, Alternative antimicrobial approach: Nano-antimicrobial materials, Evidence-Based Complementary and Alternative Medicine. 2015 (2015) 246012. https://doi.org/10.1155/2015/246012.

[97] A. Brandelli, A.C. Ritter, F.F. Veras, Antimicrobial activities of metal nanoparticles, in: M. Rai, R. Shegokar (Eds.), Metal Nanoparticles in Pharma, Springer International Publishing, Cham (2017): pp. 337–363. https://doi.org/10.1007/978-3-319-63790-7_15.

[98] V. Chiriac, D.N. Stratulat, G. Calin, S. Nichitus, V. Burlui, C. Stadolcanu, M. Popa, I.M. Popa, Antimicrobial property of zinc based nanoparticles, IOP Conference Series: Materials Science and Engineering. 133 (2016) 12055. https://doi.org/10.1088/1757-899x/133/1/012055.

[99] H. Zazo, C.I. Colino, J.M. Lanao, Current applications of nanoparticles in infectious diseases, Journal of Controlled Release. 224 (2016) 86–102. https://doi.org/10.1016/j.jconrel.2016.01.008.

[100] Y. Zhang, M. Wu, W. Dai, Y. Li, X. Wang, D. Tan, Z. Yang, S. Liu, L. Xue, Y. Lei, Gold nanoclusters for controlled insulin release and glucose regulation in diabetes, Nanoscale. 11 (2019) 6471–6479. https://doi.org/10.1039/C9NR00668K.

[101] H.J. Kim, S.-M. Lee, K.-H. Park, C.H. Mun, Y.-B. Park, K.-H. Yoo, Drug-loaded gold/iron/gold plasmonic nanoparticles for magnetic targeted chemo-photothermal treatment of rheumatoid arthritis, Biomaterials. 61 (2015) 95–102. https://doi.org/10.1016/j.biomaterials.2015.05.018.

[102] Y. Zhao, X. Zhao, Y. Cheng, X. Guo, W. Yuan, Iron oxide nanoparticles-based vaccine delivery for cancer treatment, Molecular Pharmaceutics. 15 (2018) 1791–1799. https://doi.org/10.1021/acs.molpharmaceut.7b01103.

[103] A. Sharma, A.K. Goyal, G. Rath, Recent advances in metal nanoparticles in cancer therapy, Journal of Drug Targeting. 26 (2018) 617–632. https://doi.org/10.1080/1061186X.2017.1400553.

[104] J.M. Seiffert, M.-O. Baradez, V. Nischwitz, T. Lekishvili, H. Goenaga-Infante, D. Marshall, Dynamic monitoring of metal oxide nanoparticle toxicity by label free impedance sensing, Chemical Research in Toxicology. 25 (2012) 140–152. https://doi.org/10.1021/tx200355m.

[105] S. Hussain, K. Hess, J. Gearhart, K.T. Geiss, J. Schlager, In vitro toxicity of nanoparticles in BRL 3A rat liver cells, Toxicology in Vitro: An International Journal Published in Association with BIBRA. 19 (2005) 975–983. https://doi.org/10.1016/j.tiv.2005.06.034.

[106] N.A. Hanan, H.I. Chiu, M.R. Ramachandran, W.H. Tung, N.N. Mohamad Zain, N. Yahaya, V. Lim, Cytotoxicity of plant-mediated synthesis of metallic nanoparticles: A systematic review, International Journal of Molecular Sciences. 19 (2018). https://doi.org/10.3390/ijms19061725.

[107] R.Y. Parikh, S. Singh, B.L. V Prasad, M.S. Patole, M. Sastry, Y.S. Shouche, Extracellular synthesis of crystalline silver nanoparticles and molecular evidence of silver resistance from Morganella sp.: Towards understanding biochemical synthesis mechanism, Chembiochem: A European Journal of Chemical Biology. 9 (2008) 1415–1422. https://doi.org/10.1002/cbic.200700592.

[108] P.V. AshaRani, G. Low Kah Mun, M.P. Hande, S. Valiyaveettil, Cytotoxicity and genotoxicity of silver nanoparticles in human cells, ACS Nano. 3 (2009) 279–290. https://doi.org/10.1021/nn800596w.

[109] P. Manivasagan, J. Venkatesan, K. Senthilkumar, K. Sivakumar, S.-K. Kim, Biosynthesis, antimicrobial and cytotoxic effect of silver nanoparticles using a novel nocardiopsis sp. MBRC-1, BioMed Research International. 2013 (2013) 287638. https://doi.org/10.1155/2013/287638.

[110] H.I.O. Gomes, C.S.M. Martins, J.A.V. Prior, Silver nanoparticles as carriers of anticancer drugs for efficient target treatment of cancer cells, Nanomaterials. 11 (2021). https://doi.org/10.3390/nano11040964.

[111] M. Cordani, A. Somoza, Targeting autophagy using metallic nanoparticles: A promising strategy for cancer treatment, Cellular and Molecular Life Sciences. 76 (2019). https://doi.org/10.1007/s00018-018-2973-y.

[112] S. Singla, K. Harjai, O. Katare, S. Chhibber, Encapsulation of bacteriophage in liposome accentuates its entry in to macrophage and shields it from neutralizing antibodies, PLoS One. 11 (2016) e0153777. https://doi.org/10.1371/journal.pone.0153777.

[113] S. Moghimi, A. Hunter, J. Murray, S.M. Moghimi, A.C. Hunter, J.C. Murray, Nanomedicine: Current status and future prospects, FASEB Journal: Official Publication of the Federation of American Societies for Experimental Biology. 19 (2005) 311–330. https://doi.org/10.1096/fj.04-2747rev.

[114] A. Wicki, D. Witzigmann, V. Balasubramanian, J. Huwyler, Nanomedicine in cancer therapy: Challenges, opportunities, and clinical applications, Journal of Controlled Release. 200 (2015) 138–157. https://doi.org/10.1016/j.jconrel.2014.12.030.

3 Application of MNPs in Targeted Delivery and Genetic Manipulations

Renu Kadian, Jayamanti Pandit, Charu Bharti, Rabiya, Aafrin Waziri, Puja Kumari, Arun Garg, Md Noushad Javed, Jamilur R. Ansari, and Md Sabir Alam

3.1 INTRODUCTION

Nanotherapeutics emerged as a vast field of research for medical and diagnostic purposes. Nanoparticles (NPs) are nanometric-sized particles between 1 nm and 100 nm made up of polymer or metal. Polymeric particles are prepared from natural or synthetic polymers for the delivery of therapeutics [1, 2]. Recently, scientists have focused on hybrid nanoparticles, by either coating them or combing them with other dosage carriers to achieve targeted delivery of the drug to the desired site. Nanoparticles are reported as a means to overcome the drawbacks of conventional formulations, especially in the treatment of cancer. Conventional dosage forms, such as tablets and injectables, used in cancer treatment are distributed throughout the whole body, which is responsible for the side effects or toxic effects of the drug [3]. To overcome this unspecific absorption or distribution of anticancer drugs, site-specific drug delivery carriers have been used and are still under trials. Metallic nanoparticles (MNPs) could work in this direction in combination with other drug delivery options by using smart polymers for targeted drug delivery to the tumor site specifically [4, 5].

MNPs have been used for the treatment of aliments since ancient times, and they have been reported for their medicinal properties in recent years. The optical properties of MNPs make them a better candidate for biomedical applications [6–23]. The large surface area, spatial confinement, larger pore volume, and surface stability make MNPs suitable for wider application in diverse fields [24]. The easy modification and conjugation of MNPa with a targeting ligand such as proteins, or peptides, and nucleic acid sequences is possible due to the surface chemistry of MNPs [25, 26]. Drug payload can also be enhanced to the target site in a specific cell by using a functionalized MNP, thus reducing the side effects related to the drug [27, 28]. MNPs of silver, gold, zinc, copper, iron, and other metals are used in wound dressing, food packaging, and drug delivery for their anti-microbial properties [24]. Silver nanoparticles (AgNPs) are widely reported to have antimicrobial activity and are used in the control of infectious diseases. Titanium oxide and zinc nanoparticles are well known

DOI: 10.1201/9781003317319-3

for cosmetic use, and gold nanoparticles (AuNPs) have shown promising results in imaging and the treatment of cancer [29, 30]. Functionalized gold nanoparticles made by using nucleotides have been useful for the treatment of genetic diseases [31]. Other applications of MNPs include biosensors, diagnostic agents, and water purification [32]. The phagocytic uptake of circulating nanoparticles can be decreased by surface modification of NP with PEG or other long-chain polymers, which minimizes the accumulation at non-target sites [33]. Many other possible applications of MNPs, such as magnetically responsive delivery system, photothermal therapy, neurological disease targeting, and photoimaging, are still under research [34–40].

Synthesis of metallic nanoparticles provides us with great choices, as one can go for physical methods, chemical methods, or biological methods. Today, physical and chemical methods are less considered due to environmental pollution and toxicity. Biological methods using a green synthesis approach are preferred due to their environment friendly nature, reproducibility, and easy scale-up. Plant parts such as roots, stems, leaves, flowers, and fruits are considered for the green synthesis of MNPs as it is a safe, easy, environment friendly, and stable method compared to the physicochemical method [41–44]. The success rate of nanoparticles for drug delivery and therapeutic targeting is related to the uniformity, stability, and safety of the NP. Metallic nanoparticles have gained immense attention in many research areas and are recommended as potential, stable, and easily modifiable drug targeting tools for controlled and sustained drug release [45].

This chapter discusses the synthesis and characterization of metallic nanoparticles and their potential applications in targeted drug delivery as well as genetic manifestation. It also elaborates on its implications for cell therapy and the outcome of the clinical trial under way.

3.2 SYNTHESIS OF METALLIC NANOPARTICLES

3.2.1 Chemical Methods

3.2.2 Chemical Reduction Method

By using the chemical reduction technique, Michael Faraday (1857) was able to investigate the synthesis and colloidal hues of gold. By reducing the metals with both organic and inorganic reducing solvents, it has grown to be the most used approach for the synthesis of MNPs. For instance, employing reducing chemicals such as sodium citrate, sodium borohydride ($NaBH_4$), hydrazine, and phenols, silver may be converted from the Ag+ state to the Ag0 state [46]. Copper has been observed to be the easiest metal to reduce by reducing chemicals, which results in the creation of copper NPs with an ideal surface shape [47].

3.2.3 Microwave-Assisted Synthesis Method

The microwave-assisted synthesis approach is more promising for MNP synthesis since it uses microwave heat, which is superior to conventional oil bath heating. As a result, a superior dispersity index may be obtained with the continuous manufacture of nanosized MNPs [48, 49]. The key advantages of this approach include a quick and accurate reaction time, which eventually results in less energy consumption, a higher yield of MNPs, and the elimination of cluster formation [50, 51]. For the synthesis of noble metal NPs like silver, gold, and platinum, the microwave-assisted technique is highly popular. By maximizing several physiochemical parameters, such as reaction time, concentration, temperature, and pH, the size may be readily controlled.

3.2.4 Irradiation Methods

In irradiation procedures, metal solutions are irradiated with a laser beam to produce MNPs of the ideal size. For the purpose of producing silver nanoparticles, lasers and mercury (Hg) lamps are frequently employed [52].

3.2.5 Photoinduced Reduction

A redactor is added to an electron donor species in photoinduced reduction to reduce metals [53]. It was discovered that we may synthesize many morphologies of MNPs using this approach, including spherical, triangle, cube, rod, and prism with an edge of 30–120 nm in length [54].

3.2.6 Microemulsion Techniques

For the synthesis of MNPs using microemulsion methods, a two-phase aqueous system is necessary. Its foundation is the separation of a reducing agent and a metal precursor in a two-phase aqueous environment. Thermostatically stable dispersion made from two immiscible solvents and a suitable surfactant allowed for the synthesis of MNPs [55].

3.3 PHYSICAL METHODS

3.3.1 Laser Ablation

One of the most crucial physical processes for MNP synthesis is laser ablation, which involves directing a strong laser beam onto the target substance to produce MNPs as a colloidal solution. A vacuum chamber is used throughout the whole procedure [56].

3.3.2 Mechanical Milling Method

To create MNPs, solid or thicker types of metal are milled via a mechanical process. To process the solid-state material, this approach needs a large machinery setup. The capacity and application of the machine are used to categorize the type of machine needed for the synthesis process [57].

3.4 BIOLOGICAL METHODS

3.4.1 Bacteria

The possibility of diverse microorganisms producing MNPs has been thoroughly investigated by several researchers. Bacillus licheniformis has been discovered to be capable of reducing Ag^+ ions from an aqueous solution of silver [58]. The use of *Shewanella alga* in the synthesis of gold nanoparticles was also investigated. According to a study by Nair and colleagues, it is possible to create metallic nanoparticles like gold, silver, and their alloys using the *Lacto bacillus* bacteria found in buttermilk [59].

3.4.2 Fungi

Due to their distinctive characteristics, such as metal tolerance and bioaccumulation capacity, fungi have been studied for the synthesis of MNPs. Fusarium oxysporum has been used successfully to create silver NPs [60]. The extracellular production of MNPs in filamentous fungi has been the subject of several research.

3.4.3 Algae

Several accounts claim that the sea algae *Sargassum wightii* has been used to create incredibly stable gold nanoparticles. Additionally, a similar procedure has also been used to create platinum and palladium MNPs [61].

3.4.4 Plants

In addition to microorganisms, researchers are now working to improve the process for synthesizing high-yield MNPs. In this regard, the ability of extracts from several plant sections to decrease metal salts for synthesis purposes has been investigated [62].

3.5 GREEN SYNTHESIS OF METALLIC NANOPARTICLES

It is regarded as a green method to create biogenic metallic nanoparticles by using natural resources like plants. It is now the most effective and environmentally safe method for doing away with the requirement for toxic chemicals in physical and chemical methods. There are several studies that support the biogenic production of MNPs utilizing various plant components and noble metals [63]. Different physical and chemical techniques are used to produce nanoparticles on a large scale. However, these traditional processes have some limitations, such as the need for expensive apparatus or instruments, chemicals, high production costs, and poorer yields [64]. This environmentally friendly method is particularly well-liked because of its low toxicity and green attributes. Additionally, because this approach does not require sophisticated equipment, costs are eventually reduced [65].

The bioactive substances contained in plant extracts, which primarily function as metal reducers, are what drive the entire synthesis process. Numerous metals and their derivatives have been utilized to produce Ag, Au, Pt, TiO_2, copper oxide (CuO), zinc oxide (ZnO), palladium (Pd), nickel (Ni), and iron oxide (FeO) nanoparticles over the past several years [66–73]. *Terminalia arjuna*, *Cordia dichotoma*, *Canarium ovatum*, *Dicoma tomentosa*, *Prosopis julifora*, *Cicer arietinum*, *Trigonella foenum-graecum*, and *Acacia nilotica* are some of the plants that have been utilized to make MNPs, particularly gold and silver [74–81]. In order to create gold nanoparticles, leaves extracted from a variety of plant species, including *Azadirachta indica*, *Cymbopogon* (lemon grass), *Tamarindus indica* (tamarind), and *Aloe barbadensis miller* (aloe vera), have been used [82]. In many different domains, platinum nanoparticles have been employed as catalysts. Various aqueous extracts of *X. strumarium* were used to effectively carry out the biogenic production of Pt NPs under ideal circumstances [83]. For the first time, it was revealed by Shabani et al. that sheep milk may be used to create platinum nanoparticles, which have the potential to be used in gene therapy and medication delivery [84].

3.6 CHARACTERIZATION TECHNIQUES

Researchers have given NPs a lot of attention because of their distinct physical, chemical, and mechanical characteristics. Therefore, before applying synthesized NPs in diverse fields, it is crucial to characterize their physicochemical properties. The end-use applications of the NPs were ultimately selected by the analysis of several parameters, including size, shape, surface morphology, surface area, structure, stability, elemental and mineral degradation, uniformity, intensity, etc. These methods may also be used to determine the electrical and thermal conductivity and purity of NPs. There are various methods that are used to characterize the metallic nanoparticles.

3.6.1 Spectroscopic Techniques

a. **UV–visible spectrophotometry**: The UV-visible spectroscopy method is based on the light absorption and reflection ions processes in the ultraviolet-visible electromagnetic spectrum regions, spanning from 200 nm to 800 nm, where molecules and other chemical compounds go through electronic transitions [85].
b. **Fluorescence spectroscopy**: One of the most accurate, straightforward, quick, and affordable methods for determining the concentration of NPs in the solution based on their fluorescent distinguishing characteristics is fluorescence spectroscopy (also known as spectrofluorometry). Both the kind of substance that has to be examined (the analyte) and its unknown concentration may be determined using them. This method is mostly employed to gauge the amount of fluorescent chemicals in a solution. In spectrofluorometry, a solution contained in a cuvette within the holder is passed through a beam of light with a wavelength range between 180 nm and 800 nm [86].
c. **DLS (dynamic light scattering):** There are different parameter.
   Zeta sizing: DLS is one of the most accurate methods for determining the distribution of nanoparticle sizes in aqueous media. It is sometimes referred to as photon correlation spectroscopy or quasi-elastic light scattering. It offers a general measurement of the polydispersity index (PDI) as well as the hydrodynamic diameter of the NPs [87].
   Zeta potential: Due to the presence of functional groups on the surface of the particles, the zeta potential analyzer is a complex technique used to assess the electric charge in the aqueous/colloidal media. The ionization of

the electrons causes the net charge to form on the NPs' surface. The surface groups or species that are adsorbed may be to blame. Due to the existence of a surface charge that impacts the general distribution of the ions, there is an electrical double layer close to the NPs [88].

d. **FTIR (Fourier-transform infrared spectroscopy):** The FTIR method, which is based on vibrational spectroscopy, is frequently used to characterize both organic and inorganic materials when infrared light is present. Energy is transferred or absorbed as the sample moves through electromagnetic radiation with wavelengths between 2.5 and 25 m, or wavenumbers 400 and 4000 cm^{-1}. Rising and varying rotational or vibrational motion are seen at certain wavelengths, which is mostly caused by absorption [89].
e. **NMR (nuclear magnetic resonance):** There are two types.
   CNMR: Nuclear magnetic resonance for carbon-13 (C-13) is another name for it. It aids in the identification of carbon atoms in organic molecules and produced NPs, as well as the clarification of their chemical structures [90].
   HNMR: Known further as proton NMR. It is one of the most widely utilized NMRs. Apart from tritium, the proton (hydrogen-1 nucleus) is a hypersensitive nucleus that defines the predominant signals, making proton NMR with resonance energy of 400 MHz particularly advantageous [91].
f. **XRD (X-ray diffraction):** The crystalline structure of the compound (particles) or the atomic configuration of the specified material or chemicals are both determined by this analytical approach. In terms of electromagnetic wavelength, they have a range of 0.01 to 10 nm and 120 eV to 120 KeV [92].
g. **Hydrophobicity test:** The wettability of an ideal surface may be precisely determined using the contact angle approach. The surface porosity, roughness, and heterogeneity are often described by the intrinsic property of the contact angle [93].

3.6.2 Microscopic Techniques

a. **SEM (scanning electron microscopy):** A scanning electron microscope is a type of electron microscope that uses a focused electron beam to scan over the surface of the target material to produce a picture. The required information about the surface and its composition is provided by the electrons' interactions with the secondary atoms in the sample, which result in a variety of signals based on the NPs' morphologies [94].
b. **EDX (energy-dispersive X-ray spectroscopy):** Energy-dispersive X-ray analysis (EDXA) or energy-dispersive X-ray microanalysis are frequent names for EDX (EDXMA). It is a microanalytical method used to characterize the chemical makeup of the samples or to detect and quantify the elemental compositions that are present in the samples [95].
c. **TEM with SAED (selected area electron diffraction):** In the fields of material science and biological research, TEM is a highly potent technology for analyzing the intricate details of materials. The essential components of this are viewing the morphology of the prepared NPs and the different elements involved in their creation. In essence, TEM involves placing the

entire apparatus within a vacuum chamber and allowing an electron beam with very high energy to pass through the sample that is stored on the TEM grid. As the electron beam interacts with the atoms, it is possible to see the structures, planes, dislocations, and grain boundaries that are distinctive to a certain crystal [96].

d. **AFM (atomic force microscopy):** The three-dimensional surface morphology of a sample is examined using the AFM method (3D). AFM creates the precise morphological structure of the samples by analyzing the interaction between the cantilever tip and the sample surface. The SEM and HRTEM methods are complemented by this characterization approach [97].
e. **TGA (thermo-gravimetric analysis):** TGA is a method for determining the weight fluctuations of the sample at a specific range of temperature and environment, which is utilized for the thermal characterization of the NPs. This analytical method allows for the determination of the thermal stability and bonding strength of the ligand conjugated to the nanoparticle surface [98].
f. **XPS (X-ray photoelectron spectroscopy):** For use in chemical analysis, XPS is a kind of photo-emission spectroscopy (PES) or electron spectroscopy (ESCA). It is a method for analyzing surfaces that offers data on all the components, composition, binding energies, and chemical states relevant to the surface's outermost layers, as well as quantitative and qualitative surface data [99], as shown in Figure 3.1.

3.7 TARGETED DELIVERY OF DRUG BY MNP

Cancer is the most prevailing medical problem globally, and according to a recent report by World Health Organization (WHO), cancer is reported as second major cause of death after ischemic heart disease [100]. The major cancer types in many countries are breast, lung, prostrate, ovarian, and rectum, and they have been reported as the cause of death for nearly 10 million people worldwide in 2020 [101, 102]. The drug delivery to a specific site or target site is most desired in cancer therapy. The concept of drug delivery to a therapeutic target is considered successful if most of the delivered drug reaches the target site without any significant effect on other normal cells in a controlled and sustained manner. The delivery carrier should not lead to burst drug release or delayed release at the site [103]. MNP became promising in nanomedicine for cancer treatment because of its easy functionalization, which could target drugs toward specific cancer cells. Even the same size of MNP can easily internalize into the cells compared to nonmetallic NP due to its higher surface charge density, which has been reported as a better strategy for cancer therapy [104, 105]. By using MNP, targeted delivery to the tumor cells can be achieved by using any of the strategies such as passive targeting, active targeting, and magnetic targeting [106]. The antitumor activity associated to specific MNP is related to the anti-angiogenesis and apoptosis of the NP component and the association of the metallic NP with anticancer drugs [107].

Silver and gold nanoparticles are widely reported for tumor targeting and diagnostic imaging. The in vitro study with AgNP showed cell internalization via endocytosis and caused DNA damage, mitochondrial damage, oxidative stress, and apoptosis

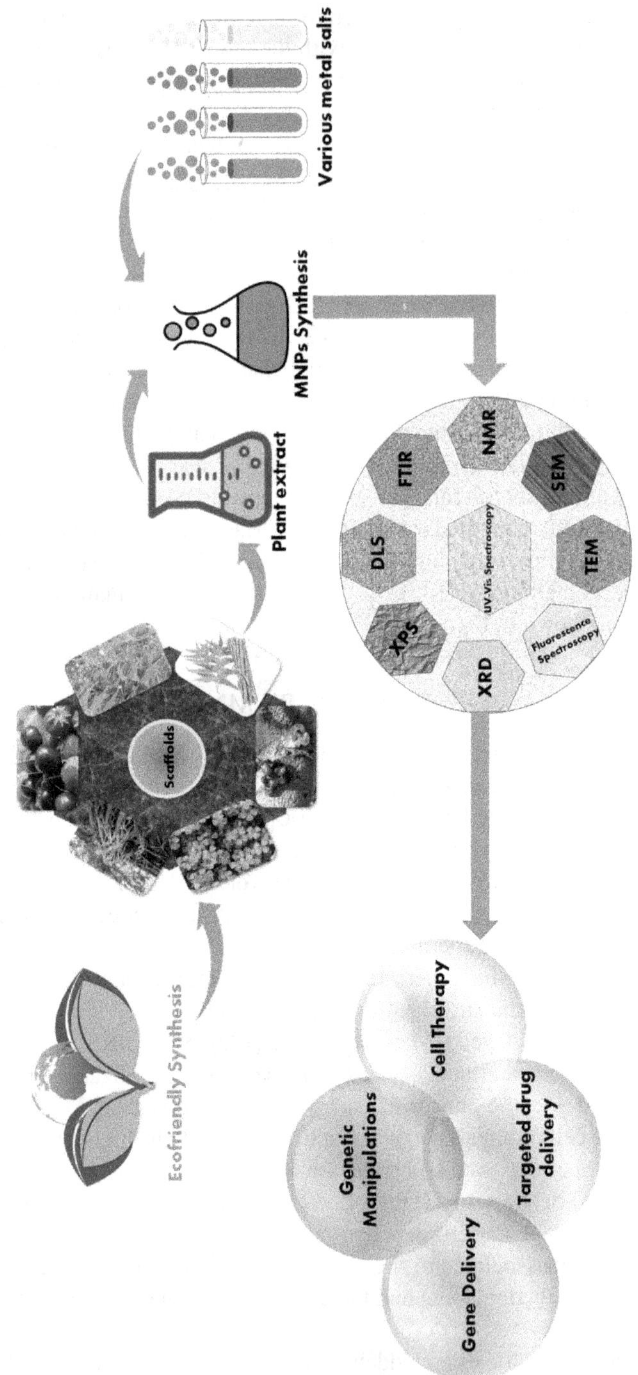

FIGURE 3.1 Synthesis, characterization, and biomedical application of MNPs.

in the cancer cell [108]. Song and the researcher reported the toxicity study of plant latex-capped AgNP against human lung cancer cells, and the findings showed dose-dependent toxicity to AS49 cells. They also reported plant latex as stabilizer for AgNP in water as well as helping in the transportation of NP into the target cell [109].

Paclitaxel-loaded silver nanospheres showed more cytotoxicity toward various cancer cell lines (MDA-MB-231, MCF-7, 4T1, Saos-2) than the bare drug and were reported to be safe for noncancerous cells [110].

Gold nanoparticles are another attractive drug delivery carrier for cancer targeting, offering efficient biodistribution of the drug along with hyperthermia [111], radiation imaging [112, 113], and photoimaging [114–116]. The delocalized nature of free electrons and the polarizability of these charges on the surface of AuNP attribute to its photothermal property. The high atomic number of AuNP is responsible for its efficiency in radiotherapy and can damage the DNA of the tumor cell [117, 118]. It has been reported that attachment of VEGF antibodies with AuNP induces apoptosis in cancer cells and is effective in the treatment of B-chronic lymphocytic leukemia [119].

In recent years, researchers have been working on metal organic framework nanocarriers for biomedical applications and reported promising results for drug delivery and disease treatment [120]. Even researchers are exploiting metallic nanoparticles from two or more metallic components for preparing nanoparticles and evaluating them for biomedical treatment. Qadri et al. (2019) investigated the efficacy of copper boron conjugated silver nanoparticles (AgCuBNP) in *Staphylococus aureus*-associated bone infection. They revealed that AgCuB NPs are able to reduce the infection in osteoblast cells due to internalized bacteria with a single dose treatment at a very small quantity (2 μg/ml) without toxicity to osteoblasts [121]. A comparative study was performed to show the effect of AgNP, CuNP, and AgCuNP on breast cancer cells and normal breast cells. The outcome of the study reported that CuNP has no effect on cancer cells or healthy cells. Both AgNP and AgCuNP have significant toxic effects on MCF-7 cells at concentrations of 20 μg/ml and 10 μg/ml, respectively. AgCu NP also showed that it is nontoxic to normal breast cells and leads to an increase in MMP-9 level, indicating that the cell death caused by NPs may be due to autophagy and oxidative stress [122].

3.7.1 Genetic Manipulations

Genetic manipulations are gene editing techniques that permit the genome to target specific locations for deletion, addition, or substitution. This can be done for both germinal and somatic genes. However, the process is associated with certain ethical, legal, and social issues [123]. The main research concern in this area is how accurately gene manipulation can be performed [124]. These days, this technique has been used to treat many human diseases. Genome editing is done to introduce certain genetic modifications in the genomic sequences of humans or other living entities [125, 126]. Somatic gene editing is done for therapeutic use and has undergone clinical trials. Examples include those used in the treatment of HIV and leukemia, where the patient is treated with modified cells from a healthy person [127–129]. The CRISPR-Cas9 technique was first used in China to edit genes for a clinical trial done for lung cancer [130]. Somatic gene editing can even be used for the treatment of beta-thalassemia and sickle-cell disease, both are single-gene disease [131]. Gene editing in the embryo or germ line can be used to correct single gene disorders permanently [132]. Although such an approach can prove a blessing for mankind by relieving destructive genetic disorders, there is still doubt whether human beings are morally responsible enough to use the technique for the betterment of mankind.

3.7.2 Gene Delivery

Gene delivery is a technology that is used in the treatment of hereditary disorders that develop after birth due to the unusual expression of genes. This technology is utilized to introduce the genetic material to the required point in the body. The delivery of genetic material cannot be done on its own as it may get destroyed by the physiological environment of the body; so, it is done using viral or other vectors. The other vectors include liposomes, nanotubes, nanoparticles, and metallic nanoparticles [133–135]. Gold nanoparticles (AuNPs), due to their stability, uniformity, and compatibility with biological environments, have gained much popularity as a gene non-vector [136]. Because of their specific electronic structure, AuNPs exhibit variable magnetic and optoelectronic capabilities that reflect a discrete electronic band known as surface plasmon resonance based on particle size, structure, interparticle spacing, and protective shell. [136–138]. Slight particle size deviation leads to a change in radical properties, which enhance its biomedical applications in the treatment of inherent and developed diseases [137, 139]. The chemistry of AuNP formulation is fairly straightforward, resulting in an appropriate size with many distinct biological molecules useful in its assimilation with the biological system. [140]. The apparent chemistry of the AuNPs is soft, which is an advantage in that it can be tailored to different biomolecules, which is not possible for other metal nanoparticles. It has a metallic core as a building block that acts as a strong support for the therapeutic material and is stable even after many dilutions, which is not possible with any other polymeric nanocarriers. AuNPs have a high exterior-to-volume ratio, which maximizes payload/carrier ratio exponentially [140]. It shows great potential in gene delivery; it can load nucleic acid by conjugation or by adsorbing with the help of regrafted cationic polymeric stabilizers [141–143]. Cationic polymers are implanted onto AuNPs on a regular basis to enhance their introduction to genetic materials that are anionic in nature. These anionic and cationic interactions form dense complexes, protecting the loaded material from any type of degradation, whether enzymatic or nonenzymatic, augmenting the cellular intake, and increasing the half-life in the cytoplasm [144, 145]. A gold nanorod has been reported for siRNA delivery to the dopaminergic signaling pathway for the therapy of drug addiction [146]. Recently, a layer-to-layer approach has been created by AuNPs for gene transport to enhance the efficacy by expanding the release kinetics of the loaded substance [147]. In the last few decades, there has been substantial research on the formulation of target-oriented gene delivery vehicles by inserting targeting moieties with the help of stabilizer ligands of AuNPs or using crosslinkers [139, 148, 149]. Although the biological functionalization of

AuNPs is a tough process because of their small size, electrostatic repulsion, colloidal instability, ineffective purifications, etc. [150]. Still, advanced bioconjugate chemistry helped in the use of AuNPs in the development of multifunctional nanoconstructs for gene therapy in cancer.

3.7.3 Cell Therapy

Cancer immunotherapy, or cell therapy, is a technique where the body's immune system is used to battle tumor cells [151]. Immunotherapy focuses on regaining the ability of the body's immune system to identify and damage cancerous cells against which the inherent immune system has been compromised [152]. Cytotoxic (CD8+) T cells act as the major immune system component responsible for removing damaged or infected cells. Antigen-presenting cells (APCs) recognize the antigens and present them to the major histocompatibility complex (MHC) [153, 154]. From here, the antigens interact with T cell receptors present on CD8+ T cells and start the immune response, leading to the activation of CD8+ T cells and their differentiation to form an army of T cells specific for that particular antigen that was presented [155]. Cancer vaccines present tumor antigen to the APCs, thus initiating the production of antigen-specific T cells by the spleen, skin, or lymph tissues [156]. Many types of different immunotherapy approaches are under investigation, such as adoptive cell therapies, monoclonal antibodies, checkpoint inhibitors [157–161]. Nanoparticles like metallic formulations act as natural transport for the spleen and lymph organs, making them good candidates for immunotherapy. These have specific properties due to which they can be designed to be used in cancer immunotherapy [162–166]. Their design can be customized as per the required application in respect of their particle size, shape, and charge [167–169]. MNPs are of specific advantage due to the perfection with which their size, structure, charge, and exterior modification can be done in a coordinated manner [170–172]. Further, these are of high density and easily taken up by the cells [173, 174]. MNPs improve the uptake of the antigen by the APCs, thus expanding the cytotoxic T cell effect [175, 176]. A significant sera antibody response was observed against the antigen delivered using gold nanoparticles of different sizes [177]. Peptide-coated AuNPs enhanced humoral response, as demonstrated by a rise in IgG exudation facilitated by the blimp/pax5 route [178]. AuNP-mediated delivery of ovalbumin antigens proved to be more significant both prophylactically and therapeutically than ovalbumin administration alone [179]. The peptide antigens alone generate a very weak immune response, which can be enhanced to a very high rate by their co-administration with adjuvant molecules incorporated on a nanoparticle carrier such as MNPs [180]. Gold, being bioinert, has inherent immune activation properties that can be used to stimulate anti-tumor immunity [176]. AuNPs coated with antigens without any adjuvant produced a strong immune response leading to T cell expansion, thus preventing cancer tumors [179]. Nude silver nanoparticles have also been reported to exhibit anti-tumor properties in lymphomas in vivo by inducing programmed cell death and slowing angiogenesis [181–183]. MNP formulations show specific advantages because of their capabilities in surface functionalization and optical or heat-centered therapeutic methods, and thus they have successfully entered clinical trials [151, 184–198] (Table 3.1).

TABLE 3.1
Biomedical Application of MNPs

MNPs	Mechanism	Disease targeting	Reference
	MNPs role in gene delivery		
AuNPs	CPIR28 peptide is the target of peptide-modified AuNPs for selective gene targeting.	CPIR28 gene targeting	[184]
AgNPs	The apoptosis caused by AgNPs damaging mitochondrial membranes is comparable to the process induced by 5-fluorouracil medicines or gene therapy treatments. As a result, AgNPs may operate as a therapeutic medication, a new chemosensitization strategy for the future.	UPRT non-transduced and transduced gene	[185]
AgNPs	AgNPs have the potential to be used as effective nonviral gene delivery vehicles with low cytotoxicity to improve the therapeutic efficiency of AgNPs used in many biomedical goods. Also used for the treatment of HeLa and A549 cells.	Nonviral gene delivery and HeLa and A549 cancer cell line treatment	[186]
AuNPs	Au-C225-p53DNA significantly reduced tumor size and targeted tumor growth in the SK-OV-3 xenograft mouse model.	Tumor suppressor gene p53 ovarian cancer	[187]
Manganese (Mn), silver (Ag), and copper (Cu) nanoparticles	According to these findings, dopaminergic neurotoxicity caused by Mn and Cu nanoparticles may be caused by similar neurodegenerative pathways.	Dopaminergic neuronal cell line, PC12	[188]
MNPs role in cell therapy			
Fe_3O_4 MNPs	After being exposed to 50 Hz electromagnetic fields, hBM-MSCs were treated with magnetic iron oxide nanoparticles (MNPs) employing RT-PCR, immunohistological examination and brain cell type-specific genes and antibodies.	Human bone marrow-derived mesenchymal stem cells (hBM-MSCs)	[189]
AuNPs	We come to the conclusion that against gram-positive pathogens, particularly L. monocytogenes, the proposed PVC modifications and gold nanoparticles are the most effective. However, it should be stressed that each strain needs a different strategy because of its unique features.	Medical device-related infections	[190]
AgNPs	We demonstrate that AgNPs reduce bacterial growth or lead to cell death by disrupting the structure and permeability of bacterial cell membrane biomolecules.	Cell surface damaged by bacterial cells	[191]
AuNPs	AuNPs are used in conjunction with fluorescence as well as reflectance measurements to detect squamous cell carcinoma of the tongue.	Squamous cell carcinoma	[192]
AuNPs	In bladder carcinoma 5637 cells, AuNPs greatly increase the generation of ROS, cause apoptosis, and limit cell migration.	Bladder cancer 5637 cells	[193]

TABLE 3.1 *(Continued)*
Biomedical Application of MNPs

MNPs role in gene delivery			
MNPs	**Mechanism**	**Disease targeting**	**Reference**
MNPs role in targeted drug delivery			
AgNPs	The results of this study demonstrate that HepG2 and Caco2 cells in culture, which are often used in in vitro models, are appropriate systems for assessing the cytotoxicity of silver nanoparticles. These well-known cell culture models and easy-to-use assays, which were employed in this study, can offer relevant toxicity and mechanistic data that can serve to better improve safety evaluations of silver nanoparticles that are utilized in food and cosmetics.	Liver HepG2 and colon Caco2 cells	[194]
AgNPs	In MCF-7 cancer cells, IMAB-AgNPs can have an inhibitory influence on viability by u*p-regulating apoptosis, which is convincing evidence that the greenly generated AgNPs are a potential drug delivery strategy.	**Breast cancer MCF-7 cells**	[195]
AgNPs	According to mechanism studies, Ag@AM can stop H1N1 from infecting host cells and can also stop DNA fragmentation, chromatin condensation, and caspase-3 activation. Reactive oxygen species (ROS) were prevented from building up, and the H1N1 virus's induction of apoptosis was countered by Ag@AM.	**H1N1 influenza virus**	[196]
AuNPs	In lung A549 carcinoma cells. Gold nanoparticles were able to trigger alternate translation of caspase-9, target oncogenes, tumor suppressor genes, and induce the apoptotic cascade in lung cancer cells.	A549 cells	[197]
AuNPs	Bleomycin and doxorubicin constitute two medications used in AuNPs-assisted combination chemotherapy that function through various pathways, reducing systemic drug toxicity, improving chemotherapy results, and decreasing the likelihood of the development of cancer treatment resistance.	HeLa cells	[198]

3.8 CONCLUSION

Metallic nanoparticle-based nanomedicine is always searching for better and more effective ways to treat illnesses. These new medicines must also be affordable, which places a strong demand on scientific research to find them. Being able to isolate the illness and spare other healthy parts of the body from injury is a critical component of any treatment. Due to their accessibility, biocompatibility, and stability, metal nanoparticles are being used as targeted drug delivery systems with great success today. The improvement in the practical potential of metallic nanoparticles emphasizes their potency as novel tools for drug delivery therapeutic modalities, particularly in the treatment of cancer, inflammation, diabetes, and antiviral therapy. Metallic nanoparticle development is quick and multidirectional.

ACKNOWLEDGMENT

Our sincere thanks to Dr. Jamilur R. Ansari and Aafrin Waziri for editing the manuscript. The authors gratefully acknowledge the contributors of the collaborators and co-workers mentioned in the cited reference. MSA gratefully acknowledges the Vice Chancellor and Director of Pharmacy at SGT University for their kind cooperation and support.

REFERENCES

1. Felice B, Prabhakaran MP, Rodriguez AP, Ramakrishna S. Drug delivery vehicles on a nano-engineering perspective. Materials Science and Engineering: C. 2014 Aug 1;41:178–195.
2. Khan I, Saeed K, Khan I. Nanoparticles: properties, applications and toxicities. Arabian Journal of Chemistry. 2019;12(7):908–931.
3. Jahangirian H, Lemraski EG, Webster TJ, Rafiee-Moghaddam R, Abdollahi Y. A review of drug delivery systems based on nanotechnology and green chemistry: green nanomedicine. International Journal of Nanomedicine. 2017;12:2957.
4. Quan Q, Zhang Y. Lab-on-a-tip (LOT): where nanotechnology can revolutionize fibre optics. Nanobiomedicine. 2015 Mar 30;2:3.
5. Bamrungsap S, Zhao Z, Chen T, Wang L, Li C, Fu T, Tan W. Nanotechnology in therapeutics: a focus on nanoparticles as a drug delivery system. Nanomedicine. 2012 Aug;7(8):1253–1271.
6. Shi J, Votruba AR, Farokhzad OC, Langer R. Nanotechnology in drug delivery and tissue engineering: from discovery to applications. Nano Letters. 2010 Sept 8;10(9):3223–3230.
7. Alam MS, Javed MN, Pottoo FH, Waziri A, Almalki FA, Hasnain MS, Garg A, Saifullah MK. QbD approached comparison of reaction mechanism in microwave synthesized gold nanoparticles and their superior catalytic role against hazardous nirto-dye. Applied Organometallic Chemistry. 2019;33(9):e5071.
8. Pandit J, Alam MS, Ansari JR, Singhal M, Gupta N, Waziri A, Sharma K, Potto FH. Multifaced applications of nanoparticles in biological science. In Nanomaterials in the Battle Against Pathogens and Disease Vectors (pp. 17–50). CRC Press, Boca Raton;2022.
9. Mishra S, Sharma S, Javed MN, Pottoo FH, Barkat MA, Alam MS, Amir M, Sarafroz M. Bioinspired nanocomposites: applications in disease diagnosis and treatment. Pharmaceutical Nanotechnology. 2019 Jun 1;7(3):206–219.
10. Alam MS, Naseh MF, Ansari JR, Waziri A, Javed MN, Ahmadi A, Saifullah MK, Garg A. Synthesis approaches for higher yields of nanoparticles. In Nanomaterials in the Battle Against Pathogens and Disease Vectors (pp. 51–82). CRC Press, Boca Raton;2022.
11. Alam MS, Garg A, Pottoo FH, Saifullah MK, Tareq AI, Manzoor O, Mohsin M, Javed MN. Gum ghatti mediated, one pot green synthesis of optimized gold nanoparticles: investigation of process-variables impact using Box-Behnken based statistical design. International Journal of Biological Macromolecules. 2017;104:758–767.
12. Javed MN, Pottoo FH, Alam MS. Metallic nanoparticle alone and/or in combination as novel agent for the treatment of uncontrolled electric conductance related disorders and/or seizure, epilepsy & convulsions. Patent Acquired on October. 2016;10:40.

13. Pottoo FH, Sharma S, Javed MN, Barkat MA, Harshita, Alam MS, Naim MJ, Alam O, Ansari MA, Barreto GE, Ashraf GM. Lipid-based nanoformulations in the treatment of neurological disorders. Drug Metabolism Reviews. 2020;52(1):185–204.
14. Javed, M.N., Alam, M.S., Waziri, A., Pottoo, F.H., Yadav, A.K., Hasnain, M.S. and Almalki, F.A., 2019. QbD applications for the development of nanopharmaceutical products. In *Pharmaceutical quality by design* (pp. 229–253). Academic Press.
15. Javed MN, Dahiya ES, Ibrahim AM, Alam M, Khan FA, Pottoo FH. Recent advancement in clinical application of nanotechnological approached targeted delivery of herbal drugs. In Nanophytomedicine (pp. 151–172). Springer, Singapore;2020.
16. Aslam M, Javed M, Deeb HH, Nicola MK, Mirza M, Alam M, Akhtar M, Waziri A. Lipid nanocarriers for neurotherapeutics: introduction, challenges, blood-brain barrier, and promises of delivery approaches. CNS & Neurological Disorders-Drug Targets (Formerly Current Drug Targets-CNS & Neurological Disorders). 2022;21.
17. Javed MN, Pottoo FH, Shamim A, Hasnain MS, Alam MS. Design of experiments for the development of nanoparticles, nanomaterials, and nanocomposites. In Design of Experiments for Pharmaceutical Product Development (pp. 151–169). Springer, Singapore;2021.
18. Singhal S, Gupta M, Alam MS, Javed MN, Ansari JR. Carbon allotropes-based nanodevices: graphene in biomedical applications. In Nanotechnology (pp. 241–269). CRC Press, Boca Raton;2022.
19. Kumar R, Dhamija G, Ansari JR, Javed MN, Alam MS. C-dot nanoparticulated devices for biomedical applications. In Nanotechnology (pp. 271–299). CRC Press, Boca Raton;2022.
20. Aslam M, Javed M, Deeb HH, Nicola MK, Mirza M, Alam M, Akhtar M, Waziri A. Lipid nanocarriers for neurotherapeutics: introduction, challenges, blood-brain barrier, and promises of delivery approaches. CNS & Neurological Disorders-Drug Targets (Formerly Current Drug Targets-CNS & Neurological Disorders). 2022;21.
21. Javed MN, Akhter MH, Taleuzzaman M, Faiyazudin M, Alam MS. Cationic nanoparticles for treatment of neurological diseases. In Fundamentals of Bionanomaterials (pp. 273–292). Elsevier, Amsterdam;2022.
22. Kumari N, Daram N, Alam MS, Verma AK. Rationalizing the use of polyphenol nano-formulations in the therapy of neurodegenerative diseases. CNS & Neurological Disorders-Drug Targets (Formerly Current Drug Targets-CNS & Neurological Disorders). 2022;21(10):966–976.
23. Naseh MF, Ansari JR, Alam MS, Javed MN. Sustainable nanotorus for biosensing and therapeutical applications. In Handbook of Green and Sustainable Nanotechnology: Fundamentals, Developments and Applications (pp. 1–21). Springer International Publishing, Cham;2022.
24. Khandel P, Yadaw RK, Soni DK, Kanwar L, Shahi SK. Biogenesis of metal nanoparticles and their pharmacological applications: present status and application prospects. Journal of Nanostructure in Chemistry. 2018 Sept;8(3):217–254.
25. Sengani M, Grumezescu AM, Rajeswari VD. Recent trends and methodologies in gold nanoparticle synthesis – a prospective review on drug delivery aspect. OpenNano. 2017 Jan 1;2:37–46.
26. Liyanage PY, Hettiarachchi SD, Zhou Y, Ouhtit A, Seven ES, Oztan CY, Celik E, Leblanc RM. Nanoparticle-mediated targeted drug delivery for breast cancer treatment. Biochimica et Biophysica Acta (BBA)-Reviews on Cancer. 2019 Apr 1;1871(2):419–433.
27. Ahmad MZ, Akhter S, Jain GK, Rahman M, Pathan SA, Ahmad FJ, Khar RK. Metallic nanoparticles: technology overview & drug delivery applications in oncology. Expert Opinion on Drug Delivery. 2010 Aug 1;7(8):927–942.
28. Alalaiwe A. The clinical pharmacokinetics impact of medical nanometals on drug delivery system. Nanomedicine: Nanotechnology, Biology and Medicine. 2019 Apr 1;17:47–61.

29. Singh K, Panghal M, Kadyan S, Yadav JP. Evaluation of antimicrobial activity of synthesized silver nanoparticles using Phyllanthus amarus and Tinospora cordifolia medicinal plants. Journal of Nanomedicine & Nanotechnology. 2014 Nov 1;5(6):1.
30. Kong FY, Zhang JW, Li RF, Wang ZX, Wang WJ, Wang W. Unique roles of gold nanoparticles in drug delivery, targeting and imaging applications. Molecules. 2017 Aug 31;22(9):1445.
31. Nejati K, Dadashpour M, Gharibi T, Mellatyar H, Akbarzadeh A. Biomedical applications of functionalized gold nanoparticles: a review. Journal of Cluster Science. 2022 Jan;33(1):1–6.
32. Parveen S, Misra R, Sahoo SK. Nanoparticles: a boon to drug delivery, therapeutics, diagnostics and imaging. Nanomedicine: Nanotechnology, Biology and Medicine. 2012 Feb 1;8(2):147–166.
33. Suk JS, Xu Q, Kim N, Hanes J, Ensign LM. PEGylation as a strategy for improving nanoparticle-based drug and gene delivery. Advanced Drug Delivery Reviews. 2016 Apr 1;99:28–51.
34. Pottoo FH, Tabassum N, Javed M, Nigar S, Rasheed R, Khan A, Barkat M, Alam M, Maqbool A, Ansari MA, Barreto GE. The synergistic effect of raloxifene, fluoxetine, and bromocriptine protects against pilocarpine-induced status epilepticus and temporal lobe epilepsy. Molecular Neurobiology. 2019;56(2):1233–1247.
35. Pottoo FH, Sharma S, Javed MN, Barkat MA, Harshita, Alam MS, Naim MJ, Alam O, Ansari MA, Barreto GE, Ashraf GM. Lipid-based nanoformulations in the treatment of neurological disorders. Drug Metabolism Reviews. 2020;52(1):185–204.
36. Pottoo FH, Javed M, Barkat M, Alam M, Nowshehri JA, Alshayban DM, Ansari MA. Estrogen and serotonin: complexity of interactions and implications for epileptic seizures and epileptogenesis. Current Neuropharmacology. 2019;17(3):214–231.
37. Pottoo FH, Tabassum N, Javed MN, Nigar S, Sharma S, Barkat MA, Alam MS, Ansari MA, Barreto GE, Ashraf GM. Raloxifene potentiates the effect of fluoxetine against maximal electroshock induced seizures in mice. European Journal of Pharmaceutical Sciences. 2020;146:105261.
38. Waziri A, Bharti C, Aslam M, Jamil P, Mirza M, Javed MN, Pottoo U, Ahmadi A, Alam MS. Probiotics for the chemoprotective role against the toxic effect of cancer chemotherapy. Anti-Cancer Agents in Medicinal Chemistry (Formerly Current Medicinal Chemistry-Anti-Cancer Agents). 2022;22(4):654–667.
39. Ibrahim AM, Chauhan L, Bhardwaj A, Sharma A, Fayaz F, Kumar B, Alhashmi M, AlHajri N, Alam MS, Pottoo FH. Brain-derived neurotropic factor in neurodegenerative disorders. Biomedicines. 2022;10(5):1143.
40. Raj S, Manchanda R, Bhandari M, Alam M. Review on natural bioactive products as radioprotective therapeutics: present and past perspective. Current Pharmaceutical Biotechnology. 2022;23.
41. Punjabi K, Choudhary P, Samant L, Mukherjee S, Vaidya S, Chowdhary A. Biosynthesis of nanoparticles: a review. International Journal of Pharmaceutical Sciences Review and Research 2015;30(1):219–226.
42. Ahmed S, Ahmad M, Swami BL, Ikram S. A review on plants extract mediated synthesis of silver nanoparticles for antimicrobial applications: a green expertise. Journal of advanced research. 2016 Jan 1;7(1):17–28.
43. Shnoudeh AJ, Hamad I, Abdo RW, Qadumii L, Jaber AY, Surchi HS, Alkelany SZ. Synthesis, characterization, and applications of metal nanoparticles. In Biomaterials and Bionanotechnology (pp. 527–612). Academic Press, London. 2019 Jan 1.
44. Iravani S, Zolfaghari B. Green synthesis of silver nanoparticles using Pinus eldarica bark extract. BioMed Research International. 2013;1–5. Available from: https://doi.org/10.1155/2013/639725.

45. Sengani M, Grumezescu AM, Rajeswari VD. Recent trends and methodologies in Au nanoparticle synthesis? A prospective review on drug delivery aspect. OpenNano. 2017;37–46. Available from: http://dx.doi.org/10.1016/j.onano.2017.07.001.
46. Sastry M, Ahmad A, Khan MI, Kumar R. Biosynthesis of metal nanoparticles using fungi and actinomycete. Current Science. 2003;162–170.
47. Korbekandi H, Iravani S, Abbasi, S. Production of nanoparticles using organisms. Critical Reviews in Biotechnology. 2009;29(4):279–306.
48. Nadagouda MN, Speth TF, Varma RS. Microwave assisted green synthesis of silver nanostructures. Accounts of Chemical Research. 2011;44(7):469–478.
49. Eustis S, Krylova G, Eremenko A, Smirnova N, Schill AW, El-Sayed M. Growth and fragmentation of silver nanoparticles in their synthesis with a FS laser and CW light by photo-sensitization with benzophenone. Photochemical and Photobiological Sciences. 2005;4(1):154–159.
50. Polshettiwar V, Nadagouda MN, Varma RS. Microwave assisted chemistry: a rapid and sustainable route to synthesis of organics and nanomaterials. Australian Journal of Chemistry. 2009;62(1):16–26.
51. Kirstein S, von Berlepsch H, Bottcher C. Photo-induced reduction of noble metal ions to metal nanoparticles on tubular J-aggregates. International Journal of Photoenergy. (2006):1–7.
52. Jin R, Charles Cao Y, Hao E, Metraux GS, Schatz GC, Mirkin CA. Controlling anisotropic nanoparticle growth through plasmon excitation. Nature. 2003;425(6957): 487–490.
53. Krutyakov YA, Olenin AY, Kudrinskii AA, Dzhurik PS, Lisichkin GV. Aggregative stability and polydispersity of silver nanoparticles prepared using two phase aqueous organic systems. Nanotechnologies in Russia 2008;3(5–6):303–310.
54. Kim M, Osone S, Kim T, Higashi H, Seto T. Synthesis of nanoparticles by laser ablation: a review. KONA Powder and Particle Journal. 2017:2017009.
55. Yadav TP, Manohar Yadav R, Pratap Singh D. Mechanical milling: a top down approach for the synthesis of nanomaterials and nanocomposites. Nanoscience and Nanotechnology. 2012;2(3):22–48.
56. Kalishwaralal K, Deepak V, Ramkumarpandian S, Nellaiah H, Sangiliyandi G. Extracellular biosynthesis of silver nanoparticles by the culture supernatant of *Bacillus licheniformis*. Materials Letters. 2008;62(29):4411–4413.
57. Nair B, Pradeep T. Preparation of gold nanoparticles from *Mirabilis jalapa* flowers. Crystal Growth and Design. 2002;2(4):293–298.
58. Ahmad A, Mukherjee P, Senapati S, Mandal D, Islam Khan M, Kumar R, Sastry M. Extracellular biosynthesis of silver nanoparticles using the fungus Fusarium oxysporum. Colloids and Surfaces, Part B: Biointerfaces. 2003;28(4):313–318.
59. Singaravelu G, Arockiamary JS, Ganesh Kumar V, Govindaraju K. A novel extracellular synthesis of monodisperse gold nanoparticles using marine alga, *Sargassum wightii* Greville. Colloids and Surfaces B: Biointerfaces. 2007;57.
60. Singaravelu G, Arockiamary JS, Ganesh Kumar V, Govindaraju K. A novel extracellular synthesis of monodisperse gold nanoparticles using marine alga, *Sargassum wightii* Greville. Colloids and Surfaces B: Biointerfaces. 2007;57:97–101.
61. Shankar SS, Rai A, Ahmad A, Sastry M. Rapid synthesis of Au, Ag, and bimetallic Au core–Ag shell nanoparticles using Neem (Azadirachta indica) leaf broth. Journal of Colloid and Interface Science. 2004;275(2):496–502.
62. Hasnain MS, Javed MM, Alam MS, Rishishwar P, Rishishwar S, Ali S, Nayak AK, Beg S. Purple heart plant leaves extract-mediated silver nanoparticle synthesis: optimization by Box-Behnken design. Materials Science and Engineering C. 2019:1105–1114.

63. Burduşel A-C, Gherasim O, Grumezescu AM, Mogoantă L, Ficai A, Andronescu E. Biomedical applications of silver nanoparticles: an up-to-date overview. Nanomaterials. 2018;8(9):681.
64. Raghupathi KR, Koodali RT, Manna AC. Size-dependent bacterial growth inhibition and mechanism of antibacterial activity of zinc oxide nanoparticles. Langmuir. 2011;27(7):4020–4028.
65. Jha AK, Prasad K. Green synthesis of silver nanoparticles using Cycas leaf. International Journal of Green Nanotechnology: Physics and Chemistry. 2010;1(2):110–117.
66. Fouda A, El-Din Hassan S, Abdo AM, El-Gamal MS. Antimicrobial, antioxidant and larvicidal activities of spherical silver nanoparticles synthesized by endophytic Streptomyces spp. Biological Trace Element Research. 2019:1–18.
67. El-Din HS, Salem S, Fouda A, Awad MA, El-Gamal MS, Abdo AM. New approach for antimicrobial activity and biocontrol of various pathogens by biosynthesized copper nanoparticles using endophytic actinomycetes. Journal of Radiation Research and Applied Sciences. 2018;11(3):262–270.
68. Kumar PV, Kala SMJ. Green synthesis, characterisation and biological activity of platinum nanoparticle using *Croton caudatus* Geisel leaf extract. International Journal of Recent Research Aspects. Special Issue: Conscientious Computing Technologies. 2018:608–612.
69. Molaei R, Farhadi K, Forough M, Hajizadeh S. Green biological fabrication and characterization of highly monodisperse palladium nanoparticles using *Pistacia atlantica* fruit broth. Journal of Nanostructures. 2018;8(1):47–54.
70. Muthukumar H, Mohammed SN, Chandrasekaran N, Sekar AD, Pugazhendhi A, Matheswaran M. Effect of iron doped zinc oxide nanoparticles coating in the anode on current generation in microbial electrochemical cells. International Journal of Hydrogen Energy. 2019;44(4):2407–2416.
71. Ahmad A, Mukherjee P, Senapati S, Mandal D, Khan MI, Kumar R, Sastry M. Extracellular biosynthesis of silver nanoparticles using the fungus Fusarium oxysporum. Colloids and surfaces B: Biointerfaces. 2003 May 1;28(4):313–318.
72. Muthukumar H, Mohammed SN, Chandrasekaran N, Sekar AD, Pugazhendhi A, Matheswaran M. Effect of iron doped Zinc oxide nanoparticles coating in the anode on current generation in microbial electrochemical cells. International Journal of Hydrogen Energy. 2019;44(4):2407–2416.
73. Goutam SP, Saxena G, Singh V, Yadav AK, Bharagava RN, Thapa KB. Green synthesis of TiO_2 nanoparticles using leaf extract of Jatropha curcas L. for photocatalytic degradation of tannery wastewater. Chemical Engineering Journal. 2018 Mar 15;336: 386–396.
74. Ahmed Q, Gupta N, Kumar A, Nimesh S. Antibacterial efficacy of silver nanoparticles synthesized employing Terminalia arjuna bark extract. Artificial Cells, Nanomedicine, and Biotechnology. 2017 Aug 18;45(6):1192–1200.
75. Kumari RM, Thapa N, Gupta N, Kumar A, Nimesh S. Antibacterial and photocatalytic degradation efficacy of silver nanoparticles biosynthesized using Cordia dichotoma leaf extract. Advances in Natural Sciences: Nanoscience and Nanotechnology. 2016 Oct 13;7(4):045009.
76. Arya G, Kumar N, Gupta N, Kumar A, Nimesh S. Antibacterial potential of silver nanoparticles biosynthesised using *Canarium ovatum* leaves extract. IET Nanobiotechnology. 2016;11(5):506–511.
77. Arya G, Mankamna Kumari R, Gupta N, Kumar A, Chandra R, Nimesh S. Green synthesis of silver nanoparticles using *Prosopis juliflora* bark extract: reaction optimization, antimicrobial and catalytic activities. Artificial Cells, Nanomedicine, and Biotechnology. 2018;46(5): 985–993.
78. Arya G, Malav AK, Gupta N, Kumar A, Nimesh S. Biosynthesis and in vitro antimicrobial potential of silver nanoparticles prepared using Dicoma tomentosa plant extract. Nanoscience and Nanotechnology-Asia. 2018;8(2):240–247.

79. Arya G, Sharma N, Ahmed J, Gupta N, Kumar A, Chandra R, Nimesh S. Degradation of anthropogenic pollutant and organic dyes by biosynthesized silver nano-catalyst from Cicer arietinum leaves. Journal of Photochemistry and Photobiology, Part B: Biology. 2017;174:90–96.
80. Goyal S, Gupta N, Kumar A, Chatterjee S, Nimesh S. Antibacterial, anticancer and antioxidant potential of silver nanoparticles engineered using *Trigonella foenum*-Graecum seed extract. IET Nanobiotechnology. 2018;12(4):526–533.
81. Arya G, Mankamna Kumari R, Pundir R, Chatterjee S, Gupta N, Kumar A, Chandra R, Nimesh S. Versatile biomedical potential of biosynthesized silver nanoparticles from *Acacia nilotica* bark. Journal of Applied Biomedicine. 2019;17(2):115–124.
82. Akhtar MS, Panwar J, Yun YS. Biogenic synthesis of metallic nanoparticles by plant extracts. ACS Sustainable Chemistry & Engineering. 2013 Jun 3;1(6):591–602.
83. Naseer A, Ali A, Ali S, Mahmood A, Kusuma HS, Nazir A, Yaseen M, Khan MI, Ghaffar A, Abbas M, Iqbal M. Biogenic and eco-benign synthesis of platinum nanoparticles (Pt NPs) using plants aqueous extracts and biological derivatives: environmental, biological and catalytic applications. Journal of Materials Research and Technology. 2020 Jul 1;9(4):9093–9107.
84. Gholami-Shabani M, Gholami-Shabani Z, Shams-Ghahfarokhi M, Akbarzadeh A, Riazi G, Razzaghi-Abyaneh M. Biogenic approach using sheep milk for the synthesis of platinum nanoparticles: the role of milk protein in platinum reduction and stabilization. International Journal of Nanoscience and Nanotechnology. 2016;12(4):199–206.
85. Coates AR, Halls G, Hu Y. Novel classes of antibiotics or more of the same? British Journal of Pharmacology. 2011 May 163(1):184–194.
86. Amendola V, Meneghetti M. Size evaluation of gold nanoparticles by UV-vis spectroscopy. The Journal of Physical Chemistry C. 2009;113(11):4277–4285.
87. Parang Z, Keshavarz A, Farahi S, Elahi SM, Ghoranneviss M, Parhoodeh S. Fluorescence emission spectra of silver and silver/cobalt nanoparticles. Scientia Iranica. 2012;19(3):943–947.
88. Lim J, Yeap S, Che H, Low S. Characterization of magnetic nanoparticle by dynamic light scattering. Nanoscale Research Letters. 2013;8(1):381.
89. Dougherty GM, Rose KA, Tok JB, Pannu SS, Chuang FY, Sha MY, Chakarova G, Penn SG. The zeta potential of surface-functionalized metallic nanorod particles in aqueous solution. Electrophoresis. 2008;29(5):1131–1139.
90. Faghihzadeh F, Anaya NM, Schifman LA, Oyanedel-Craver V. Fourier transform infrared spectroscopy to assess molecular-level changes in microorganisms exposed to nanoparticles. Nanotechnology for Environmental Engineering. 2016;1(1):1.
91. Marbella LE, Millstone JE. NMR techniques for noble metal nanoparticles. Chemistry of Materials. 2015;27(8):2721–2739.
92. Guo C, Yarger JL. Characterizing gold nanoparticles by NMR spectroscopy. Magnetic Resonance in Chemistry. 2018;56(11):1074–1082.
93. Marbella LE, Millstone JE. NMR techniques for noble metal nanoparticles. Chemistry of Materials. 2015;27(8):2721–2739.
94. Ingham B. X-ray scattering characterisation of nanoparticles. Crystallography Reviews. 2015;21(4):229–303.
95. Gao N, Yan Y. Characterisation of surface wettability based on nanoparticles. Nanoscale; 2012;4(7):2202–2218.
96. Buhr E, Senftleben N, Klein T, Bergmann D, Gnieser D, Frase CG, Bosse H. Characterization of nanoparticles by scanning electron microscopy in transmission mode. Measurement Science and Technology. 2009;20(8):084025.
97. Herzing AA, Watanabe M, Edwards JK, Conte M, Tang Z-R, Hutchings GJ, Kiely CJ. Energy dispersive X-ray spectroscopy of bimetallic nanoparticles in an aberration corrected scanning transmission electron microscope. Faraday Discussions. 2008;138:337–351.

98. Asadi Asadabad M, Jafari Eskandari M. Transmission electron microscopy as best technique for characterization in nanotechnology. Synthesis and Reactivity in Inorganic, Metal-Organic, and Nano-Metal Chemistry. 2015 Mar 4;45(3):323–326.
99. Sublemontier O, Nicolas C, Aureau D, Patanen M, Kintz H, Liu X, Gaveau MA, Le Garrec JL, Robert E, Barreda FA, Etcheberry A. X-ray photoelectron spectroscopy of isolated nanoparticles. The journal of physical chemistry letters. 2014 Oct 2;5(19):3399–3403.
100. Darwich S, Mougin K, Rao A, Gnecco E, Jayaraman S, Haidara H. Manipulation of gold colloidal nanoparticles with atomic force microscopy in dynamic mode: influence of particle–substrate chemistry and morphology, and of operating conditions Beilstein Journal of Nanotechnology. 2011;2(1):85–98.
101. Mansfield E, Tyner KM, Poling CM, Blacklock JL. Determination of nanoparticle surface coatings and nanoparticle purity using microscale thermogravimetric analysis. Analytical Chemistry. 2014;86(3):1478–1484.
102. Mattiuzzi C, Lippi G. Current cancer epidemiology. Journal of Epidemiology and Global Health. 2019 Dec;9(4):217.
103. Ferlay J, Ervik M, Lam F, Colombet M, Mery L, Piñeros M, Znaor A, Soerjomataram I, Bray F. Global Cancer Observatory: Cancer Today. International Agency for Research on Cancer, Lyon;2020 (https://gco.iarc.fr/today, accessed February 2021).
104. Mandal D, Maran A, Yaszemski MJ, Bolander ME, Sarkar G. Cellular uptake of gold nanoparticles directly cross-linked with carrier peptides by osteosarcoma cells. Journal of Materials Science: Materials in Medicine. 2009;20(1):347–350.
105. Evans ER, Bugga P, Asthana V, Drezek R. Metallic nanoparticles for cancer immunotherapy. Materials Today. 2018;21:673–685.
106. Shang L, Zhou X, Zhang J, Shi Y, Zhong L. Metal nanoparticles for photodynamic therapy: a potential treatment for breast cancer. Molecules. 2021;26:6532.
107. Alphandéry E. Biodistribution and targeting properties of iron oxide nanoparticles for treatments of cancer and iron anemia disease. Nanotoxicology. 2019;13:573–596.
108. Alphandéry E. Natural metallic nanoparticles for application in nano-oncology. International Journal of Molecular Sciences. 2020 Jun 21;21(12):4412.
109. Yesilot S, Aydin C. Silver nanoparticles; a new hope in cancer therapy? Eastern Journal of Medicine. 2019;24(1):111–116.
110. Song JY, Jang HK, Kim BS. Biological synthesis of gold nanoparticles using Magnolia kobus and Diopyros kaki leaf extracts. Process Biochemistry. 2009 Oct 1;44(10):1133–1138.
111. Danışman-Kalındemirtaş F, Kariper İA, Hepokur C, Erdem-Kuruca S. Selective cytotoxicity of paclitaxel bonded silver nanoparticle on different cancer cells. Journal of Drug Delivery Science and Technology. 2021 Feb 1;61:102265.
112. Lal S, Clare SE, Halas NJ. Nanoshell-enabled photothermal cancer therapy: impending clinical impact. Accounts of Chemical Research. 2008 Dec 16;41(12):1842–1851.
113. Roa W, Zhang X, Guo L, Shaw A, Hu X, Xiong Y, Gulavita S, Patel S, Sun X, Chen J, Moore R. Gold nanoparticle sensitize radiotherapy of prostate cancer cells by regulation of the cell cycle. Nanotechnology. 2009 Aug 26;20(37):375101.
114. Hainfeld JF, Slatkin DN, Smilowitz HM. The use of gold nanoparticles to enhance radiotherapy in mice. Physics in Medicine & Biology. 2004 Sep 3;49(18):N309.
115. Jung Y, Reif R, Zeng Y, Wang RK. Three-dimensional high-resolution imaging of gold nanorods uptake in sentinel lymph nodes. Nano letters. 2011 Jul 13;11(7):2938–2943.
116. Chen J, Glaus C, Laforest R, Zhang Q, Yang M, Gidding M, Welch MJ, Xia Y. Gold nanocages as photothermal transducers for cancer treatment. Small. 2010 Apr 9;6(7):811–817.
117. Wang H, Huff TB, Zweifel DA, He W, Low PS, Wei A, Cheng JX. In vitro and in vivo two-photon luminescence imaging of single gold nanorods. Proceedings of the National Academy of Sciences. 2005 Nov 1;102(44):15752–15756.

118. Hainfeld JF, Dilmanian A, Zhong Z, Slatkin DN, Kalef-Ezra JA, Smilowitz HM. Gold nanoparticles enhance the radiation therapy of a murine squamous cell carcinoma. Cancer Research. 2011 Apr 15;71(8_Supplement):2679.
119. Dreaden EC, Austin LA, Mackey MA, El-Sayed MA. Size matters: gold nanoparticles in targeted cancer drug delivery. Therapeutic delivery. 2012 Apr;3(4):457–478.
120. Mukherjee P, Bhattacharya R, Bone N, Lee YK, Patra CR, Wang S, Lu L, Secreto C, Banerjee PC, Yaszemski MJ, Kay NE. Potential therapeutic application of gold nanoparticles in B-chronic lymphocytic leukemia (BCLL): enhancing apoptosis. Journal of Nanobiotechnology. 2007 Dec;5(1):1–3.
121. Sun Y, Zheng L, Yang Y, Qian X, Fu T, Li X, Yang Z, Yan H, Cui C, Tan W. Metal–organic framework nanocarriers for drug delivery in biomedical applications. Nano-Micro Letters. 2020 Dec;12(1):1–29.
122. Qadri S, Abdulrehman T, Azzi J, Mansour S, Haik Y. AgCuB nanoparticle eradicates intracellular S. aureus infection in bone cells: in vitro. Emergent Materials. 2019;2:219–31.
123. Al Tamimi S, Ashraf S, Abdulrehman T, Parray A, Mansour SA, Haik Y, Qadri S. Synthesis and analysis of silver–copper alloy nanoparticles of different ratios manifest anticancer activity in breast cancer cells. Cancer Nanotechnology. 2020;11:1–16.
124. Howard HC, van El CG, Forzano F, Radojkovic D, Rial-Sebbag E, de Wert G, Borry P, Cornel MC; Public and professional policy committee of the European society of human genetics: One small edit for humans, one giant edit for humankind? Points and questions to consider for a responsible way forward for gene editing in humans. European Journal of Human Genetics. 2018 Jan;26(1):1–11.
125. Lecuona Ramírez ID, Casado M, Marfany i Nadal G, López Baroni MJ, Escarrabill M. Gene editing in humans: Towards a global and inclusive debate for responsible research. Yale Journal of Biology and Medicine. 2017 Dec 19;90(4):673–681.
126. Marraffini LA, Sontheimer EJ. CRISPR interference: RNA-directed adaptive immunity in bacteria and archaea. Nature Reviews Genetics. 2010;11(3):181–190.
127. Garneau JE, Dupuis MÈ, Villion M, Romero DA, Barrangou R, Boyaval P, Fremaux C, Horvath P, Magadán AH, Moineau S. The CRISPR/Cas bacterial immune system cleaves bacteriophage and plasmid DNA. Nature. 2010;468(7320):67.
128. Maeder ML, Gersbach CA. Genome-editing technologies for gene and cell therapy. Molecular Therapy. 2016;24:430–446.
129. Reardon S. First CRISPR clinical trial gets green light from US panel. Nature 2016 Jun;22.
130. Reardon S. Leukaemia success heralds wave of gene-editing therapies. Nature. 2015;527:146–147.
131. Cyranoski D. CRISPR gene-editing tested in a person for the first time. Nature. 2016;539:479.
132. Maeder ML, Gersbach CA. Genome-editing technologies for gene and cell therapy. Molecular Therapy. 2016;24:430–446.
133. Liang P, Xu Y, Zhang X, Ding C, Huang R, Zhang Z, Lv J, Xie X, Chen Y, Li Y, Sun Y. CRISPR/Cas9-mediated gene editing in human tripronuclear zygotes. Protein Cell. 2015;6:363–372.
134. Putnam D. Polymers for gene delivery across length scales. Nature Materials. 2006;5: 439–451.
135. Pack DW, Hoffman AS, Pun S, Stayton PS. Design and development of polymers for gene delivery. Nature Reviews Drug Discovery. 2005;4:581–593.
136. Morille M, Passirani C, Vonarbourg A, Clavreul A, Benoit JP. Progress in developing cationic vectors for non-viral systemic gene therapy against cancer. Biomaterials. 2008;29:3477–3496.
137. Feldheim DL. Colby AF Jr., Eds., Metal Nanoparticles Synthesis, Characterization and Applications, Marcel Dekker, New York, 2002.

138. Saha K, Agasti SS, Kim C, Li X, Rotello VM. Gold nanoparticles in chemical and biological sensing. Chemical Reviews. 2012;112:2739–2779.
139. Jain PK, Lee KS, El-Sayed IH, El-Sayed MA. Calculated absorption and scattering properties of gold nanoparticles of different size, shape, and composition: applications in biological imaging and biomedicine. The Journal of Physical Chemistry B. 2006;110:7238–7248.
140. Daniel MC, Astruc D. Gold nanoparticles: assembly, supramolecular chemistry, quantum-size-related properties, and applications toward biology, catalysis, and nanotechnology. Chemical Reviews. 2004;104:293–346.
141. Hostetler NJ, Wingate JE, Zhong CJ, Harris JE, Vachet RW, Clark MR, Londono JD, Green SJ, Stokes JJ, Wignall GD, Glish GL, Porter MD, Evans ND, Murray RW. Alkanethiolate gold cluster molecules with core diameters from 1.5 to 5.2 nm: core and monolayer properties as a function of core size. Langmuir. 1998;14:17–30.
142. Rosi NL, Giljohann DA, Thaxton CS, Lytton-Jean AK, Han MS, Mirkin CA. Oligonucleotide-modified gold nanoparticles for intracellular gene regulation. Science 2006;312:1027–1030.
143. Kong WH, Bae KH, Jo SD, Kim JS, Park TG. Cationic lipid-coated gold nanoparticles as efficient and noncytotoxic intracellular siRNA delivery vehicles. Pharmaceutical Research. 2012;29:362–374.
144. Ghosh P, Han G, De M, Kim CK, Rotello VM. Gold nanoparticles in delivery applications. Advanced Drug Delivery Reviews. 2008;60:1307–1315.
145. Alexis F, Pridgen E, Molnar LK, Farokhzad OC. Factors affecting the clearance and biodistribution of polymeric nanoparticles. Molecular Pharmaceutics. 2008;5:505–515.
146. Dykxhoorn DM, Palliser D, Lieberman J. The silent treatment: siRNAs as small molecule drugs. Gene Therapy. 2006;13:541–552.
147. Bonoiu AC, Mahajan SD, Ding H, Roy I, Yong KY, Kumar R, Hu R, Bergey EJ, Schwartz SA, Prasad PN. Nanotechnology approach for drug addiction therapy: gene silencing using delivery of gold nanorod-siRNA nanoplex in dopaminergic neurons. Proceedings of the National Academy of Sciences of the United States of America. 2009;106:5546–5550.
148. Tang Z, Wang Y, Podsiadlo P, Kotov NA. Biomedical applications of layer-by-layer assembly: from biomimetics to tissue engineering. Advanced Materials. 2006;18: 3203–3224.
149. Radwan SH, Azzazy HME. Gold nanoparticles for molecular diagnostics. Expert Review of Molecular Diagnostics. 2009;9:511–524.
150. Zayats M, Baron R, Popov I, Willner I. Biocatalytic growth of Au nanoparticles: from mechanistic aspects to biosensor design. Nano Letters. 2005;5:21–25.
151. Bahadur KCR, Thapa B, Bhattarai N. Gold nanoparticle-based gene delivery: promises and challenges. Nanotechnology Reviews. 2014 Jun 1;3(3):269–280.
152. Evans ER, Bugga P, Asthana V, Drezek R. Metallic Nanoparticles for Cancer Immunotherapy. Mater Today (Kidlington). 2018 Jul–Aug;21(6):673–685.
153. Farkona S, Diamandis EP, Blasutig IM. Cancer immunotherapy: the beginning of the end of cancer? BMC Medicine. 2016 May 5;14:73.
154. Krummel MF, Bartumeus F, Gerard A. T cell migration, search strategies and mechanisms. Nature Reviews Immunology. 2016 Mar;16:193–201.
155. Tagliamonte M, Petrizzo A, Tornesello ML, Buonaguro FM, Buonaguro L. Antigen-specific vaccines for cancer treatment. Human Vaccines & Immunotherapeutics. 2014;10:3332–3346.
156. Kreiter S, Vormehr M, Van de Roemer N, Diken M, Löwer M, Diekmann J, Boegel S, Schrörs B, Vascotto F, Castle JC, Tadmor AD. Mutant MHC class II epitopes drive therapeutic immune responses to cancer. Nature. 2015 Apr 30;520:692–696.
157. Melief CJ, van Hall T, Arens R, Ossendorp F, van der Burg SH. Therapeutic cancer vaccines. Journal of Clinical Investigation. 2015 Sep;125:3401–3412.

158. Gattinoni L, Powell Jr DJ, Rosenberg SA, Restifo NP. Adoptive immunotherapy for cancer: building on success. Nature Reviews Immunology. 2006 May;6:383–393.
159. Klebanoff CA, Acquavella N, Yu Z, Restifo NP. Therapeutic cancer vaccines: are we there yet? Immunological Reviews. 2011 Jan;239:27–44.
160. Rosenberg SA, Restifo NP, Yang JC, Morgan RA, Dudley ME. Adoptive cell transfer: a clinical path to effective cancer immunotherapy. Nature Reviews Cancer. 2008 Apr;8:299–308.
161. Rosenberg SA, Yang JC, Restifo NP. Cancer immunotherapy: moving beyond current vaccines. Nature Medicine. 2004 Sep;10:909–915.
162. Weiner LM, Dhodapkar MV, Ferrone S. Monoclonal antibodies for cancer immunotherapy. Lancet. 2009 Mar 21;373:1033–1040.
163. Almeida JP, Figueroa ER, Drezek RA. Gold nanoparticle mediated cancer immunotherapy. Nanomedicine. 2014 Apr;10:503–514.
164. Park YM, Lee SJ, Kim YS, Lee MH, Cha GS, Jung ID, Kang TH, Han HD. Nanoparticle-based vaccine delivery for cancer immunotherapy. Immune Network. 2013 Oct;13:177–183.
165. Shao K, Singha S, Clemente-Casares X, Tsai S, Yang Y, Santamaria P. Nanoparticle-based immunotherapy for cancer. ACS Nano. 2015 Jan 27;9:16–30.
166. Pelaz B, Alexiou C, Alvarez-Puebla RA, Alves F, Andrews AM, Ashraf S, Balogh LP, Ballerini L, Bestetti A, Brendel C, Bosi S. Diverse Applications of Nanomedicine. ACS Nano. 2017 Mar 28;11:2313–2381.
167. Ramos AP, Cruz MAE, Tovani CB, Ciancaglini P. Biomedical applications of nanotechnology. Biophysical Reviews. 2017 Apr;9:79–89.
168. Bachmann MF, Jennings GT. Vaccine delivery: a matter of size, geometry, kinetics and molecular patterns. Nature Reviews Immunology. 2010 Nov;10:787–796.
169. Niikura K, Matsunaga T, Suzuki T, Kobayashi S, Yamaguchi H, Orba Y, Kawaguchi A, Hasegawa H, Kajino K, Ninomiya T, Ijiro K. Gold nanoparticles as a vaccine platform: influence of size and shape on immunological responses in vitro and in vivo. ACS Nano. 2013 May 28;7:3926–3938.
170. Salatin S, Maleki Dizaj S, Yari Khosroushahi A. Effect of the surface modification, size, and shape on cellular uptake of nanoparticles. Cell Biology International. 2015 Aug;39:881–890.
171. Sperling RA, Parak WJ. Surface modification, functionalization and bioconjugation of colloidal inorganic nanoparticles. Philosophical Transactions of the Royal Society A. 2010 Mar 28;368:1333–1383.
172. Almeida JP, Chen AL, Foster A, Drezek R. In vivo biodistribution of nanoparticles. Nanomedicine (Lond) 2011 Jul;6:815–835.
173. Barnaby SN, Lee A, Mirkin CA. Probing the inherent stability of siRNA immobilized on nanoparticle constructs. Proceedings of the National Academy of Sciences of the United States of America. 2014 Jul 08;111:9739–9744.
174. Almeida JP, Figueroa ER, Drezek RA. Gold nanoparticle mediated cancer immunotherapy. Nanomedicine. 2014 Apr;10:503–514.
175. Shao K, Singha S, Clemente-Casares X, Tsai S, Yang Y, Santamaria P. Nanoparticle-based immunotherapy for cancer. ACS Nano. 2015 Jan 27;9:16–30.
176. Chen YS, Hung YC, Lin WH, Huang GS. Assessment of gold nanoparticles as a size-dependent vaccine carrier for enhancing the antibody response against synthetic foot-and-mouth disease virus peptide. Nanotechnology. 2010 May 14;21:195101.
177. Lee CH, Syu SH, Chen YS, Hussain SM, Onischuk AA, Chen WL, Huang GS. Gold nanoparticles regulate the blimp1/pax5 pathway and enhance antibody secretion in B-cells. Nanotechnology. 2014 Mar 28;25:125103.
178. Almeida JP, Lin AY, Figueroa ER, Foster AE, Drezek RA. In vivo gold nanoparticle delivery of peptide vaccine induces anti-tumor immune response in prophylactic and therapeutic tumor models. Small. 2015 Mar;11:1453–1459.

179. Bode C, Zhao G, Steinhagen F, Kinjo T, Klinman DM. CpG DNA as a vaccine adjuvant. Expert Review of Vaccines. 2011 Apr;10:499–511.
180. Arvizo RR, Bhattacharyya S, Kudgus RA, Giri K, Bhattacharya R, Mukherjee P. Intrinsic therapeutic applications of noble metal nanoparticles: past, present and future. Chemical Society Reviews. 2012 Apr 7;41:2943–2970.
181. Sriram MI, Kanth SB, Kalishwaralal K, Gurunathan S. Antitumor activity of silver nanoparticles in Dalton's lymphoma ascites tumor model. International Journal of Nanomedicine. 2010 Oct 05;5:753–762.
182. Jacob JA, Shanmugam A. Silver nanoparticles provoke apoptosis of Dalton's ascites lymphoma in vivo by mitochondria dependent and independent pathways. Colloids Surf B Biointerfaces. 2015 Dec 1;136:1011–1016.
183. Antony JJ, Sithika MA, Joseph TA, Suriyakalaa U, Sankarganesh A, Siva D, Kalaiselvi S, Achiraman S. In vivo antitumor activity of biosynthesized silver nanoparticles using Ficus religiosa as a nanofactory in DAL induced mice model. Colloids Surf B Biointerfaces. 2013 Aug 1;108:185–190.
184. Xu X, Liu Y, Yang Y, Wu J, Cao M, Sun L. One-pot synthesis of functional peptide-modified gold nanoparticles for gene delivery. Colloids and Surfaces A: Physicochemical and Engineering Aspects. 2022 May 5;640:128491.
185. Gopinath P, Gogoi SK, Chattopadhyay A, Ghosh SS. Implications of silver nanoparticle induced cell apoptosis for in vitro gene therapy. Nanotechnology. 2008 Jan 29;19(7):075104.
186. Sarkar K, Banerjee SL, Kundu PP, Madras G, Chatterjee K. Biofunctionalized surface-modified silver nanoparticles for gene delivery. Journal of Materials Chemistry B. 2015;3(26):5266–5276.
187. Kotcherlakota R, Vydiam K, Jeyalakshmi Srinivasan D, Mukherjee S, Roy A, Kuncha M, Rao TN, Sistla R, Gopal V, Patra CR. Restoration of p53 function in ovarian cancer mediated by gold nanoparticle-based EGFR targeted gene delivery system. ACS Biomaterials Science & Engineering. 2019 Jun 5;5(7):3631–3644.
188. Wang J, Rahman MF, Duhart HM, Newport GD, Patterson TA, Murdock RC, Hussain SM, Schlager JJ, Ali SF. Expression changes of dopaminergic system-related genes in PC12 cells induced by manganese, silver, or copper nanoparticles. Neurotoxicology. 2009 Nov 1;30(6):926–933.
189. Choi YK, Lee DH, Seo YK, Jung H, Park JK, Cho H. Stimulation of neural differentiation in human bone marrow mesenchymal stem cells by extremely low-frequency electromagnetic fields incorporated with MNPs. Applied Biochemistry and Biotechnology. 2014 Oct;174(4):1233–1245.
190. Kwiatkowska A, Granicka LH, Grzeczkowicz A, Stachowiak R, Bącal P, Sobczak K, Darowski M, Kozarski M, Bielecki J. Gold nanoparticle-modified poly (vinyl chloride) surface with improved antimicrobial properties for medical devices. Journal of Biomedical Nanotechnology. 2018 May 1;14(5):922–932.
191. Ansari MA, Khan HM, Khan AA, Ahmad MK, Mahdi AA, Pal R, Cameotra SS. Interaction of silver nanoparticles with Escherichia coli and their cell envelope biomolecules. Journal of Basic Microbiology. 2014 Sep;54(9):905–915.
192. Nour M, Hamdy O, Faid AH, Eltayeb EA, Zaky AA. Utilization of gold nanoparticles for the detection of squamous cell carcinoma of the tongue based on laser-induced fluorescence and diffuse reflectance characteristics: an in vitro study. Lasers in Medical Science. 2022 Aug 24:1.
193. Daei S, Ziamajidi N, Abbasalipourkabir R, Khanaki K, Bahreini F. Anticancer effects of gold nanoparticles by inducing apoptosis in bladder cancer 5637 cells. Biological Trace Element Research. 2022 Jun;200(6):2673–2683.
194. Sahu SC, Zheng J, Graham L, Chen L, Ihrie J, Yourick JJ, Sprando RL. Comparative cytotoxicity of nanosilver in human liver HepG2 and colon Caco2 cells in culture. Journal of Applied Toxicology. 2014 Nov;34(11):1155–1166.

195. Sadat Shandiz SA, Shafiee Ardestani M, Shahbazzadeh D, Assadi A, Ahangari Cohan R, Asgary V, Salehi S. Novel imatinib-loaded silver nanoparticles for enhanced apoptosis of human breast cancer MCF-7 cells. Artificial cells, nanomedicine, and biotechnology. 2017 Aug 18;45(6):1082–1091.
196. Li Y, Lin Z, Zhao M, Guo M, Xu T, Wang C, Xia H, Zhu B. Reversal of H1N1 influenza virus-induced apoptosis by silver nanoparticles functionalized with amantadine. RSC Advances. 2016;6(92):89679–89686.
197. Tiloke C, Phulukdaree A, Anand K, Gengan RM, Chuturgoon AA. Moringa oleifera gold nanoparticles modulate oncogenes, tumor suppressor genes, and caspase-9 splice variants in a549 cells. Journal of cellular biochemistry. 2016 Oct;117(10):2302–2314.
198. Farooq MU, Novosad V, Rozhkova EA, Wali H, Ali A, Fateh AA, Neogi PB, Neogi A, Wang Z. Gold nanoparticles-enabled efficient dual delivery of anticancer therapeutics to HeLa cells. Scientific Reports. 2018 Feb 13;8.

4 Emerging Trends in Anti-infectious Metallic Nanoparticles

Hayrunnisa Nadaroglu and Azize Alayli

4.1 INTRODUCTION

It would be right to say that nanotechnology is among the most innovative values of our age. These technological developments include a long way to use the products obtained by the design of substances that start their journey at the atomic scale in different fields. While working with materials of this small size and knowing that matter obeys the rules of atomic physics, the excitement of benefiting from the unique properties it acquires offers a new perspective to these researchers every day. Nanotechnology is a word that combines the concepts of nano and technology. It can be defined as the technology of small-sized substances. Nanotechnology includes the following areas within the meaning of the word: Nanotechnology is a research area developed by different disciplines. Some of the disciplines that make up nanotechnology are given in the following. With the developments in nanotechnology, it is seen that it finds use in many places and enters almost every part of your life. Some of these areas are as follows (Figures 4.1 and 4.2) [1, 2].

Nanotechnology is a research area developed by different disciplines. Some of the disciplines that make up nanotechnology are given in the following.

With the developments in nanotechnology, it has been noticed that it finds use in many places and enters almost every part of your life [3–9]. Some of these areas are as follows (Figure 4.3):

i. Electricity: With the development of nanomaterials, more conductive and lighter materials with higher storage capacity have been created, and thus, superior properties have been performed. Thanks to nanomaterials used in electronics, it has been possible to produce higher-resolution displays, detectors, and processors.
ii. Energy: There are different studies in this field. One of these examples can exhibited as the use of nanomaterials that increase heat conduction in factory or city heating systems. In addition, nanoadditives that increase performance by adding them to fuels can be given as an example of nanomaterials used in the field of energy.

 DOI: 10.1201/9781003317319-4

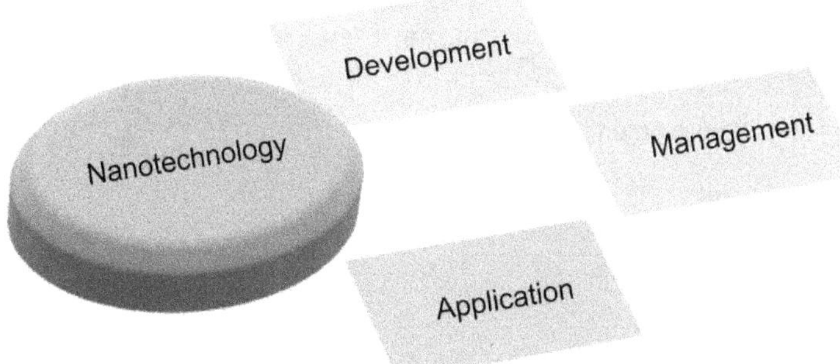

FIGURE 4.1 Strategies of nanotechnology.

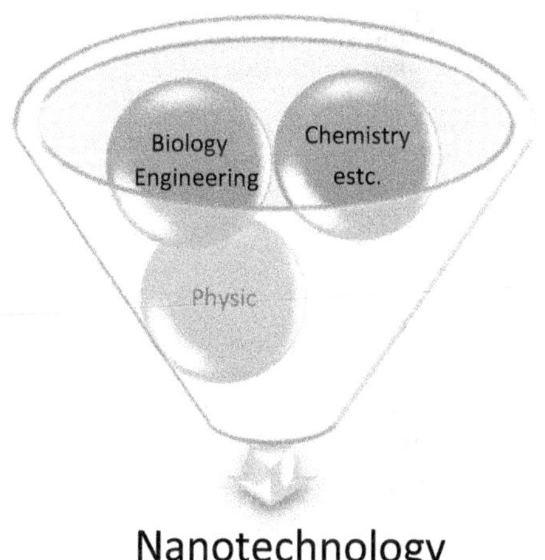

FIGURE 4.2 Science branches that make up nanotechnology.

iii. Artificial intelligence: It has entered this field in nanoapplications with electricity and electronics as a multidisciplinary field. Thanks to nanosensors, it is possible to run processors.
iv. In the field of the environment: Nanomaterial applications are quite common, one of the most striking is the use of nanomaterials in the removal of environmentally harmful components.
v. Health field: It is an area where nanotechnology gets into a lot. Today, we see more and more that nanomaterials can be used separately in the diagnosis, follow-up, and treatment steps in the health field. Nanomaterials are often used in drug compositions used for cancer treatment, from rapid test kits used for this purpose.
vi. In the pharmaceutical industry: The use of nanomaterials brings to mind innovative pharmaceutical applications. In addition, nanomaterials are

FIGURE 4.3 Usage areas of nanotechnology.

employed in the development of controlled-release drugs and target-sensitive drugs.

vii. In the agricultural sector: There are further areas where nanocomponents are utilized, some of them are the product of innovative nanomaterials used instead of pesticides and the use of nano fertilizers for the development of plants.

viii. Food sector: Today, it is a different sector into which nanotechnology enters. In this sector, different areas, such as smart packaging and foods with increased nutritional value, are studied.

ix. The cosmetics sector: It is one of the fastest-growing sectors for nanotechnology as well as being new. In this area, nanocomponents are included in some sunscreens, hair dyes, and colored cosmetics.

x. Biomedical: When examined, it is a very common practice to include nanoparticles due to the development of biomedical such as orthoses and prostheses compatible with tissues, increasing the sensitivity of biomedical used for diagnosis. The replacement of biosensors by nanosensors should not be strange in the future. It can be thought that alternatives to biosensors will be designed by utilizing the biomimetic properties of structurally more durable nanocomponents.

xi. In the defense industry: Nanotechnology has taken its place in the creation of light equipment, the production of strong armor, and the production of more durable ammunition.

Protection: for this purpose, radiation-resistant nanomaterials were produced, and nanoproducts in buildings and ships were used to protect the exterior and components.

Nanoparticles, which find such widespread use, are synthesized by different methods. These methods are based on two basic principles. From this point of view, the bottom-up is the synthesis of nano-sized materials using atomic-sized building blocks. For this purpose, it is tried to build the desired structure between 1 and 100 nm by designing it one by one with nanobuilding blocks. The chemical synthesis method is among the examples to be given of this principle.

The top-down method, on the other hand, is a different approach used to obtain nanostructures. In this principle, nanolevels are reached by different methods by leaving the bulk material. The methods used for this can vary from physical etching methods to laser etching methods as shown in Figure 4.4 [10].

Each of the nanostructures synthesized by different methods has its own physical and chemical structure. In fact, many properties such as chemical, physical, biological, electrical, and biomimetic properties of nanostructures have the same chemical structure and only differ in size differ. For these reasons, the shape and size of the nanostructure obtained are important. From this point of view, nanostructures can be seen in Figure 4.5.

These structures, which were synthesized in different sizes and shapes, are used in many different fields, as shown in Figure 4.5. When we look at this table, metal nanoparticles with ^{0}D, which is one of the basic nanostructures, draw attention to their small sizes. Metal nanoparticles can be synthesized by different methods in the

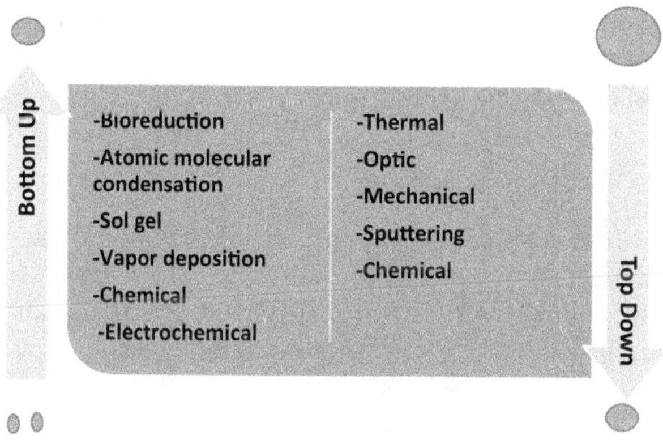

FIGURE 4.4 Nanomaterial synthesis methods.

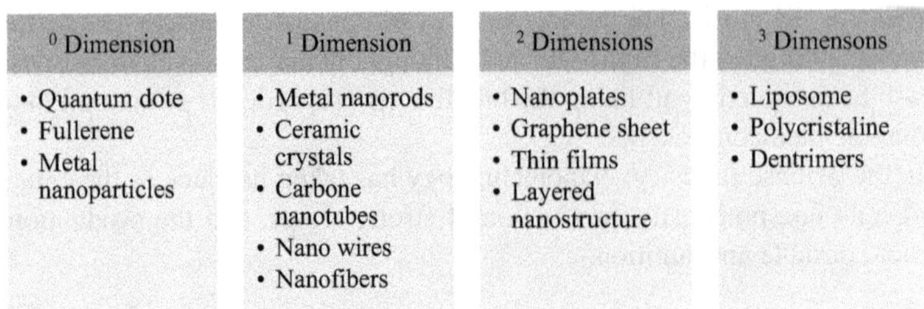

FIGURE 4.5 Classification of nanostructures by dimensions.

aforementioned ways. These synthesized metal nanoparticles are used in many fields for the production of nanomaterials. We will discuss their antimicrobial, antiviral, and anti-infective properties, which are important for human health, in detail in the following sections [4, 11, 12].

4.2 METALLIC NANOPARTICLES

Metallic nanoparticles (M-NPs) can be found in different salts (Ag_2S, etc.) or oxides (Ag_2O) structures of metals, as well as containing metals in their core or surface. In these materials, metal nanoparticles can be located in or on an organic or inorganic structure (Figure 4.6).

This type of metal nanoparticle, which was synthesized by the different methods mentioned earlier, has found many uses today. Some of these areas are shown in Table 4.1 [13].

Metal nanoparticles used in different fields can also be categorized as metal nanoparticles (Ag, Au, Cu, Pt, and Ti) according to their structures. These types of nanoparticle are used in many different fields. In addition, metal sulfide-type nanoparticles are among the most studied structures today. They are compounds in the class of metal-containing nanoparticles in nano metal frameworks, where nano-metals and organic structures are utilized, which is an even more current area. In addition, doped metal-metal or metal-metal oxides are classified as metal nano particular structures. It is seen that these types of nanostructures are widely studied due to their drug delivery and antimicrobial effects in Figure 4.7 [24].

4.3 ANTI-INFECTIVE M-NPs

It has been reported that one out of every four deaths worldwide is due to problems caused by infection. The elderly, children, and chronic patients with weakened immune systems are more affected by this type of infection. This type of infectious disease can be caused by bacteria, viruses, and protozoa. Treatment processes are interrupted by infectious pathogens caused by these factors, either because the existing drugs are not sufficient or because the agent develops resistance against the existing ones. The reason why bacterial infections are so common is the formation of bacterial species that develop resistance to antibiotics and the formation of different

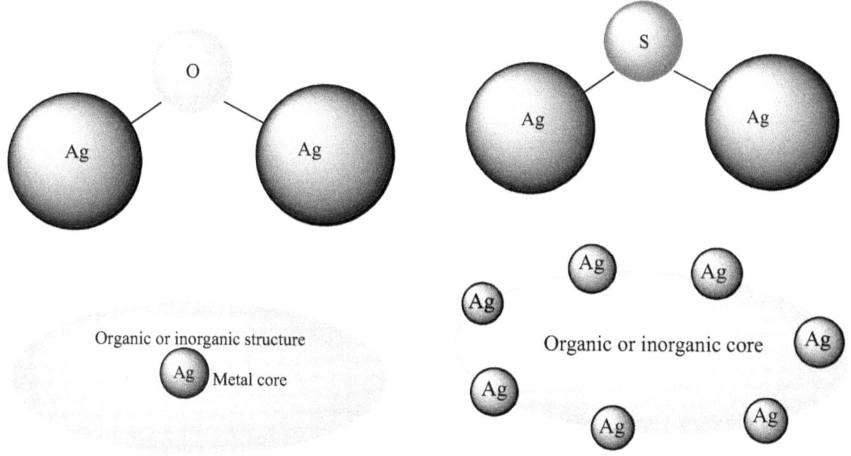

FIGURE 4.6 Types of metal nanoparticles.

TABLE 4.1
Commonly Used Metal Nanoparticles

Metal nanoparticles	Usage	Reference
Aluminum	Aluminum nanoparticles are used in drug delivery, materials, surface coatings, explosives, and as fuel additives.	[14]
Iron	Iron nanoparticles are involved in many drug formulations. These drugs are used in iron replacement therapy for chronic renal failure. They are also used in different ways such as CNS imaging, macrophage imaging, and blood pooling agents, cellular labeling, and lymph node imaging.	[15]
Gold	Gold nanoparticles are employed as an imaging agent for the determination of microorganisms and their metabolites, the development of biosensors, the imaging of tumor cells, endocytosis and receptor identification, the examination of metabolism, and different types of microscopy.	[16]
Silver	Silver nanoparticles are widely used in many fields such as medicine, food, and health especially because of their antimicrobial effects.	[17]
Copper	Copper nanoparticles (Cu NPs) are used because of their antibacterial properties and antifungal activity, in addition to their catalytic, optical, and electrical properties.	[18]
Cerium	Cerium nanoparticles are used for bone defects, osteosarcoma, tooth treatment, implant production, and antimicrobial effects.	[19]
Manganese	Manganese nanoparticles are used in fields such as wastewater treatment, catalysis, sensors, supercapacitors, and alkaline and rechargeable batteries.	[20]
Nickel	Nickel nanoparticles are used in a variety of catalytic applications because they have antibacterial, antifungal, cytotoxic, anticancer, antioxidant, healing, and enzyme inhibition properties.	[21]
Titanium	Titanium nanoparticles are utilized in the therapy of cancer as well as in the photodynamic inactivation of antibiotic-resistant bacteria and in sunscreens.	[22]
Zinc	Zinc nanoparticles are widely used in biosensors, cosmetics, and drug delivery systems, as well as due to their antimicrobial effects.	[23]

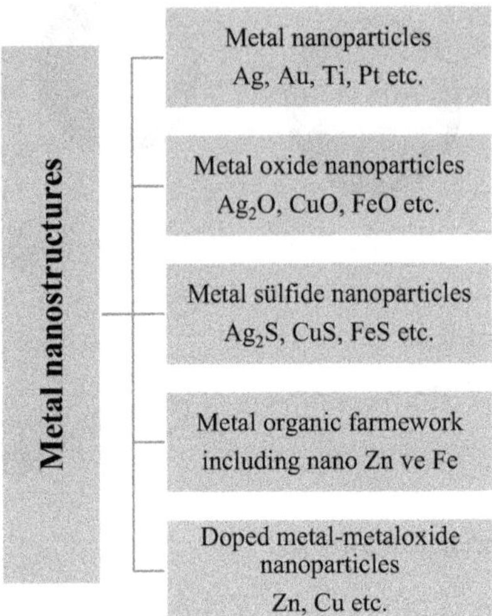

FIGURE 4.7 Metal nanostructures [24].

microorganisms due to mutation [11, 25, 26]. The rapid development of new antibiotics is needed to control and neutralize these microorganisms. This process is quite costly and slow. In addition, the synthesized new antibiotic drug active ingredients may have different side effects. Because of this, as well as the various stages in the process, the commercial introduction and sale of a synthesized novel antibiotic medicine can take a long time. Studies have shown that most metal nanoparticles have antimicrobial effects. Silver and zinc metal nanoparticles, the most widely known and used, exhibit anti-infective properties against pathogenic microorganisms [27, 28]. For this reason, a large number of studies on their use are ongoing. Thus, the use of nanometals in the development of new-generation drugs by developing new strategies is an important research area. The anti-infective properties of metal nanoparticles have been investigated by different research groups in pathogenic microorganisms as well as in infectious viruses and protozoa. The data obtained from this research will be given in more detail in the following sections [29–34].

4.3.1 Antibacterial M-NPs

As mentioned in the previous section, one of the most encountered problems in the field of health today is antibiotic resistance. The active ingredients of many drugs are used to lessen the effects on pathogens over time, or they do not have enough effect on microorganisms as a result of misusage, and they can contribute to the formation of resistant strains by mutating microorganisms. For these reasons, alternative strategies are required to combat the new pathogenic microorganism-borne diseases that will be encountered. As a result of the research, it has been shown in many studies by different groups that metal and metal-based (metal oxide, metal sulfide, metal-doped, and metal-organic framework) structures have antimicrobial effects. Among these, in the research made with metal oxides, ZnO, Ag_2O, FeO, MnO_2, etc, it is remarked that some metal nanoparticles, such as those, can be used in some medical applications due to their antimicrobial activity [35]. With the materials developed through the use of metal nanoparticles with antimicrobial properties, treatments that shorten the healing process are being devised. In addition, these properties of silver nanoparticles can be used in burn treatments (Tablo 1). Numerous studies have been performed in which silver nanoparticles have anti-inflammatory, antiangiogenesis, antiplatelet, antifungal, anticancer, and antibacterial activities. The use of TiO_2 nanoparticles is also advised to prevent the transmission of infectious diseases [36–40].

Zinc oxide (bio-safe) nanoparticles are an attractive antimicrobial source of choice, mainly because the zinc on which they are founded has an important place in our immune system. The antimicrobial effect of ZnO nanoparticles, which was investigated, was seen to have a strong effect in the study of Sirelkhatim and his group [41]. In another study, aluminum oxide and copper oxide nanometals were used, and it was understood that both had antimicrobial effects [42]. It is understood that metal nanoparticles can be used alone, or synergy can be reached by employing more than one metal nanoparticle jointly.

In addition, there are different studies on the anticancer and proangiogenic properties of metal nanometals. In another study, CdS was analyzed as an antimicrobial agent [43]. It has been supported by the results of studies that it is effective on microorganisms such as *Aspergillus flavus* and *Pseudomonas aeruginosa*. In different studies, the effect of cadmium nanoparticles on *Escherichia coli* was investigated. In addition to these studies, the anticancer effects of Cd nanoparticles are mentioned [24].

4.3.2 Antiviral M-NPs

Being able to develop drugs against viral infections, the effects of which we face and struggle with has been one of the most researched areas of health systems. Studies show that antiviral drugs become ineffective very quickly due to the mutation rate of viruses. In addition, the side effects of the chemicals in these drugs are not exactly understood. It can have side effects that can be much more dangerous, especially for the elderly and children. Accordingly, it is necessary to develop different antiviral treatments. For this purpose, many studies are taken out. While vaccines are effective in the period until the virus infects the metabolism, different methods should be used in the treatment of the affected organism. For this purpose, nanoparticles have been applied as new methods. Nanometals have antiviral effects by preventing the replication of viruses. Recently, the use of silver nanoparticles has been studied in many areas due to their easy synthesis. However, it has been determined that silver nanoparticles can be used as anti-infectives due to their antiviral effects.

A study by Chen and Liang indicated that silver nanoparticles show antiviral activities against the influenza A virus, hepatitis B virus, human parainfluenza virus, herpes simplex virus, and human immunodeficiency virus in Table 4.2 [44, 45]. In addition, it has been marked that metal nanoparticles employed with silver nanoparticles increase the synergy in the therapy process [45]. In another study, gold nanoparticles were operated together with antiviral medicines, and it was shown that they prevented viral infection by inhibiting the binding of the virus to the cell.

TABLE 4.2
Nanometals Effective on Viruses [44]

Viruses	Nanometals
Herpes virus	Silver, gold, tin, zinc, gold
Herpes simplex virus	Copper
Hepatitis	Copper, silver, gold, iron
Corona	Silver sulfide
Plant virus	Nickel
HIV	Silver, gallium, gold
Influenza	Gold, silver

In addition, it was expected that gold nanoparticles could be used in the transport of different drugs and to increase their effectiveness and it prevents the spread of infected virus as well. In another study, gold nanoparticles were used for the cure procedure jointly with hyaluronic acid against the hepatitis C virus [46, 47]. In Table 4.2, viruses and nanometals with antiviral effects are shown.

4.3.3 Anti-protozoal M-NPs

Infections related to protozoa have historically been associated with tropical or subtropical regions. The reason for this is that the vectors living in the mentioned environments are directly connected to the climate. However, nowadays, it has been noticed that it is effective in carrying disease vectors and factors related to parasites more easily and quickly due to the factors in which global warming, migration, and easy travel are interested. The emergence of a parasite-borne infection that has never been seen before in a place far from the source is quite common for the reasons mentioned earlier.

More expected in the tropics, malaria is an illness caused by protozoan parasites of the Plasmodium genus transferred by female Anopheles mosquitoes. It has been noticed that this disease has caused significant epidemics and deaths in the areas where it is seen. For the reasons listed, the fight against malaria is an important research topic. In this reported study, it has been exhibited that nanometals can be used not only in the fight against infections but also in diagnosis. *Entamoeba histolytica* and *Cryptosporidium parvum* vectors, which are the most important parasitic infections that cause diarrhea among individuals with weakened immune systems, are most effective among children as they are effective on the gastrointestinal system. Studies against these two parasites have shown that silver and copper metal nanoparticles show anti-protozoan activity. As a result of this research, it was comprehended that there was a significant decrease in cyst viability for CuO nanoparticles against *Entamoeba histolytica* cysts and for Ag nanoparticles against *Cryptosporidium parvum* oocysts.

TABLE 4.3
Effects of Nanometals on Protozoa

Parasites	Metal nanoparticles
Malaria	Iron (III) oxide, magnesium (II) oxide, zinc oxide, aluminum (III) oxide, cerium (IV) oxide, gold, and silver
Leishmaniasis	Gold and silver
Helminth infections	Gold, silver, zinc, and iron
Toxoplasma Gondii	Gold, silver, and platinum
Trypanosoma brucei	Gold and silver
Entamoeba histolytica cysts	Copper (II) oxide
Cryptosporidium parvum oocysts	Silver

In addition, metal nanoparticles can be employed to design delivery systems responsible for the intracellular entry of antiparasitic drugs [48, 49]. In addition, due to the lack of a vaccine used in the treatment of parasitic diseases, the development of drugs used against protozoa has gained vital importance. The activity of antiparasitic drugs is rather limited. In addition, the high side effects appear to be a huge disadvantage.

In another study, the antiprotozoal activities of gold, silver, and platinum metals against *Toxoplasma gondii* and *Trypanosoma brucei* were examined. As a result of the research, it was noted that gold, silver, and platinum metal nanoparticles were effective against *T. gondii*, and gold and silver nanoparticles were effective against *T. brucei* Table 4.3 [50, 51].

4.4 MECHANISMS OF ACTION ANTI-INFECTIVE M-NPs

Enzymes are biological catalysts used in diagnosis, treatment, and follow-up processes in the field of health. They are usually formed from proteins and coenzymes. Although structurally quite complex, their presence is essential for the processes occurring in vivo and in vitro. Generally, their reactions occur at maximum rates under mild conditions. Many mechanisms by which metals in the structure of enzymes, which are biological catalysts, participate in the reaction one-to-one, have been exhibited in the research done so far. It is also comprehended that some of the metals have catalytic properties. It is clear that these data have led scientists to investigate the enzyme properties of nanoparticles. With the rapid advancement of nanotechnology, the concept of nanozymes has emerged. It has been marked that many nanoparticles gaining nanosize exhibit properties similar to enzymes. These enzyme-mimicking nanostructures can also be used in non-moderate reaction conditions. This gives them the advantage of use [52, 53].

Another study was shown to explain the anti-infective effect of nanoparticles. It is understood that some enzymes have antibiotic-like properties, and such enzymes are named enzymobiotics. In this way, it will be possible to produce nanozymobiotics from nanozymes. Compared with enzymobiotics, nanozymobiotics have a broad spectrum and high durability, a feature that will help combat the problem of antibiotic resistance in bacteria. Now, we will explain in the following section which properties of metal nanoparticles mimic enzymes and provide an antimicrobial effect.

Nanometals with peroxidase-like properties; such nanozymes catalyze the decomposition of H_2O_2 and produce free hydroxyl radicals by strong oxidation. These radicals are known as one of the most destructive reactive oxygen species. It not only breaks down cell components, nucleotides, and other bacterial biofilms but also exerts an antimicrobial effect by disrupting the structural integrity of bacteria and causing their death. Normally, hydrogen peroxide is also used for sterilization. In some studies investigating these properties, many nanozymes with peroxidase-like activity have been reported as promising antibacterial nanozymbiotics. For example, research using iron-oxide nanozymes produced nanozymobiotics and repressed the survival of *Salmonella enteritidis* [54].

Another aspect of nanozymes that mimics enzymes is that they can show similar properties to oxidase enzymes. In oxidase-catalyzed reactions, oxygen (O_2) along with water is oxidized to H_2O_2, and in some cases, to superoxide radicals (O^{2-}). The radicals formed during this process show antibacterial activity. It has been comprehended that radicals can oxidize the outer surface of the microorganism, damage the cell membrane, and disrupt the cell structure. In another study, when the properties of Au nanozymobiotics were concerned, it was seen that MSN (mesoporous silica)-AuNPs exhibited antibacterial properties against both Gram-negative (*E. coli*) and Gram-positive (*S. aureus*) bacteria, disrupted the biofilm, and prevented its re-formation [33, 37, 55].

It has been observed that metal nanoparticles also catalyze reactions similar to those of another enzyme, deoxyribonucleases. Due to their small size, nanoparticles can easily enter the cell and even contact the DNA of the microorganism. Thanks to these metal-nanoparticles, which have a disintegrating effect on the DNA of the microorganism, both microorganisms will be destroyed and the formation of resistant organisms will be prevented [56, 57].

Methods such as ultraviolet, chlorine, ozone normally used for sterilization cannot fully decompose the DNA of the microorganism. For this reason, when nanometals with DNAase features are employed, it will be possible to completely kill the microorganism and to remove the antibiotic-resistant genes released during extinction. Recently, it has been reported that metals such as Cu(II), Cr(III), Zn(II), and Ce(IV) can show DNAase activity by breaking the phosphodiester bond in the structure of DNA.

In addition to these, different strategies are being designed for the control and prevention of microorganisms. It is created especially in studies where synergy is created by using the nanonzymobiotic products of more than one nanoparticle or much stronger antimicrobial effects are likely by using both antibiotics and nanometals (Figure 4.8) [54].

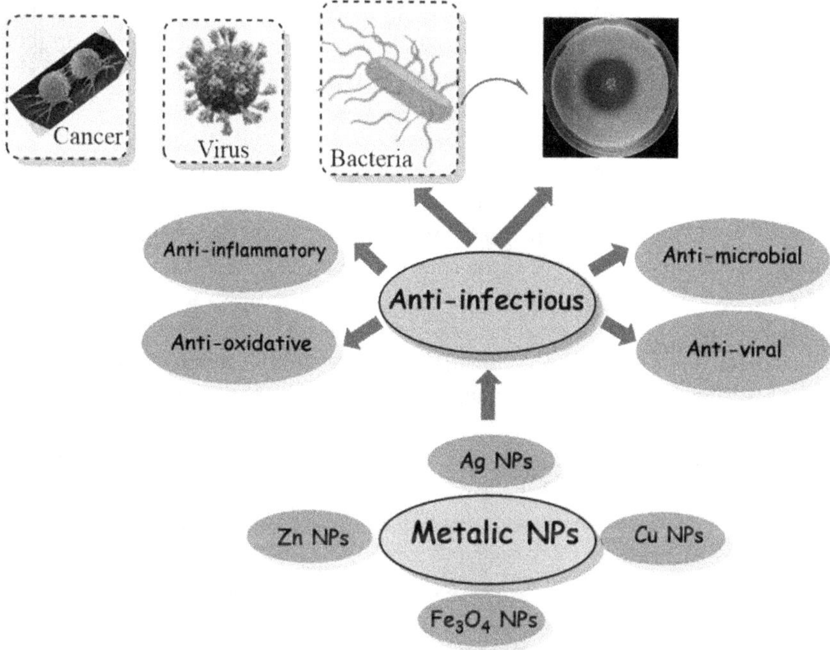

FIGURE 4.8 Antimicrobial, anti-virus activity, anti-oxidative, and anti-inflammatory activities of metal NPs.

4.5 ANTIMICROBIAL SUSCEPTIBILITY TESTING OF ANTI-INFECTIVE M-NPs

There are articles examining the effects of metal nanoparticles on infection diagnosis, antibiotic delivery, and the immune systems of living things. However, articles in the field of medicine are restricted in number among all articles, and they are mostly on the synthesis and characterization of new nanoparticles. There are a limited number of studies on the effect of M-NPs in the fields of infection and immunity. This literature is mainly aimed at diagnosing infectious diseases. In in vitro studies, it is desired to detect the DNA sequence or protein that enables the diagnosis of the disease agent with M-NPs. The development of diagnostic methods using M-NPs in immunoassays is due to their suitability in terms of their structural properties. For this reason, immune gold labeling of antibodies has been used for a long time, especially in diagnostics. Au-NPs used in diagnosis are obtained by reducing the starting material to tetrachloroauric acid ($HAuCl_4$), and it is founded on the principle that the AuNPs formed in the colloidal form are attached to the antibody proteins and the imaging takes place [58, 59]. The immunoassay method is a bioanalytical method based on the selective interaction of the antigen with the antibody. It is observed by measuring the activity of a substance in the medium during the reaction of the immunoanalytical reagents present during the binding of the antigen to the antibody. The most important reason for the application of the immunoassay method is its high specificity and sensitivity [60]. The fact that NPs have different characteristics unique to them has enabled them to be used in the immunoanalysis method. The non-specific binding of NPs to proteins has limited application. In order to prevent this, thanks to the functional groups attached to the support material, the properties of NPs are enhanced and their more functional use is delivered [61]. Thus, it was chosen to investigate a new class of catalysts known to have high catalytic activity. Enzymes are not the only catalysts that convert the substrate to the product. In addition, the enzyme shows mimetic properties in metal and metal oxide NPs and catalyzes the related reaction. With these features, biosensors with low cost, new designable features, high catalytic activity, stability in a wider pH range, improved resistance to high substrate concentrations, and low concentration sensing capability can be designed in laboratory conditions by using metal and metal oxide NPs [62, 63]. Metals and metal NPs have great potential to replace conventional methods for detection against a variety of molecules, especially when modified with specific biorecognition molecules.

One of the metal oxides, cerium oxide NPs (nanoceria), has been utilized for decades in fuel additives, catalysts, glass polishing, solid oxide fuel cells, solar cells, and as a potential pharmacological material [64]. Cerium oxide NPs are metal oxide NPs that have the ability to perform oxidation-reduction reactions due to their Ce^{3+}/Ce^{4+} redox features [65, 66]. Due to the tendency of nanoceria to coagulate after synthesis, this important problem can be eliminated by modifying their surfaces. For this purpose, the surface of nanoceria was defunctionalized with polyacrylic acid (PAA), polyethylene glycol (PEG), dextran, polyethyleneimine, cyclodextrin, glucose, and folic acid. In this way, the surface of the nanoceria is desensitized to non-specific interactions, and its toxicity is decreased [67]. The redox activity in the nanoceria structure is due to the presence of the Ce^{3+}/Ce^{4+} ion pairs. The activity of nanoceria enables it to display similar mimetic activity to oxidoreductase enzymes (catalase, superoxide dismutase (SOD), and oxidases) [68]. In this way, the detection of H_2O_2, which is composed of a harmful oxidizing agent in the body, is brought out even at low concentrations by mimetic enabling the activity of nanoceria on SOD [69]. In addition to the properties described, it was decided that nanoceria had superior antioxidant activity as well as in vivo anti-inflammatory activity [69].

It has been determined that magnetic iron oxide nanoparticles play a role in electrochemical sensor systems, in the curing process of different cancer types, and in the detection and removal of different compounds [70, 71]. Iron oxide nanoparticles, which are believed to be chemically and biologically inert, can be covered with antibodies and enzymes to increase their functional properties [72–74]. It was determined that Fe_3O_4 magnetic nanoparticles also showed peroxidase enzyme activity as an enzyme mimetic and it was apprehended that they interacted strongly with some compounds. In this way, the disadvantage of the enzyme being affected by adverse environmental conditions is prevented, and the efficiency of Fe_3O_4 magnetic nanoparticles is improved [75]. It was determined that iron oxide nanoparticles mimic the enzyme activities and show the properties of peroxidase and catalase enzymes [76].

Other enzyme-mimic metal NPs include gold, silver, and platinum. AuNPs have been used as immunolabels and immunosensors. Due to their optical, chemical, electrical, and catalytic properties, AuNPs have a user area where colorimetric, plasmonic, and electrochemical measurements are created [77].

The antibody immobilized to the surface can be determined by colloidal AuNPs, which recognize its structure and offer high specific binding capacity. The binding of enzymes and related compounds to surfaces modified with mercapto and amino groups is promoted. Therefore, bioconjugated AuNPs can be easily connected to nanoprobes. AuNPs show sensitive binding at picomolar concentrations. AuNPs mimetic show glucose oxidase enzyme activity [78].

AgNPs, one of the other metal oxide NPs with commercial importance, are used in many fields, especially medicine, cosmetics, and the textile industry, due to their strong antibacterial properties [79, 80]. It was determined that AgNPs displayed strong immunoeffective activity as well as antibacterial activity. In particular, it has been written that they have mimetic peroxidase enzyme activity and show specific effects with human C-reactive protein (CRP) [81].

4.6 CLINICAL APPLICATIONS OF M-NPS

4.6.1 Drug Delivery

When drugs need to reach specific organs or tissues directly, certain challenges arise when it comes to ensuring their targeting. For this reason, it has been attempted to overcome these difficulties by doing a lot of work on drug distribution systems. Unlike existing systems, newly designed drug delivery systems can overcome challenges such as drug solubility and pH [82]. Thanks to newly designed drug delivery systems, it has become possible to overcome the blood-brain barrier. The use of metal NPs in drug delivery systems has ensured that NPs have superior properties compared to other drug delivery systems due to all their physical, chemical, optical, etc. properties. A large capacity has been recognized for AuNPs and metal NPs to play a role in drug delivery systems.

It is a great advantage that AuNPs can be easily adjusted and functionalized with different types of molecules. As a result of coating AuNPs with polyethylene glycol (PEG), their toxicity can be changed and directed to the desired target [83]. In the studies, it has been determined that AuNPs coated with PEG can bind to tumor structures specifically through the -NH_2 group [84]. When AuNPs were directed to tumor tissues jointly with Doxorubicin (DOX) after being coated with PEG, it was observed that the drug concentration in tumor tissues was much higher compared to other free doxorubicin applications. This has made it possible to develop a more effective cancer treatment method for liver cancer, which is known to be a very aggressive type of cancer [85].

The Au-DOX used was loaded in a single stage. In nanohybrid structures obtained from this species, drug toxicity has been reduced, and it has also contributed to reducing the possibility of resistance development as a cancer drug [86].

It was determined that AgNPs obtained by the green synthesis method showed cytotoxic effects. Again, it has been determined that anticancer drugs such as cisplatin modified with palladium nanoparticles (PdNPs), which are widely employed in chemotherapy, are highly influential against A549, SKOV-3, and HepG2 cell lines [87].

4.6.2 Nanoparticles for Gene Delivery

Nucleic acids are transported by biological and synthetic vectors. Biological vectors provide efficient transport. However, immunogenicity, carcinogenicity, and inflammation may pose problems for clinical applications. By using nanoparticle-based vectors, the distribution of nucleic acids can be supplied, and the solution to biological-based vector problems can be provided [88]. The gene therapy method is founded on the exogenous transfer of DNA or RNA for the treatment or prevention of diseases. When the immune system is active during viral infections, the effectiveness of gene therapy decreases. Problems are solved by using non-viral systems such as metallic NPs. In some studies, it has been determined that AuNPs protect the nucleic acid structure by preventing the degradation of DNA and RNA by the nuclease enzyme [89]. Studies of this kind have been very promising for the development of gene distribution systems [90].

4.6.3 Protein Delivery

Recent studies on the use of nanoparticles in the transport of proteins have shown very promising results. Modified AuNPs can be used for protein binding and morphological structure analysis. The efficacy of insulin-modified AuNPs in the treatment of diabetes has been tried, and positive results have been reported in the rat model [91]. In addition, it was decided that AuNPs coated with non-toxic polysaccharides such as chitosan had a positive effect on insulin release. After binding AuNPs with Herceptin, they showed strong binding to receptors (SK-BR3) on the breast cancer surface [92]. AgNPs have been used in protein carrier systems. It was determined that AgNPs increased the penetration into cell lines and the efficiency of the transport systems in the cancer cell lines [93].

With the obtained findings, it has been demonstrated that metal NPs have very promising results in cancer treatment and are useful in solving problems [94].

4.7 METABOLIC RESPONSES OF M-NPs

The harmful effects of nanosized materials have not been fully determined today. Adequate research, duration, knowledge, the path of the substance in nature, accumulation, etc., which are the necessary criteria for the examination of this subject, are not understood exactly. The effects on the environment and humans of processes such as the production, transportation, use, and destruction of nanomaterials are not fully known. The potential risk assessment of nanomaterials takes place in four stages (Figures 4.9 and 4.10).

The potential risk assessment of nanomaterials takes place in four stages;

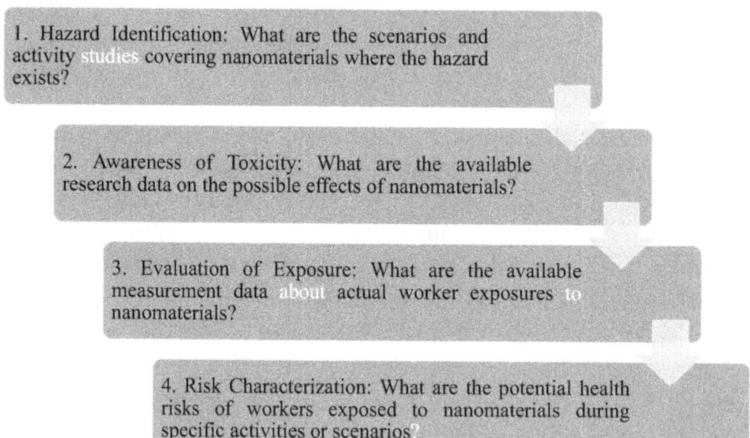

FIGURE 4.9 Risk assessment of nanomaterials takes place in four phases.

The 5D rule applies when examining the toxicology of nanoparticles.

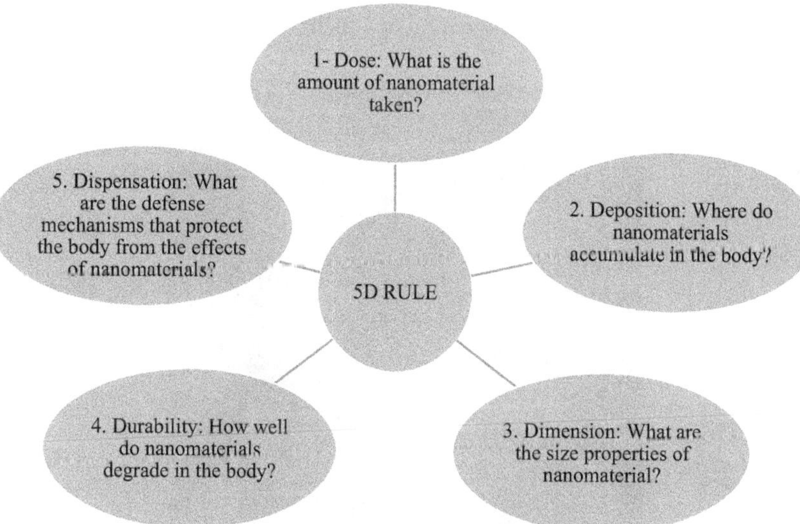

FIGURE 4.10 5D rule for the toxicology of nanoparticles.

4.7.1 Biologic Barriers

The entry of nanoparticles into the living body occurs in four ways. These are inhalation, digestion, injection, and dermal penetration (Figure 4.11).

4.7.1.1 Inhalation

For inhaled nanomaterials, the affected target site is usually the lungs. Nanomaterials have been demonstrated to induce inflammatory responses in the lungs. However, the factors that determine the severity of the reactions are not fully understood. Also, the long-term health consequences of many cases of repeated exposure are unknown. Epidemiological studies examining the health effects of ambient air pollution have revealed that, in addition to the effects on the lungs, people who breathe air containing a high percentage of nanoparticles are more likely to develop cardiovascular disease. However, the relevance of these findings to workers exposed to nanomaterials in the workplace has not been fully proven.

4.7.1.2 Dermal

In addition to respiration, nanomaterials are also connected to the skin and digestive tract as a result of workplace exposure. Except for nanomaterials used in cosmetic products, research on nanomaterials affecting the skin is very restricted. The effects that occur after skin contact are predicted to be in the contact area. Research into the skin absorption potential of nanomaterials has revealed that any skin absorption that will occur will be in very low quantities.

4.7.1.3 Injection

There is no information that can reveal general conclusions about the fate of nanomaterials entering the digestive tract. Unlike larger particles, nanoparticles easily pass through the skin and similar biological membranes, thus penetrating various cells,

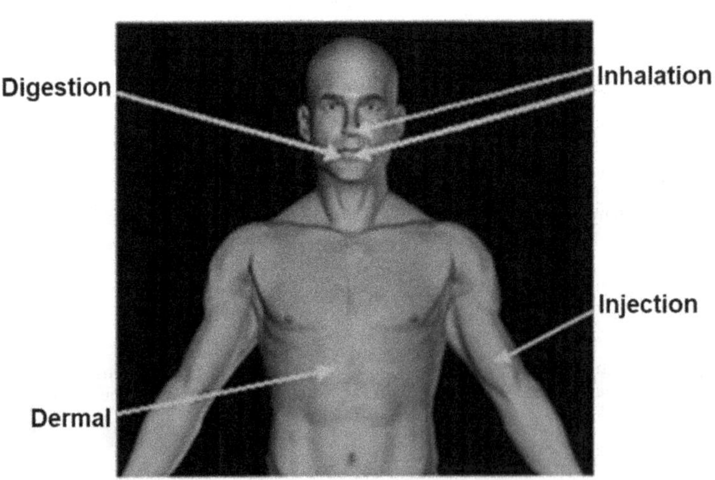

FIGURE 4.11 Exposure pathways to nanomaterials.

tissues, and organs. When mixed with blood, they are moved to other vital internal organs and tissues.

4.7.2 Immune Response

To investigate the effects of exposure to nanoparticles on the immune system, it is necessary to study organs and cells at the immune level. In the immune system, nanoparticles have effects on primary and secondary lymphoid cells. The primary lymphoid organs are the bone marrow and the thymus. These organs are concerned with the production of mature T- and B-lymphocytes and myeloid cells such as macrophages [95]. Secondary lymphoid organs include the spleen, lymph nodes and tonsils, and mucosa-associated lymphoid tissues including Peyer's patches [96]. These organs work together in an organized manner.

The defense behavior against a foreign particle taken into a living organism is carried out by immune cells in metabolism. The center where the primary immunity in the body is brought out is believed to be leukocytes. In addition, there are cells that provide signal transduction to lymphocytes and appear against cytokines released by T lymphocytes and macrophages. Leukocytes are ordered into three different groups: lymphocytes, phagocytes, and other helper cells. Lymphocytes are mainly formed of T cells, B cells, and large granular lymphocytes. Phagocytes consist of mononuclear phagocytes, neutrophils, and eosinophil systems. Helper cells are basophils, mast cells, and platelets. Lymphocytes play an important role in the specific immune recognition of pathogens in living things [97].

Antibodies are created by B cells against particles taken into the organism from the outside and bind to a specific surface receptor. Lymphocytes consist of two different types of T cells: helper T cells and cytotoxic T cells. While T cells are involved in cell division, differentiation, and antibody production by interacting with B cells, cytotoxic T cells interact with mononuclear phagocytes to assist in the elimination of antigens. Phagocytes eliminate particles inside the cell by phagocytosis [98].

Cells mainly involved in detoxification activity are cytotoxic T cells, LGLs, or eosinophils. Lymphocytes recognize tumors and infected surface antigens. The spleen consists of both B and T lymphocytes (65% and 25%), which make up the cell immune system. Macrophages comprise around 4–5% of the spleen immune cell population. The rest of the cells are mostly neutrophils and eosinophils [96].

It is worrying that nanoparticles act in the direction of stimulating or suppressing the immune system when exposed to living things. It is important to understand how nanoparticle-based new-generation drugs developed, especially in drug delivery systems, affect the immune system in living things. Studies have been documented on which drug systems designed specifically to target the immune system and their use in the treatment of inflammatory and autoimmune disorders arc beneficial [99].

Due to the small size of nanoparticles, it is possible to penetrate all organs in the body. When the effects of SiO_2, TiO_2, ZrO_2, and CoNPs on human macrophages were studied, it was determined that inflammatory cytokine production took place [100].

4.7.3 Targeting of M-NPs into Infected Organs

The main cause of infectious diseases is believed to be the growth of pathogenic agents such as bacteria, fungi, and viruses in host organisms. In particular, it is very difficult to prevent the negative effects of infectious diseases on public health. Especially in geographies where dirty drinking water and unsanitary conditions prevail, it is not possible to prevent the spread of endemic infectious diseases. Studies have shown that these diseases affect 500 million people annually and cause 1–3 million deaths [101].

It is possible to use nanoparticle-based drug systems in the therapy process of diseases with drug loading, controlled drug release, and the use of specific receptor-targeted drugs [102]. It is potential that drugs developed with nanotechnology can be an alternative to antibiotics. It is possible for nanosized drug systems to pass through cell membranes and reach tissues and organs. After inhalation, ingestion, or skin contact with nano-drug systems, they can efficiently reach the bloodstream. In this way, it is achievable to effectively provide therapeutic agents to targeted organs and to supply time-dependent drug departure by using nano-drug designs [102].

The surfaces of synthesized NPs are altered with substances such as DNA and protein and used by converting them to a biocompatible and less toxic form. NPs, whose surfaces are adjusted with biomaterials such as proteins, lipids, nucleic acids, and carbohydrates, are antimicrobially useful. In addition, the selection of therapeutically selected bioactive molecules increases their effectiveness against bacteria, fungi, and viruses that cause infectious diseases.

Metal-NPs or drug-bound metal-NPs can help the body's key components (antibodies and cytokines) to fight disease and strengthen the immune system. In the research, it was studied that metal NPs such as silver, zinc, gold, and cerium inactivate microorganisms [79, 103]. Properties can be used in various drug or biomolecule nanotherapy applications by transporting metal-NPs with liposomes or C-based nanomaterials.

Due to the antibacterial activities of silver NPs, they have been utilized in the treatment of burns, wounds, and teeth. In addition, the use of silver NPs in mouthwashes has been declared recently. It has been determined that silver NPs bind to S-containing proteins present in biological molecules and have the ability to bind to the phosphorus in DNA. Drug strategies changed with AgNPs can be efficiently targeted to target cells/organs. AgNPs have been written to interact with and destroy the HIV virus. It has been informed that the means of activity are via the electron transfer mechanism [104].

By coating the metal NPs with silica NPs, it is possible to create surfaces that can be changed more easily, and it has the potential to reduce the toxicity of metal NPs. In this way, metal NPs coated with silica NPs are required for their target [105].

Titanium dioxide NPs bind to Ca^{2+} ions, which are affected in bridging lipopolysaccharides in metabolism. Thus, the generation of proinflammatory signals by phagocytes occurs [106]. The most important features of TiO_2 NPs are that they have photocatalytic activity, and it has been said that TiO_2 NPs kill bacteria and tumor cells in culture in vitro with UV irradiation. Taking advantage of these properties, it has been determined that single and drug-formulated forms of TiO_2 NPs are targeted to tumor tissues and give successful results in therapy. In in vivo studies on rats, it has been determined that it has successful effects in the treatment of the disease by transiting the blood-brain barrier and opposing brain tumors [107, 108].

ZnO NPs and CuO NPs have high antibacterial activity. ZnO NPs act as bacteriostatic agents in a light environment. In this sense, it has been determined that the protective response against carcinogenesis can be activated by the stimulation of DNA damage-inducing agents in ZnO NPs [109, 110]. It has been determined that iron oxide NPs take part in imaging systems. In addition, superparamagnetic iron oxide NPs have played an important role as theranostic agents in ultrasound-enhanced brain drug delivery. In addition, magnetic iron NPs are also employed to target tumor cells [111, 112].

4.7.4 Elimination

As a result of developing nanotechnological applications, the most important concern is the elimination of nanoparticles. The United States Food and Drug Administration (FDA) stipulates that nanomaterials can be metabolized in the body after their use as medical and pharmaceutical agents [113]. The proposed strategy for the elimination of nanomaterials relates to their biodegradation and elimination by the kidneys. Elimination of intravenous nanoparticles occurs by renal and hepatobiliary elimination [114]. The mechanism of elimination by the renal route is well understood. In general, it has been said that nanoparticles less than 5.5 nm experience efficient urinary excretion due to the pore size limit of glomerular filtration in the kidneys. For this reason, aspects such as nanoparticle size, core density, surface charge, and surface chemistry are regarded in applications. However, the details of hepatobiliary elimination are not well known. In some studies, it has been conveyed that nanoparticles undergo hepatobiliary elimination via hepatocytes, via transcytosis, then pass into the bile vent, pass into the digestive system, and are destroyed through feces. It has been reported that this elimination takes a long time, occasionally taking months [115, 116].

4.7.5 Toxicology

The development of nanotechnology and the synthesis of new nanomaterials with each passing day make M-NPs think about their negative effects on humans, animals, and the environment. The limited information about the effects of nanomaterials on living things causes even more concern. In the studies, it was determined that when a large number of nanoparticles enter the body, they trigger stress reactions by creating an overload on the phagocytes. This causes inflammation and weakens the body's defense mechanisms. Due to their increased chemical reactivity, they increase the formation of free radicals, which can cause damage to vital proteins, membranes, and even DNA. Like other materials that can cause DNA damage, they can cause mutations in cells and, afterward, cancers [117].

In an in vitro study on human epithelial cell cultures with fluorescently labeled SiO_2 nanoparticles, they showed that particles smaller than 70 nm could penetrate the cell nucleus. In the same study, the accumulation of proteins known to be indicators of damage in DNA replication and transcription was also marked in nuclei penetrated by nanoparticles [118]. Some nanoparticles penetrating through the skin layer are then carried to various lymph nodes through the lymph channel, especially nanoparticles smaller than 30 nm inhaled can be transported to almost every organ of the body by passing through the blood circulation system, and then to the liver, spleen, bone marrow, and heart. There are also some studies showing that they can even accumulate in the brain. In another study, polystyrene nanoparticles with a diameter of 50 and 100 nm were provided orally to female mice for 10 days. At autopsies performed at the end of the experiment, an average of 34% and 26% of these particles were absorbed in various tissues of the bodies of the mice, whereas particles were larger than 300 nm. It has been documented that they are not seen in blood, heart, and lung tissues when given. In another study conducted with hydrophobic latex nanoparticles of 14 and 415 nm sizes, it was observed that these particles were able to penetrate the mucous gel layer of the small intestine in 2 and 30 minutes, respectively, and it was deduced that as the size of the particles decreased, the penetration rate of the mucosal barrier increased. Principle 15 of the Declaration of the United Nations Conference on Environment and Development (Rio) (1992) states that "In the case of a serious and irreversible hazard, the absence of full scientific certainty should not be an excuse for delaying measures that can prevent environmental degradation". When an action or policy decision is likely to have dubious consequences for the environment and people and there is no consensus among scientists on the issue, the stance should be in favor of inaction. That is, a risk to health should never be accepted [119].

4.8 CONCLUSION

Due to their superior properties and size advantages, M-NPs have high advantages both in the cell and in their interactions on the cell surface. In addition, M-NPs can be prepared with desired properties and attached to specific cells to improve their therapeutic efficacy. For this reason, many studies have focused on the development of therapeutics by targeting diseases. Most of the data on the use of M-NPs developed for the treatment of diseases are from preclinical considerations. Serious research is required on their clinical usability. There is also a need to examine the toxicological properties and pharmacokinetics of compounds based on metal NPs. Common to M-NPs is that they have the potential to overcome drug resistance. There is no doubt about the potential of metal NPs to be promising therapeutics for the treatment of different diseases.

ACKNOWLEDGMENTS

We would like to thank Hayrunnisa Nadaroglu, Ataturk University, Erzurum/Turkey, and Azize Alayli, Sakarya Applied Science University, Sakarya/Turkey, for all their support. The authors are also thankful to Dr. Jamilur R. Ansari, Dronacharya College of Engineering, Gurgaon, India, for the needful editing in the chapter. The study was not funded by any institution.

REFERENCES

[1] O.C. Farokhzad, R. Langer, Nanomedicine: developing smarter therapeutic and diagnostic modalities, Advanced Drug Delivery Reviews. 58 (2006) 1456–1459. https://doi.org/10.1016/J.ADDR.2006.09.011.

[2] A. Brandelli, A.C. Ritter, F.F. Veras, Antimicrobial activities of metal nanoparticles, in: Metal Nanoparticles in Pharma, Springer International Publishing, Cham, 2017: pp. 337–363. https://doi.org/10.1007/978-3-319-63790-7_15.

[3] A.M. Ibrahim, L. Chauhan, A. Bhardwaj, A. Sharma, F. Fayaz, B. Kumar, M. Alhashmi, N. AlHajri, M.S. Alam, F.H. Pottoo, Brain-Derived neurotropic factor in neurodegenerative disorders, Biomedicines. 10 (2022) 1143. https://doi.org/10.3390/biomedicines 10051143.

[4] M.N. Javed, F.H. Pottoo, A. Shamim, M.S. Hasnain, M.S. Alam, Design of experiments for the development of nanoparticles, nanomaterials, and nanocomposites, in: S. Beg (Ed.), Design of Experiments for Pharmaceutical Product Development, Springer, Singapore, 2021: pp. 151–169. https://doi.org/10.1007/978-981-33-4351-1_9.

[5] F.H. Pottoo, M.N. Javed, M.A. Barkat, M.S. Alam, J.A. Nowshehri, D.M. Alshayban, M.A. Ansari, Estrogen and serotonin: complexity of interactions and implications for epileptic seizures and epileptogenesis, Current Neuropharmacology. 17 (2019) 214–231. https://doi.org/10.2174/1570159X16666180628164432.

[6] M. Aslam, M.N. Javed, H.H. Deeb, M.K. Nicola, M.A. Mirza, M.S. Alam, M.H. Akhtar, A. Waziri, Lipid nanocarriers for neurotherapeutics: introduction, challenges, blood-brain barrier, and promises of delivery approaches, CNS & Neurological Disorders-Drug Targets (Formerly Current Drug Targets-CNS & Neurological Disorders). 21 (2022) 952–965. https://doi.org/10.2174/1871527320666210706104240.

[7] F.H. Pottoo, S. Sharma, M.N. Javed, M.A. Barkat, Harshita, M.S. Alam, M.J. Naim, O. Alam, M.A. Ansari, G.E. Barreto, G.M. Ashraf, Lipid-based nanoformulations in the treatment of neurological disorders, Drug Metabolism Reviews. 52 (2020) 185–204. https://doi.org/10.1080/03602532.2020.1726942.

[8] A. Waziri, C. Bharti, M. Aslam, P. Jamil, Mohd. A. Mirza, M.N. Javed, U. Pottoo, A. Ahmadi, M.S. Alam, Probiotics for the chemoprotective role against the toxic effect of cancer chemotherapy, Anti-Cancer Agents in Medicinal Chemistry (Formerly Current Medicinal Chemistry-Anti-Cancer Agents). 22 (2022) 654–667. https://doi.org/10.2174/1871520621666210514000615.

[9] M.S. Hasnain, M.N. Javed, M.S. Alam, P. Rishishwar, S. Rishishwar, S. Ali, A.K. Nayak, S. Beg, Purple heart plant leaves extract-mediated silver nanoparticle synthesis: optimization by Box-Behnken design, Materials Science and Engineering: C. 99 (2019) 1105–1114. https://doi.org/10.1016/j.msec.2019.02.061.

[10] S. Bayda, M. Adeel, T. Tuccinardi, M. Cordani, F. Rizzolio, The history of nanoscience and nanotechnology: from chemical–physical applications to nanomedicine, Molecules. 25 (2019) 112. https://doi.org/10.3390/molecules25010112.

[11] M.S. Alam, M.N. Javed, F.H. Pottoo, A. Waziri, F.A. Almalki, M.S. Hasnain, A. Garg, M.K. Saifullah, QbD approached comparison of reaction mechanism in microwave synthesized gold nanoparticles and their superior catalytic role against hazardous nirto-dye, Applied Organometallic Chemistry. (2019). https://doi.org/10.1002/aoc.5071.

[12] M.N. Javed, E.S. Dahiya, A.M. Ibrahim, M.S. Alam, F.A. Khan, F.H. Pottoo, Recent advancement in clinical application of nanotechnological approached targeted delivery of herbal drugs, in: S. Beg, M.A. Barkat, F.J. Ahmad (Eds.), Nanophytomedicine, Springer, Singapore, 2020: pp. 151–172. https://doi.org/10.1007/978-981-15-4909-0_9.

[13] A.M. Schrand, M.F. Rahman, S.M. Hussain, J.J. Schlager, D.A. Smith, A.F. Syed, Metal-based nanoparticles and their toxicity assessment, Wiley Interdisciplinary Reviews. Nanomedicine and Nanobiotechnology. 2 (2010) 544–568. https://doi.org/10.1002/WNAN.103.

[14] N.A. Monteiro-Riviere, S.J. Oldenburg, A.O. Inman, Interactions of aluminum nanoparticles with human epidermal keratinocytes, Journal of Applied Toxicology. 30 (2010) 276–285. https://doi.org/10.1002/jat.1494.

[15] S.M. Dadfar, K. Roemhild, N.I. Drude, S. von Stillfried, R. Knüchel, F. Kiessling, T. Lammers, Iron oxide nanoparticles: diagnostic, therapeutic and theranostic applications, Advanced Drug Delivery Reviews. 138 (2019) 302–325. https://doi.org/10.1016/j.addr.2019.01.005.

[16] L.A. Dykman, N.G. Khlebtsov, Gold nanoparticles in biology and medicine: recent advances and prospects, Acta Naturae. 3 (2011) 34. https://doi.org/10.32607/20758251-2011-3-2-34-56.

[17] X.-F. Zhang, Z.-G. Liu, W. Shen, S. Gurunathan, Silver nanoparticles: synthesis, characterization, properties, applications, and therapeutic approaches, International Journal of Molecular Sciences. 17 (2016) 1534. https://doi.org/10.3390/ijms17091534.

[18] B.A. Camacho-Flores, O. Martínez-Álvarez, M.C. Arenas-Arrocena, R. Garcia-Contreras, L. Argueta-Figueroa, J. de la Fuente-Hernández, L.S. Acosta-Torres, Copper: synthesis techniques in nanoscale and powerful application as an antimicrobial agent, Journal of Nanomaterials. 2015 (2015) 1–10. https://doi.org/10.1155/2015/415238.

[19] H. Li, P. Xia, S. Pan, Z. Qi, C. Fu, Z. Yu, W. Kong, Y. Chang, K. Wang, D. Wu, X. Yang, The advances of ceria nanoparticles for biomedical applications in orthopaedics, International Journal of Nanomedicine. 15 (2020) 7199–7214. https://doi.org/10.2147/IJN.S270229.

[20] X. Liu, C. Chen, Y. Zhao, B. Jia, A review on the synthesis of manganese oxide nanomaterials and their applications on lithium-ion batteries, Journal of Nanomaterials. 2013 (2013) 1–7. https://doi.org/10.1155/2013/736375.

[21] S.S. Sana, R.P. Singh, M. Sharma, A.K. Srivastava, G. Manchanda, A.R. Rai, Z.-J. Zhang, Biogenesis and application of nickel nanoparticles: a review, Current Pharmaceutical Biotechnology. 22 (2021) 808–822. https://doi.org/10.2174/138920102299921 0101235233.

[22] D. Ziental, B. Czarczynska-Goslinska, D.T. Mlynarczyk, A. Glowacka-Sobotta, B. Stanisz, T. Goslinski, L. Sobotta, Titanium dioxide nanoparticles: prospects and applications in medicine, Nanomaterials. 10 (2020) 387. https://doi.org/10.3390/nano10020387.

[23] S. Sabir, M. Arshad, S.K. Chaudhari, Zinc oxide nanoparticles for revolutionizing agriculture: synthesis and applications, The Scientific World Journal. 2014 (2014) 1–8. https://doi.org/10.1155/2014/925494.

[24] A.A. Yaqoob, H. Ahmad, T. Parveen, A. Ahmad, M. Oves, I.M.I. Ismail, H.A. Qari, K. Umar, M.N. Mohamad Ibrahim, Recent advances in metal decorated nanomaterials and their various biological applications: a review, Frontiers in Chemistry. 8 (2020). https://doi.org/10.3389/fchem.2020.00341.

[25] F.H. Pottoo, N. Tabassum, M.N. Javed, S. Nigar, S. Sharma, M.A. Barkat, Harshita, M.S. Alam, M.A. Ansari, G.E. Barreto, G.M. Ashraf, Raloxifene potentiates the effect of fluoxetine against maximal electroshock induced seizures in mice, European Journal of Pharmaceutical Sciences. 146 (2020) 105261. https://doi.org/10.1016/j.ejps.2020.105261.

[26] N. Kumari, N. Daram, M.S. Alam, A.K. Verma, Rationalizing the use of polyphenol nano-formulations in the therapy of neurodegenerative Diseases, CNS & Neurological Disorders-Drug Targets (Formerly Current Drug Targets-CNS & Neurological Disorders). 21 (2022) 966–976. https://doi.org/10.2174/1871527321666220512153854.

[27] S. Raj, R. Manchanda, M. Bhandari, M.S. Alam, Review on natural bioactive products as radioprotective therapeutics: present and past perspective, Current Pharmaceutical Biotechnology. 23 (2022) 1721–1738. https://doi.org/10.2174/13892010236662201101 04645.

[28] F.H. Pottoo, N. Tabassum, M.N. Javed, S. Nigar, R. Rasheed, A. Khan, M.A. Barkat, M.S. Alam, A. Maqbool, M.A. Ansari, G.E. Barreto, G.M. Ashraf, The synergistic effect

of raloxifene, fluoxetine, and bromocriptine protects against pilocarpine-induced status epilepticus and temporal lobe epilepsy, Molecular Neurobiology. 56 (2019) 1233–1247. https://doi.org/10.1007/s12035-018-1121-x.

[29] M.S. Alam, M.F. Naseh, J.R. Ansari, A. Waziri, M.N. Javed, A. Ahmadi, M.K. Saifullah, A. Garg, Synthesis approaches for higher yields of nanoparticles, in: Nanomaterials in the Battle Against Pathogens and Disease Vectors, 1st ed., CRC Press, Boca Raton, 2022: pp. 51–82. https://doi.org/10.1201/9781003126256-3.

[30] M.F. Naseh, J.R. Ansari, M.S. Alam, M.N. Javed, Sustainable nanotorus for biosensing and therapeutical applications, in: U. Shanker, C.M. Hussain, M. Rani (Eds.), Handbook of Green and Sustainable Nanotechnology, Springer International Publishing, Cham, 2022: pp. 1–21. https://doi.org/10.1007/978-3-030-69023-6_47-1.

[31] M. Kumar, Madhavi, J.R. Ansari, Studies on the heating ability by varying the size of Fe_3O_4 magnetic nanoparticles for hyperthermia, Nanoscience and Technology: An International Journal. 13 (2022) 33–45. https://doi.org/10.1615/NanoSciTechnolIntJ.2022040075.

[32] C.A. Sunilbhai, M.S. Alam, K.K. Sadasivuni, J.R. Ansari, SPR assisted diabetes detection, in: K.K. Sadasivuni, J.-J. Cabibihan, A.K. A M Al-Ali, R.A. Malik (Eds.), Advanced Bioscience and Biosystems for Detection and Management of Diabetes, Springer International Publishing, Cham, 2022: pp. 91–131. https://doi.org/10.1007/978-3-030-99728-1_6.

[33] J.R. Ansari, S.M. Hegazy, M.T. Houkan, K. Kannan, A. Aly, K.K. Sadasivuni, Nanocellulose-based materials/composites for sensors, in: Nanocellulose Based Composites for Electronics, Elsevier, Amsterdam, 2021: pp. 185–214. https://doi.org/10.1016/B978-0-12-822350-5.00008-4.

[34] J. Pandit, M.S. Alam, J.R. Ansari, M. Singhal, N. Gupta, A. Waziri, K. Sharma, F. Hyder Potto, Multifaced applications of nanoparticles in biological science, in: Nanomaterials in the Battle Against Pathogens and Disease Vectors, 1st ed., CRC Press, Boca Raton, 2022: pp. 17–50. https://doi.org/10.1201/9781003126256-2.

[35] O. Janson, S. Gururaj, S. Pujari-Palmer, M. Karlsson Ott, M. Strømme, H. Engqvist, K. Welch, Titanium surface modification to enhance antibacterial and bioactive properties while retaining biocompatibility, Materials Science and Engineering: C. 96 (2019) 272–279. https://doi.org/10.1016/j.msec.2018.11.021.

[36] M. Kumar, M. Na, J. R Ansari, Analysis of the heating ability by varying the size of Fe_3O_4 magnetic nanoparticles for hyperthermia, Nanoscience and Technology: An International Journal. (2022). https://doi.org/10.1615/NanoSciTechnolIntJ.2022040075.

[37] S. Mishra, S. Sharma, M.N. Javed, F.H. Pottoo, M.A. Barkat, Harshita, M.S. Alam, M. Amir, M. Sarafroz, Bioinspired nanocomposites: applications in disease diagnosis and treatment, Pharmaceutical nanotechnology. 7 (2019) 206–219. https://doi.org/10.2174/2211738507666190425121509.

[38] R. Kumar, G. Dhamija, J.R. Ansari, Md.N. Javed, Md.S. Alam, C-Dot nanoparticulated devices for biomedical applications, in: Nanotechnology, 1st ed., CRC Press, Boca Raton, 2022: pp. 271–299. https://doi.org/10.1201/9781003220350-15.

[39] S. Singhal, M. Gupta, M.S. Alam, M.N. Javed, J.R. Ansari, Carbon allotropes-based nanodevices, in: Nanotechnology, 1st ed., CRC Press, Boca Raton, 2022: pp. 241–269. https://doi.org/10.1201/9781003220350-14.

[40] N. Singh, J.R. Ansari, M. Pal, A. Das, D. Sen, D. Chattopadhyay, A. Datta, Enhanced blue photoluminescence of cobalt-reduced graphene oxide hybrid material and observation of rare plasmonic response by tailoring morphology, Applied Physics A. 127 (2021) 568. https://doi.org/10.1007/s00339-021-04697-1.

[41] A. Sirelkhatim, S. Mahmud, A. Seeni, N.H.M. Kaus, L.C. Ann, S.K.M. Bakhori, H. Hasan, D. Mohamad, Review on zinc oxide nanoparticles: antibacterial activity and toxicity mechanism, Nano-Micro Letters. 7 (2015) 219–242. https://doi.org/10.1007/s40820-015-0040-x.

[42] R. Sankar, R. Maheswari, S. Karthik, K.S. Shivashangari, V. Ravikumar, Anticancer activity of Ficus religiosa engineered copper oxide nanoparticles, Materials Science and Engineering: C. 44 (2014) 234–239. https://doi.org/10.1016/j.msec.2014.08.030.
[43] W. Cai, J. Wang, C. Chu, W. Chen, C. Wu, G. Liu, Metal-Organic framework-based stimuli-responsive systems for drug delivery, Advanced Science. 6 (2019) 1801526. https://doi.org/10.1002/advs.201801526.
[44] L. Chen, J. Liang, An overview of functional nanoparticles as novel emerging antiviral therapeutic agents, Materials Science and Engineering: C. 112 (2020) 110924. https://doi.org/10.1016/j.msec.2020.110924.
[45] T.Q. Huy, N.T. Hien Thanh, N.T. Thuy, P. Van Chung, P.N. Hung, A.T. Le, N.T. Hong Hanh, Cytotoxicity and antiviral activity of electrochemical—synthesized silver nanoparticles against poliovirus, Journal of Virological Methods. 241 (2017) 52–57. https://doi.org/10.1016/J.JVIROMET.2016.12.015.
[46] M.Y. Lee, J.A. Yang, H.S. Jung, S. Beack, J.E. Choi, W. Hur, H. Koo, K. Kim, S.K. Yoon, S.K. Hahn, Hyaluronic acid-gold nanoparticle/interferon α complex for targeted treatment of hepatitis C virus infection, ACS Nano. 6 (2012) 9522–9531. https://doi.org/10.1021/NN302538Y.
[47] B. Aderibigbe, Metal-based nanoparticles for the treatment of infectious diseases, Molecules. 22 (2017) 1370. https://doi.org/10.3390/molecules22081370.
[48] W.M. Hikal, A. Bratovcic, R.S. Baeshen, K.G. Tkachenko, H.A.H. Said-Al Ahl, Nanobiotechnology for the detection and control of waterborne parasites, Open Journal of Ecology. 11 (2021) 203–223. https://doi.org/10.4236/oje.2021.113016.
[49] G. Volpedo, L. Costa, N. Ryan, G. Halsey, A. Satoskar, S. Oghumu, Nanoparticulate drug delivery systems for the treatment of neglected tropical protozoan diseases, The Journal of Venomous Animals and Toxins Including Tropical Diseases. 25 (2019). https://doi.org/10.1590/1678-9199-JVATITD-1441-18.
[50] O.S. Adeyemi, N.I. Molefe, O.J. Awakan, C.O. Nwonuma, O.O. Alejolowo, T. Olaolu, R.F. Maimako, K. Suganuma, Y. Han, K. Kato, Metal nanoparticles restrict the growth of protozoan parasites, Artificial Cells, Nanomedicine, and Biotechnology. 46 (2018) S86–S94. https://doi.org/10.1080/21691401.2018.1489267.
[51] O.S. Adeyemi, C.G. Whiteley, Interaction of nanoparticles with arginine kinase from Trypanosoma brucei: kinetic and mechanistic evaluation, International Journal of Biological Macromolecules. 62 (2013) 450–456. https://doi.org/10.1016/j.ijbiomac.2013.09.008.
[52] D. V. Laurents, R.L. Baldwin, Characterization of the unfolding pathway of hen egg white lysozyme, Biochemistry. 36 (1997) 1496–1504. https://doi.org/10.1021/BI962198Z/ASSET/IMAGES/LARGE/BI962198ZF00007.JPEG.
[53] F. Vatansever, W.C.M.A. de Melo, P. Avci, D. Vecchio, M. Sadasivam, A. Gupta, R. Chandran, M. Karimi, N.A. Parizotto, R. Yin, G.P. Tegos, M.R. Hamblin, Antimicrobial strategies centered around reactive oxygen species-bactericidal antibiotics, photodynamic therapy, and beyond, FEMS Microbiology Reviews. 37 (2013) 955–989. https://doi.org/10.1111/1574-6976.12026.
[54] C. Zhou, Q. Wang, J. Jiang, L. Gao, Nanozybiotics: nanozyme-based antibacterials against bacterial resistance, Antibiotics. 11 (2022) 390. https://doi.org/10.3390/antibiotics11030390.
[55] J.R. Ansari, N. Singh, S. Anwar, S. Mohapatra, A. Datta, Silver nanoparticles decorated two dimensional MoS2 nanosheets for enhanced photocatalytic activity, Colloids and Surfaces A: Physicochemical and Engineering Aspects. 635 (2022) 128102. https://doi.org/10.1016/j.colsurfa.2021.128102.
[56] M.E. Allentoft, M. Collins, D. Harker, J. Haile, C.L. Oskam, M.L. Hale, P.F. Campos, J.A. Samaniego, T.P.M. Gilbert, E. Willerslev, G. Zhang, R.P. Scofield, R.N. Holdaway, M. Bunce, The half-life of DNA in bone: measuring decay kinetics in 158 dated fossils, Proceedings: Biological Sciences. 279 (2012) 4724–4733. https://doi.org/10.1098/RSPB.2012.1745.

[57] T.K.N. Luong, I. Govaerts, J. Robben, P. Shestakova, T.N. Parac-Vogt, Polyoxometalates as artificial nucleases: hydrolytic cleavage of DNA promoted by a highly negatively charged ZrIV-substituted Keggin polyanion, Chemical Communications. 53 (2017) 617–620. https://doi.org/10.1039/C6CC08555E.
[58] N. Panté, B. Fahrenkrog, Exploring nuclear pore complex molecular architecture by immuno-electron microscopy using Xenopus oocytes, Methods in Cell Biology. 122 (2014) 81–98. https://doi.org/10.1016/B978-0-12-417160-2.00004-7.
[59] N. Panté, Use of intact xenopus oocytes in nucleocytoplasmic transport studies, Methods in Molecular Biology (Clifton, NJ). 322 (2006) 301–314. https://doi.org/10.1007/978-1-59745-000-3_21.
[60] X. Chuanlai, P. Cifang, H. Kai, J. Zhengyu, W. Wukang, Chemiluminescence enzyme immunoassay (CLEIA) for the determination of chloramphenicol residues in aquatic tissues, Luminescence. 21 (2006) 126–128. https://doi.org/10.1002/BIO.892.
[61] J. Jiang, G. Oberdörster, A. Elder, R. Gelein, P. Mercer, P. Biswas, Does nanoparticle activity depend upon size and crystal phase? Nanotoxicology. 2 (2008) 33–42. https://doi.org/10.1080/17435390701882478.
[62] I. Karaduman, A.A. Güngör, H. Nadaroğlu, A. Altundaş, S. Acar, Green synthesis of γ-Fe_2O_3 nanoparticles for methane gas sensing, Journal of Materials Science: Materials in Electronics. 28 (2017). https://doi.org/10.1007/s10854-017-7510-5.
[63] H. Nadaroğlu, A. Alayli Güngör, S. İnce, Synthesis of nanoparticles by green synthesis method, International Journal of Innovative Research and Reviews. 1 (2017) 6–9.
[64] K. Reed, A. Cormack, A. Kulkarni, M. Mayton, D. Sayle, F. Klaessig, B. Stadler, Exploring the properties and applications of nanoceria: is there still plenty of room at the bottom? Environmental Science: Nano. 1 (2014) 390–405. https://doi.org/10.1039/C4EN00079J.
[65] S. Tsunekawa, R. Sivamohan, S. Ito, A. Kasuya, T. Fukuda, Structural study on monosize CeO2-x nano-particles, Nanostructured Materials. 11 (1999) 141–147. https://doi.org/10.1016/S0965-9773(99)00027-6.
[66] H. Nadaroglu, H. Onem, A.A. Gungor, Green synthesis of Ce_2O_3 NPs and determination of its antioxidant activity, IET Nanobiotechnology. 11 (2017). https://doi.org/10.1049/iet-nbt.2016.0138.
[67] A. Asati, S. Santra, C. Kaittanis, S. Nath, J.M. Perez, Oxidase-Like activity of polymer-coated cerium oxide nanoparticles, Angewandte Chemie International Edition. 48 (2009) 2308–2312. https://doi.org/10.1002/anie.200805279.
[68] T. Pirmohamed, J.M. Dowding, S. Singh, B. Wasserman, E. Heckert, A.S. Karakoti, J.E.S. King, S. Seal, W.T. Self, Nanoceria exhibit redox state-dependent catalase mimetic activity, Chemical Communications. 46 (2010) 2736. https://doi.org/10.1039/b922024k.
[69] C. Xu, X. Qu, Cerium oxide nanoparticle: a remarkably versatile rare earth nanomaterial for biological applications, NPG Asia Materials. 6 (2014) e90. https://doi.org/10.1038/am.2013.88.
[70] H. Nadaroglu, A. Alayli, S. Ceker, H. Ogutcu, G. Agar, Investigation of antimicrobial and genotoxic effects of Fe_2O_3, NiO and CoO NPs synthesized by green synthesis, Journal of Nanoanalysis. (2020). https://doi.org/10.22034/JNA.2020.1894035.1190.
[71] I. Karaduman, A.A. Güngör, A. Nadaroğlu, A. Altundaş, A. Acar, Green synthesis of γ-Fe_2O_3 nanoparticles for methane gas sensing, Journal of Materials Science: Materials in Electronics. 28 (2017) 16094–16105.
[72] H. Nadaroglu, G. Mosber, A.A. Gungor, G. Adıguzel, A. Adiguzel, Biodegradation of some azo dyes from wastewater with laccase from Weissella viridescens LB37 immobilized on magnetic chitosan nanoparticles, Journal of Water Process Engineering. 31 (2019). https://doi.org/10.1016/j.jwpe.2019.100866.
[73] H. Nadaroglu, Z. Sonmez, Purification of an endo-beta 1,4-mannanase from Clitocybe Geotropa and immobilization on chitosan-coated magnetite nanoparticles: application for fruit juices, Digest Journal of Nanomaterials and Biostructures. 11 (2016).

[74] A. Alayli Gungor, H. Nadaroglu, E. Kalkan, N. Celebi, Fenton process-driven decolorization of Allura Red AC in wastewater using apolaccase-modified or native nanomagnetite immobilized on silica fume, Desalination and Water Treatment. 57 (2016). https://doi.org/10.1080/19443994.2015.1074620.

[75] L. Gao, J. Zhuang, L. Nie, J. Zhang, Y. Zhang, N. Gu, T. Wang, J. Feng, D. Yang, S. Perrett, X. Yan, Intrinsic peroxidase-like activity of ferromagnetic nanoparticles, Nature Nanotechnology. 2 (2007) 577–583. https://doi.org/10.1038/nnano.2007.260.

[76] H. Wei, E. Wang, Fe_3O_4 magnetic nanoparticles as peroxidase mimetics and their applications in H2O2 and glucose detection, Analytical Chemistry. 80 (2008) 2250–2254. https://doi.org/10.1021/ac702203f.

[77] D. Kim, Y. Kim, S. Hong, J. Kim, N. Heo, M.-K. Lee, S. Lee, B. Kim, I. Kim, Y. Huh, B. Choi, Development of lateral flow assay based on size-controlled gold nanoparticles for detection of hepatitis B surface antigen, Sensors. 16 (2016) 2154. https://doi.org/10.3390/s16122154.

[78] H. Vaisocherová-Lísalová, I. Víšová, M.L. Ermini, T. Špringer, X.C. Song, J. Mrázek, J. Lamačová, N. Scott Lynn, P. Šedivák, J. Homola, Low-fouling surface plasmon resonance biosensor for multi-step detection of foodborne bacterial pathogens in complex food samples, Biosensors and Bioelectronics. 80 (2016) 84–90. https://doi.org/10.1016/j.bios.2016.01.040.

[79] S. Cicek, A.A. Gungor, A. Adiguzel, H. Nadaroglu, Biochemical evaluation and green synthesis of nano silver using peroxidase from Euphorbia (Euphorbia amygdaloides) and its antibacterial activity, Journal of Chemistry. 2015 (2015). https://doi.org/10.1155/2015/486948.

[80] H. Nadaroglu, A. Alayli, S. Ceker, H. Ogutcu, G. Agar, Biosynthesis of silver nanoparticles and investigation of genotoxic effects and antimicrobial activity, International Journal of Nano Dimension. (2020) 158–167.

[81] S. Sloan-Dennison, S. Laing, N.C. Shand, D. Graham, K. Faulds, A novel nanozyme assay utilising the catalytic activity of silver nanoparticles and SERRS, The Analyst. 142 (2017) 2484–2490. https://doi.org/10.1039/C7AN00887B.

[82] R.B. Rigon, M.H. Oyafuso, A.T. Fujimura, M.L. Gonçalez, A.H. Do Prado, M.P.D. Gremião, M. Chorilli, Nanotechnology-based drug delivery systems for melanoma antitumoral therapy: a review, BioMed Research International. 2015 (2015) 1–22. https://doi.org/10.1155/2015/841817.

[83] Y.J. Gu, J. Cheng, C.C. Lin, Y.W. Lam, S.H. Cheng, W.T. Wong, Nuclear penetration of surface functionalized gold nanoparticles, Toxicology and Applied Pharmacology. 237 (2009) 196–204. https://doi.org/10.1016/J.TAAP.2009.03.009.

[84] R. Bhattacharya, C.R. Patra, A. Earl, S. Wang, A. Katarya, L. Lu, J.N. Kizhakkedathu, M.J. Yaszemski, P.R. Greipp, D. Mukhopadhyay, P. Mukherjee, Attaching folic acid on gold nanoparticles using noncovalent interaction via different polyethylene glycol backbones and targeting of cancer cells, Nanomedicine: Nanotechnology, Biology and Medicine. 3 (2007) 224–238. https://doi.org/10.1016/J.NANO.2007.07.001.

[85] T. Cui, J.-J. Liang, H. Chen, D.-D. Geng, L. Jiao, J.-Y. Yang, H. Qian, C. Zhang, Y. Ding, Performance of doxorubicin-conjugated gold nanoparticles: regulation of drug location, ACS Applied Materials & Interfaces. 9 (2017) 8569–8580. https://doi.org/10.1021/acsami.6b16669.

[86] M.U. Farooq, V. Novosad, E.A. Rozhkova, H. Wali, A. Ali, A.A. Fateh, P.B. Neogi, A. Neogi, Z. Wang, Gold Nanoparticles-enabled efficient dual delivery of anticancer therapeutics to HeLa cells, Scientific Reports. 8 (2018) 2907. https://doi.org/10.1038/s41598-018-21331-y.

[87] A. Schmidt, V. Molano, M. Hollering, A. Pöthig, A. Casini, F.E. Kühn, Evaluation of new palladium cages as potential delivery systems for the anticancer drug cisplatin, Chemistry: A European Journal. 22 (2016) 2253–2256. https://doi.org/10.1002/chem.201504930.

[88] C. Dufes, I. Uchegbu, A. Schatzlein, Dendrimers in gene delivery, Advanced Drug Delivery Reviews. 57 (2005) 2177–2202. https://doi.org/10.1016/j.addr.2005.09.017.
[89] Y. Ding, Z. Jiang, K. Saha, C.S. Kim, S.T. Kim, R.F. Landis, V.M. Rotello, Gold nanoparticles for nucleic acid delivery, Molecular Therapy. 22 (2014) 1075–1083. https://doi.org/10.1038/mt.2014.30.
[90] E.-Y. Kim, R. Schulz, P. Swantek, K. Kunstman, M.H. Malim, S.M. Wolinsky, Gold nanoparticle-mediated gene delivery induces widespread changes in the expression of innate immunity genes, Gene Therapy. 19 (2012) 347–353. https://doi.org/10.1038/gt.2011.95.
[91] H.M. Joshi, D.R. Bhumkar, K. Joshi, V. Pokharkar, M. Sastry, Gold nanoparticles as carriers for efficient transmucosal insulin delivery, Langmuir. 22 (2006) 300–305. https://doi.org/10.1021/la051982u.
[92] P. Rathinaraj, A. Al-Jumaily, D.S. Huh, Internalization: acute apoptosis of breast cancer cells using herceptin-immobilized gold nanoparticles, Breast Cancer: Targets and Therapy. (2015) 51. https://doi.org/10.2147/BCTT.S69834.
[93] P. Di Pietro, L. Zaccaro, D. Comegna, A. Del Gatto, M. Saviano, R. Snyders, D. Cossement, C. Satriano, E. Rizzarelli, Silver nanoparticles functionalized with a fluorescent cyclic RGD peptide: a versatile integrin targeting platform for cells and bacteria, RSC Advances. 6 (2016) 112381–112392. https://doi.org/10.1039/C6RA21568H.
[94] P.P.P. Kumar, D.-K. Lim, Gold-Polymer nanocomposites for future therapeutic and tissue engineering applications, Pharmaceutics. 14 (2021) 70. https://doi.org/10.3390/pharmaceutics14010070.
[95] J.J. Muñoz, J. García-Ceca, D. Alfaro, M.A. Stimamiglio, T. Cejalvo, E. Jiménez, A.G. Zapata, Organizing the thymus gland, Annals of the New York Academy of Sciences. 1153 (2009) 14–19. https://doi.org/10.1111/j.1749-6632.2008.03965.x.
[96] V. Bronte, M.J. Pittet, The spleen in local and systemic regulation of immunity, Immunity. 39 (2013) 806–818. https://doi.org/10.1016/j.immuni.2013.10.010.
[97] Y.D. Mahnke, T.M. Brodie, F. Sallusto, M. Roederer, E. Lugli, The who's who of T-cell differentiation: human memory T-cell subsets, European Journal of Immunology. 43 (2013) 2797–2809. https://doi.org/10.1002/eji.201343751.
[98] G.J. Tobón, J.H. Izquierdo, C.A. Cañas, B lymphocytes: development, tolerance, and their role in autoimmunity—focus on systemic lupus erythematosus, Autoimmune Diseases. 2013 (2013) 1–17. https://doi.org/10.1155/2013/827254.
[99] S.T. Stern, S.E. McNeil, Nanotechnology safety concerns revisited, Toxicological Sciences: An Official Journal of the Society of Toxicology. 101 (2008) 4–21. https://doi.org/10.1093/TOXSCI/KFM169.
[100] M. Lucarelli, A.M. Gatti, G. Savarino, P. Quattroni, L. Martinelli, E. Monari, D. Boraschi, Innate defence functions of macrophages can be biased by nano-sized ceramic and metallic particles, European Cytokine Network. 15 (2004) 339–346.
[101] S. Sundar, V.K. Prajapati, Drug targeting to infectious diseases by nanoparticles surface functionalized with special biomolecules, Current Medicinal Chemistry. 19 (n.d.).
[102] R.P. Allaker, G. Ren, Potential impact of nanotechnology on the control of infectious diseases, Transactions of the Royal Society of Tropical Medicine and Hygiene. 102 (2008) 1–2. https://doi.org/10.1016/j.trstmh.2007.07.003.
[103] H. Nadaroglu, H. Onem, A.A. Gungor, Green synthesis of Ce_2O_3 NPs and determination of its antioxidant activity, IET Nanobiotechnology. 11 (2017). https://doi.org/10.1049/iet-nbt.2016.0138.
[104] M. Mahato, P. Pal, B. Tah, M. Ghosh, G.B. Talapatra, Study of silver nanoparticle–hemoglobin interaction and composite formation, Colloids and Surfaces B: Biointerfaces. 88 (2011) 141–149. https://doi.org/10.1016/j.colsurfb.2011.06.024.
[105] L. Cai, Z.-Z. Chen, X.-M. Dong, H.-W. Tang, D.-W. Pang, Silica nanoparticles based label-free aptamer hybridization for ATP detection using hoechst33258 as the signal reporter, Biosensors & Bioelectronics. 29 (2011) 46–52. https://doi.org/10.1016/j.bios.2011.07.064.

[106] B. Jovanović, T. Ji, D. Palić, Gene expression of zebrafish embryos exposed to titanium dioxide nanoparticles and hydroxylated fullerenes, Ecotoxicology and Environmental Safety. 74 (2011) 1518–25. https://doi.org/10.1016/j.ecoenv.2011.04.012.

[107] X. Gao, S. Yin, M. Tang, J. Chen, Z. Yang, W. Zhang, L. Chen, B. Yang, Z. Li, Y. Zha, D. Ruan, M. Wang, Effects of developmental exposure to TiO_2 nanoparticles on synaptic plasticity in hippocampal dentate gyrus area: an in vivo study in anesthetized rats, Biological Trace Element Research. 143 (2011) 1616–1628. https://doi.org/10.1007/S12011-011-8990-4.

[108] A. Mohammadipour, A. Fazel, H. Haghir, F. Motejaded, H. Rafatpanah, H. Zabihi, M. Hosseini, A.E. Bideskan, Maternal exposure to titanium dioxide nanoparticles during pregnancy; impaired memory and decreased hippocampal cell proliferation in rat offspring, Environmental Toxicology and Pharmacology. 37 (2014) 617–625. https://doi.org/10.1016/J.ETAP.2014.01.014.

[109] K.W. Ng, S.P.K. Khoo, B.C. Heng, M.I. Setyawati, E.C. Tan, X. Zhao, S. Xiong, W. Fang, D.T. Leong, J.S.C. Loo, The role of the tumor suppressor p53 pathway in the cellular DNA damage response to zinc oxide nanoparticles, Biomaterials. 32 (2011) 8218–8225. https://doi.org/10.1016/J.BIOMATERIALS.2011.07.036.

[110] M. Raffi, S. Mehrwan, T.M. Bhatti, J.I. Akhter, A. Hameed, W. Yawar, M.M. Ul Hasan, Investigations into the antibacterial behavior of copper nanoparticles against Escherichia coli, Annals of Microbiology. 60 (2010) 75–80. https://doi.org/10.1007/S13213-010-0015-6/FIGURES/6.

[111] H.L. Liu, P.Y. Chen, H.W. Yang, J.S. Wu, I.C. Tseng, Y.J. Ma, C.Y. Huang, H.C. Tsai, S.M. Chen, Y.J. Lu, C.Y. Huang, M.Y. Hua, Y.H. Ma, T.C. Yen, K.C. Wei, In vivo MR quantification of superparamagnetic iron oxide nanoparticle leakage during low-frequency-ultrasound-induced blood-brain barrier opening in swine, Journal of Magnetic Resonance Imaging: JMRI. 34 (2011) 1313–1324. https://doi.org/10.1002/JMRI.22697.

[112] S. Sundar, V.K. Prajapati, Drug targeting to infectious diseases by nanoparticles surface functionalized with special biomolecules, Current Medicinal Chemistry. 19 (2012) 3196. https://doi.org/10.2174/092986712800784630.

[113] H.S. Choi, J.V. Frangioni, Nanoparticles for biomedical imaging: fundamentals of clinical translation, Molecular Imaging. 9 (2010) 291. https://doi.org/10.2310/7290.2010.00031.

[114] Y.N. Zhang, W. Poon, A.J. Tavares, I.D. McGilvray, W.C.W. Chan, Nanoparticle-liver interactions: cellular uptake and hepatobiliary elimination, Journal of Controlled Release: Official Journal of the Controlled Release Society. 240 (2016) 332–348. https://doi.org/10.1016/J.JCONREL.2016.01.020.

[115] T. Soji, Y. Murata, A. Ohira, H. Nishizono, M. Tanaka, D.C. Herbert, Evidence that hepatocytes can phagocytize exogenous substances, The Anatomical Record. 233 (1992) 543–546. https://doi.org/10.1002/AR.1092330408.

[116] K.I. Ogawara, M. Yoshida, K. Furumoto, Y. Takakura, M. Hashida, K. Higaki, T. Kimura, Uptake by hepatocytes and biliary excretion of intravenously administered polystyrene microspheres in rats, Journal of Drug Targeting. 7 (2009) 213–221. https://doi.org/10.3109/10611869909085504.

[117] K.F. Soto, A. Carrasco, T.G. Powell, K.M. Garza, L.E. Murr, Comparative in vitro cytotoxicity assessment of some manufactured nanoparticulate materials characterized by transmission electron microscopy, Journal of Nanoparticle Research. 2.7 (2005) 145–169. https://doi.org/10.1007/S11051-005-3473-1.

[118] M. Chen, A. Von Mikecz, Formation of nucleoplasmic protein aggregates impairs nuclear function in response to SiO_2 nanoparticles, Experimental Cell Research. 305 (2005) 51–62. https://doi.org/10.1016/J.YEXCR.2004.12.021.

[119] United Nations, A/CONF.151/26/Vol. I: Rio Declaration on Environment and Development (1992). https://cil.nus.edu.sg/databasecil/1992-rio-declaration-on-environment-and-development/

5 Anti-inflammatory Metallic NPs

Type, Role, and Mechanisms

Mohit Sanduja, Tinku Gupta, Vikas Jogpal, and Reena Badhwar

5.1 INTRODUCTION

The principal process by which the body heals tissue damage and protects itself from external stressors is inflammation. It is a common and protective reaction to injury, infection, or other harmful stimuli, and it is connected to a number of disorders in humans. It is the outcome of a series of closely controlled processes that are activated in response to a stimulus [1] (Lawrence et al., 2002). Swelling, redness, pain, and other symptoms in the inflamed area of the body are caused by infection or injury. The pathology is believed to be built on that basis. The restoration of cell homeostasis, as well as tissue shape and function, relies on a good host response [2]. Primarily, without inflammatory feedback, infections, wounds, and tissue damage cannot recover. The two intrinsic components of host defenses that mediate this response are innate and adaptive immune responses [3]. The host immune system is the body's first line of protection against foreign substances, while adaptive immunity is formed by granulocytes, phagocytes, etc. [4]. The inflammatory response is mediated by immune cells (mast cells, neutrophils, dendritic cells, and macrophages) and non-invasive cells like endothelium and fibroblasts [5]. Adaptive or acquired immunity is usually defined as specificity and helps eliminate infections and establish immune memory in subsequent stages [6]. Acute inflammation is an essential defense mechanism to eliminate foreign particles, damaged tissue, and prevent injury in the future. It results from extrinsic, transient, chemical, mechanical, or infectious stimulation (i.e., a few hours to a few days). External stimulation is not necessary for abnormal/chronic inflammation, which may cause various unpleasant and debilitating symptoms [7].

In both acute and chronic inflammation, systemic TNF generation by macrophages activates the primary microglia cells, which are the core component of the innate immune system. In cases, where microglial cells have previously been triggered by chronic neurodegenerative alterations, acute inflammation stimulates the innate immune response, which results in the secretion of cytotoxic chemical mediators that further induce inflammation and aggravate neurodegeneration [8]. A host with a functional immune system often experiences an encounter with foreign stimuli

DOI: 10.1201/9781003317319-5

within a few minutes. Different inflammatory pathways and specific tissues are triggered in response to the stimulus. According to the harm the virus has produced, inflammation lasts for a while, and prolonged inflammation has systemic effects. Hepatocytes create prostaglandins (PGs) and basic proteins such as C-reactive protein as a consequence of excess cytokines and coagulation factors, which impact the central nervous system and result in fever, exhaustion, and discomfort [9]. Some of the most significant mechanisms for modulating ignition are the NF-kB and COX-2 signaling pathways.

5.2 NANOPARTICLES (NPs)

In recent decades, nanotechnology has played a vital role in biomedicine, diagnosis, treatment, industry, scientific research probes, and environmental protection [10–13]. Several synthetic processes and procedures have been used to create a diverse range of nanomaterials of different morphologies [14]. They are found in dimensions from 1 to 100 nm or in particles less than 100 nm [15]. Additionally, it offers several uses; particularly in the biomedical industry, such as medication release systems, dental dentures and artificial teeth; in-vitro diagnostics, bioluminescence imaging, opto-electronics and sensing devices [16]. Surprisingly, the material in the nanometer dimension has several distinct physiochemical and physiological properties that are fundamentally unique from the material in its original state and are quite beneficial as per scientific and research contexts [17]. The current trend is to develop NPs that confer better therapeutic properties in addition to their environment. NPs have several advantages, such as a high surface-to-volume ratio, surface plasmon coherence, the ability to conjugate them with chelators to create customized materials with specific features, toxicity toward pathogens, potent anticancer action against malignant cells, and catalytic uses [18].

In the past, ultra-fine NPs were produced by using metals, polymers, silica, phospholipid bilayers, liposomes, and inorganic dyes. Although there are many solid phase preparation techniques, including grinding, laser pyrolysis, mechanical alloying, aerosol processes, atomic condensation, vapor deposition, and biological processes, bottom-up preparation refers to the agglomeration of atoms to form stable molecular clusters, so although top-down preparation involves the disintegration of a large macro-material in to smaller ones [19]. The different types, roles, and anti-inflammatory capabilities based on NP mechanisms are described in detail in this review document.

5.3 TYPES AND ROLES OF METALLIC NANOPARTICLES

Because of their unique physicochemical and biological properties, metallic NPs (i.e., stability, universal adaptability, biocompatibility, adhering capabilities, medicinal properties) have been intensively researched and exploited for a variety of scientific objectives [20–22]. Various metallic NPs or their oxides [23] were stated to possess anti-inflammatory effects like silver [24], gold [25], selenium [26], copper [27], nickel [28], zinc oxide [29], zinc peroxide [30], magnesium oxide [31], cerium oxide [32], iron oxide [33], and titanium dioxide [34]. In recent times, various roles of metallic nanoparticles in the field of bio nanotechnology, medical science, drug delivery, and various disease targeting [35–44].

5.3.1 Zinc Oxide Nanoparticles

Zinc oxide NPs exhibit excellent biological characteristics [45] due to strong ionic nature. It is a yellow amorphous substance with a large surface area, size, and crystal composition [46]. Compared to other NPs, ZnO NPs are more favorable to nano-antibiotics because of their economical, whitish appearance, antibacterial properties, and UV shielding. Interestingly, a few studies have shown that ZnO NPs have high selective toxicity to bacteria, but have some side effects on living cells [47]. Using the aqueous extract of Camellia sinensis, 16-nm hexagonal wurtzite biomaterials were produced during the green synthesis of ZnO NPs. Compared to market antibiotics, NP ZnO is superior in antibacterial efficacy against bacteria and fungi [48]. According to a study, folic acid - loaded polyethylene glycol ZnO NPs also demonstrated anti-tumor and anti-inflammatory qualities [49]. In accordance with findings from a different investigation, mesoporus ZnO NPs combined with doxorubicin hydrochloride (DOX) had anti-inflammatory and anticancer properties [50]. Therefore, in vitro/vivo pharmacological activities are required to completely comprehend the various merits of ZnO as a possible therapeutic delivery service. Table 5.1 [51–64] and 5.2 [65–69] summarize the list of various ZNO-NPs anti-inflammatory nanoformulations.

5.3.2 Silver Nanoparticles

AgNPs are distinguished from a number of the noble metallic NPs and nano-composites due to their great capacity and enormous packages with inside the fabric and meals industries, water purification industry [70], environmental pollutants safety, biomedical devices/pharmacological study (anticancer, antibacterial, anti-angiogenic, comparison marketers in illustration strategies for forecasting illnesses [71]and problem-solving probes in organic structures for the detection of numerous dreadful sicknesses [72]. To increase the effectiveness and healing index of pills, as well as to overcome any resistance or hurdles posed by their green shipment [73]. Nobel silver NPs, which are counterparts to ZnO NPs, operate as an anti-inflammatory agent in the medical business. It acts on four basic principles: (1) lowering VEGF levels; (2) preventing mucin hypersecretion; (3) lowering HIF-1 expression; and (4) inhibiting pro-inflammatory cytokine generation. Researchers have shown that epithelial VEGF increases the sensitivity of antigen, which is crucial for the regulation of physiological dysfunction, and the promotion of T-cell (TH-2), cell-mediated inflammation pathway, formation of pro-inflammatory cytokines (IL-4, IL-5, IL-9, and IL-13) [74]. The phosphorylation of Src at Y419 by VEGF and IL-1 stimulates endothelial permeability via the Src kinase pathway. AgNPs disrupt Y419 phosphorylation in a dose-dependent manner and inactivate the Src kinase pathway, lowering VEGF and IL-1-induced vascular endothelial permeability [75]. By mediating bacterial death, AgNPs lower the production of hypoxia-inducible factor (HIF) -1 and limit pro-inflammatory expression of genes. As per current research, less O_2 induces an increase in TNF-α, IL-1, and IL-6 levels in macrophages and Kupffer cells in

TABLE 5.1
Phytometallic-based anti-inflammatory nanoparticles with mechanisms and inflammatory model

S. No	Metallic nanoparticles	Plant extract	Inflammation Model	Mechanism of action	References
1	Silver (Ag)	Leucasaspera	Carrageenan-induced paw edema	Edema (↓)	[50]
2		*Viburnum opulus fruit*	HaCaT cell line, Wistar rats	Cytokine production (↓),	[51]
3		Calophyllum-tomentosum Leaf	Bovine albumin	Inhibition of albumin denaturation	[52]
4		*Salvia officinalis* leaf	MCF-7 cells	Suppress COX-2 expression	[53]
5		Syzygiuma-romaticum	Bovine serum albumin	Inhibition of protein denaturation, down-regulation of cytokines	[54]
6		Chamaeme-lumnobile	Carrageenan-induced paw edema	cytokine production (↓)	[55]
7	Gold (Au)	*Litchi chinensis*	Canageenan-induced paw edema model	Edema (↓)	[56]
8		Pnmusdomestica gum			[57]
9		Prunusserrulata fruit	LPS-induced RAW264.7 cell line	Pro-inflammatory cytokines and inflammatory mediators (↓)	[59]
10	Zinc (Zn)	Trianthemaportulac astrum Linn. Leaf	Bovine albumin = human RBCs	Membrane stabilization, inhibition of albumin denaturation, proteinase inhibition	[59]
11		Andrographi-spaniculata leaf	Bovine serum albumin	Protein denaturation inhibition	[60]
12	Selenium (Se)	Spermacocehispida leaf			[61]
13	Titanium Dioxide (TiO_2)	Grape seed extarct	Bovine serum albumin	Protein denaturation inhibition	[62]
14	Copper oxide	Seed extract of Bacopamonnieri	Carrageean-Induced Paw	Inflammatory mediator	[63]

TABLE 5.2
Summary of Anti-inflammatory Potential of ZnO-NP.

ZnO-NPs of	Size	Mode of evaluation	Target/model used	Anti-Inflammatory Prolife	Reference
Delphinium uncinatum root extract	16–28 nm	In vitro study	COX-1, COX-2, 15-LOX, and sPLA2	Among all, ZnO NP showed best inhibitory potential against sPLA2 (33.2%).	[65]
Etoricoxib + Montelukast	217–298 nm	In vivo study	Mice model and response were recorded using plethysmometer	Maximum inhibition (81.67%) was observed after 2 hr by ZnO -NP of combination of Etoricoxib (1.49 mg) + Montelukast (4.2 mg)	[66]
Kalanchoe pinnata leaf extract	24 nm	In vitro study	IL-1β, IL-6, TNF-α, and COX-2 inflammatory gene expressions	ZnO NP suppress the all the targeted expression in dose-dependent manner.	[67]
Tabernaemontana heyneana Wall leaf and stem extract	6.89–8.14 nm	In vitro study	HRBC membrane stabilizing method	ZnO-NP produced similar membrane stabilization as standard (diclofenac).	[68]
Vernonia amygdalina leaf powder extract	20–40 nm	In vivo study	Air Sack Assessment method and cytokines release study	ZnO-NP suppresses the release of pro-inflammatory cytokines and desensitized the nociceptors by reducing the production of PGs.	[69]

the liver. Adipokines which was observed to be more prevalent in hypoxic tissue was intimately associated with the lymphangiogenesis [76]. Angiogenesis is aided by HIF-1, VEGF, and anecdotal data shows that AgNPs limit angiogenesis progression through in vitro activity [77]. AgNPs in lung tissues limit the overproduction of mucins, that is, Muc5ac, which causes airway blockage and chronic inflammation [78]. It has also been demonstrated to reduce perivascular and peribronchial inflammation and glycoproteins [79]. At greater doses, it reduces the construction of inflammatory cytokines as well as COX-2 gene expression [80]. A list of different silver metal-based NPs is summarized in Table 5.3.

5.3.3 Gold Nanoparticles

Gold NPs have special features that can help in treatment. They are primarily controlled by three mechanisms:

1. Falling reactive oxygen species production;
2. Diminishing LPS-induced cytokine;
3. Transformation (MAPK and PI3K pathways) in liver cells

TABLE 5.3
Summary of Anti-inflammatory Potential of AgNPs.

Ag-NPs of	Size	Mode of evaluation	Target/model used	Anti-inflammatory prolife	Reference
Cotyledon orbiculata leaves	106–137 nm	In-vitro	Determination of cytokines response by the stimulation of macrophages	At the dose of 5 µg/ml, Ag-NP tremendously decrease the level of TNF-α, IL-1β and IL-6,	[81]
Prunus serrulata fruit extract	66 nm	In-vitro	Pro-inflammatory cytokines release response in RAW264.7 cells.	Ag-NPs decrease the production NO and PGE2 level and also suppress the level of iNOS and COX expressions in RAW264.7 cells	[82]
Soft Coral Nephthea sp. extract	5–11 nm	In-vitro	COX inhibitor screening kit	Ag-NP total extract showed more potency toward COX-1 with IC_{50} of 33.72 µg/ml, while Petroleum ether extract was found to be more selective toward COX-II with IC_{50} of 3.34 µg/ml	[83]
Black Currant extract suspension	95 nm and 215 nm	In-vitro and in-vivo	DCC induce colitis animal study and study on LPS stimulated macrophage of RAW264.7 (Griess Test)	AgNP significantly decreases the NO production up to 47.95% at 1 ppm administration. Inhibit the colon shortening at the dose of 2 mg/kg	[84]
Phyllospongia lamellose extract	2.47–27.55 nm	In vitro	COX inhibitory assay	Ag-NPs showed moderate inhibitory effect on both COX-I and COX-II.	[85]

AuNPs may be employed as anti-inflammatory medications because it has been demonstrated that they stimulate ROS formation in phagocytes in a dose-dependent way. Several authors have reported that AuNPs diagnose various disease and targets anticancer, antibacterial, neurological diseases, etc. [86–96]. By oxidizing lipids and the cell—signaling proteins tyrosine phosphatases, it damages DNA [97]. LPS enhances splenocyte secretion of pro-inflammatory cytokines (IL-1, IL-17, TNF-α) and is inhibited by AuNP through IL-1-induced epithelial cell proliferation down-regulation phenomena [98].

The MAPK pathway is activated when LPS combines with toll-like receptors on the surface of cell. The PI3K pathway regulates gene expression, cell growth, protein production, and cytokine levels [99]. Gao et al. 2019 discovered P12 (G20), a peptide hybrid with a 20-nanometer core that demonstrated strong action in decreasing

TABLE 5.4
Summary of Anti-inflammatory Potential of AuNPs.

Au-NPs of	Size	Mode of evaluation	Target/model used	Anti-inflammatory Prolife	Reference
Prunus serrulata fruit extract	65 nm	In-vitro	Western Blotting analysis for protein associated with for NF-kB and MPAK pathways	Au-NPs suppress the induction of NF-kB signaling pathway and reduced the expressions of p-IkB, NF-kB, p-38, p-JNK and p-ERK proteins in dose-dependent manner.	[100]
Metal oxide (TiO_2)	8 5 nm	In vitro	Quantification of cytokines and pre-inflammatory responses on TNF-α and IL-1β LPS-induced inflammation model	Au-TiO_2 reduces the concentration of TNF-α and IL-1β. Au-TiO_2 also suppresses the inflammation induced by LPS.	[101]
Chaetomium globosum extract	23 nm	In-vivo and in-vitro	Carrageenan-induced Paw edema model and western blotting (LOX & COX-II)	AuNPs were found to suppress LOX and COX-II expression. In in vivo study, it also reduces the production of PGs in dose-dependent manner.	[102]
Citrus sinensis (Orange) peel extract	52.44 nm	In-vitro	Western blotting on expressions of iNOS and COX-II proteins and LPS induced RAW264.7 cells	AuNPs diminish the expressions of iNOS and COX-II in dose-dependent manner and reduces the production of NO in LPS - induced RAW264.7 cell.	[103]
Sambucus wightiana powdered extract	14.2 nm	In vivo and in-vitro	Carrageenan-induced Paw edema model. COX and LOX assay	AuNPs potentially inhibited both COX and LOX enzymes with IC_{50} values 17.27 and 30.64 µg/ml respectively. At 20 mg/kg dose, AuNPs inhibited up to 74.49 % inflammation.	[104]

TLR signaling, which is a major component in the inflammatory pathways leading to acute lung damage. Numerous different types of gold NPs are presented in Table 5.4 [100–104].

5.3.4 Titanium Dioxide Nanoparticles

In our civilization, titanium dioxide (TiO_2) seems to have become embedded. It may be found in a variety of consumer products and everyday objects, including paving stones, paper, and plastics. Global production of titanium dioxide in 2009 reached 4.68 million tones. It works by

1. Reducing platelet numbers
2. Increasing thrombin-anti-thrombin levels

PSG1 (P selectin Glycol protein) is produced by neutrophils and monocytes, and when it interacts with P selectin, it produces free radicals that cause oxidative stress in macrophages. According to studies, blood incubated with TiO_2NPs resulted in a decrease in platelet numbers [105]. Thrombin increases inflammation via signaling through protease-activated receptors or downstream mediators (PARs). P-selectin and cytokines are all induced when PAR is activated. It is a cell adhesion protein that helps activated platelets clump together on leukocytes (Table 5.5). According to research, TiO_2 NPs raise TAT levels, which mean a more thrombin inactivation by anti-thrombin occurs, and inflammation is decreased as a result of blocking the PAR pathway [106].

5.3.5 Copper Oxide Nanoparticles

Tyrosinase and SOD are copper-metal-based enzymes are helpful in various physiological processes [107]. Copper-based nanomaterials can be employed to neutralize free radicals, as per a recent research study. The anti-inflammatory effect of copper oxide NPs, according to Rajesh Kumar and coworkers (2021), stabilizes the membrane, which releases lysosomal enzymes that cause inflammation (Table 5.5) [108–113].

TABLE 5.5
Summary of the Anti-inflammatory Potential of TiO_2, CuO, Platinum and Selenium.

Nanoparticles	Size	Mode of evaluation	Target/model used	Anti-inflammatory prolife	Reference
TiO_2 nanoparticle of *Alpinia calcarata* rhizomes extract	60–130 nm	In vitro	Protein denaturation bioassay (egg and bovine albumin)	TiO_2 NPs prevents the denaturation of protein in dose-dependent manner.	[108]
TiO_2 nanoparticle of *Mucuna pruriens* plant extract	≤ 50 nm	In-vitro	Albumin denaturation assay	NPs inhibits the denaturation of protein in dose-dependent manner and at dose of 50 μL, 89% inhibition was reported.	[109]
CuO nanoparticle of *Capsicum frutescens* leaves	25–35 nm	In-vitro	Albumin denaturation assay	CuO NPs inhibit the denaturation up to 90% at 150 μg/ml concentration.	[110]

TABLE 5.5 ***(Continued)***
Summary of the Anti-inflammatory Potential of TiO_2, CuO, Platinum and Selenium.

Nanoparticles	Size	Mode of evaluation	Target/model used	Anti-inflammatory prolife	Reference
CuO nanoparticle of (Etoricoxib + Montelukast)	87 nm	In vivo	Carrageenan-induced hind paw edema method	At 5 mg/kg dose, NP suppresses the inflammation up to 89.6%.	[111]
Se nanoparticle of *Enterococcus durans* A8–1	Not mentioned	In-vitro	Lipopolysaccharide -induced Caco-2 cells	Se NPs suppressed the expressions of inflammatory cytokines IL and TNF-α in dose-dependent manner.	[112]
Platinum nanoparticles	20–100 nm	In-vivo and In-vitro	Lipopolysaccharide - induced RAW 264.7 cells. Western blotting on protein expressions of cytokines	PtNPs inhibits the production of COX-II and PGE_2 in LPS induced RAW 264.7 cells. Also suppress the expression of cytokines IL and TNF-α in dose-dependent manner.	[113]

5.3.6 Selenium Nanoparticles

Due to its inclusion in the catalytic sites of several antioxidant enzymes, including glutathione peroxidase (GPx) and thioredoxin reductase (TrxR), selenium (Se) is a necessary component of the healthy cell [114] and has anti-inflammatory activity [115]. Nano-Se has the ability to block lymphocyte adherence to endothelial cells by downregulating pro-inflammatory genes via a metalloproteinase-dependent method that forms a soluble L-selectin by cleaving its receptor [116]. Se-NPs function by blocking the NF-B pathway and the production of PG-E2. By triggering pro-inflammatory cytokines, the cytokine NF-β encourages the inflammatory reaction. Se-NPs are known to suppress the release of NF-β by inhibiting the phosphorylation and appearance of iNOS and COX-2 [117].

5.3.7 Lanthanide Nanoparticles

Lanthanide-based NPs (LNPs) have a lot of potential in medicine. Through activation of the NLRP-3 inflammasome, a range of nanocrystals, including LNPs, trigger a powerful inflammatory response. The capacity of peptide surface coatings to control the inflammatory response of nanocrystals should be very useful for pharmacological applications of LNPs and other tailored nanomaterials [118]. A typical lanthanide member can survive temperatures much beyond 700 degrees Celsius. In nuclear medicine, lanthanide radionuclides are well-known for their ability to detect and treat malignant cancers [119].

5.3.8 Platinum Nanoparticles

Platinum is a silvery-white metal that is more expensive than gold and is prized for its malleability, strength, and ductility. LPS-produced inflammatory mediators and intracellular ROS were prevented from forming by Pt-NPs. Pt NPs also decreased IB phosphorylation/degradation, ERK1/2 and AKT phosphorylation, as well as nuclear factor kappa-B transcriptional activity (NFB). Due to the down regulation of the NFB signaling pathway in macrophages, Pt NPs exhibit anti-inflammatory properties and may have potential as an anti-inflammatory medication.

5.4 MECHANISM OF METALLIC NANOPARTICLE ON INFLAMMATORY DISORDER

An abnormality in the regulatory signals that control the inflammatory process triggers inflammation by causing cellular damage [120]. The sequence of destruction indicates that an inflammatory response is triggered by immune cells that are activated to assist in repressing the reaction [121]. They are large, heterogeneous; mononucleated cells that are created in the bone marrow and found in the bloodstream as monocytes [122]. Two types of cells are involved, such as: pro-inflammatory (Figure 5.1) M1 macrophages (which increases inflammation) and anti-inflammatory M2 macrophages, which are alternately activated in response to anti-inflammatory responses and stimulate tissue and organ remodeling [123].

Depending on the state of the response, it can initiate, control, and uphold the inflammatory reactions by switching between these two phenotypes. It also engages in phagocytosis, which engulfs cellular and tissue debris, and promotes inflammation by producing signals that further trigger macrophages through membrane proteins, lipopolysaccharides, and cytokines (interleukins, interferons, chemokines, lymphokines), and tumor necrosis growth factor [124]. Neutrophils move to the inflammatory site as a result of inflammation, and they release pro-inflammatory substances that draw macrophages to the area [125].

Inhibition of the electron transport chain; enzyme and protein disruption; ROS production; cell signaling inhibition; cell membrane damage; oxidative damage of cell components; ribosome binding; DNA fragmentation; replication inhibition; cell wall damage [126]; protein corona formation; nanoparticles cell communications and uptake by cells are some of the mechanisms of action of various metallic NPs (Figure 5.2).

Metal NPs can enter the body through the nose, mouth, or skin. Due to their compact size, they may readily pass through the majority of biological membranes, particularly mucosal linings, and even reach the sense organs. Metal NPs enter the circulatory system as soon as they enter the body and inevitably bind with proteins in the blood plasma [127]. A protein corona forms around the nanoparticle as a result of this interaction. IgG, IgM, and fibrinogen are among of the most prevalent

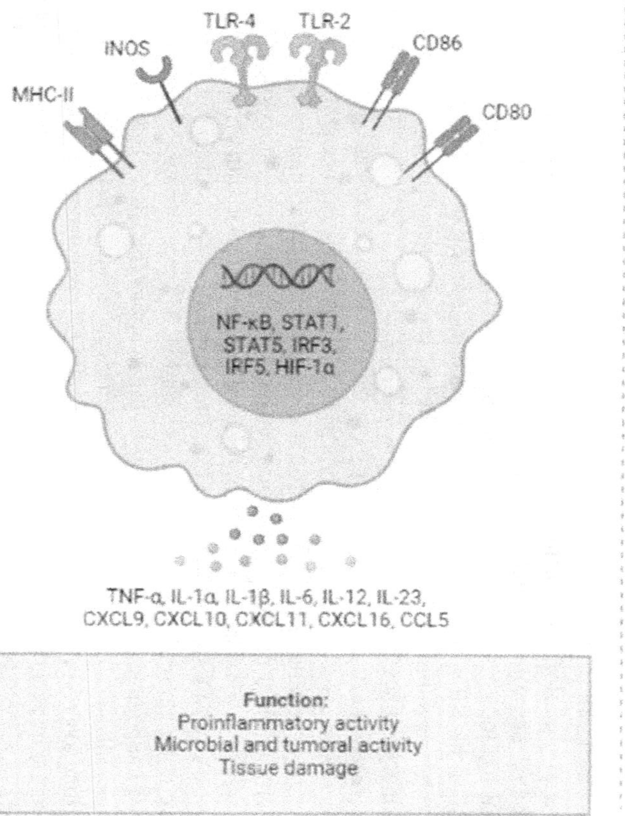

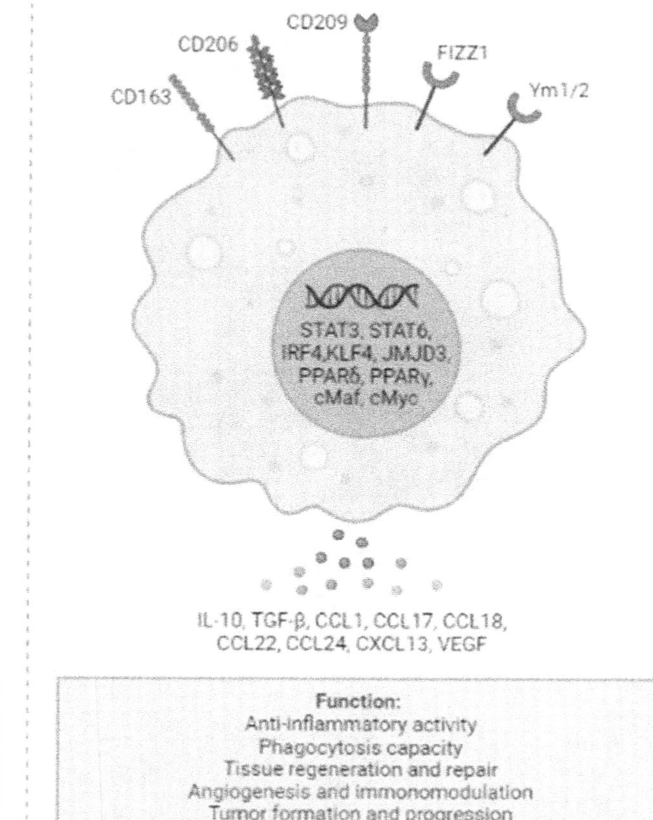

FIGURE 5.1 Mechanism of action of different macrophages (M1 and M2) on the inflammation phenomena.

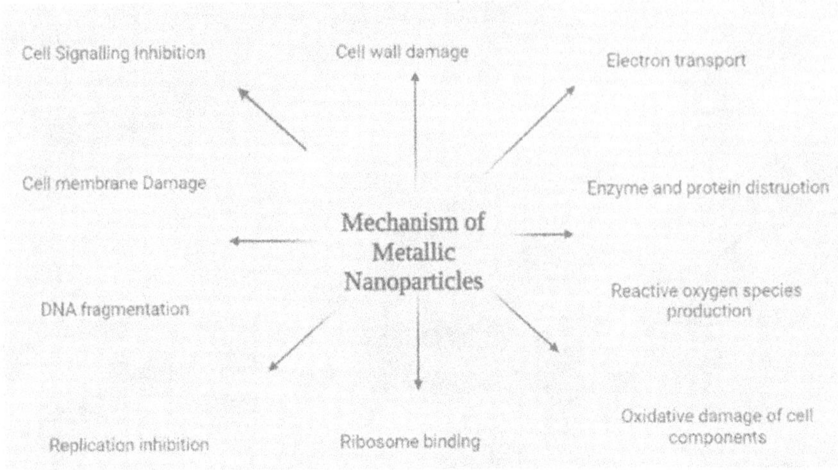

FIGURE 5.2 Schematic diagram of the mechanism of action of metallic nanoparticles.

proteins found in NPs. According to studies, both types of immunoglobulins participate in the inflammatory process [128], which is influenced by the physical features of the NPs, including surface charge, size, geometric shape, degree of hydrophobicity [129], surface roughness and curvature, as well as the chemical composition of the NPs [130]. Although serum proteins are less numerous in blood plasma, they have a strong affinity for metal NPs. As a result, serum proteins make up a large portion of the protein corona that surrounds NPs [131]. The NP's exterior shape is altered by this protein, giving it a biological identity [132]. The transit and interaction of the NP with numerous chemical reactions are determined by its identity. NPs enter cells through cell membrane pores or through ion channels. The size of the NP determines this sort of ingression. Although NP uptake does not need membrane receptors, it does require sticky contacts such as electrostatic interactions, Van der Waals forces, and steric interactions [133]. At higher concentration levels, the majority of cellular vesicles easily contain endocyte specific small metallic NPs. Macrophages and neutrophils perform phagocytosis and macro-pinocytosis. First, macrophages or neutrophils in inflammatory regions interact with protein-coated metal NPs in M2 macrophages [134]. M2 macrophages exhibit faster nanoparticle uptake in the presence of serum proteins than M1. M2 macrophages contain much more immunoglobulin and complement factor receptors (FCGR2B and CD36) than M1 macrophages, suggesting that M2-induced receptors attach to the protein corona, according to a phagocytosis gene arrays study [135]. This suggests that serum protein adsorption (especially immunoglobulins and complement components) is important for M2 macrophage NP absorption. Neutrophils create extracellular traps (NET) around themselves in response to intrinsic and environmental stimuli such as uric acid, cholesterol, pathogenic germs, and foreign particles [136]. Mitosis is the process by which these NETs develop, and is regulated by microbes, and inflammatory stimuli. NET synthesis requires the protein kinase-3 and reactive oxygen species that are arise from lipid peroxide when the membrane, is damaged. As a result, of the increased surface area of the cell membrane, more O_2 may be absorbed on the surface, and more ROS can subsequently be created [137]. Due to the DNA and antimicrobial proteins that make up the majority of these NETs, gold NPs may be readily bound inside of them.

5.5 CONCLUSION

A notable and comprehensive report on the kind, function, and methods employed by numerous metallic NPs is provided in this review. The NPs' ability to be ingested by phagocytic cells is facilitated by the corona protein that forms around them. Pro-inflammatory cytokines, ROS mechanisms, and NF-B and COX-2 pathways play a crucial role in the anti-inflammatory phenomenon. A review of the literature has been provided, with particular attention paid to the several clean sources for producing different NPs, the NPs' unique morphologies, the inflammatory paradigms used in the experiment, and the nanoparticles' alleged mechanisms of action. The anti-inflammatory properties of different NPs can be exploited in drug development, healthcare, and pharmaceutical sectors, and they may offer a minimally invasive therapy option for a variety of inflammatory conditions.

CONFLICT OF INTEREST

The authors have no financial conflicts to report.

ACKNOWLEDGMENT

I want to express my thanks to the pharmacy department at GD Goenka University in Gurugram, Haryana, India, for equipping me with all of the research resources.

REFERENCES

1. Lawrence W. Willoughby DA, and Gilroy DW. Anti-inflammatory lipid mediators and insights into the resolution of inflammation. Nature Reviews Immunology. 2002; 2:787–795.
2. Akire S. Pathogen recognition and immune system. Cell. 2006, 124:783–801.
3. Fujiwara N, Kobayashi K. Macrophages in inflammation. Current Drug Targets-Inflammation & Allergy. 2005 Jun 1;4(3):281–286.
4. Ahmed AU. An overview of inflammation: mechanism and consequences. Frontiers in Biology. 2011 Aug;6(4):274–281.
5. Beckmann N, Cannet C, Babin AL, Blé FX, Zurbruegg S, Kneuer R, Dousset V. In vivo visualization of macrophage infiltration and activity in inflammation using magnetic resonance imaging. Wiley Interdisciplinary Reviews: Nanomedicine and Nanobiotechnology. 2009 May;1(3):272–298.
6. Netea MG. Training innate immunity: the changing concept of immunological memory in innate host defence. European Journal of Clinical Investigation. 2013 Aug;43(8):881–884.
7. Medzhitov R. Inflammation 2010: New adventures of an old flame. Cell. 2010 Mar 19;140(6):771–776.

8. Holmes C, Cunningham C, Zotova E, Woolford J, Dean C, Kerr SU, Culliford D, Perry VH. Systemic inflammation and disease progression in Alzheimer disease. Neurology. 2009 Sep 8;73(10):768–774.
9. Kahya MC, Nazıroğlu M, Çiğ B. Melatonin and selenium reduce plasma cytokine and brain oxidative stress levels in diabetic rats. Brain Injury. 2015 Oct 15;29(12):1490–1496.
10. Guerra FD, Attia MF, Whitehead DC, Alexis F. Nanotechnology for environmental remediation: materials and applications. Molecules. 2018 Jul 18;23(7):1760.
11. Mishra S, Sharma S, Javed MN, Pottoo FH, Barkat MA, Alam MS, Amir M, Sarafroz M. Bioinspired nanocomposites: applications in disease diagnosis and treatment. Pharmaceutical Nanotechnology. 2019 Jun 1;7(3):206–219.
12. Javed MN, Dahiya ES, Ibrahim AM, Alam M, Khan FA, Pottoo FH. Recent advancement in clinical application of nanotechnological approached targeted delivery of herbal drugs. In Nanophytomedicine (pp. 151–172). Springer. 2020.
13. Javed MN, Pottoo FH, Shamim A, Hasnain MS, Alam MS. Design of experiments for the development of nanoparticles, nanomaterials, and nanocomposites. In Design of Experiments for Pharmaceutical Product Development (pp. 151–169). Springer. 2021.
14. Aygün A, Gülbağça F, Nas MS, Alma MH, Çalımlı MH, Ustaoglu B, Altunoglu YC, Baloğlu MC, Cellat K, Şen F. Biological synthesis of silver nanoparticles using Rheum ribes and evaluation of their anticarcinogenic and antimicrobial potential: a novel approach in phytonanotechnology. Journal of Pharmaceutical and Biomedical Analysis. 2020 Feb 5;179:113012.
15. Farokhzad OC, Langer R. Impact of nanotechnology on drug delivery. ACS Nano. 2009 Jan 27;3(1):16–20.
16. Ramos AP, Cruz MA, Tovani CB, Ciancaglini P. Biomedical applications of nanotechnology. Biophysical Reviews. 2017 Apr;9(2):79–89.
17. Gatoo MA, Naseem S, Arfat MY, Mahmood Dar A, Qasim K, Zubair S. Physicochemical properties of nanomaterials: implication in associated toxic manifestations. BioMed Research International. 2014 Oct;2014:498420.
18. Akter M, Sikder MT, Rahman MM, Ullah AA, Hossain KF, Banik S, Hosokawa T, Saito T, Kurasaki M. A systematic review on silver nanoparticles-induced cytotoxicity: physicochemical properties and perspectives. Journal of Advanced Research. 2018 Jan 1;9:1–6.
19. Richards R, Bönnemann H. Synthetic approaches to metallic nanomaterials. Nanofabrication Towards Biomedical Applications: Techniques, Tools, Applications, and Impact. 2005 Jan 20:1–32.
20. De Crozals G, Bonnet R, Farre C, Chaix C. Nanoparticles with multiple properties for biomedical applications: a strategic guide. Nano Today. 2016 Aug 1;11(4):435–463.
21. Hasnain MS, Javed MN, Alam MS, Rishishwar P, Rishishwar S, Ali S, Nayak AK, Beg S. Purple heart plant leaves extract-mediated silver nanoparticle synthesis: optimization by Box-Behnken design. Materials Science and Engineering: C. 2019 Jun 1;99:1105–1114.
22. Javed MN, Pottoo FH, Alam MS. Metallic nanoparticle alone and/or in combination as novel agent for the treatment of uncontrolled electric conductance related disorders and/or seizure, epilepsy & convulsions. Patent Acquired on October. 2016;10:40.
23. Piñón-Segundo E, Mendoza-Muñoz N, Quintanar-Guerrero D. Nanoparticles as dental drug-delivery systems. In Nanobiomaterials in Clinical Dentistry (pp. 475–495). William Andrew Publishing. 2013 Jan 1.
24. Del Turco S, Ciofani G, Cappello V, Navarra T, Caselli C, Gemmi M, Mattoli V, Basta G. Anti-inflammatory and antioxidant effects of cerium oxide nanoparticles in human endothelial cells. European Heart Journal. 2013 Aug 1;34(suppl_1).
25. Uchiyama MK, Deda DK, Rodrigues SF, Drewes CC, Bolonheis SM, Kiyohara PK, Toledo SP, Colli W, Araki K, Farsky SH. In vivo and in vitro toxicity and anti-inflammatory properties of gold nanoparticle bioconjugates to the vascular system. Toxicological Sciences. 2014 Dec 1;142(2):497–507.

26. Lu KY, Lin PY, Chuang EY, Shih CM, Cheng TM, Lin TY, Sung HW, Mi FL. H2O2-depleting and O_2-generating selenium nanoparticles for fluorescence imaging and photodynamic treatment of proinflammatory-activated macrophages. ACS Applied Materials & Interfaces. 2017 Feb 15;9(6):5158–5172.
27. Angajala G, Pavan P, Subashini R. One-step biofabrication of copper nanoparticles from Aegle marmelos correa aqueous leaf extract and evaluation of its anti-inflammatory and mosquito larvicidal efficacy. RSC Advances. 2014;4(93):51459–51470.
28. Spoorthy HP, Archna M, Rekha N, Satish S. Synthesis of nickel nanoparticles via biological entity and their anti-inflammatory activity. Journal of Microbiology and Biotechnology. 2017;7:1–6.
29. Ilves M, Palomäki J, Vippola M, Lehto M, Savolainen K, Savinko T, Alenius H. Topically applied ZnO nanoparticles suppress allergen induced skin inflammation but induce vigorous IgE production in the atopic dermatitis mouse model. Particle and Fibre Toxicology. 2014 Dec;11(1):1–2.
30. Ali K, Dwivedi S, Azam A, Saquib Q, Al-Said MS, Alkhedhairy AA, Musarrat J. Aloe vera extract functionalized zinc oxide nanoparticles as nanoantibiotics against multi-drug resistant clinical bacterial isolates. Journal of Colloid and Interface Science. 2016 Jun 15;472:145–156.
31. Jahangiri L, Kesmati M, Najafzadeh H. Evaluation of analgesic and anti-inflammatory effect of nanoparticles of magnesium oxide in mice with and without ketamine. European Review for Medical and Pharmacological Sciences. 2013 Oct 1;17(20):2706–2710.
32. Del Turco S, Ciofani G, Cappello V, Navarra T, Caselli C, Gemmi M, Mattoli V, Basta G. Anti-inflammatory and antioxidant effects of cerium oxide nanoparticles in human endothelial cells. European Heart Journal. 2013 Aug 1;34(Suppl_1); Wu HY, Chung MC, Wang CC, Huang CH, Liang HJ, Jan TR. Iron oxide nanoparticles suppress the production of IL-1beta via the secretory lysosomal pathway in murine microglial cells. Particle and Fibre Toxicology. 2013 Dec;10(1):1.
33. Wu PC, Hsiao HT, Lin YC, Shieh DB, Liu YC. The analgesia efficiency of ultrasmall magnetic iron oxide nanoparticles in mice chronic inflammatory pain model. Nanomedicine: Nanotechnology, Biology and Medicine. 2017 Aug 1;13(6):1975–1981.
34. Agarwal H, Shanmugam V. A review on anti-inflammatory activity of green synthesized zinc oxide nanoparticle: mechanism-based approach. Bioorganic Chemistry. 2020 Jan 1;94:103423.
35. Javed MN, Alam MS, Waziri A, Pottoo FH, Yadav AK, Hasnain MS, Almalki FA. QbD applications for the development of nanopharmaceutical products. In Pharmaceutical Quality by Design (pp. 229–253). Academic Press. 2019 Jan 1.
36. Javed MN, Pottoo FH, Shamim A, Hasnain MS, Alam MS. Design of experiments for the development of nanoparticles, nanomaterials, and nanocomposites. In Design of Experiments for Pharmaceutical Product Development (pp. 151–169). Springer. 2021.
37. Pandit J, Alam MS, Ansari JR, Singhal M, Gupta N, Waziri A, Sharma K, Potto FH. Multifaced Applications of Nanoparticles in Biological Science. In Nanomaterials in the Battle Against Pathogens and Disease Vectors (pp. 17–50). CRC Press. 2022.
38. Kumar R, Dhamija G, Ansari JR, Javed MN, Alam MS. C-dot nanoparticulated devices for biomedical applications. In Nanotechnology (pp. 271–299). CRC Press. 2022.
39. Alam MS, Naseh MF, Ansari JR, Waziri A, Javed MN, Ahmadi A, Saifullah MK, Garg A. Synthesis approaches for higher yields of nanoparticles. In Nanomaterials in the Battle Against Pathogens and Disease Vectors (pp. 51–82). CRC Press. 2022.
40. Bharti C, Alam MS, Javed MN, Khalid M, Saifullah FA, Manchanda R. Silica based nanomaterial for drug delivery. Nanomaterials: Evolution and Advancement Towards Therapeutic Drug Delivery (Part II). 2021 Jun 2:57.
41. Sangeet Kumar Mall SK, Yadav T, Waziri A, Alam MS. Treatment opportunities with *Fernandoa adenophylla* and recent novel approaches for natural medicinal phytochemicals as a drug delivery system. Exploration of Medicine. 2022;3:516–539.

42. Naseh MF, Ansari JR, Alam MS, Javed MN. Sustainable nanotorus for biosensing and therapeutical applications. In Handbook of Green and Sustainable Nanotechnology: Fundamentals, Developments and Applications (pp. 1–21). Springer International Publishing. 2022 Aug 19.
43. Sunilbhai CA, Alam M, Sadasivuni KK, Ansari JR. SPR Assisted diabetes detection. In Advanced Bioscience and Biosystems for Detection and Management of Diabetes (pp. 91–131). Springer. 2022.
44. Singhal S, Gupta M, Alam MS, Javed MN, Ansari JR. Carbon Allotropes-based nanodevices: Graphene in biomedical applications. In Nanotechnology (pp. 241–269). CRC Press. 2022.
45. Basnet P, Chanu TI, Samanta D, Chatterjee S. A review on bio-synthesized zinc oxide nanoparticles using plant extracts as reductants and stabilizing agents. Journal of Photochemistry and Photobiology B: Biology. 2018 Jun 1;183:201–221.
46. Ali K, Dwivedi S, Azam A, Saquib Q, Al-Said MS, Alkhedhairy AA, Musarrat J. Aloe vera extract functionalized zinc oxide nanoparticles as nanoantibiotics against multi-drug resistant clinical bacterial isolates. Journal of Colloid and Interface Science. 2016 Jun 15;472:145–156.
47. Akbarian M, Mahjoub S, Elahi SM, Zabihi E, Tashakkorian H. Appraisal of the biological aspect of Zinc oxide nanoparticles prepared using extract of Camellia sinensis L. Materials Research Express. 2019 Jul 3;6(9):095022.
48. Vimala K, Soundarapandian K. Erbitux conjugated zinc oxide nanoparticles to enhance antitumor efficiency via targeted drug delivery system for breast cancer therapy. Annals of Oncology. 2017 Nov 1;28:x41.
49. Barick KC, Nigam S, Bahadur D. Nanoscale assembly of mesoporous ZnO: A potential drug carrier. Journal of Materials Chemistry. 2010;20(31):6446–6452.
50. Kumaran N, Vijayaraj R, Swarnakala. Biosynthesis of silver nano particles from Leucas aspera (Willd.) link and its anti-inflammatory potential against carrageen induced paw edema in rats. International Journal of Pharmaceutical Sciences and Research. 2017 Jun 1;8(6):2588–2593.
51. Moldovan B, David L, Vulcu A, Olenic L, Perde-Schrepler M, Fischer-Fodor E, Baldea I, Clichici S, Filip GA. In vitro and in vivo anti-inflammatory properties of green synthesized silver nanoparticles using *Viburnum opulus* L. fruits extract. Materials Science and Engineering: C. 2017 Oct 1;79:720–727.
52. Govindappa M, Hemashekhar B, Arthikala MK, Rai VR, Ramachandra YL. Characterization, antibacterial, antioxidant, antidiabetic, anti-inflammatory and antityrosinase activity of green synthesized silver nanoparticles using Calophyllum tomentosum leaves extract. Results in Physics. 2018 Jun 1;9:400–408.
53. Baharara J, Ramezani T, Mousavi M, Asadi-Samani M. Antioxidant and anti-inflammatory activity of green synthesized silver nanoparticles using Salvia officinalis extract. Annals of Tropical Medicine and Public Health. 2017 Sep 1;10(5).
54. Varghese RE, Ragavan D, Sivaraj S, Gayathri D, Kannayiram G. Anti-inflammatory activity of Syzygium aromaticum silver nanoparticles: in vitro and in silico study. Asian Journal of Pharmaceutical and Clinical Research 2017;10:370–373.
55. Erjaee H, Nazifi S, Rajaian H. Effect of Ag-NPs synthesised by Chamaemelum nobile extract on the inflammation and oxidative stress induced by carrageenan in mice paw. IET Nanobiotechnology. 2017 Sep;11(6):695–701.
56. Murad U, Khan SA, Ibrar M, Ullah S, Khattak U. Synthesis of silver and gold nanoparticles from leaf of Litchi chinensis and its biological activities. Asian Pacific Journal of Tropical Biomedicine. 2018 Mar 1;8(3):142.
57. Islam NU, Amin R, Shahid M, Amin M, Zaib S, Iqbal J. A multi-target therapeutic potential of Prunus domestica gum stabilized nanoparticles exhibited prospective anticancer, antibacterial, urease-inhibition, anti-inflammatory and analgesic properties. BMC Complementary and Alternative Medicine. 2017 Dec;17(1):1–7.

58. Singh P, Ahn S, Kang JP, Veronika S, Huo Y, Singh H, Chokkaligam M, El-Agamy Farh M, Aceituno VC, Kim YJ, Yang DC. In vitro anti-inflammatory activity of spherical silver NPs and monodisperse hexagonal gold NPs by fruit extract of Prunus serrulata: a green synthetic approach. Artificial Cells, Nanomedicine, and Biotechnology. 2018 Dec;46(8):2022–2032.
59. Yadav E, Singh D, Yadav P, Verma A. Ameliorative effect of biofabricated ZnO nanoparticles of Trianthema portulacastrum Linn. on dermal wounds via removal of oxidative stress and inflammation. RSC Advances. 2018 Jun 13;8(38):21621–21635.
60. Rajakumar G, Thiruvengadam M, Mydhili G, Gomathi T, Chung IM. Green approach for synthesis of zinc oxide nanoparticles from Andrographis paniculata leaf extract and evaluation of their antioxidant, anti-diabetic, and anti-inflammatory activities. Bioprocess and Biosystems Engineering. 2018 Jan;41(1):21–30.
61. Vennila K, Chitra L, Balagurunathan R, Palvannan T. Comparison of biological activities of selenium and silver nanoparticles attached with bioactive phytoconstituents: green synthesized using Spermacoce hispida extract. Advances in Natural Sciences: Nanoscience and Nanotechnology. 2018 Jan 5;9(1):015005.
62. Labh AK, Rajasekar A, Rajeshkumar S. Anti-inflammatory activity of titanium dioxide nanoparticles synthesised using grape seed extract: an in vitro study. Plant Cell Biotechnology and Molecular Biology. 2020 Aug 24:24–31.
63. Khuda F, Haq ZU, Ilahi I, Ullah R, Khan A, Fouad H, Khalil AA, Ullah Z, Sahibzada MU, Shah Y, Abbas M. Synthesis of gold nanoparticles using Sambucus wightiana extract and investigation of its antimicrobial, anti-inflammatory, antioxidant and analgesic activities. Arabian Journal of Chemistry. 2021 Oct 1;14(10):103343.
64. Rehman H, Ali W, Khan NZ, Aasim M, Khan T, Khan AA. Delphinium uncinatum mediated biosynthesis of zinc oxide nanoparticles and in-vitro evaluation of their antioxidant, cytotoxic, antimicrobial, anti-diabetic, anti-inflammatory, and anti-aging activities. Saudi Journal of Biological Sciences. 2023 Jan 1;30(1):103485.
65. Sulaiman S, Ahmad S, Naz SS, Qaisar S, Muhammad S, Ullah R, Al-Sadoon MK, Gulnaz A. Synthesis of zinc oxide based etoricoxib and montelukast nanoformulations and their evaluation through analgesic, anti-inflammatory, anti-pyretic and acute toxicity activities. Journal of King Saud University-Science. 2022 Jun 1;34(4):101938.
66. Agarwal H, Nakara A, Shanmugam VK. Anti-inflammatory mechanism of various metal and metal oxide nanoparticles synthesized using plant extracts: a review. Biomedicine & Pharmacotherapy. 2019 Jan 1;109:2561–2572.
67. Manasa DJ, Chandrashekar KR, Kumar MP, Suresh D, Kumar DM, Ravikumar CR, Bhattacharya T, Murthy HA. Proficient synthesis of zinc oxide nanoparticles from Tabernaemontana heyneana wall. via green combustion method: antioxidant, anti-inflammatory, antidiabetic, anticancer and photocatalytic activities. Results in Chemistry. 2021 Jan 1;3:100178.
68. Liu H, Kang P, Liu Y, An Y, Hu Y, Jin X, Cao X, Qi Y, Ramesh T, Wang X. Zinc oxide NPs synthesised from the Vernonia amygdalina shows the anti-inflammatory and antinociceptive activities in the mice model. Artificial Cells, Nanomedicine, and Biotechnology. 2020 Dec;48(1):1068–1078.
69. Barkalina N, Charalambous C, Jones C, Coward K. Nanotechnology in reproductive medicine: emerging applications of nanomaterials. Nanomedicine: Nanotechnology, Biology and Medicine. 2014 Jul 1;10(5):e921–e938.
70. Butola BS, Mohammad F. Silver nanomaterials as future colorants and potential antimicrobial agents for natural and synthetic textile materials. RSC Advances. 2016;6(50):44232–44247.
71. Yetisen AK, Qu H, Manbachi A, Butt H, Dokmeci MR, Hinestroza JP, Skorobogatiy M, Khademhosseini A, Yun SH. Nanotechnology in textiles. ACS Nano. 2016 Mar 22;10(3):3042–3068.

72. Deshmukh SP, Patil SM, Mullani SB, Delekar SD. Silver nanoparticles as an effective disinfectant: a review. Materials Science and Engineering: C. 2019 Apr 1;97:954–965.
73. Lee CG, Link H, Baluk P, Homer RJ, Chapoval S, Bhandari V, Kang MJ, Cohn L, Kim YK, McDonald DM, Elias JA. Vascular endothelial growth factor (VEGF) induces remodeling and enhances TH2-mediated sensitization and inflammation in the lung. Nature Medicine. 2004 Oct;10(10):1095–1103.
74. Sheikpranbabu S, Kalishwaralal K, Venkataraman D, Eom SH, Park J, Gurunathan S. Silver nanoparticles inhibit VEGF-and IL-1β-induced vascular permeability via Src dependent pathway in porcine retinal endothelial cells. Journal of Nanobiotechnology. 2009 Oct;7(1):1–2.
75. Imtiyaz HZ, Simon MC. Hypoxia-inducible factors as essential regulators of inflammation. Diverse Effects of Hypoxia on Tumor Progression. 2010:105–120.
76. Lin N, Simon MC. Hypoxia-inducible factors: key regulators of myeloid cells during inflammation. The Journal of Clinical Investigation. 2016 Oct 3;126(10):3661–3671.
77. Yang T, Yao Q, Cao F, Liu Q, Liu B, Wang XH. Silver nanoparticles inhibit the function of hypoxia-inducible factor-1 and target genes: insight into the cytotoxicity and antiangiogenesis. International Journal of Nanomedicine. 2016;11:6679.
78. Jang S, Park JW, Cha HR, Jung SY, Lee JE, Jung SS, Kim JO, Kim SY, Lee CS, Park HS. Silver nanoparticles modify VEGF signaling pathway and mucus hypersecretion in allergic airway inflammation. International Journal of Nanomedicine. 2012;7:1329–1343.
79. Franková J, Pivodová V, Vágnerová H, Juráňová J, Ulrichová J. Effects of silver nanoparticles on primary cell cultures of fibroblasts and keratinocytes in a wound-healing model. Journal of Applied Biomaterials & Functional Materials. 2016 Apr;14(2):137–142.
80. Tyavambiza C, Elbagory AM, Madiehe AM, Meyer M, Meyer S. The antimicrobial and anti-inflammatory effects of silver nanoparticles synthesised from Cotyledon orbiculata aqueous extract. Nanomaterials. 2021 May;11(5):1343.
81. Singh P, Ahn S, Kang JP, Veronika S, Huo Y, Singh H, Chokkaligam M, El-Agamy Farh M, Aceituno VC, Kim YJ, Yang DC. In vitro anti-inflammatory activity of spherical silver NPs and monodisperse hexagonal gold NPs by fruit extract of Prunus serrulata: a green synthetic approach. Artificial Cells, Nanomedicine, and Biotechnology. 2018 Dec;46(8):2022–2032.
82. Abdelhafez OH, Ali TFS, Fahim JR, Desoukey SY, Ahmed S, Behery FA, Kamel MS, Gulder TAM, Abdelmohsen UR. Anti-Inflammatory potential of green synthesized silver NPs of the soft coral Nephthea Sp. supported by metabolomics analysis and docking studies. International Journal of Nanomedicine. 2020;15:5345–5360.
83. Krajewska JB, Długosz O, Sałaga M, Banach M, Fichna J. Silver nanoparticles based on blackcurrant extract show potent anti-inflammatory effect in vitro and in DSS-induced colitis in mice. International Journal of Pharmaceutics. 2020 Jul 30;585:119549.
84. Al-Khalaf AA, Hassan HM, Alrajhi AM, Mohamed RAEH, Hozzein WN. Anti-cancer and anti-inflammatory potential of the green synthesized silver NPs of the red sea sponge Phyllospongia lamellosa supported by metabolomics analysis and docking study. Antibiotics. 2021;10(10):1155.
85. Alam MS, Garg A, Pottoo FH, Saifullah MK, Tareq AI, Manzoor O, Mohsin M, Javed MN. Gum ghatti mediated, one pot green synthesis of optimized gold nanoparticles: Investigation of process-variables impact using Box-Behnken based statistical design. International Journal of Biological Macromolecules. 2017 Nov 1;104:758–767.
86. Alam MS, Javed MN, Pottoo FH, Waziri A, Almalki FA, Hasnain MS, Garg A, Saifullah MK. QbD approached comparison of reaction mechanism in microwave synthesized gold nanoparticles and their superior catalytic role against hazardous nirto-dye. Applied Organometallic Chemistry. 2019 Sep;33(9):e5071.
87. Pottoo FH, Tabassum N, Javed M, Nigar S, Rasheed R, Khan A, Barkat M, Alam M, Maqbool A, Ansari MA, Barreto GE. The synergistic effect of raloxifene, fluoxetine, and

bromocriptine protects against pilocarpine-induced status epilepticus and temporal lobe epilepsy. Molecular Neurobiology. 2019 Feb;56(2):1233–1247.

88. Pottoo FH, Sharma S, Javed MN, Barkat MA, Harshita, Alam MS, Naim MJ, Alam O, Ansari MA, Barreto GE, Ashraf GM. Lipid-based nanoformulations in the treatment of neurological disorders. Drug Metabolism Reviews. 2020 Jan 2;52(1):185–204.
89. Pottoo FH, Javed M, Barkat M, Alam M, Nowshehri JA, Alshayban DM, Ansari MA. Estrogen and serotonin: complexity of interactions and implications for epileptic seizures and epileptogenesis. Current Neuropharmacology. 2019 Mar 1;17(3):214–231.
90. Pottoo FH, Tabassum N, Javed MN, Nigar S, Sharma S, Barkat MA, Alam MS, Ansari MA, Barreto GE, Ashraf GM. Raloxifene potentiates the effect of fluoxetine against maximal electroshock induced seizures in mice. European Journal of Pharmaceutical Sciences. 2020 Apr 15;146:105261.
91. Aslam M, Javed M, Deeb HH, Nicola MK, Mirza M, Alam M, Akhtar M, Waziri A. Lipid nanocarriers for neurotherapeutics: introduction, challenges, blood-brain barrier, and promises of delivery approaches. CNS & Neurological Disorders-Drug Targets (Formerly Current Drug Targets-CNS & Neurological Disorders). 2022;21.
92. Waziri A, Bharti C, Aslam M, Jamil P, Mirza M, Javed MN, Pottoo U, Ahmadi A, Alam MS. Probiotics for the chemoprotective role against the toxic effect of cancer chemotherapy. Anti-Cancer Agents in Medicinal Chemistry (Formerly Current Medicinal Chemistry-Anti-Cancer Agents). 2022 Feb 1;22(4):654–667.
93. Javed MN, Akhter MH, Taleuzzaman M, Faiyazudin M, Alam MS. Cationic nanoparticles for treatment of neurological diseases. In Fundamentals of Bionanomaterials (pp. 273–292). Elsevier. 2022 Jan 1.
94. Kumari N, Daram N, Alam MS, Verma AK. Rationalizing the use of polyphenol nano-formulations in the therapy of neurodegenerative diseases. CNS & Neurological Disorders-Drug Targets (Formerly Current Drug Targets-CNS & Neurological Disorders). 2022 Dec 1;21(10):966–976.
95. Raj S, Manchanda R, Bhandari M, Alam M. Review on natural bioactive products as radioprotective therapeutics: present and past perspective. Current Pharmaceutical Biotechnology. 2022;23(14).
96. Ibrahim AM, Chauhan L, Bhardwaj A, Sharma A, Fayaz F, Kumar B, Alhashmi M, AlHajri N, Alam MS, Pottoo FH. Brain-derived neurotropic factor in neurodegenerative disorders. Biomedicines. 2022 May;10(5):1143.
97. Kingston M, Pfau JC, Gilmer J, Brey R. Selective inhibitory effects of 50-nm gold nanoparticles on mouse macrophage and spleen cells. Journal of Immunotoxicology. 2016 Mar 3;13(2):198–208.
98. Zhang J, Wang X, Vikash V, Ye Q, Wu D, Liu Y, Dong W. ROS and ROS-mediated cellular signaling. Oxidative Medicine and Cellular Longevity. 2016:4350965.
99. de Carvalho TG, Garcia VB, de Araújo AA, da Silva Gasparotto LH, Silva H, Guerra GC, de Castro Miguel E, de Carvalho Leitão RF, da Silva Costa DV, Cruz LJ, Chan AB. Spherical neutral gold nanoparticles improve anti-inflammatory response, oxidative stress and fibrosis in alcohol-methamphetamine-induced liver injury in rats. International Journal of Pharmaceutics. 2018 Sep 5;548(1):1–4.
100. Gao W, Wang Y, Xiong Y, Sun L, Wang L, Wang K, Lu HY, Bao A, Turvey SE, Li Q, Yang H. Size-dependent anti-inflammatory activity of a peptide-gold nanoparticle hybrid in vitro and in a mouse model of acute lung injury. Acta Biomaterialia. 2019 Feb 1;85:203–217.
101. Singh P, Ahn S, Kang JP, Veronika S, Huo Y, Singh H, Chokkaligam M, El-Agamy Farh M, Aceituno VC, Kim YJ, Yang DC. In vitro anti-inflammatory activity of spherical silver NPs and monodisperse hexagonal gold NPs by fruit extract of Prunus serrulata: a green synthetic approach. Artificial Cells, Nanomedicine, and Biotechnology. 2018 Dec;46(8):2022–2032.

102. Fujita T, Zysman M, Elgrabli D, Murayama T, Haruta M, Lanone S, Ishida T, Boczkowski J. Anti-inflammatory effect of gold nanoparticles supported on metal oxides. Scientific Reports. 2021 Nov 30;11(1):1.
103. Gao L, Mei S, Ma H, Chen X. Ultrasound-assisted green synthesis of gold nanoparticles using citrus peel extract and their enhanced anti-inflammatory activity. Ultrasonics Sonochemistry. 2022 Feb 1;83:105940.
104. Khuda F, Haq ZU, Ilahi I, Ullah R, Khan A, Fouad H, Khalil AA, Ullah Z, Sahibzada MU, Shah Y, Abbas M. Synthesis of gold nanoparticles using Sambucus wightiana extract and investigation of its antimicrobial, anti-inflammatory, antioxidant and analgesic activities. Arabian Journal of Chemistry. 2021 Oct 1;14(10):103343.
105. Sonmez O, Sonmez M. Role of platelets in immune system and inflammation. Porto Biomedical Journal. 2017 Nov 1;2(6):311–314.
106. Seisenbaeva GA, Fromell K, Vinogradov VV, Terekhov AN, Pakhomov AV, Nilsson B, Ekdahl KN, Vinogradov VV, Kessler VG. Dispersion of TiO_2 nanoparticles improves burn wound healing and tissue regeneration through specific interaction with blood serum proteins. Scientific Reports. 2017 Nov 13;7(1):1.
107. Vanathi P, Rajiv P, Sivaraj R. Synthesis and characterization of Eichhornia-mediated copper oxide nanoparticles and assessing their antifungal activity against plant pathogens. Bulletin of Materials Science. 2016 Sep;39(5):1165–1170.
108. Pratheema P, Gurupriya S, Ramesh J, Cathrine L, Pratheema P. Anti-inflammatory and anti-bacterial activity of titanium nanoparticles synthesized from rhizomes of Alpinia calcarata. International Journal for Research in Applied Science and Engineering Technology. 2018;6:2472–2477.
109. Thangavelu L, Rajeshkumar S, Arivarasu L, Aditya BS. Antioxidant and antiinflammatory activity of titanium dioxide nanoparticles synthesised using Mucuna pruriens. Journal of Pharmaceutical Research International. 2021 Dec 28:414–422.
110. Velsankar K, Suganya S, Muthumari P, Mohandoss S, Sudhahar S. Ecofriendly green synthesis, characterization and biomedical applications of CuO nanoparticles synthesized using leaf extract of Capsicum frutescens. Journal of Environmental Chemical Engineering. 2021 Oct 1;9(5):106299.
111. Sulaiman S, Ahmad S, Naz SS, Qaisar S, Muhammad S, Alotaibi A, Ullah R. Synthesis of copper oxide-based nanoformulations of etoricoxib and montelukast and their evaluation through analgesic, anti-inflammatory, anti-pyretic, and acute toxicity activities. Molecules. 2022 Feb 21;27(4):1433.
112. Liu J, Shi L, Tuo X, Ma X, Hou X, Jiang S, Lv J, Cheng Y, Guo D, Han B. Preparation, characteristic and anti-inflammatory effect of selenium nanoparticle-enriched probiotic strain Enterococcus durans A8–1. Journal of Trace Elements in Medicine and Biology. 2022 Dec 1;74:127056.
113. Rehman H, Ali W, Khan NZ, Aasim M, Khan T, Khan AA. Delphinium uncinatum mediated biosynthesis of zinc oxide nanoparticles and in-vitro evaluation of their antioxidant, cytotoxic, antimicrobial, anti-diabetic, anti-inflammatory, and anti-aging activities. Saudi Journal of Biological Sciences. 2023 Jan 1;30(1):103485.
114. El-Ghazaly MA, Fadel N, Rashed E, El-Batal A, Kenawy SA. Anti-inflammatory effect of selenium NPs on the inflammation induced in irradiated rats. Canadian Journal of Physiology and Pharmacology. 2017;95(2):101–110.
115. Kahya MC, Naziroğlu M, Çiğ B. Melatonin and selenium reduce plasma cytokine and brain oxidative stress levels in diabetic rats. Brain Injury. 2015 Oct 15;29(12):1490–1496.
116. Ahrens I, Ellwanger C, Smith BK, Bassler N, Chen YC, Neudorfer I, Ludwig A, Bode C, Peter K. Selenium supplementation induces metalloproteinase-dependent L-selectin shedding from monocytes. Journal of Leukocyte Biology. 2008 Jun;83(6):1388–1395.

117. Zhu C, Zhang S, Song C, Zhang Y, Ling Q, Hoffmann PR, Li J, Chen T, Zheng W, Huang Z. Selenium nanoparticles decorated with Ulva lactuca polysaccharide potentially attenuate colitis by inhibiting NF-κB mediated hyper inflammation. Journal of Nanobiotechnology. 2017 Dec;15(1):1–5.
118. Yao H, Zhang Y, Liu L, Xu Y, Liu X, Lin J, Zhou W, Wei P, Jin P, Wen LP. Inhibition of lanthanide nanocrystal-induced inflammasome activation in macrophages by a surface coating peptide through abrogation of ROS production and TRPM2-mediated Ca^{2+} influx. Biomaterials. 2016;108:143–156.
119. Liu H, Kang P, Liu Y, An Y, Hu Y, Jin X, Cao X, Qi Y, Ramesh T, Wang X. Zinc oxide NPs synthesised from the Vernonia amygdalina shows the anti-inflammatory and antinociceptive activities in the mice model. Artificial Cells, Nanomedicine, and Biotechnology. 2020 Dec;48(1):1068–1078.
120. Bianchi ME, Manfredi AA. How macrophages ring the inflammation alarm. Proceedings of the National Academy of Sciences. 2014 Feb 25;111(8):2866–2867.
121. Wynn TA, Vannella KM. Macrophages in tissue repair, regeneration, and fibrosis. Immunity. 2016 Mar 15;44(3):450–462.
122. Epelman S, Lavine KJ, Randolph GJ. Origin and functions of tissue macrophages. Immunity. 2014 Jul 17;41(1):21–35.
123. Liu YC, Zou XB, Chai YF, Yao YM. Macrophage polarization in inflammatory diseases. International Journal of Biological Sciences. 2014;10(5):520.
124. Fujiwawa N, Kobayashi K. Macrophage in inflammation. Current drug Targets Inflammation & Allergy. 2005;4:281–286.
125. Bahadar H, Maqbool F, Niaz K, Abdollahi M. Toxicity of nanoparticles and an overview of current experimental models. Iranian Biomedical Journal. 2016 Jan;20(1):1.
126. Begum SJP, Pratibha S, Rawat JM, Venugopal D, Sahu P, Gowda A, Qureshi KA, Jaremko M. Recent advances in green synthesis, characterization, and applications of bioactive metallic NPs. Pharmaceuticals. 2022;15(4):455.
127. Bahadar H, Maqbool F, Niaz K, Abdollahi M. Toxicity of nanoparticles and an overview of current experimental models. Iranian Biomedical Journal. 2016 Jan;20(1):1.
128. Zhang J, Wang X, Vikash V, Ye Q, Wu D, Liu Y, Dong W. ROS and ROS-mediated cellular signaling. Oxidative Medicine and Cellular Longevity. 2016:4350965.
129. Walkey CD, Chan WC. Understanding and controlling the interaction of nanomaterials with proteins in a physiological environment. Chemical Society Reviews. 2012;41(7):2780–2799.
130. Schwartz-Albiez R, Monteiro RC, Rodriguez M, Binder CJ, Shoenfeld Y. Natural antibodies, intravenous immunoglobulin and their role in autoimmunity, cancer and inflammation. Clinical & Experimental Immunology. 2009 Dec;158(Supplement_1):43–50.
131. Aschermann S, Lux A, Baerenwaldt A, Biburger M, Nimmerjahn F. The other side of immunoglobulin G: suppressor of inflammation. Clinical & Experimental Immunology. 2010 May;160(2):161–167.
132. Angajala G, Pavan P, Subashini R. One-step biofabrication of copper nanoparticles from Aegle marmelos correa aqueous leaf extract and evaluation of its anti-inflammatory and mosquito larvicidal efficacy. RSC Advances. 2014;4(93):51459–51470.
133. Simkó M, Fiedeler U, Gazsó A, Nentwich M. The Impact of Nanoparticles on Cellular Functions (NanoTrust Dossier No. 007). Institut für Technikfolgen-Abschätzung. 2011 Jan.
134. Monopoli MP, Bombelli FB, Laurent S. Protein – Nanoparticle interactions: opportunities and challenges. Chemical Reviews. 2011;111:5610–5637.
135. Kuhn A, Wozniacka A, Szepietowski JC, Gläser R, Lehmann P, Haust M, Sysa-Jedrzejowska A, Reich A, Oke V, Hügel R, Calderon C. Photoprovocation in cutaneous

lupus erythematosus: a multicenter study evaluating a standardized protocol. Journal of Investigative Dermatology. 2011 Aug 1;131(8):1622–1630.

136. Binnemars-Postma KA, Ten Hoopen HW, Storm G, Prakash J. Differential uptake of nanoparticles by human M1 and M2 polarized macrophages: protein corona as a critical determinant. Nanomedicine. 2016 Nov;11(22):2889–28902.
137. Muñoz LE, Bilyy R, Biermann MH, Kienhöfer D, Maueröder C, Hahn J, Brauner JM, Weidner D, Chen J, Scharin-Mehlmann M, Janko C. Nanoparticles size-dependently initiate self-limiting NETosis-driven inflammation. Proceedings of the National Academy of Sciences. 2016 Oct 4;113(40): E5856–E5865.

6 Trends in Theranostic Applications of Metallic Nanoparticles

Md Sabir Alam, Mansi Garg, Vijay Bhalla, Puja Kumari, Renu Kadyan, Md Meraj Anjum, Tejpal Yadav, Madhu Yadav, Mukesh Kumar, Aafrin Waziri, Syed Muzammil Munawar, Dhandayuthabani Rajendiran, Khaleel Basha Sabjan, and Jamilur R. Ansari

6.1 INTRODUCTION

A complicated set of illnesses known as cancer are responsible for one-third of all morbidity and mortality worldwide. One of the worst diseases in the world today is cancer. According to World Health Organization (WHO) statistics, cancer claims 8.97 million lives each year, placing it second on the list of causes of death behind ischemic heart disease [1]. Cancer is mostly brought on by driver mutations that turn on proto-oncogenes and turn off tumor suppressor genes. Cancer is characterized by changes in cell activity, which include resistance to growth-inhibiting signals, an unrestricted capacity for reproduction, evasion of apoptosis, continuing angiogenesis, self-sufficiency in tissue invasion, signal growth, and tumors [2]. Additionally, malignant tumors and neoplasms have features that enable tumor-induced inflammation and genome mutation, in addition to avoiding immune detection and rewiring the energy metabolism. Rapidly expanding aberrant cells have the ability to spread to surrounding bodily regions when they cross their typical boundaries, body parts and migrate to other body tissues from their original site; cancer metastasis is the term for the later process [3]. It takes several stages for cancer to develop from a precancerous lesion to a malignant tumor. As it does, healthy cells are transformed into tumor cells. These changes result from a person's genetic traits interacting with different categories of stress factors, including: (a) biological cancer-causing elements, such as viruses or bacteria, environmental pollutants, and physical stress; (b) chemical carcinogens like arsenic, aflatoxin, and cigarette smoke; (c) physical carcinogens like UV and ionizing radiation [4].

As conventional therapeutic techniques (such as chemotherapy, radiation, and surgeries) could have adverse effects. Due to inadequate therapy outcomes, interest has shifted to introducing nanotechnology in cancer management. However,

DOI: 10.1201/9781003317319-6

correct diagnosis is necessary for the effective and appropriate treatment of cancer since each type of tumor involves a unique therapeutic regimen [5]. Chemotherapy, radiation therapy, and/or surgery are the most popular forms of cancer treatment. However, malignancy-focused nanomedicines have the potential to be an effective, non-invasive therapeutic alternative for the management of a variety of cancer types and may even have an anti-tumor effect [6]. Nanocarriers like MNPs are excellent candidates for cancer therapy because they can simultaneously encapsulate and deliver a variety of anticancer medications with different therapeutic mechanisms [7]. These MNPs are novel carriers and imaging agents in the treatment of cancer and may be able to overcome difficulties brought on by traditional chemotherapy. These are said to play a crucial role in cancer therapy and can thereby increase medication transport, silencing of genes, and targeting [8]. MNPs with functionalized targeting ligands exhibit superior energy eposition control in malignancies. In addition to their therapeutic advantages, these MNPs are also used as a diagnostic tool for tumor cell imaging by passive and active targeting, and they recently paved the way for precision drug administration and site-specific targeting [9]. In comparison to traditional cancer treatments, nanomedicine has a number of benefits, including multifunctionality, effective drug transport, and controlled release of chemotherapeutic drugs. These benefits are made possible by the unique morphological and physical properties of nanoparticles (NPs), such as their small size, chemical composition, enormous surface area, and customized form [10].

Regardless of whether they are based on polymeric, liposomal, or inorganic metallic formulations, NPs are great candidates for delivering immunotherapeutic medicines since they naturally migrate to different organs via the spleen and lymph organs. Additionally, by reducing side effects, extending the duration that drugs spend in the bloodstream, and preventing drug breakdown before it reaches the target site, nanomaterials can be utilized as cytotoxics and/or enhancers of conventional chemotherapies [11]. Because of their concentrated surface functionalization, extended action, restricted size and shape distribution, and ability to be utilized in optical or heat-based treatment procedures, metal-based nanoparticles (NPs) are particularly interesting in the field of nanomedicine [12].

The increased density of metallic NPs enables them to be absorbed by cells more quickly when compared to non-metallic NPs of equivalent size, which makes them useful for cancer management measures. Furthermore, it was said that metal NPs, particularly when functionalized with specific ligands that offer regulated deposition inside cancer cells, could deliver medications more efficiently while also providing better targeting and gene silencing. By converting unfavorable conditions into ones that can be used therapeutically, metallic nanoconstructs can change the tumor microenvironment (TME). For instance, the ability of metallic nanoparticles to target biological systems may be enhanced by external stimuli including light, heat, ultrasonic radiation, and magnetic fields by modifying their redox potential and producing reactive oxygen species (ROS), which further sensitize target tissues [13]. The intrinsic properties of tumor tissues, such as pH, redox potential, and hypoxia, operate as additional viable triggers for the activity of metal-based NPs and the release of drugs, enhancing the effectiveness of treatment. Furthermore, it has been demonstrated that some metallic NPs can cause oxidative stress in cancer cells even in the absence of external stimulation [14]. It is also thought to be a great stabilizing strategy to surface functionalize metallic NPs with various organic compounds, macromolecules, or noble metal coatings [15].

6.1.1 Applications in Cancer Theranostic

MNPs have consistently captured the interest of scientists and are currently extensively used in the biological field. They have drawn attention in nanotechnology due to their immense potential [16]. These MNPs can now be produced and conjugated with various functional groups, allowing them to form new derivatives using ligands and pharmaceuticals. In biotechnology, magnetic separation, target analytes, carriers for drug and gene delivery, diagnostic imaging, and pre-concentration of personalized medication administration aging open up a wide variety of potential [17]. Additionally, a variety of imaging techniques have been developed over a long period of time for the imaging of various disease states, including magnetic resonance imaging (MRI), optical imaging, ultrasound, computed tomography (CT), and surface enhanced Raman spectroscopy (SERS) [14]. These imaging modalities demand a contrast agent with particular physiochemical properties more urgently because of the differences in their instrumentation and processes. For application in various imaging modalities, a number of nanoparticulated contrast agents, such as magnetic NPs, AuNPs, and AgNPs, were consequently developed [18]. To exploit multiple imaging modalities simultaneously, new multifunctional nanocages and nano-shells have also been developed. MNPs can be applied in a range of biological contexts, including photoablation, bioimaging, hyperthermia, medicine administration, and tumor targeting [19].

6.1.2 Biosensors

In the medical industry, notably in oncology research, biosensors can be employed successfully. An analytical tool used to examine biological material is called a biosensor. An electrical signal is produced by converting a biological, chemical, or biochemical response [20]. A biosensor is made up of three fundamental components: (I) a bioelement, which is often made up of nucleic acids, enzymes, cells, or antibodies. (III) the electronic device, which consists of a processor, a display, and an amplifier. (II) the transducer, which may be optical, electrical, piezoelectric, pyroelectric, or gravimetric [21]. In light of the fact that there are more cancer cases worldwide every year, surveys on early cancer diagnosis are crucial. Moreover, the potential for biosensor-based cancer treatment monitoring offers hope for a customized treatment. For this reason, there is still a need for more affordable, sensitive, and streamlined methods that can provide even more information on a given disease [22]. For these advanced works, some recent research has focused on employing MNPs as biosensors in cancer therapy. Because MNPs may easily engage with recognition molecules on the surface of biomolecules, such as antibodies, for detection, they are used in biosensing applications. MNPs have a high sensitivity for detecting cancer cells. Additionally, MNPs cause strong electromagnetic fields to form on the surface of the particles, which in turn improves radiative characteristics like scattering and absorption. MNPs have robust and easily-tunable optical characteristics as a result. This is applicable to optical imaging [23]. SPR, colorimetric, fluorescent, electrical, electrochemical, and biobarcode assay sensors are among the many types of sensors that can be created using these distinctive qualities of MNPs [24].

6.1.3 Bioimaging

Medical imaging techniques are currently essential for disease diagnosis and therapy. The rapidly developing science of nanomedicine has substantially influenced bioimaging, which includes the process of imaging tumor cells [25]. This imaging process could involve NPs directly or indirectly. When taking direct action, NPs can use specialized tools like heat analyzers, MRIs, and fluorescence microscopy to find the tumor cells. Au, Ag, and paramagnetic iron NPs are a few examples of NPs that can exhibit chromogen or fluorescence in a biological system [26]. These NPs do not, however, exhibit their own chromogenic or fluorescent behavior in the biological system. In their place, they transport substrate, reactive chromogen, enzymes, antibodies, or substrate. Indirect action is the term for this. This interaction between these NPs and the specific biological system they enter is facilitated by that environment. MNPs have widely been used during the past few years because of their exceptional optical and chemical properties [27]. Due to the surface plasmon absorption (SPA) phenomenon, MNPs' optical characteristics can be altered by a small change in size. The size and form of NPs affect the substance's conductive and valence band energy differentials as well. It might be able to modify the absorption-emission profile for bioimaging properties because of how form and size influence it [28].

6.1.4 Photoablation

Photothermal therapy (PTT) and photodynamic therapy (PDT) are the two types of photoablation therapy (PAT). PDT makes use of non-toxic light-sensitive compounds (photosensitizers), which, when exposed to light of a specific wavelength, turn highly toxic. The goal of this therapy is to target tumor cells [29]. When different photosensitizers, like TiO_2 NPs, are exposed to light at a specific wavelength during PDT, holes and photo-induced electrons are created. Reactive oxygen species, also known as oxidative radicals, are created by these electrons' subsequent interactions with hydroxyl ions or water (ROS). The ROS species are what induce the expected cell death. In PTT, a close NIR light source is held in front of cancer cells. When this light energy is further transformed into heat energy, which speeds up cell death, hyperthermia is created [30].

6.1.5 Hyperthermia

The medical sciences have defined the idea of employing heat to obtain therapeutic advantages. Typically, heat has been used to destroy tumor cells in a manner similar to surgically removing tumors [13]. It is sometimes referred to as thermoablation, a process that uses dangerous heat levels to cause irreversible protein coagulation and, as a result, cell death. The collateral damage to normal cells/tissues outside the ablation zone could be reduced by a little temperature rise that does not cause cell apoptosis on its own. This could still be helpful in therapy. Temperatures between 41 and 50 °C are considered hyperthermia. The term "thermoablation" is used to describe a process where the temperature exceeds 50 °C [31]. Notably, the results of this temperature rise depend on the passage of time, the uniformity of tissue temperature, and the kind of tissue. Necrosis can occur even at temperatures as low as 42 °C that are maintained for longer than 60 minutes. Hyperthermia improves blood circulation over time in cancer cells (but not in healthy cells), which improves oxygenation and perfusion of the hypoxic malignant core, which is minimally vascularized and mainly radiation resistant, as well as the delivery of anticancer drugs. The cells are more acidotic inside the hypoxic core of the tumor, which makes them more sensitive to heat but less sensitive to radiation. These activities, when combined, result in an altered extracellular milieu that is sensitive to heat-induced cell injury [32]. When different photosensitizers, like TiO_2 NPs, are exposed to light at a specific wavelength during PDT, holes and photo-induced electrons are created. Reactive oxygen species, also known as oxidative radicals, are created by these electrons' subsequent interactions with hydroxyl ions or water (ROS). The ROS species are what induce the expected cell death. In PTT, a close NIR light source is held in front of cancer cells. When this light energy is further transformed into heat energy, hyperthermia is created, which speeds up cell death [6].

6.1.6 Drug Delivery and Tumor Targeting

Different cancer therapies have very low molecular weights, are uniformly dispersed throughout the body, have a short half-life in the systemic circulation, and have a higher clearance rate [33]. As a result, a very modest proportion of the active medicine actually reaches the area it is intended to target, and its diffusion into healthy tissues causes several harmful side effects. Poor delivery at the intended site causes several issues, including multidrug resistance. The best chances of delivering therapeutic benefits are believed to come from combining a variety of targeting technologies that have a synergistic effect, the ability to deliver a variety of therapeutic medications, and the ability to successfully target them while avoiding biological barriers in organisms [34]. NPs with an active and passive targeting focus are used for targeting. MNPs can enter tumor tissues through vascular fenestrations in the case of passive targeting, indicating targeting through permeability and retention while accumulating at the tumor site. This is because different tumors have inadequate lymphatic drainage and faulty vasculature as a result of their fast growth into solid tumors. Additionally, adding hydrophilic moieties, like PEG, to the surface of NPs can help with solubility, inhibit macrophage absorption, keep the carriers in the bloodstream, and guard against enzymatic oxidation while doing in vivo research [35]. To achieve active targeting, other moieties, such as antibodies, can be employed to functionalize NPs. According to well-established evidence, the functionalized NPs efficiently target particular receptors or surface proteins on tumor cells and enhance their anticancer action, triggering cell apoptosis with minimum harm to healthy tissues, as shown in Figure 6.1 [36].

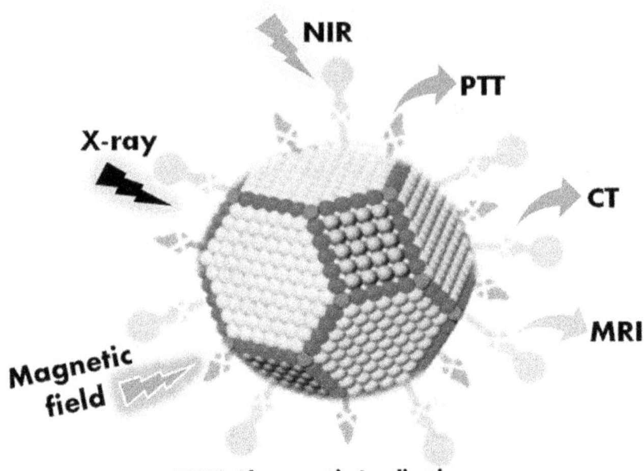

FIGURE 6.1 MNPs theranostics applications.

6.2 IMAGING TECHNIQUE FOR METALLIC NANOPARTICLES

6.2.1 Optical Imaging

In general, optical imaging techniques have a variety of significant benefits for developing medical equipment for POC diagnostics. For a rapid and accurate diagnosis, optical imaging can deliver high-resolution microscopic and macroscopic data. Recent developments in a variety of optical technologies, including optoelectronics, optical fibers, and micro-optics, have made it possible to further miniaturize and lower the price of optical imaging platforms [37–42]. For many years, optical imaging has been an effective technique for research in biomedical as well as clinical applications. Microscopic inspection of specimens (such as tissue, sputum, or blood film) is still regarded as the gold standard for the diagnosis of many ailments, notably infectious diseases like malaria and tuberculosis. Fluorescence microscopy, one of many optical imaging platforms, is particularly significant because of its great sensitivity and specificity [43, 44].

6.2.2 Ultrasound Imaging

Ultrasonic (US) imaging is one of the most commonly used techniques in clinical practice because of its low cost, wide diagnostic usefulness, and ease of handling [45]. It is used by medical professionals with a variety of specializations, including cardiologists, radiologists, gynecologists, surgeons, and many others involved in the medical field, for quick follow-up checks as well as an initial screening tool. US imaging is also advised for assessing the stage of carotid artery stenosis [46], for examining the perfusion of tumors [47], and for examining organs after their transplantation [48] due to its capacity to visualize blood vessels, blood velocity, and blood flow by Power and Color Doppler. In addition to these diagnostic uses, high-intensity focused ultrasound (HIFU) is gaining popularity for its use in the removal of ureteric stones [49], as well as benign uterine myomas and malignant tumors [50]. The diseased tissue is shifted over in this situation by the focused sonic energy. The diseased tissue is destroyed because of the acoustic energy being absorbed, which leads to an increase in the local temperature.

6.2.3 Raman Spectroscopy

Raman spectroscopy can serve as an in vivo modality for image formation because of its high sensitivity and minimal autofluorescence, which can be enhanced by surface-enhanced Raman scattering (SERS) nanoparticles. Nanoparticles with SERS have been extensively used in biological applications like biomarker identification and image-directed theragnostic. In Raman spectroscopy, the Rayleigh effect causes most photons to scatter elastically when light interacts with matter [51]. A tiny percentage of photons, however, experience inelastic scattering, in which energy transfer takes place when the input photon interacts with the scattering material [52]. The Raman spectroscopic technique is considered one of the vibrational spectroscopic techniques that relies on the inelastic scattering of photons through the materials of investigation, which leads to the fingerprint of distinctive characteristics [53].

6.2.4 Magnetic Resonance

Magnetic Resonance Imaging (MRI) is a crucial technique for diagnosis in the medical field and is also used as an in vivo imaging technique in the field of diagnostics [54]. The capacity of MRI to obtain 3D tomographic data in entire tissue samples with extremely high spatial resolution is one of the technology's most promising aspects. The benefit of MRI is that it may gather pictures non-invasively and without the use of radiotracers (PET or SPECT) or ionizing radiation (X-ray or CT), which lowers the danger of harm to delicate core organs like the central nervous system as well as the brain.

Protons relax under the influence of an externally applied magnetic field and a radiofrequency pulse, which is the basis of MRI [55]. When exposed to a radio frequency pulse, proton magnetic moments become disrupted and revert to their initial condition, which is also known as the relaxation process. When present, proton magnetic moments align in one direction and process around an external magnetic field. The two categories used to classify relaxation processes are transverse (T1, spin-lattice relaxation) and longitudinal (T2, spin-spin relaxation). MRI creates contrast pictures using these relaxation mechanisms. A brighter contrast in T1-weighted images has been produced by a quicker longitudinal relaxation process, whereas a darker contrast in T2-weighted images has been produced by a faster transverse relaxation process. Despite being a widely common and effective imaging method, MRI has many drawbacks, including the following: First, because relaxing protons experience the same environment throughout their relaxation, it is not easy to distinguish between normal and sick cells in the same tissue.

Second, it can be used to acquire anatomical data; however, it does not provide comprehensive molecular data on the provided tissue [56, 57].

6.2.5 Nuclear Imaging

Because of their high sensitivity, nuclear imaging techniques like positron emission tomography and single photon computed tomography have had exceptional success. Radionuclide imaging is the most widely used imaging technique for molecular imaging. It has been discovered that there is a great deal of interest in developing novel nanomaterials for this use in clinics, particularly for the detection and treatment of cancer by studying illnesses at the molecular level using non-invasive imaging techniques, despite the fact that a wide variety of molecules have been used for radionuclide-based imaging techniques. According to the society of nuclear medicine, cellular as well as molecular activity in individuals and other living things can be imaged, defined, and interpreted using molecular imaging [58, 59].

6.2.6 Computer Tomography

Another imaging technique that has become highly popular in clinics is CT. It is largely utilized in clinics for tumor detection, monitoring, and screening. Most radionuclides used in CT are iodine-based. The biggest drawback of CT is that it employs X-rays to create contrast, which is dangerous for patients because of concerns about the development of cancer from prolonged exposure to this kind of radiation. But according to current studies in the sector, low-dose CT has helped to increase early tumor identification while also lowering the risk of subsequent cancers [60].

6.2.7 Role of Magnetic Resonance Imaging in MNPs

Magnetic Resonance Imaging (MRI) works upon the same principle as Nuclear Magnetic Resonance (NMR) and is an important technique for the assessment of information about tumor size, location, and its biological characteristics [61]. It is useful both preclinically and clinically [62–68]. All these assessments proved to be helpful in improving the therapeutic management and targeting the recognizable biological feature of the tumor [69]. MRI, because of its high spatial resolution and specificity for tissue, has an important place among all clinical imaging modalities. Magnetic nanoparticles (MNPs) are nanomaterials that are enthusiastically being used in several technological utilities such as MRI, site-specific drug delivery, and magnetic hyperthermia [70]. It has been considered one of the most useful diagnostic techniques for tumor imaging and gives excellent results [71, 72]. It is safer and more economical compared to other imaging techniques because it does not utilize radioactive agents and ionizing radiation. MRI is perfect for providing a detailed image of soft tissues; however, it is not capable of discriminating normal tissues from those with lesions due to its low intrinsic sensitivity [73, 74]. To overcome this shortcoming, contrast agents were introduced, which enhanced the contrast effect by accelerating the magnetic relaxation [73, 74]. Based upon their mechanism of action, these contrast agents are divided into three categories: T1 are those that act by precising the longitudinal duration of relaxation of surrounding water molecules and known as T1, second are those that act by precising the horizontal relaxation time of water protons and called T2, third is a new approach based on chemical exchange saturation transfer [75]. In clinical applications, T1 is used as it creates the highest resolution images and produces a bright signal [76]. Magnetic nanoparticles (MNPs) are characteristically fabricated NPs from pure metals, for example, iron, cobalt, nickel, or a mixture of the metals with polymers [77]. The MNPs have been extensively used in many biological applications, like cancer diagnosis and treatment [78], and MRI [79]. The main reason for their enhanced use is that they are capable of being magnetically manipulated by an exogenous magnetic field [80]. In addition to this, their magnetic properties can be further improved by the application of biocompatible coatings, which make them more acceptable for a definite target in the human body [81]. MNPs are used in MRI as they expand the image contrast of the beleaguered tissues. MNPs are concentrated in specific tissues and enhance relaxation of proton ions to enhance their visibility [82]. MNPs, which are used for imaging purposes, are nanocrystalline particles whose function can be further improved by the application of biocompatible coatings and ligands. It is important to measure the biomarkers and cells more efficiently and accurately for an early and fast disease investigation and prevention of cancer progression. Early diagnosis can prevent tumor growth and metastasis. More effective and accurate quantification of biomarkers and cells is very important for early disease identification and to prevent cancer spread and growth to other tissues [83].

6.2.8 Role of Raman Spectroscopy in MNPs

Surface enhanced Raman spectroscopy (SERS) is an effective analytical technique that, in combination with nanotechnology, is widely used for molecular sensing of environmental pollutants, biological markers, pesticides, drug products for bioimaging and monitoring of diagnostic reactions [84–92]. It is a unique fingerprinting spectroscopy where the data acquisition is non-destructive, and further, it shows single-molecule sensitivity [93]. Despite the fact that Raman spectroscopy encounters several challenges in analytical practice, the complex matrix of the real sample interferes with the spectrum. Its cost is high because the substrate it uses is noble metal and not recyclable. The concentration impact of the substrate is limited for the target. The technique for the fabrication of hot spots is complicated. All these problems can be overcome by manipulating the magnetic force through the use of magnetic nanoparticles. The analyte can be clearly and quickly separated from the sample matrix, which reduces interference in the result. Magnetic substrate is recyclable and thus can be reused [94]. They are helpful in the detection of low-abundance analytes due to their efficient and instantaneous concentrating abilities. The sequence of substrates due to magnetic attraction creates a constructive density of "hot spots" for consequent spectra enhancements. The most popular nanoparticles used in SARS are gold and silver [95]. Due to their size-dependent specifications, ease of surface modification, low toxicity to living cells, diversity of size and shape, and ease of functionalization by conjugation with antibodies and other tumor-targeting biomolecules, gold nanoparticles (AuNPs) have been used for imaging for a long time [96,

97]. Tumor location can be the reason for the variation of AuNPs that can be formed for SERS [98, 99]. AuNPs can reach malignant tissues by permeating tumor vessels during circulation. Second, it can also be transferred by attachment to specific ligands, which can target specific receptors [100–103]. Silver nanoparticles (AgNPs) are also being effectively used and are a thousand times more efficient than gold nanoparticles [104]. RS in vivo has been explored for cancer detection; although it is not a new technique, cancer detection is its new application [105]. It is suitable to be used with a fiber-optic probe for the detection of colorectal cancer [106]. As a result, intelligently planned MNP-plasmonic nanostructures and successful coupling with SERS might significantly enhance its sensing capability and expand its application sectors.

6.2.9 Computer Tomography and Magnetic Nanoparticle

Computed tomography (CT), an imaging technique using X-rays applicable to the whole body, has an X-ray source and a detector array, which are the two main components of this imaging technique. The source sends the X-rays into the patient's body, where some of the rays get absorbed. After passing through the patient, the rays are detected by the detector array, and the X-ray flux is recorded [107]. Barium suspension and iodinated small molecules are clinically approved contrast agents for CT. There has been a lot of excitement in recent years about the development and research of nanoparticles as CT contrast agents. These have a number of advantages over the small-molecule CT contrast agents now in use, including longer blood dwell times and a lack of need for nanoparticle-based contrast. Furthermore, there is a need for nanoparticle-based contrast agents in the case of renally impaired people and for those who show hypersensitivity to iodinated contrast [108]. In the past decade, for biomedical applications, many different types of nanoparticles have been used, such as emulsions and solid metal nanoparticles [109–115]. Gold nanoparticles, due to certain qualities such as a well-developed synthetic approach, are observed as biocompatible [116] and are under assessment for their beneficial pharmacological qualities in several clinical trials [117–120]. Hainfeld, in a 2006 report, reported a contrast agent in which he used a 1.9 nm-gold nanoparticle formulation [121]. He injected 2.7 g of gold/kg of rats with these nanoparticles. This high dose is not economical, but no toxicity of any kind was observed. Further, an experiment was conducted using long-circulating CT contrast agents. Gold nanoparticle with 10 nm core coated with PEG-2000 was used, and several parameters showed no toxicity signs except that the liver retention time was longer after three days of post-injection [122]. Another experiment was done using a 5 nm core, where the mice were given 493 mg gold/kg, and showed excellent results due to small nanoparticle diameter easily excreted by the kidney and no retention in the bloodstream [123]. The dose of 493 mg gold/kg led to a circulation half-life of 14.6 hours, and 100 HU contrast was seen in the aorta for quite a long time. Small-diameter nanoparticles provide exceptional contrast for the blood pool, so tumor neovasculature gets resolved [122]. Also, small-diameter nanoparticles are easily excreted by the kidney as compared to larger ones, which get entrapped in the blood stream and are too large to be excreted by the kidney.

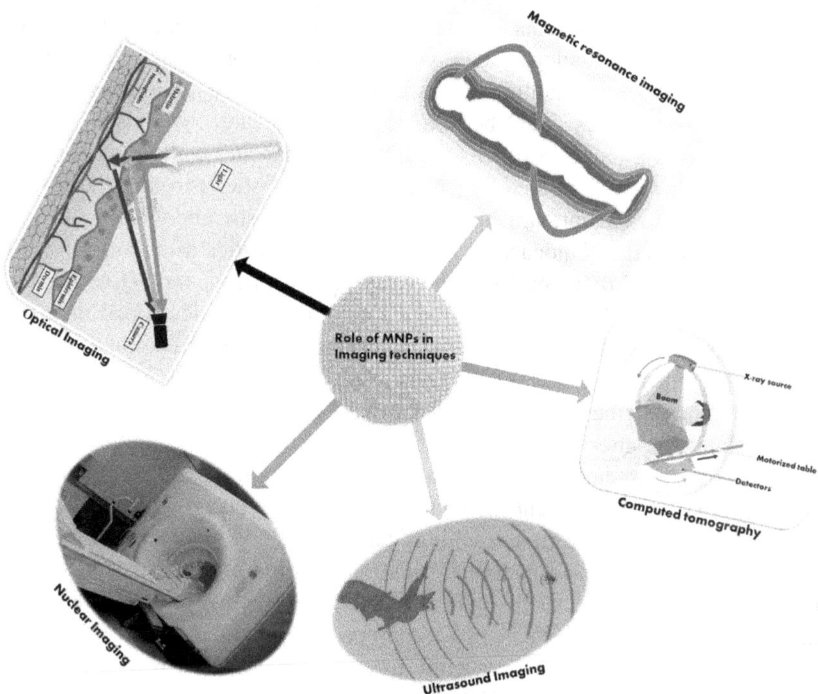

FIGURE 6.2 Role of MNPs in imaging technique.

By using different stabilizers many different gold nanoparticles were developed. Stabilizers used are dendrimers, thioctic acid, gum Arabic, PLGA (124), or heparin as shown in Figure 6.2 [124–133].

6.3 APPLICATION OF THERANOSTIC METALLIC NANOPARTICLES

The word theranostic, which combines the Greek terms for healing and diagnostic, allows for the simultaneous diagnosis, treatment, and non-forestall examination of a problem. Rapid advancements in bioscience and technology have created new treatment options for a number of diseases, moving from "standard remedy" to "customized remedy" approaches. In a news release from the organization Cardiovascular Diagnostic in August 1998, US consultant John Funkhouser used the term "theranostics" for the first time. The development of structures for the simultaneous switch of lively materials and imaging of cancerous regions has additionally garnered good-sized study interest. This new class of nanosystems is known as theranostics [134, 135]. In current years, with the improvement of nanotechnology, new horizons have been opened in treatment and prognosis of cancer. Application of nanotechnology to increase strategies and approaches for focused shipping of anticancer capsules and discounting of aspect results has been significantly studied in the closing decade [136]. Because of developments in nanotechnology, a wide range of nanomaterials, including magnetic nanoparticles, polymer conjugations, silica nanoparticles, micelles, dendrimers, and quantum dots, are being created at a rapid rate. With these nanoparticles, theranostics could be used to achieve greater synergistic outcomes with fewer adverse effects. In order to attain molecular and cellular performance, recent research has concentrated on enhancing theranostics through the conjugation, absorption, entrapment, and encapsulation of diagnostic and therapeutic agents in polymeric materials. Nanocarriers are used in theranostics as a platform for the incorporation of novel therapeutics such as capsules, peptides, and DNA, as well as imaging agents like radio metals. A unique method is being developed in which a single injection will enable imaging while also controlling the activation of theranostics for the treatment of most malignancies. Nanootheranostics is projected to be a successful nanomedicine in the future because of its distinctive features, including target specificity, prognosis/imaging, and the capacity to bundle the recovery medicament inside one nanoplatform. Electrical, optical, biological, and magnetic homes are all combined with different merchants for nanotheranostics. Nanotheranostics might be preferable to the first teachings on materials, which are inorganic and herbal. The most often used nanoplatforms in medicine are natural nanoparticles and specific biopolymers, such as dendrimers, lipoproteins, and liposome-based NPs. Nonmetallic nanoparticles can be appropriately characterize in the intended region, are biocompatible, and frequently form a covalent bond with the opposing agent. However, quantum dots (QDs) and superparamagnetic iron oxide nanoparticles (SPIONs) found in steel nanoparticles are widely studied and used effectively as imaging materials. By placing painkillers on or inside the nanoplatform, the aforementioned nanoparticles are successfully converted into nanotheranostics. A complicated group of diseases known as cancer are caused by DNA mutations that affect the processes of mobile growth and the mobile cycle.

6.3.1 Theranostic Agents

Nanotechnology advancements have enabled the creation of several nanoparticles that can be used as molecular imaging and/or diagnostic tools for the identification of cancer. Due to their capacity for chemotherapy, some nanocarriers are used in medication delivery and have received clinical approval. The nanoparticles' properties in this area open up the possibility of using them as theranostic agents.

6.3.2 Metallic Nanoparticles

Metallic nanoparticles have recently attracted significant interest in the field of most cancer treatments due to their vast physicochemical features, such as plasmonic resonance, fluorescence, and catalytic activity. Due to their unique characteristics of excessive ground region to amount ratio, metallic nanoparticles also permit excessive loading potential of therapeutic compounds, focused on ligands, and as imaging agents. Additionally, the majority of metallic nanoparticles are favored by the strict requirements in drug delivery methods for the most effective cancer, neurological, antibacterial, antifungal treatment, including reducing systemic toxicity and concentrating just on the diseased tissues [137–162].

6.3.3 Application of Theranostic Nanomedicine

Numerous biological difficulties, including cancer, respiratory disorders, cardiovascular diseases, kidney disorders, and neurodegenerative diseases, are treated with theranostic nanomedicine. This chapter reviews recent advancements in theranostic nanomedicine, which is used to treat serious illnesses.

6.3.4 Cancer

One of the deadliest and most difficult clinical issues is cancer. The lack of affordable tablets and capacity restoration vendors for the most effective cancer treatment results in increased mortality costs worldwide every year. Metastasis and tumors that are multidrug-resistant make treating most malignancies difficult. The likelihood of successfully treating most malignancies will increase with the early identification of most tumors. However, certain cancer types are difficult to detect in their early stages. Most cancers can now be addressed and visualized in great detail because of recent advancements in nanotheranostics. For the most effective cancer treatment, many nanotheranostic materials have been developed that can diagnose and deliver healing elements to malignant regions. Additionally, it keeps track of medicine's effectiveness. The majority of cancer treatments involve magnetic nanoparticles, which are functionalized with anticancer drugs and/or antibodies and stabilized by way of a polymer coating. Here, a magnetic nanoparticle enables the imaging of malignant tissues, while coated polymers improve the nanoparticles' biocompatibility. To improve the target selectivity, nanoparticles are functionalized with antibodies, and restoration sellers are tagged to boost the drug accumulation on the targeted spot.

6.3.5 Photothermal and Molecular Imaging

Molecular imaging is a technique for identifying, describing, and measuring natural processes at the cellular and subcellular levels within intact live animals. Nuclear therapy molecular imaging (PET, SPECT) has a high sensitivity but some resolution, rapid half-life of tracers, high instrumentation cost, and complexity restrictions. Endoscopic operations can make use of optical imaging since it is very sensitive, reasonably priced, and capable of producing incredibly high resolution. Contrary to unique imaging modalities like CT and MRI, clinical programs are constrained by their shallow depth of penetration, making it difficult to image deep malignancies. By reducing photon absorption through tissue components, NIR wavelengths (800–1000 nm) can be employed to improve tissue penetration in optical imaging, enabling in vivo optical imaging programs [163, 164]. With the discovery of creative fluorescent merchants, interest in optical imaging programs for maximum cancer diagnosis has expanded. Targeted fluorescence contrast chemicals can be used as an adjuvant to help identify the line dividing the tumor from healthy tissue and to guide surgical resection.

6.3.7 Ultrasonic Image-Guided Therapy

Through the EPR effect, nanobubble-shaped ultrasound assessment shops can be activated for controlled shipping and imaging reasons and can preferentially extravasate into tumour tissue. Through cavitation, nanobubbles can enhance molecule permeability and hence raise the cytotoxicity of delivered products [165]. According to Gao et al. method, perfluorocarbon nanodroplets stabilized with the aid of biodegradable block-copolymer micelles were used to create DOX-containing nanoemulsions [166]. This group demonstrated the use of real-time ultrasonography imaging together with the capacity to release encapsulated capsules through a cavitation effect that is distinctive to the tumor location and may improve the uptake of medications tailored to the tumor [167]. In a subsequent work from the same team, Rapoport et al. created a second-generation technique using perfluoro-15-crown-5-ether (PFCE) nanodroplets loaded with paclitaxel. In consideration of multimodal monitoring of shipping and biodistribution using ultrasonography and MRI in mice, these nanoagents demonstrated all ultrasound and fluorine 19F MR assessment properties [168]. The application of this method to the therapy and imaging of the majority of malignancies still faces several challenging issues, particularly those related to the nonuniform vascularization of the tumor and the inhomogeneity of nanodroplet distribution there. This problem, which may potentially have an effect on various forms of nanotherapy, appears to be responsible for the emergence of drug resistance in some tumor-related regions [169].

6.4 CLINICAL APPLICATION OF METALLIC NANOPARTICLES

Many innovative applications for inorganic nanomaterials fall short of regulatory approval while having excellent clinical outcomes because of issues with endpoint design, clinical comparator selection, and clinical data analysis methodologies that aren't up to date with governmental standards. For instance, the FDA or EMA had granted clinical approval for a number of SPION-based MRI contrast agents, including Resovist®/Cliavist® and Feridex I.V.®/Endorem®. They were later discontinued or withdrawn inside the United States and/or the European Union due to their low sensitivity since they were less competitive in the medical and commercial markets than gadolinium-based contrast agents [170].

There are currently 18 new nanocarriers in the clinical testing phase in 2016, according to recent publications. Out of the 18 liposomes, 12 nanocarriers were chosen, and 17 of them are suggested for the treatment of cancer (15 for therapy and 2 for imaging). The only non-cancer indication is mRNA-1944, which is made up of two mRNAs that code for the heavy and light chains of an anti-Chikungunya antibody and is generated in lipid nanoparticles to ward off the infection of the Chikungunya virus. It should be noted that there are more clinical trials researching the use of nanocarriers to deliver mRNA, but since the majority of these trials used intradermal or other delivery methods, they won't be covered here. We recommend reading a current review on mRNA delivery methods, which has a strong emphasis on the most recent clinical trials and delivery methods [171].

There are currently 563 nanomedicines in various phases of clinical development and 100 nanomedicines on the market. The majority of these nanomedicines are in phase I (33%), phase II (22%) and clinical trials (21%), with the majority of attention going to cancer (53%), and infection (14%) [172]. As a result, clinical trials and metallic nanocarriers are interesting issues that are addressed and reviewed in this study [173]. We provide an update on the level of inorganic nanoparticles' usage as therapeutic, defensive, and research agents, as well as their advancement in clinically useful applications.

6.4.1 Therapeutic Agents

In the fields of therapeutic agents like cancer treatment, iron replacement therapy, antimicrobial agents, bone substitutes, and antidotes for heavy metal toxicity, they have great clinical applications and translations. Various forms of synthesized inorganic nanoparticles exhibit exceptional performance.

6.4.2 Antitumor Therapy

Over the past few decades, significant advancements have been made in the study of nanotechnology, the quick development of medications, and nano-drugs that fight cancer. More than 70 cancer nanomedicines are now being studied in clinical trials, and there are at least 16 tumor nanomedicines on the market [174]. The principal inorganic nanoparticle-based nanomedicines for cancer treatment that have received clinical approval or are now undergoing clinical studies are iron, gold, and hafnium.

6.4.3 Iron-Based Inorganic Nanoparticles

Iron-based nanoparticles (IONPs) have greatly influenced medicine and bioimaging over the past 80 years. They are used extensively in bioimaging [175] and localized magnetic heating and targeting [176, 177]. Localized heating is particularly advantageous for tumor ablation and therapies using hyperthermia [178, 179]. Several superparamagnetic iron oxide nanoparticles (SPIONs) have shown promise in preclinical and clinical studies as hyperthermia anticancer drugs. The use of Nanotherm™ therapy in clinical trials for glioblastoma tumors increased overall survival by up to 12 months [180]. However, metastatic lesions—which are challenging to individually target with local treatments—are what lead to the majority of cancer-related deaths. Additionally, sufficiently high nanomaterial concentrations in tumor areas are necessary for thermal ablation or hyperthermia (Table 6.1) [181].

IONPs might also be used to provide therapeutic doses to target areas, augment the affected area with targeted magnetic field modulation, and deliver targeted combination therapy with external stimulation, all of which would boost anticancer treatment. This is because there is a great opportunity to improve the resistance of the material and to offer a magnetic target [182–201].

6.4.4 Gold-Based Inorganic Nanoparticles

There is a lot of literature describing novel and unusual medicinal formulations based on AuNPs; however, only a few gold nanoconstructs have reached the stage of

TABLE 6.1
Biomedical Application of Metallic Nanoparticles (MNPs)

MNPs	Mechanism	Disease	Reference
Cancer Disease Targeting			
AuNPs	Chemotherapy to kill MCF7 breast cancer cells while sparing healthy cells is a good option for folate-based tumor targeting using CurAu-PVP NCs.	Breast cancer MCF7 cell line	[182]
	Oxidative stress caused by gold nanoparticles increased paclitaxel's ability to kill cancer cells by two times. These discoveries gave clear structural guidelines for creating biocompatible nanocarriers for future nanomedicine.	Lung cancer cell line A549	[183]
AgNPs	Tamarindus indica-derived Silver Nanoparticles were found to inhibit the growth of MCF-7 breast cancer cells in a dose-dependent manner using the MTT test, and their anticancer potential was revealed utilizing live and dead assays (Ao/EtBr), ROS, and Rho123 assays.	MCF-7 human breast cancer cell line	[184]
FeNPs	In this investigation, we report that the DDS's tumor cell-targeting and pH-responsive releasing capabilities are confirmed by iron nanoparticles utilized in U-87 MG malignant glioma cells.	U-87 MG malignant glioma cells	[185]
ZnNPs	It has been discovered that PBA-ZnO-Q lessens tumor-related toxicity in the liver, kidney, and spleen. It is believed that the combination of cytotoxic effects of quercetin and ZnO on cancer cells is what gives the nanohybrid its cytotoxic potential. Overall, the data demonstrated the unique nanohybrid PBA-ZnO-chemotherapeutic Q's potential, which might be taken into consideration for clinical cancer treatment.	Human breast cancer cells (MCF-7)	[186]
Microbial Disease Targeting			
AgNPs	In this research we found plant based silver nanoparticles not the metabolites of the aqueous extract that were present on their surface, was responsible for the antibacterial activity. The plant extract modulated cytotoxic and immunological responses as well as being crucial in the creation of stable AgNPs.	Skin and soft tissue infectious agents	[187]
CuNPs	Green synthesize CuNPs were using RAFE and GJLE that shows the antibacterial activity against *S. aureus* and *E. coli*.	Antibacterial activity	[188]
ZnNPs	aqueous extract of the macroalgae *Ulva fasciata* based zinc nanoparticles was synthesize and we recommend the usage of our product to potentially control the growth of pathogenic bacteria (gram-positives as well as gram-negative bacteria) and prevent the possible environmental pollution with the dyes and heavy metals.	Marine Macroalgae, *U. fasciata* Delile, characterization, antibacterial activity, photocatalysis, and tanning wastewater treatment	[189]
AuNPs	*Panchagavya*-based AuNPs were discovered to have strong antibacterial action against gram-negative bacteria and moderate antibacterial activity against gram-positive bacteria.	Antibacterial activity	[190]
FeNPs	*S. aureus* and *B. subtilis* were more resistant to *Purpureocillium lilacinum*-based iron nanoparticles than gram-negative bacteria (*E. coli* and *P. aeruginosa*).	Antibacterial activity	[191]

TABLE 6.1 *(Continued)*
Biomedical Application of Metallic Nanoparticles (MNPs)

MNPs	Mechanism	Disease	Reference
Neurological Disease Targeting			
AuNPs	Treatment with LCM-GNP mainly in the cerebral cortex suggests that this area of the brain may have functions related to the control of neuronal activity.	Antiepileptic activity (seizures)	[192]
AuNPs	Animal behavior, oxidative stress, neurotransmitter levels, and the cholinergic system may all be modulated by Au@TPM. This method might be connected to the Nrf2/Keap1 signalling pathway's mediating role.	Alzheimer's disease	[193]
AuNPs	The goal of this study was to develop a sensitive and selective immunosensor for the quick identification of the α-syncline protein in real samples for early stage Parkinson's disease diagnosis using an AuNP-Gr bioconjugated-specific antibody.	Parkinson's disease	[194]
AgNPs	PS composites with thorny Au particles prepared via dopamine reduction proved to be the most efficient substrate for SERS measurement purposes, and this substrate was used and tested for specific and selective detection of protein A, a marker of *S. aureus* (in the model sample and in a real joint knee fluid sample), as well as for the detection of Tau protein.	Alzheimer's disease	[195]
AgNPs	Following KA (Kainic acid) treatment, THSN (Tualang honey silver nanoparticles) improved seizures, locomotor activity, and memory performance. The ability of THSN to enhance the latency to seizure and the frequency of line crossings, as well as the higher recognition index, indicate this.	Antiepileptic activity (seizures)	[196]
Anti-viral Disease Targeting			
AgNPs	The potential use of silver-based nanoparticles, which are about 10 nm in size, as an antiviral treatment is supported by the fact that they reduce MPV (monkeypox virus) infection in vitro.	Monkeypox virus Plaque	[197]
MNPs	Anti-viral nanotheranostic compounds made of metal. Metal-based NPs may serve either as carriers for the sustained and precise administration of active antiviral compounds or as diagnostic tools for the quick and accurate detection of viral infections.	COVID-19	[198]
AuNPs	Sulfated ligands and other active compounds that target several stages of the HIV life cycle can be found on the gold nanoplatform, which makes gold nanoparticles an interesting scaffold for the creation of novel, multifunctional anti-HIV systems.	Anti-HIV agents	[199]
ZnNPs	In the current study, acyclovir (ACV) was analyzed and the results were applied for the determination of ACV in the analysis of urine and pharmaceutical dosage form. Calcium doped zinc oxide (Ca-ZnO) nanoparticles were developed as a novel electroanalytical sensing tool for the detection of antiviral drug.	Anti-viral disease	[200]
Ag, Cu, ZnNPs	The antiviral activity of nanosilver particles as a coating is the strongest of the three nanomaterials tested, while copper oxide has modest activity and zinc oxide does not appear to significantly diminish virus infectivity. As a result, nanosilver and copper oxide have the potential to act as antiviral coatings on solid surfaces and filter media, reducing transmission and superspreading events while also providing crucial information for present and future pandemic mitigation efforts.	SARS-CoV-2	[201]

clinical trials/testing. The majority of clinical research in materials is science-driven, with the goal being to build multifunctional NPs with new properties rather than focus on enhancing features needed in the clinic, which results in this discrepancy between preclinical and clinical investigations. Since, for instance, both inorganic and organic NPs that have reached the clinic tend to be simple and only include a few elements, it is simpler to predict their in vivo behavior and scale up their synthesis [202]. A PEGylated-AuNP called CYT-6091 contains recombinant tumor necrosis factor alpha (TNF), a pro-inflammatory cytokine generated by activated leukocytes [203]. Due to its capacity to regulate the immune response and exert an anti-tumor effect, TNF has been thoroughly investigated in both preclinical and clinical cancer studies [204]. The right therapeutic effectiveness may be achieved with local delivery of TNF without experiencing any systemic adverse effects. Without the need for additional reagents, AuNP and TNF can be connected directly via thiol residues. Because PEGylation reduces particle absorption and clearance by the reticuloendothelial system, TNF-AuNP circulation duration is greatly lengthened. TNF-AuNP also accumulates in tumors more quickly than free TNF and TNF-AuNP that hasn't been PEGylated [205]. Patients with advanced solid tumors received intravenous injections of 27-nm AuNP conjugated with TNF and thiolated PEG in the first clinical trial (NCT00356980), which began in 2006 [206]. Two injections separated by 14 days make up one cycle for each patient. All of the doses—which varied from 50 to 600 g/m2—were accepted without any problems. The maximum acceptable dose of TNF could be increased when given as a nanoformulation without suffering any negative side effects (in contrast to treatment with parental-free protein). Electron microscopy was used to confirm the existence of gold nanoparticles in tumor tissue samples twenty-four hours after treatment. The tissue distribution and toxicity of intravenously given CYT-6091 were evaluated in a phase 0 clinical investigation (NCT00436410). The trial's participants were receiving surgery for either primary or metastatic cancer when it began shortly after in 2006 and is now ongoing. The National Cancer Institute and Cytimmune, the biotechnology business that created CYT-6091, reportedly secured a clinical trial agreement in July 2020 to carry out a phase II clinical trial of the medication. The commencement date for this study has not yet been determined.

6.4.5 Iron Supplements

A lack of iron, especially when combined with anemia (iron deficiency anemia) (IDA), is a serious issue for public health. Oral and intravenous conventional clinical iron replacement therapy include the primary reasons why iron compositions are not appropriate: tolerability and safety concerns, respectively [207]. hence, an active field of research into treatments IDA is an investigation into manufacturer iron nanoformulations. Considering the pharmaceutical industry's progress, numerous technologies, nanotechnology, and IONP-based goods FDA- or EMA-approved for use as an iron substitute in clinical trials are still overwhelmingly used to treat anemia with respect to surface modification [208], controllable size, and significant biocompatibility and biodegradability [209]. One cutting-edge application of IONPs is seen in Feraheme®/Rienso®, which uses SPIONs coated with polyglucose sorbitol carboxymethylether as an iron supplement [210]. Feraheme®, an alternative to conventional iron treatments, showed clear superiority, including enhanced pharmacokinetic features, simpler delivery techniques, and favorable tolerance in both healthy people and those with disabilities, even when bolus injections are used [211]. Following clinical trials, Feraheme® was shown to be safe and effective for anemic patients with CKD, increasing iron reserves and hemoglobin response without major adverse effects [212]. Notably, Feraheme® made managing iron deficiency anemia in individuals with chronic renal failure easier in a Phase III trial by delivering up to 1 g of iron with two intravenous injections rather than roughly five to eight dosages of slow infusion [213]. Additionally, the ability to properly provide high doses of Feraheme® nursing time as a result of several injections. Additionally, Feraheme® could be used for routine phlebotomy, resulting in fewer venipunctures being necessary and the insertion of intravenous catheters to protect veins for potential hemodialysis access.

6.4.6 Antibacterial Agents

Another therapeutic area that is being thoroughly investigated for the advancement of inorganic nanomedicine is antibacterial therapy. Numerous metal ions, particularly silver, have the potential to be antibacterial substances. AgNPs have been added to a variety of cleaning solutions because they easily penetrate bacteria and induce toxicity through ion release [214]. Utilizing the antimicrobial properties of AgNPs to create synergistic effects when combined with other treatments is a significant application [215]. There are currently 5 clinical trials using nanosilver fluoride (AgNPs mixed with fluor varnish) and 9 clinical trials using silver nanoparticles (AgNPs) as an antibacterial treatment for various ailments. Additionally, a phase III trial is being conducted to compare the efficiency of Purell, the industry-standard antibacterial hand gel, and SilvaSorb, a type of antimicrobial silver nanoparticle gel, by examining the surrogate biomarker *S. Marcescens* survival rate on the patients' hands after a single application of each gel. Although these AgNPs have received increased attention, further research is still needed to fully understand their biosafety [216]. AgNPs have a long history of usage in medicine, particularly for antimicrobial purposes. However, prolonged contact with Ag-based nanocarriers may cause the permanent coloring of the skin or eyes that is thought to be caused by silver deposits in vivo [217, 218]. Therefore, research into the underlying biological impact, biodistribution, and pharmacokinetics of Ag-based nanocarriers is essential before they are employed in therapeutic situations.

6.4.7 Antidote for Heavy Metal Poisoning

Radiocesium and thallium toxicity are two examples of the heavy metal burden that persists globally. With a radiation half-life of 30 years, 137Cs is most likely the radioactive cesium isotope that poses the greatest concern to human health and the environment. Clinical symptoms associated with 137Cs exposure are most likely caused by ionizing radiation. In addition, long-term exposure to 137Cs causes bone marrow suppression, bleeding, infection, and mortality in humans [219]. Acute gastrointestinal symptoms of thallium toxicity were first noticed, and thereafter came neurological problems. Due to the rarity of thallium poisoning, the clinical licensing of Radiogardase® (Prussian blue) as an antitoxin for heavy metal poisoning has been hurried. A "dirty bomb" that may cause widespread 137Cs poisoning could be the reason. Prussian blue, which is contained in the crystal lattice of Prussian blue, will attach to released Cs or Th in the gastrointestinal system and exchange them for K [220]. Prussian blue also demonstrates increased safety and toleration [221]. The risk of hyperkalemia should be considered when maintaining potassium levels during Prussian blue therapy.

6.4.8 Composite Inorganic Nanoparticles

The performances of a large variety of inorganic metals work in concert with one another to create synergistic effects. As a result, the overall achievement of synthetic inorganic nanocarriers is exceptional compared to that of the authentic component material when definite requirements are met. Presently, five items made from composite metallic nanocarriers are undergoing clinical studies. The great majority of these items have metal components that are antibacterial in nature, as shown in Figure 6.3 [222].

6.5 TARGETING ELEMENTS OF THERAPEUTIC METALLIC NANOPARTICLES

Over the course of the past century, scientists have been captivated by metallic nanoparticles, which are now finding widespread application in the field of

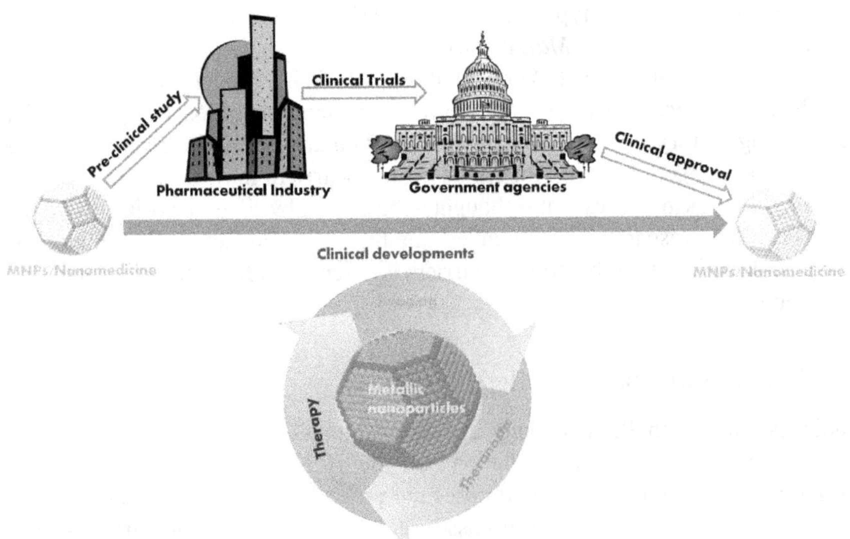

FIGURE 6.3 Metallic nanoparticles in clinical applications.

biomedicine. They are of particular interest because of the enormous impact that nanotechnology could have on various fields [223–225]. The significance of nanotechnology in many scientific domains is dependent on the manufacture and modification of NPs, more especially metals, which requires a significant alteration of the characteristics of those nanoparticles [226]. For the treatment of a wide variety of infections, medicine has traditionally made use of noble metals and the compounds derived from them as medicinal agents since ancient times. Recent advances in nanobiotechnology have led to the creation of numerous nanomaterials with a variety of uses, including the treatment of cancer, the labeling of cells, hyperthermia, target-specific and immunoassays, tissue regeneration, magnetic resonance imaging (MRI), and the identification of genetic disorders, angiogenesis, hereditary condition diagnosis, tumor detection, photothermal therapy, photoimaging, and drug delivery [227, 228], antimicrobial, anti-inflammatory, and anticancer applications [229]. All of the aspects of metallic nanoparticles that are intended to be targeted for various diseases, conditions, detection/diagnosis, and microbial infections are discussed in more detail in the following in a more specific fashion.

6.5.1 Tumor Targeted by Magnetic Hyperthermia Technique

Patients with recurrent glioblastoma have seen a 7-month improvement in survival after undergoing magnetic hyperthermia. Treatment of mice with an alternating magnetic field of 202 kHz and 27 mT for 27 sessions of 30 minutes each and administration of 500 to 700 g of iron from megnetosomes with poly-L-lysine (M-PLL) to intracranial U87-Luc tumors of 1.5 mm^3 indicated the improved anticancer activity of M-PLL. In the first magnetic session, we set the temperature to that of a standard hyperthermia treatment, which is 42 degrees Celsius. Bioluminescence from live glioblastoma cells completely vanished 68 days after tumor cell implantation in all treated animals (D68). At day 350, all of these mice were still active and healthy. Tumor cells were not detected in brain tissue histology, indicating complete recovery [230].

6.5.2 Role of Gold Nanoparticles in Tumor Targeting

In a recent study, the cytotoxic effects of gold nanoparticles on human colorectal carcinoma (HCT-116) and breast cancer (Michigan Cancer Foundation-7) cells were assessed (AuNPs). An anticancer study found that cancer cells treated with AuNPs had much fewer cells than control cells. The AuNPs-hibiscus sample, with its microscopic size and uniform dispersion, outperformed the AuNPs-curcumin control [231].

In the current work, gold nanoparticles were created by using cetyltrimethylammonium bromide (CTAB) and were then loaded with fluorouracil (5-FU). The ex vivo permeability of these formulations was examined in the dorsal skin of mice after adding 5-FU/CTAB-GNPs to gel and cream bases. A431 tumor-bearing mice were used to test the identical formulations' antitumor effectiveness in vivo. In comparison to free 5-FU gel and cream formulations, the permeability of 5-FU GNP gel and cream through mouse skin was approximately two times greater. In vivo tests using a mouse model in which A431 skin cancer cells were injected beneath the skin reduced tumor volume by 6.8 and 18.4 times, respectively, compared to the untreated control. These findings support the hypothesis that topical 5-FU/CTAB-GNPs may improve therapeutic effectiveness against skin cancer [232].

In this research we developed a polyethylene glycol (PEG) functionalized AuNPs loaded with docetaxel (Dtx) for use in prostate cancer targeting. Prostate cancer cell lines were tested to determine the efficacy of docetaxel-encapsulated AuNPs in combating the disease (PC3). When treated with a 40 M concentration of the drug-encapsulated nanoformulations for 24, 48, and 72 hours, cell viability decreased to about 40%. Optical microscopy was used to detect damage to prostate cancer cells after exposure to Dtx-encapsulated AuNPs. Due to its potent cytotoxic action against prostate cancer cell lines, the proposed nanoformulation holds promise as a tool for prostate cancer medication administration [233].

6.5.3 Role of Gold Nanoparticles in MDR Bacteria

In this research, we investigate that carbapenem-loaded AuNPs have been shown to be a viable nanosize delivery vehicle for enhancing therapeutic effectiveness and eradicating carbapenem-resistant bacteria. The MIC of AuNPs loaded showed significant effectiveness compared to alone antibiotics (Imipenem (Ipm) and Meropenem) [234].

6.5.4 CeO_2 and PbO MNPs for Colon Cancer Targeting

In this study, MNPs were synthesized using a green synthesis method that produced lead oxide (PbO) and cerium oxide (CeO_2) from an aqueous fruit extract of Prosopis fracta. PbO and CeO2 nanoparticles were tested for their ability to kill HT-29 cell lines. The cytotoxic results showed that at greater dosages, CeO_2 nanoparticles were less dangerous than PbO nanoparticles. Since CeO_2 nanoparticles offer both industrial and biomedical potential, they might be regarded as an appropriate material for biological and biomedical applications [235].

6.5.5 Role of Iron Oxide Nanoparticles for Antimicrobial and Anticancer Activity

In the current study, Fe_2O_3-NPs are synthesized for their antibacterial and anticancer properties. Using an agar diffusion method, we looked for antibacterial activity against *B. subtilis*, *E. coli*, *C. albicans*, *P. aeruginosa*, and *S. aureus*. In human cancer cell lines HeLa and MCF7, the anticancer efficacy of the iron NPs was assessed. Zones of inhibition of *S. aureus* and *B. subtilis*, respectively, measured 21 and 22 mm, according to antimicrobial tests. Similar to this, the Candida albicans fungus under investigation is successfully inhibited by Fe2O3-NPs, with a zone of inhibition of 24 mm. When the 2,5-diphenyl-2H-tetrazolium bromide (MTT) assay was used to test the NPs against both cell lines, it was shown that they had significant anticancer activity [236].

6.5.6 Role of Inorganic Nanoparticles for Surface Properties and Antibacterial Activity

Titanate nanotubes with nano-heterostructures were created in this study using a microwave-assisted The development of AgNPs, Ag_2O, CeO_2, and TiO_2 (anatase), as well as the intercalation of Ag^+ and Ce^{4+} into the interlayer gaps of the nanotubes, reveal a complex structure revealed by the hydrothermal technique and ion exchange processes. Testing was done with two different species of bacteria, and the resulting nano-heterostructure exhibits good antibacterial activity, minimum cytotoxicity, and high cell adhesion, making it a viable material for future health applications (gram-positive and gram-negative) [237].

6.5.7 Antibacterial Bacterial and Cancer Applications of AgNPs

In this research, AgNPs were synthesized by *Carica papaya* latex extract, which a was highly stable silver nanoparticle in the current investigation (CP AgNPs). Human pathogens such as *Enterococcus faecalis*, *Bacillus subtilis*, *Vibrio cholerae*, *Escherichia coli*, *Proteus mirabilis*, and *Klebsiella pneumoniae* are easily defeated by the CP AgNPs' potent antibacterial action. However, the chemotherapeutic potential of AgNPs was tested in in vitro anticancer research using a human breast carcinoma cell line (MCF-7). The IC50 value for the anticancer impact of synthetic CP AgNPs was determined to be 19.88 μg/ml, indicating a dose-dependent response. These findings strongly suggested the potential use of biogenic CP AgNPs as a nanodrug formulation for the treatment of bacterial infections and chemotherapeutics for breast cancer [238].

6.5.8 Green Synthesis AgNPs for Anticandidal Activity

In this study, collected leaves from the *Syngonium podophyllum* plant were used to create AgNPs. AgNPs made from *S. podophyllum* were found to have substantial antifungal effects against *Candida albicans*, and the broth dilution method was used to determine their minimum inhibitory concentration values. The antifungal effectiveness was confirmed by measurements of the diameter of the zone of inhibition created by the disc diffusion technique [239].

6.5.9 Plant-Based Gold Nanoparticles for Colon Cancer and Inflammation Manifestation

In the current investigation, *Terminalia bellirica* was used for the synthesis of gold nanoparticles. The zebrafish model for colorectal cancer cell line (HT29) produced by synthesized gold nanoparticles demonstrates their anticancer and antiinflammatory potential. To further understand the underlying biological mechanism of anticancer activity, quantitative real-time polymerase chain reaction (qRT-PCR) was used to evaluate the expression of cancer markers (qPCR). Alterations in apoptosis and the activity of antioxidant enzymes were also studied. All of the cancer

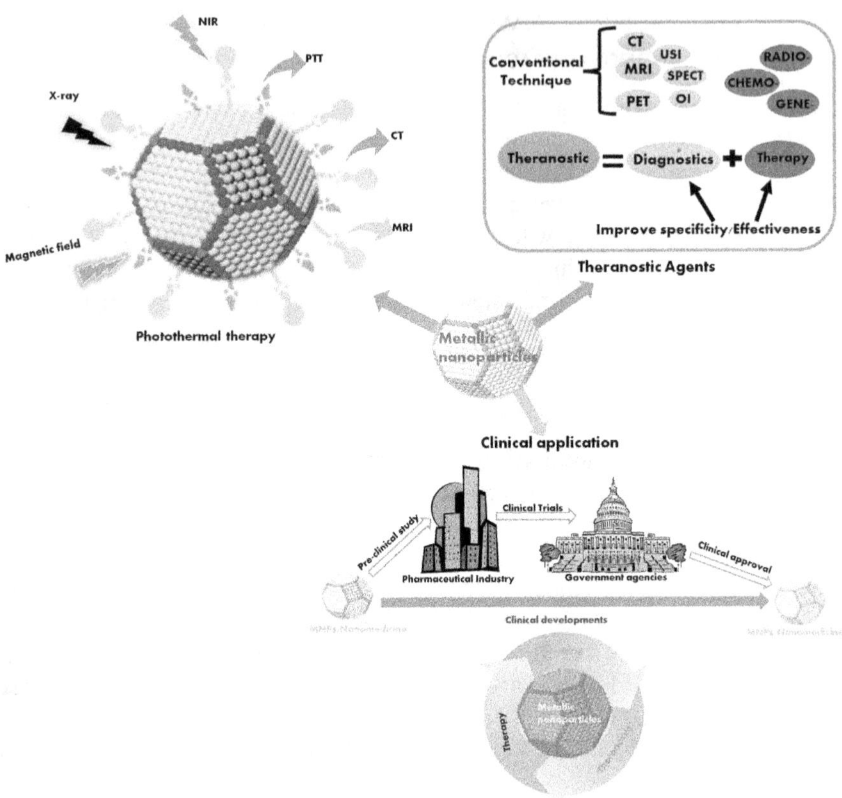

FIGURE 6.4 Metallic nanoparticles in biomedical applications.

indicators evaluated showed a significant decrease in expression after being exposed to the synthesized nanoparticles.

Treatment with nanoparticles was associated with significant alterations in the activity of antioxidative enzymes, an encouraging marker of apoptosis. This was validated by measuring the expression of TNF-α and iNOS (induced nitric oxide synthase) and histological inspection in a TNBS-induced zebrafish model showing anti-inflammatory efficacy. The expression of inflammatory markers was drastically reduced after treatment with nanoparticles, as shown in Figure 6.4 [240].

6.6 CONCLUSIONS

Several nanomaterials have developed during the past several years as promising tools in theranostic applications, aiding efficient diagnostic imaging and therapy. This is due to their inherent molecular properties and multifunctionality. It is essential for both diagnosis and treatment that theranostic MNPs are targeted effectively. Theranostic MNPs made chemically, however, have issues with biocompatibility, effectiveness in in vivo tumor targeting, and cost. A thorough overview of the trends in MNPs' theranostic applications has been discussed. Modern diagnostic imaging techniques that use metallic nanoparticles include, for instance, optical imaging, CT, MRI, Raman spectroscopy, ultrasonic imaging, and nuclear imaging (MNPs). Along with clinical trials, their therapeutic components are also mentioned, such as anticancer medications and targeted agents. The difficulties and potential uses oftheranostic MNPs in therapeutic and diagnostic procedures are discussed. These results in enhanced therapeutic efficacy and fewer medication adverse effects. This chapter covers the mechanics of drug transport as well as a brief overview of nanotechnological methods for the efficient administration of theranostic drugs. It also covers the pathophysiologies and etiologies of several brain-related disorders.

CONFLICT OF INTEREST

The authors declare that there is no conflict of interest.

ACKNOWLEDGMENTS

The authors are thankful to their respective organizations for their cooperation and support. MSA is thankful to SGT University for their cooperation and support. JRA gratefully acknowledges the HoD, Applied Science and Director Sir, DCE for cooperation and financial support.

REFERENCES

1. Ferlay J, Colombet M, Soerjomataram I, Parkin DM, Piñeros M, Znaor A, Bray F. Cancer statistics for the year 2020: An overview. International Journal of Cancer. 2021;149(4):778–789.
2. Aghebati-Maleki A, Dolati S, Ahmadi M, Baghbanzhadeh A, Asadi M, Fotouhi A, Yousefi M, Aghebati-Maleki L. Nanoparticles and cancer therapy: Perspectives for application of nanoparticles in the treatment of cancers. Journal of Cellular Physiology. 2020;235(3):1962–1972.
3. Ali MA, Ahmed T, Wu W, Hossain A, Hafeez R, Islam Masum MM, Wang Y, An Q, Sun G, Li B. Advancements in plant and microbe-based synthesis of metallic nanoparticles and their antimicrobial activity against plant pathogens. Nanomaterials. 2020;10(6):1146.
4. Evans ER, Bugga P, Asthana V, Drezek R. Metallic nanoparticles for cancer immunotherapy. Materials Today. 2018;21(6):673–685.
5. Chandra H, Kumari P, Bontempi E, Yadav S. Medicinal plants: Treasure trove for green synthesis of metallic nanoparticles and their biomedical applications. Biocatalysis and Agricultural Biotechnology. 2020;24:101518.
6. Cordani M, Somoza A. Targeting autophagy using metallic nanoparticles: A promising strategy for cancer treatment. Cellular and Molecular Life Sciences. 2019;76(7):1215–1242.
7. Farzin A, Etesami SA, Quint J, Memic A, Tamayol A. Magnetic nanoparticles in cancer therapy and diagnosis. Advanced Healthcare Materials. 2020;9(9):1901058.
8. Hanan NA, Chiu HI, Ramachandran MR, Tung WH, Mohamad Zain NN, Yahaya N, Lim V. Cytotoxicity of plant-mediated synthesis of metallic nanoparticles: A systematic review. International Journal of Molecular Sciences. 2018;19(6):1725.

9. Madamsetty VS, Mukherjee A, Mukherjee S. Recent trends of the bio-inspired nanoparticles in cancer theranostics. Frontiers in Pharmacology. 2019;10:1264.
10. Rai M, Ingle AP, Birla S, Yadav A, Santos CA. Strategic role of selected noble metal nanoparticles in medicine. Critical Reviews in Microbiology. 2016;42(5):696–719.
11. Raj S, Khurana S, Choudhari R, Kesari KK, Kamal MA, Garg N, Ruokolainen J, Das BC, Kumar D. Specific targeting cancer cells with nanoparticles and drug delivery in cancer therapy. In Seminars in Cancer Biology. Elsevier. 2021.
12. Schabath MB, Cote ML. Cancer progress and priorities: Lung cancer. Cancer Epidemiology, Biomarkers & Prevention. 2019;28(10):1563–1579.
13. Sharma A, Goyal AK, Rath G. Recent advances in metal nanoparticles in cancer therapy. Journal of Drug Targeting. 2018;26(8):617–632.
14. Sharma H, Mishra PK, Talegaonkar S, Vaidya B. Metal nanoparticles: A theranostic nanotool against cancer. Drug Discovery Today. 2015;20(9):1143–1151.
15. Singh P, Pandit S, Mokkapati VR, Garg A, Ravikumar V, Mijakovic I. Gold nanoparticles in diagnostics and therapeutics for human cancer. International Journal of Molecular Sciences. 2018;19(7):1979.
16. Sztandera K, Gorzkiewicz M, Klajnert-Maculewicz B. Gold nanoparticles in cancer treatment. Molecular Pharmaceutics. 2018;16(1):1–23.
17. Zhang D, Ma XL, Gu Y, Huang H, Zhang GW. Green synthesis of metallic nanoparticles and their potential applications to treat cancer. Frontiers in Chemistry. 2020;8:799.
18. Mukherjee S, Patra CR. Biologically synthesized metal nanoparticles: Recent advancement and future perspectives in cancer theranostics. Future Science. 2017;3(3):FSO203.
19. Sivasankarapillai VS, Somakumar AK, Joseph J, Nikazar S, Rahdar A, Kyzas GZ. Cancer theranostic applications of MXene nanomaterials: Recent updates. Nano-Structures & Nano-Objects. 2020;22:100457.
20. Fathi F, Rashidi M-R, Omidi Y. Ultra-sensitive detection by metal nanoparticles-mediated enhanced SPR biosensors. Talanta. 2019;192:118–127.
21. Wang B, Akiba U, Anzai, J-I. Recent progress in nanomaterial-based electrochemical biosensors for cancer biomarkers: A review. Molecules. 2017;22(7):1048.
22. Malekzad H, Sahandi Zangabad P, Mirshekari H, Karimi M, Hamblin MR. Noble metal nanoparticles in biosensors: Recent studies and applications. Nanotechnology Reviews. 2017;6(3):301–329.
23. Devi RV, Doble M, Verma R.S. Nanomaterials for early detection of cancer biomarker with special emphasis on gold nanoparticles in immunoassays/sensors. Biosensors and Bioelectronics. 2015;68:688–698.
24. Dong W, Ren Y, Bai Z, Yang Y, Chen Q. Fabrication of hexahedral Au-Pd/graphene nanocomposites biosensor and its application in cancer cell H_2O_2 detection. Bioelectrochemistry. 2019;128:274–282.
25. Mirabello V, Calatayud DG, Arrowsmith RL, Ge H, Pascu SI. Metallic nanoparticles as synthetic building blocks for cancer diagnostics: From materials design to molecular imaging applications. Journal of Materials Chemistry B. 2015;3(28):5657–5672.
26. Lim WQ, Phua SZ, Xu HV, Sreejith S, Zhao Y. Recent advances in multifunctional silica-based hybrid nanocarriers for bioimaging and cancer therapy. Nanoscale. 2016; 8(25):12510–12519.
27. Nejati K, Dadashpour M, Gharibi T, Mellatyar H, Akbarzadeh A. Biomedical applications of functionalized gold nanoparticles: A review. Journal of Cluster Science. 2021:1–16.
28. Sankar R, Rahman PK, Varunkumar K, Anusha C, Kalaiarasi A, Shivashangari KS, Ravikumar V. Facile synthesis of Curcuma longa tuber powder engineered metal nanoparticles for bioimaging applications. Journal of Molecular Structure. 2017;1129:8–16.
29. Jaswal R, Kaliannagounder VK, Kumar D, Park CH, Kim CS. Modulated plasmonic nanofibrous scaffold reinforced breast cancer photo-ablation and breast neurotization with resensation. Composites Part B: Engineering. 2022;243:110129.

30. Margheri G, Zoppi A, Olmi R, Trigari S, Traversi R, Severi M, Bani D, Bianchini F, Torre E, Margheri F, Chillà A. Tumor-tropic endothelial colony forming cells (ECFCs) loaded with near-infrared sensitive Au nanoparticles: A "cellular stove" approach to the photoablation of melanoma. Oncotarget. 2016. 7(26):39846.
31. Pinel S, Thomas N, Boura C, Barberi-Heyob M. Approaches to physical stimulation of metallic nanoparticles for glioblastoma treatment. Advanced Drug Delivery Reviews. 2019;138:344–357.
32. Caizer C. Magnetic hyperthermia-using magnetic metal/oxide nanoparticles with potential in cancer therapy. In Metal Nanoparticles in Pharma (pp. 193–218). Springer. 2017.
33. Rai M, Ingle AP, Gupta I, Brandelli A. Bioactivity of noble metal nanoparticles decorated with biopolymers and their application in drug delivery. International Journal of Pharmaceutics. 2015 Dec 30;496(2):159–172.
34. Paramasivam G, Kayambu N, Rabel AM, Sundramoorthy AK, Sundaramurthy A. Anisotropic noble metal nanoparticles: Synthesis, surface functionalization and applications in biosensing, bioimaging, drug delivery and theranostics. Acta Biomaterialia. 2017;49:45–65.
35. Sharma P, Mehta M, Dhanjal DS, Kaur S, Gupta G, Singh H, Thangavelu L, Rajeshkumar S, Tambuwala M, Bakshi HA, Chellappan DK. Emerging trends in the novel drug delivery approaches for the treatment of lung cancer. Chemico-Biological Interactions. 2019;309:108720.
36. Kumari P, Ghosh B, Biswas S. Nanocarriers for cancer-targeted drug delivery. Journal of Drug Targeting. 2016;24(3):179–191.
37. Wang J. Electrochemical biosensors: Towards point-of-care cancer diagnostics. Biosensors and Bioelectronics. 2006 Apr 15;21(10):1887–1892.
38. Myers FB, Lee LP. Innovations in optical microfluidic technologies for point-of-care diagnostics. Lab on a Chip. 2008;8(12):2015–2031.
39. Boyle DS, Hawkins KR, Steele MS, Singhal M, Cheng X. Emerging technologies for point-of-care CD4 T-lymphocyte counting. Trends in Biotechnology. 2012 Jan 1;30(1):45–54.
40. Weigl B, Domingo G, LaBarre P, Gerlach J. Towards non-and minimally instrumented, microfluidics-based diagnostic devices. Lab on a Chip. 2008;8(12):1999–2014.
41. Chin CD, Linder V, Sia SK. Lab-on-a-chip devices for global health: Past studies and future opportunities. Lab on a Chip. 2007;7(1):41–57.
42. Linder V. Microfluidics at the crossroad with point-of-care diagnostics. Analyst. 2007;132(12):1186–1192.
43. Yang M, Sun S, Kostov Y, Rasooly A. A simple 96-well microfluidic chip combined with visual and densitometry detection for resource-poor point of care testing. Sensors and Actuators B: Chemical. 2011 Mar 31;153(1):176–181.
44. Makler MT, Palmer CJ, Ager AL. A review of practical techniques for the diagnosis of malaria. Annals of Tropical Medicine and Parasitology. 1998 Jun 1;92(4):419–434.
45. Shung KK. Diagnostic ultrasound: Past, present, and future. Journal of Medical and Biological Engineering. 2011 Jan 1;31(6):371–374.
46. von Reutern GM, Goertler MW, Bornstein NM, Sette MD, Evans DH, Goertler MW, Hetzel A, Kaps M, Perren F, Razumovky A, Shiogai T. Grading carotid stenosis using ultrasonic methods. Stroke. 2012 Mar;43(3):916–921.
47. Baker JA, Soo MS. The evolving role of sonography in evaluating solid breast masses. In Seminars in Ultrasound, CT and MRI (Vol. 21, No. 4, pp. 286–296). WB Saunders. 2000 Aug 1.
48. Cosgrove DO, Chan KE. Renal transplants: What ultrasound can and cannot do. Ultrasound Quarterly. 2008 Jun 1;24(2):77–87.
49. Coleman AJ, Saunders JE. A review of the physical properties and biological effects of the high amplitude acoustic fields used in extracorporeal lithotripsy. Ultrasonics. 1993 Jan 1;31(2):75–89.

50. Al-Bataineh O, Jenne J, Huber P. Clinical and future applications of high intensity focused ultrasound in cancer. Cancer Treatment Reviews. 2012 Aug 1;38(5): 346–353.
51. Nie S, Emory SR. Probing single molecules and single nanoparticles by surface-enhanced Raman Scattering Science. 1997 Feb 21;275(5303):1102–1106.
52. Reguera J, Langer J, de Aberasturi DJ, Liz-Marzán LM. Anisotropic metal nanoparticles for surface-enhanced Raman scattering. Colloidal Synthesis of Plasmonic Nanometals. 2020 Apr 23:713–754.
53. Lyon LA, Keating CD, Fox AP, Baker BE, He L, Nicewarner SR, Mulvaney SP, Natan MJ. Raman spectroscopy. Analytical Chemistry. 1998 Jun 15;70(12):341–362.
54. Merbach AS, Helm L, Toth E. The Chemistry of Contrast Agents in Medical Magnetic Resonance Imaging. John Wiley & Sons. 2013 Feb 19.
55. Hu X, Norris DG. Advances in high-field magnetic resonance imaging. Annual Review of Biomedical Engineering. 2004 Aug 15;6:157–184.
56. Sun C, Lee JS, Zhang M. Magnetic nanoparticles in MR imaging and drug delivery. Advanced Drug Delivery Reviews. 2008 Aug 17;60(11):1252–1265.
57. Jun YW, Huh YM, Choi JS, Lee JH, Song HT, Kim S, Yoon S, Kim KS, Shin JS, Suh JS, Cheon J. Nanoscale size effect of magnetic nanocrystals and their utilization for cancer diagnosis via magnetic resonance imaging. Journal of the American Chemical Society. 2005 Apr 27;127(16):5732–5733.
58. Hoffman JM, Gambhir SS. Molecular imaging: The vision and opportunity for radiology in the future. Radiology. 2007 Jul;244(1):39–47.
59. Margolis DJ, Hoffman JM, Herfkens RJ, Jeffrey RB, Quon A, Gambhir SS. Molecular imaging techniques in body imaging. Radiology. 2007 Nov;245(2):333–356.
60. Iyer AK, He J, M Amiji M. Image-guided nanosystems for targeted delivery in cancer therapy. Current Medicinal Chemistry. 2012 Jul 1;19(19):3230–3240.
61. Crimì F, Capelli G, Spolverato G, Bao QR, Florio A, Milite Rossi S, Cecchin D, Albertoni L, Campi C, Pucciarelli S, Stramare R. MRI T2-weighted sequences-based texture analysis (TA) as a predictor of response to neoadjuvant chemo-radiotherapy (nCRT) in patients with locally advanced rectal cancer (LARC). La Radiologia Medica. 2020 Dec;125(12):1216–1224.
62. Petralia G, Summers PE, Agostini A, Ambrosini R, Cianci R, Cristel G, Calistri L, Colagrande S. Dynamic contrast-enhanced MRI in oncology: How we do it. La Radiologia Medica. 2020 Dec;125(12):1288–1300.
63. Bragg A, Candelaria R, Adrada B, Huang M, Rauch G, Santiago L, Scoggins M, Whitman G. Imaging of noncalcified ductal carcinoma in situ. Journal of Clinical Imaging Science. 2021;11.
64. Tamada T, Ueda Y, Ueno Y, Kojima Y, Kido A, Yamamoto A. Diffusion-weighted imaging in prostate cancer. Magnetic Resonance Materials in Physics, Biology and Medicine. 2021 Sep 7:1–5.
65. Ye C, Lin Q, Jin Z, Zheng C, Ma S. Predictive effect of DCE-MRI and DWI in brain metastases from NSCLC. Open Medicine. 2021 Jan 1;16(1):1265–1275.
66. Assadsangabi R, Babaei R, Songco C, Ivanovic V, Bobinski M, Chen YJ, Nabavizadeh SA. Multimodality oncologic evaluation of superficial neck and facial lymph nodes. La Radiologia Medica. 2021 Aug;126(8):1074–1084.
67. Chianca V, Albano D, Messina C, Vincenzo G, Rizzo S, Del Grande F, Sconfienza LM. An update in musculoskeletal tumors: From quantitative imaging to radiomics. La Radiologia Medica. 2021 Aug;126(8):1095–1105.
68. Cusumano D, Meijer G, Lenkowicz J, Chiloiro G, Boldrini L, Masciocchi C, Dinapoli N, Gatta R, Casà C, Damiani A, Barbaro B. A field strength independent MR radiomics model to predict pathological complete response in locally advanced rectal cancer. La Radiologia Medica. 2021 Mar;126(3):421–429.

69. Haris M, Yadav SK, Rizwan A, Singh A, Wang E, Hariharan H, Reddy R, Marincola FM. Molecular magnetic resonance imaging in cancer. Journal of Translational Medicine. 2015 Dec;13(1):1–6.
70. Mikhaylov G, Mikac U, Butinar M, Turk V, Turk B, Psakhie S, Vasiljeva O. Theranostic applications of an ultra-sensitive T_1 and T_2 magnetic resonance contrast agent based on cobalt ferrite spinel nanoparticles. Cancers (Basel). 2022 Aug 20;14(16):4026. doi: 10.3390/cancers14164026.
71. Tang H, Wu EX, Ma QY, Gallagher D, Perera GM, Zhuang T. MRI brain image segmentation by multi-resolution edge detection and region selection. Computerized Medical Imaging and Graphics. 2000; 24:349–357. doi: 10.1016/s0895-6111(00)00037-9.
72. Sipkins DA, Cheresh DA, Kazemi MR, Nevin LM, Bednarski MD, Li KCP. Detection of tumor angiogenesis in vivo by αvβ3-targeted magnetic resonance imaging. Nature Medicine. 1998;4:623–626. doi: 10.1038/nm0598-623.
73. Hung AH, Duch MC, Parigi G, Rotz MW, Manus LM, Mastarone DJ, Dam KT, Gits CC, MacRenaris KW, Luchinat C, Hersam MC. Mechanisms of gadographene-mediated proton spin relaxation. The Journal of Physical Chemistry C: Nanomater Interfaces. 2013;117 doi: 10.1021/jp406909b.
74. Matosziuk LM, Leibowitz JH, Heffern MC, MacRenaris KW, Ratner MA, Meade TJ. Structural optimization of Zn(II)-activated magnetic resonance imaging probes. Inorganic Chemistry. 2013;52:12250–12261. doi: 10.1021/ic400681j.
75. Mao X, Xu J, Cui H. Functional nanoparticles for magnetic resonance imaging. Wiley Interdisciplinary Reviews: Nanomedicine and Nanobiotechnology. 2016 Nov;8(6): 814–841. doi: 10.1002/wnan.1400.
76. Ahmed HU, Kirkham A, Arya M, Illing R, Freeman A, Allen C, Emberton M. Is it time to consider a role for MRI before prostate biopsy? Nature Reviews Clinical Oncology. 2009 Apr;6(4):197–206.
77. Alagiri M, Muthamizhchelvan C, Ponnusamy S. Structural and magnetic properties of iron, cobalt and nickel nanoparticles. Synthetic Metals. 2011 Aug 1;161(15–16):1776–1780.
78. Kallumadil M, Tada M, Nakagawa T, Abe M, Southern P, Pankhurst QA. Corrigendum to "Suitability of commercial colloids for magnetic hyperthermia" (Journal of Magnetism and Magnetic Materials. 321 (2009) 1509–1513). Journal of Magnetism and Magnetic Materials. 2009 Nov;321(21):3650–3651.
79. Lee N, Hyeon T. Designed synthesis of uniformly sized iron oxide nanoparticles for efficient magnetic resonance imaging contrast agents. Chemical Society Reviews. 2012;41(7):2575–2589.
80. Tran N, Webster TJ. Magnetic nanoparticles: biomedical applications and challenges. Journal of Materials Chemistry. 2010;20(40):8760–8767.
81. Gupta AK, Gupta M. Synthesis and surface engineering of iron oxide nanoparticles for biomedical applications. Biomaterials. 2005 Jun 1;26(18):3995–4021.
82. Sun C, Lee JS, Zhang M. Magnetic nanoparticles in MR imaging and drug delivery. Advanced Drug Delivery Reviews. 2008 Aug 17;60(11):1252–1265.
83. Redig AJ, McAllister SS. Breast cancer as a systemic disease: A view of metastasis. Journal of Internal Medicine. 2013 Aug;274(2):113–126.
84. Wang H, Jiang X, Lee ST, He Y. Silicon nanohybridbased surface-enhanced Raman scattering sensors. Small. 2014; 10:4455–4468.
85. Tang S, Li Y, Huang H, Li P, Guo Z, Luo Q, Wang Z, Chu PK, Li J, Yu XF. Efficient enrichment and self-assembly of hybrid nanoparticles into removable and magnetic SERS substrates for sensitive detection of environmental pollutants. ACS Applied Materials & Interfaces. 2017 Mar 1;9(8):7472–7480.
86. Guo M, Dong J, Xie W, Tao L, Lu W, Wang Y, Qian W. SERS tags-based novel monodispersed hollow gold nanospheres for highly sensitive immunoassay of CEA. Journal of Materials Science. 2015 May;50(9):3329–3336.

87. Rubira RJ, Camacho SA, Aoki PH, Paulovich FV, Oliveira ON, Constantino CJ. Probing trace levels of prometryn solutions: From test samples in the lab toward real samples with tap water. Journal of Materials Science. 2016 Mar;51(6):3182–3190.
88. Bao ZY, Liu X, Chen Y, Wu Y, Chan HL, Dai J, Lei DY. Quantitative SERS detection of low-concentration aromatic polychlorinated biphenyl-77 and 2, 4, 6-trinitrotoluene. Journal of Hazardous Materials. 2014 Sep 15;280:706–712.
89. Aoki PH, Furini LN, Alessio P, Aliaga AE, Constantino CJ. Surface-Enhanced Raman scattering (SERS) applied to cancer diagnosis and detection of pesticides, explosives, and drugs. Reviews in Analytical Chemistry. 2013 Feb 1;32(1):55–76.
90. Ben-Jaber S, Peveler WJ, Quesada-Cabrera R, Cortés E, Sotelo-Vazquez C, Abdul-Karim N, Maier SA, Parkin IP. Photo-induced enhanced Raman spectroscopy for universal ultra-trace detection of explosives, pollutants and biomolecules. Nature Communications. 2016 Jul 14;7(1):1–6.
91. Han Y, Lei SL, Lu JH, He Y, Chen ZW, Ren L, Zhou X. Potential use of SERS-assisted theranostic strategy based on Fe_3O_4/Au cluster/shell nanocomposites for bio-detection, MRI, and magnetic hyperthermia. Materials Science and Engineering: C. 2016 Jul 1;64:199–207.
92. Ngo HT, Gandra N, Fales AM, Taylor SM, Vo-Dinh T. Sensitive DNA detection and SNP discrimination using ultrabright SERS nanorattles and magnetic beads for malaria diagnostics. Biosensors and Bioelectronics. 2016 Jul 15;81:8–14.
93. Chen P, Zhao A, Wang J, He Q, Sun H, Wang D, Sun M, Guo H. In-situ monitoring reversible redox reaction and circulating detection of nitrite via an ultrasensitive magnetic Au@ Ag SERS substrate. Sensors and Actuators B: Chemical. 2018 Mar 1;256:107–116.
94. Schlücker S. Surface-Enhanced Raman spectroscopy: Concepts and chemical applications. Angewandte Chemie International Edition. 2014 May 5;53(19):4756–4795.
95. Lai H, Xu F, Wang L. A review of the preparation and application of magnetic nanoparticles for surface-enhanced Raman scattering. Journal of Materials Science. 2018 Jun;53(12):8677–8698.
96. Zhang L, Wang T, Yang L, Liu C, Wang C, Liu H, Wang YA, Su Z. General route to multifunctional uniform yolk/mesoporous silica shell nanocapsules: A platform for simultaneous cancer-targeted imaging and magnetically guided drug delivery. Chemistry–A European Journal. 2012 Sep 24;18(39):12512–12521.
97. Kalmodia S, Harjwani J, Rajeswari R, Yang W, Barrow CJ, Ramaprabhu S, Krishnakumar S, Elchuri SV. Synthesis and characterization of surface-enhanced Raman-scattered gold nanoparticles. International Journal of Nanomedicine. 2013;8:4327.
98. Paciotti GF, Myer L, Weinreich D, Goia D, Pavel N, McLaughlin RE, Tamarkin L. Colloidal gold: A novel nanoparticle vector for tumor directed drug delivery. Drug Delivery. 2004 Jan 1;11(3):169–183.
99. Lal S, Clare SE, Halas NJ. Nanoshell-enabled photothermal cancer therapy: impending clinical impact. Accounts of Chemical Research. 2008 Dec 16;41(12):1842–1851.
100. Jain PK, Huang X, El-Sayed IH, El-Sayed MA. Noble metals on the nanoscale: optical and photothermal properties and some applications in imaging, sensing, biology, and medicine. Accounts of Chemical Research. 2008 Dec 16;41(12):1578–1586.
101. Vo-Dinh T, Wang HN, Scaffidi J. Plasmonic nanoprobes for SERS biosensing and bioimaging. Journal of Biophotonics. 2010 Jan;3(1–2):89–102.
102. Qiu Y, Deng D, Deng Q, Wu P, Zhang H, Cai C. Synthesis of magnetic Fe_3O_4–Au hybrids for sensitive SERS detection of cancer cells at low abundance. Journal of Materials Chemistry B. 2015;3(22):4487–4495.
103. El-Sayed IH. Nanotechnology in head and neck cancer: The race is on. Current Oncology Reports. 2010 Mar;12(2):121–128.

104. Maeda H, Wu J, Sawa T, Matsumura Y, Hori K. Tumor vascular permeability and the EPR effect in macromolecular therapeutics: A review. Journal of Controlled Release. 2000 Mar 1;65(1–2):271–284.
105. Baranska M, editor. Optical Spectroscopy and Computational Methods in Biology and Medicine. Springer Science & Business Media. 2013 Dec 5.
106. Ravanshad R, Karimi Zadeh A, Amani AM, Mousavi SM, Hashemi SA, Savar Dashtaki A, Mirzaei E, Zare B. Application of nanoparticles in cancer detection by Raman scattering based techniques. Nano Reviews & Experiments. 2018 Jan 1;9(1):1373551.
107. Winawer SJ, Fletcher RH, Miller L, Godlee F, Stolar MH, Mulrow CD, Wooli SH, Glick SN, Ganiars TG, Bond JH, Rosen L. Erratum: Colorectal cancer screening: Clinical guidelines and rationale (Gastroenterology (1997) 112 (594–642)). Gastroenterology. 1997;112(3):1060.
108. Kulkarni NM, Uppot RN, Eisner BH, Sahani DV. Radiation dose reduction at multidetector CT with adaptive statistical iterative reconstruction for evaluation of urolithiasis: How low can we go? Radiology. 2012;265(1):158–166.
109. Cormode DP, Naha PC, Fayad ZA. Nanoparticle contrast agents for computed tomography: a focus on micelles. Contrast Media & Molecular Imaging. 2014 Jan–Feb;9(1):37–52. doi: 10.1002/cmmi.1551.
110. Cormode DP, Skajaa T, Fayad ZA, Mulder WJM. Nanotechnology in medical imaging: Probe design and applications. Arteriosclerosis, Thrombosis, and Vascular Biology. 2009;29:992–1000.
111. Mulder WJM, Strijkers GJ, Van Tilborg GAF, Cormode DP, Fayad ZA, Nicolay K. Nanoparticulate assemblies of amphiphiles and diagnostically active materials for multimodality imaging. Accounts of Chemical Research. 2009;42(7):904–914.
112. Slowing II, Trewyn BG, Lin VSY. Mesoporous silica nanoparticles for intracellular delivery of membrane-impermeable proteins. Journal of the American Chemical Society. 2007;129(28):8845–8849.
113. Lanza GM, Winter PM, Caruthers SD, Hughes MS, Cyrus T, Marsh JN, Neubauer AM, Partlow KC, Wickline SA. Nanomedicine opportunities for cardiovascular disease with perfluorocarbon nanoparticles. Nanomedicine. 2006;1(3):321–329.
114. Cormode DP, Jarzyna PA, Mulder WJM, Fayad ZA. Modified natural nanoparticles as contrast agents for medical imaging. Advanced Drug Delivery Reviews. 2010;62(3):329–338.
115. Cherukuri P, Gannon CJ, Leeuw TK, Schmidt HK, Smalley RE, Curley SA, Weisman RB. Mammalian pharmacokinetics of carbon nanotubes using intrinsic near-infrared fluorescence. Proceedings of the National Academy of Sciences of the United States of America. 2006;103(50):18882–18886.
116. Yang K, Zhang S, Zhang G, Sun X, Lee S-T, Liu Z. Graphene in mice: ultrahigh in vivo tumor uptake and efficient photothermal therapy. Nano Letters. 2010;10(9):3318–3323.
117. Thakor AS, Jokerst J, Zavaleta C, Massoud TF, Gambhir SS. Gold nanoparticles: A revival in precious metal administration to patients. Nano Letters. 2011;11:4029–4036.
118. Libutti SK, Paciotti GF, Byrnes AA, Alexander HR, Gannon WE, Walker M, Seidel GD, Yuldasheva N, Tamarkin L. Phase I and pharmacokinetic studies of CYT-6091, a novel PEGylated colloidal gold-rhTNF nanomedicine. Clinical Cancer Research. 2010;16(24):6139–6149.
119. Staves B. Pilot Study of AurolaseTM Therapy in Refractory and/or Recurrent Tumors of the Head and Neck. Clinical Trials Identifier: NCT00848042.2010. https://clinicaltrials.gov/ct2/show/results/NCT00848042.
120. Plasmonic Nanophotothermic Therapy of Atherosclerosis (NANOM). Clinical Trial Identifier: NCT01270139. http://clinicaltrialsgov/ct2/show/NCT01270139.

121. Kharlamov A. Plasmonic Photothermal and Stem Cell Therapy of Atherosclerosis Versus Stenting (NANOM PCI). Clinical Trial Identifier: NCT01436123.2011. https://clinicaltrials.gov/ct2/show/NCT01436123.
122. Hainfeld Clearance JF, Slatkin DN, Focella TM, Smilowitz HM. Gold nanoparticles: A new X-ray contrast agent. The British Journal of Radiology. 2006;79:248–253.
123. Cai QY, Kim SH, Choi KS, Kim SY, Byun SJ, Kim KW, Park SH, Juhng SK, Yoon KH. Colloidal gold nanoparticles as a blood-pool contrast agent for X-ray computed tomography in mice. Investigative Radiology. 2007;42(12):797–806.
124. Choi HS, Liu W, Misra P, Tanaka E, Zimmer JP, Ipe BI, Bawendi MG, Frangioni JV. Renal of quantum dots. Nature Biotechnology. 2007;25(10):1165–1170.
125. Guo R, Wang H, Peng C, Shen MW, Pan MJ, Cao XY, Zhang GX, Shi XY. X-ray attenuation property of dendrimer-entrapped gold nanoparticles. The Journal of Physical Chemistry C 2010;114(1):50–56.
126. Wang H, Zheng LF, Peng C, Guo R, Shen MW, Shi XY, Zhang GX. Computed tomography imaging of cancer cells using acetylated dendrimer-entrapped gold nanoparticles. Biomaterials. 2011;32(11):2979–2988.
127. Peng C, Zheng LF, Chen Q, Shen MW, Guo R, Wang H, Cao XY, Zhang GX, Shi XY. PEGylated dendrimer-entrapped gold nanoparticles for in vivo blood pool and tumor imaging by computed tomography. Biomaterials. 2012;33(4):1107–1119.
128. Kojima C, Umeda Y, Ogawa M, Harada A, Magata Y, Kono K. X-ray computed tomography contrast agents prepared by seeded growth of gold nanoparticles in PEGylated dendrimer. Nanotechnology. 2010;21(24).
129. Ghann WE, Aras O, Fleiter T, Daniel MC. Syntheses and characterization of lisinopril-coated gold nanoparticles as highly stable targeted CT contrast agents in cardiovascular diseases. Langmuir. 2012;28(28):10398–10408.
130. Kattumuri V, Katti K, Bhaskaran S, Boote EJ, Casteel SW, Fent GM, Robertson DJ, Chandrasekhar M, Kannan R, Katti KV. Gum Arabic as a phytochemical construct for the stabilization of gold nanoparticles: In vivo pharmacokinetics and X-ray-contrast-imaging studies. Small. 2007;3(2):333–341.
131. Boote E, Fent G, Kattumuri V, Casteel S, Katti K, Chanda N, Kannan R, Churchill R. Gold nanoparticle contrast in a phantom and juvenile swine: models for molecular imaging of human organs using X-ray computed tomography. Academic Radiology. 2010;17(4):410–417.
132. Mieszawska AJ, Gianella A, Cormode DP, Zhao Y, Meijerink A, Langer R, Farokhzad OC, Fayad ZA, Mulder WJM. Engineering of lipid-coated PLGA nanoparticles with a tunable payload of diagnostically active nanocrystals for medical imaging. Chemical Communications. 2012;48:5835–5837.
133. Sun IC, Eun DK, Na JH, Lee S, Kim I-J, Youn I-C, Ko C-Y, Kim H-S, Lim D, Choi K, Messersmith PB, Park TG, Kim SY, Kwon IC, Kim K, Ahn C-H. Heparin-coated gold nanoparticles for liver-specific CT imaging. Chemistry: A European Journal. 2009;15(48):13341–13347.
134. Mohan P, Rapoport N. Doxorubicin as a molecular nanotheranostic agent: Effect of doxorubicin encapsulation in micelles or nanoemulsions on the ultrasound-mediated intracellular delivery and nuclear trafficking. Molecular Pharmaceutics. 2010 Dec 6;7(6):1959–1973.
135. Timko BP, Whitehead K, Gao W, Kohane DS, Farokhzad O, Anderson D, Langer R. Advances in drug delivery. Annual Review of Materials Research. 2011 Aug 4;41:1–20.
136. Clawson C, Ton L, Aryal S, Fu V, Esener S, Zhang L. Synthesis and characterization of lipid–polymer hybrid nanoparticles with pH-triggered poly (ethylene glycol) shedding. Langmuir. 2011 Sep 6;27(17):10556–10561.
137. Javed MN, Dahiya ES, Ibrahim AM, Alam M, Khan FA, Pottoo FH. Recent advancement in clinical application of nanotechnological approached targeted delivery of herbal drugs. In Nanophytomedicine (pp. 151–172). Springer. 2020.

138. Mishra S, Sharma S, Javed MN, Pottoo FH, Barkat MA, Alam MS, Amir M, Sarafroz M. Bioinspired nanocomposites: applications in disease diagnosis and treatment. Pharmaceutical Nanotechnology. 2019 Jun 1;7(3):206–219.
139. Javed MN, Dahiya ES, Ibrahim AM, Alam M, Khan FA, Pottoo FH. Recent advancement in clinical application of nanotechnological approached targeted delivery of herbal drugs. In Nanophytomedicine (pp. 151–172). Springer. 2020.
140. Aslam M, Javed M, Deeb HH, Nicola MK, Mirza M, Alam M, Akhtar M, Waziri A. Lipid nanocarriers for neurotherapeutics: Introduction, challenges, blood-brain barrier, and promises of delivery approaches. CNS & Neurological Disorders-Drug Targets (Formerly Current Drug Targets-CNS & Neurological Disorders). 2022;21.
141. Javed MN, Pottoo FH, Shamim A, Hasnain MS, Alam MS. Design of experiments for the development of nanoparticles, nanomaterials, and nanocomposites. In Design of Experiments for Pharmaceutical Product Development (pp. 151–169). Springer. 2021.
142. Javed MN, Alam MS, Waziri A, Pottoo FH, Yadav AK, Hasnain MS, Almalki FA. QbD applications for the development of nanopharmaceutical products. In Pharmaceutical quality by design (pp. 229–253). Academic Press. 2019 Jan 1.
143. Kumar R, Dhamija G, Ansari JR, Javed MN, Alam MS. C-Dot nanoparticulated devices for biomedical applications. In Nanotechnology (pp. 271–299). CRC Press. 2022.
144. Bharti C, Alam MS, Javed MN, Khalid M, Saifullah FA, Manchanda R. Silica based nanomaterial for drug delivery. Nanomaterials: Evolution and Advancement Towards Therapeutic Drug Delivery (Part II). 2021 Jun 2:57.
145. Mall SK, Yadav T, Waziri A, Alam MS. Treatment opportunities with *Fernandoa adenophylla* and recent novel approaches for natural medicinal phytochemicals as a drug delivery system. Exploration of Medicine. 2022;3:516–539.
146. Naseh MF, Ansari JR, Alam MS, Javed MN. Sustainable nanotorus for biosensing and therapeutical applications. In Handbook of Green and Sustainable Nanotechnology: Fundamentals, Developments and Applications (pp. 1–21). Springer International Publishing. 2022 Aug 19.
147. Sunilbhai CA, Alam M, Sadasivuni KK, Ansari JR. SPR assisted diabetes detection. In Advanced Bioscience and Biosystems for Detection and Management of Diabetes (pp. 91–131). Springer. 2022.
148. Singhal S, Gupta M, Alam MS, Javed MN, Ansari JR. Carbon allotropes-based nanodevices: Graphene in biomedical applications. In Nanotechnology (pp. 241–269). CRC Press. 2022.
149. Pottoo FH, Tabassum N, Javed M, Nigar S, Rasheed R, Khan A, Barkat M, Alam M, Maqbool A, Ansari MA, Barreto GE. The synergistic effect of raloxifene, fluoxetine, and bromocriptine protects against pilocarpine-induced status epilepticus and temporal lobe epilepsy. Molecular Neurobiology. 2019 Feb;56(2):1233–1247.
150. Pottoo FH, Sharma S, Javed MN, Barkat MA, Harshita, Alam MS, Naim MJ, Alam O, Ansari MA, Barreto GE, Ashraf GM. Lipid-based nanoformulations in the treatment of neurological disorders. Drug Metabolism Reviews. 2020 Jan 2;52(1):185–204.
151. Pottoo FH, Javed M, Barkat M, Alam M, Nowshehri JA, Alshayban DM, Ansari MA. Estrogen and serotonin: Complexity of interactions and implications for epileptic seizures and epileptogenesis. Current Neuropharmacology. 2019 Mar 1;17(3):214–231.
152. Pottoo FH, Tabassum N, Javed MN, Nigar S, Sharma S, Barkat MA, Alam MS, Ansari MA, Barreto GE, Ashraf GM. Raloxifene potentiates the effect of fluoxetine against maximal electroshock induced seizures in mice. European Journal of Pharmaceutical Sciences. 2020 Apr 15;146:105261.
153. Waziri A, Bharti C, Aslam M, Jamil P, Mirza M, Javed MN, Pottoo U, Ahmadi A, Alam MS. Probiotics for the chemoprotective role against the toxic effect of cancer chemotherapy. Anti-Cancer Agents in Medicinal Chemistry (Formerly Current Medicinal Chemistry-Anti-Cancer Agents). 2022 Feb 1;22(4):654–667.

154. Javed MN, Akhter MH, Taleuzzaman M, Faiyazudin M, Alam MS. Cationic nanoparticles for treatment of neurological diseases. In Fundamentals of Bionanomaterials (pp. 273–292). Elsevier. 2022 Jan 1.
155. Kumari N, Daram N, Alam MS, Verma AK. Rationalizing the use of polyphenol nano-formulations in the therapy of neurodegenerative diseases. CNS & Neurological Disorders-Drug Targets (Formerly Current Drug Targets-CNS & Neurological Disorders). 2022 Dec 1;21(10):966–976.
156. Raj S, Manchanda R, Bhandari M, Alam M. Review on natural bioactive products as radioprotective therapeutics: Present and past perspective. Current Pharmaceutical Biotechnology. 2022;23(14).
157. Ibrahim AM, Chauhan L, Bhardwaj A, Sharma A, Fayaz F, Kumar B, Alhashmi M, AlHajri N, Alam MS, Pottoo FH. Brain-Derived neurotropic factor in neurodegenerative disorders. Biomedicines. 2022 May;10(5):1143.
158. Alam MS, Garg A, Pottoo FH, Saifullah MK, Tareq AI, Manzoor O, Mohsin M, Javed MN. Gum ghatti mediated, one pot green synthesis of optimized gold nanoparticles: Investigation of process-variables impact using Box-Behnken based statistical design. International Journal of Biological Macromolecules. 2017 Nov 1;104:758–767.
159. Alam MS, Javed MN, Pottoo FH, Waziri A, Almalki FA, Hasnain MS, Garg A, Saifullah MK. QbD approached comparison of reaction mechanism in microwave synthesized gold nanoparticles and their superior catalytic role against hazardous nirto-dye. Applied Organometallic Chemistry. 2019 Sep;33(9):e5071.
160. Pandit J, Alam MS, Ansari JR, Singhal M, Gupta N, Waziri A, Sharma K, Potto FH. Multifaced applications of nanoparticles in biological science. In Nanomaterials in the Battle Against Pathogens and Disease Vectors (pp. 17–50). CRC Press. 2022.
161. Alam MS, Naseh MF, Ansari JR, Waziri A, Javed MN, Ahmadi A, Saifullah MK, Garg A. Synthesis approaches for higher yields of nanoparticles. In Nanomaterials in the Battle Against Pathogens and Disease Vectors (pp. 51–82). CRC Press. 2022.
162. Javed MN, Pottoo FH, Alam MS. Metallic nanoparticle alone and/or in combination as novel agent for the treatment of uncontrolled electric conductance related disorders and/or seizure, epilepsy & convulsions. Patent Acquired on October. 2016;10:40.
163. Krasia-Christoforou T, Georgiou TK. Polymeric theranostics: Using polymer-based systems for simultaneous imaging and therapy. Journal of Materials Chemistry B. 2013;1(24):3002–3025.
164. Juillerat-Jeanneret L. The targeted delivery of cancer drugs across the blood–brain barrier: chemical modifications of drugs or drug-nanoparticles? Drug Discovery Today. 2008 Dec 1;13(23–24):1099–1106.
165. Lammers T, Subr V, Ulbrich K, Hennink WE, Storm G, Kiessling F. Polymeric nanomedicines for image-guided drug delivery and tumor-targeted combination therapy. Nano Today. 2010 Jun 1;5(3):197–212.
166. Xiao Y, Hong H, Matson VZ, Javadi A, Xu W, Yang Y, Zhang Y, Engle JW, Nickles RJ, Cai W, Steeber DA. Gold nanorods conjugated with doxorubicin and cRGD for combined anticancer drug delivery and PET imaging. Theranostics. 2012;2(8):757.
167. Lammers T, Kiessling F, Hennink WE, Storm G. Nanotheranostics and image-guided drug delivery: Current concepts and future directions. Molecular Pharmaceutics. 2010 Dec 6;7(6):1899–1912.
168. Mashal A, Sitharaman B, Li X, Avti PK, Sahakian AV, Booske JH, Hagness SC. Toward carbon-nanotube-based theranostic agents for microwave detection and treatment of breast cancer: Enhanced dielectric and heating response of tissue-mimicking materials. IEEE Transactions on Biomedical Engineering. 2010 Feb 18;57(8):1831–1834.
169. Kang H, Mintri S, Menon AV, Lee HY, Choi HS, Kim J. Pharmacokinetics, pharmacodynamics and toxicology of theranostic nanoparticles. Nanoscale. 2015;7(45):18848–18862.

170. Dreaden EC, Austin LA, Mackey MA, El-Sayed MA. Size matters: Gold nanoparticles in targeted cancer drug delivery. Therapeutic Delivery. 2012 Apr;3(4):457–478.
171. Kowalski PS, Rudra A, Miao L, Anderson DG. Delivering the messenger: Advances in technologies for therapeutic mRNA delivery. Molecular Therapy. 2019 Apr 10;27(4):710–728.
172. Shan X, Gong X, Li J, Wen J, Li Y, Zhang Z. Current approaches of nanomedicines in the market and various stage of clinical translation. Acta Pharmaceutica Sinica B. 2022 Mar 1;12(7).
173. Huang H, Feng W, Chen Y, Shi J. Inorganic nanoparticles in clinical trials and translations. Nano Today. 2020 Dec 1;35:100972.
174. He H, Liu L, Morin EE, Liu M, Schwendeman A. Survey of clinical translation of cancer nanomedicines—lessons learned from successes and failures. Accounts of Chemical Research. 2019 Aug 19;52(9):2445–2461.
175. Thermoablation-Retention MN. Maintenance in the Prostate: A Phase 0 Study in Men. www.ClinicalTrials.gov.
176. Nardecchia S, Sánchez-Moreno P, de Vicente J, Marchal JA, Boulaiz H. Clinical trials of thermosensitive nanomaterials: An overview. Nanomaterials. 2019 Feb 2;9(2):191.
177. Thomsen LB, Thomsen MS, Moos T. Targeted drug delivery to the brain using magnetic nanoparticles. Therapeutic Delivery. 2015 Oct;6(10):1145–1155.
178. Fernández-Barahona I, Muñoz-Hernando M, Ruiz-Cabello J, Herranz F, Pellico J. Iron oxide nanoparticles: An alternative for positive contrast in magnetic resonance imaging. Inorganics. 2020 Apr 10;8(4):28.
179. Tietze R, Zaloga J, Unterweger H, Lyer S, Friedrich RP, Janko C, Pöttler M, Dürr S, Alexiou C. Magnetic nanoparticle-based drug delivery for cancer therapy. Biochemical and Biophysical Research Communications. 2015 Dec 18;468(3):463–470.
180. Huang H, Feng W, Chen Y, Shi J. Inorganic nanoparticles in clinical trials and translations. Nano Today. 2020 Dec 1;35:100972.
181. Maier-Hauff K, Ulrich F, Nestler D, Niehoff H, Wust P, Thiesen B, Orawa H, Budach V, Jordan A. Efficacy and safety of intratumoral thermotherapy using magnetic iron-oxide nanoparticles combined with external beam radiotherapy on patients with recurrent glioblastoma multiforme. Journal of Neuro-Oncology. 2011 Jun;103(2):317–324.
182. Mahalunkar S, Yadav AS, Gorain M, Pawar V, Braathen R, Weiss S, Bogen B, Gosavi SW, Kundu GC. Functional design of pH-responsive folate-targeted polymer-coated gold nanoparticles for drug delivery and in vivo therapy in breast cancer. International Journal of Nanomedicine. 2019;14:8285.
183. Sun H, Liu Y, Bai X, Zhou X, Zhou H, Liu S, Yan B. Induction of oxidative stress and sensitization of cancer cells to paclitaxel by gold nanoparticles with different charge densities and hydrophobicities. Journal of Materials Chemistry B. 2018;6(11):1633–1639.
184. Gomathi AC, Rajarathinam SX, Sadiq AM, Rajeshkumar S. Anticancer activity of silver nanoparticles synthesized using aqueous fruit shell extract of Tamarindus indica on MCF-7 human breast cancer cell line. Journal of Drug Delivery Science and Technology. 2020 Feb 1;55:101376.
185. Song H, Wang C, Zhang H, Yao L, Zhang J, Gao R, Tang X, Chong T, Liu W, Tang Y. A high-loading drug delivery system based on magnetic nanomaterials modified by hyperbranched phenylboronic acid for tumor-targeting treatment with pH response. Colloids and Surfaces B: Biointerfaces. 2019 Oct 1;182:110375.
186. Sadhukhan P, Kundu M, Chatterjee S, Ghosh N, Manna P, Das J, Sil PC. Targeted delivery of quercetin via pH-responsive zinc oxide nanoparticles for breast cancer therapy. Materials Science and Engineering: C. 2019 Jul 1;100:129–140.
187. Mussin J, Robles-Botero V, Casañas-Pimentel R, Rojas F, Angiolella L, Martín-Martínez S, Giusiano G. Antimicrobial and cytotoxic activity of green synthesis silver

nanoparticles targeting skin and soft tissue infectious agents. Scientific Reports. 2021 Jul 15;11(1):1–2.

188. Nieto-Maldonado A, Bustos-Guadarrama S, Espinoza-Gomez H, Flores-López LZ, Ramirez-Acosta K, Alonso-Nuñez G, Cadena-Nava RD. Green synthesis of copper nanoparticles using different plant extracts and their antibacterial activity. Journal of Environmental Chemical Engineering. 2022 Apr 1;10(2):107130.
189. Fouda A, Eid AM, Abdelkareem A, Said HA, El-Belely EF, Alkhalifah DH, Alshallash KS, Hassan SE. Phyco-synthesized zinc oxide nanoparticles using marine macroalgae, ulva fasciata delile, characterization, antibacterial activity, photocatalysis, and tanning wastewater treatment. Catalysts. 2022 Jul 8;12(7):756.
190. Sathiyaraj S, Suriyakala G, Gandhi AD, Babujanarthanam R, Almaary KS, Chen TW, Kaviyarasu K. Biosynthesis, characterization, and antibacterial activity of gold nanoparticles. Journal of Infection and Public Health. 2021 Dec 1;14(12):1842–1847.
191. Hammad EN, Salem SS, Mohamed AA, El-Dougdoug W. Environmental impacts of ecofriendly iron oxide nanoparticles on dyes removal and antibacterial activity. Applied Biochemistry and Biotechnology. 2022 Jul 26:1–5.
192. Temizyürek A, Yılmaz CU, Emik S, Akcan U, Atış M, Orhan N, Arıcan N, Ahishali B, Tüzün E, Küçük M, Gürses C. Blood-brain barrier targeted delivery of lacosamide-conjugated gold nanoparticles: Improving outcomes in absence seizures. Epilepsy Research. 2022 May 6:106939.
193. Zhang J, Liu R, Zhang D, Zhang Z, Zhu J, Xu L, Guo Y. Neuroprotective effects of maize tetrapeptide-anchored gold nanoparticles in Alzheimer's disease. Colloids and Surfaces B: Biointerfaces. 2021 Apr 1;200:111584.
194. Aminabad ED, Mobed A, Hasanzadeh M, Feizi MA, Safaralizadeh R, Seidi F. Sensitive immunosensing of α-synuclein protein in human plasma samples using gold nanoparticles conjugated with graphene: an innovative immuno-platform towards early stage identification of Parkinson's disease using point of care (POC) analysis. RSC Advances. 2022;12(7):4346–4357.
195. Prucek R, Panáček A, Gajdová Ž, Večeřová R, Kvítek L, Gallo J, Kolář M. Specific detection of Staphylococcus aureus infection and marker for Alzheimer disease by surface enhanced Raman spectroscopy using silver and gold nanoparticle-coated magnetic polystyrene beads. Scientific Reports. 2021 Mar 18;11(1):1.
196. Hasim H, Salam SK, Rao PV, Muthuraju S, Muzaimi M, Asari MA. Silver nanoparticles synthesized using Tualang honey ameliorate seizures, locomotor activity, and memory function in KA-induced status epilepticus in male rats. Biomedical Research and Therapy. 2022 Sep 30;9(9):5291–5300.
197. Rogers JV, Parkinson CV, Choi YW, Speshock JL, Hussain SM. A preliminary assessment of silver nanoparticle inhibition of monkeypox virus plaque formation. Nanoscale Research Letters. 2008 Apr;3(4):129–133.
198. Ibrahim Fouad G. A proposed insight into the anti-viral potential of metallic nanoparticles against novel coronavirus disease-19 (COVID-19). Bulletin of the National Research Centre. 2021 Dec;45(1):1–22.
199. Di Gianvincenzo P, Marradi M, Martínez-Ávila OM, Bedoya LM, Alcamí J, Penadés S. Gold nanoparticles capped with sulfate-ended ligands as anti-HIV agents. Bioorganic & Medicinal Chemistry Letters. 2010 May 1;20(9):2718–2721.
200. Ilager D, Shetti NP, Malladi RS, Shetty NS, Reddy KR, Aminabhavi TM. Synthesis of Ca-doped ZnO nanoparticles and its application as highly efficient electrochemical sensor for the determination of anti-viral drug, acyclovir. Journal of Molecular Liquids. 2021 Jan 15;322:114552.
201. Merkl P, Long S, McInerney GM, Sotiriou GA. Antiviral activity of silver, copper oxide and zinc oxide nanoparticle coatings against SARS-CoV-2. Nanomaterials. 2021 May;11(5):1312.

202. Cervadoro A, Giverso C, Pande R, Sarangi S, Preziosi L, Wosik J, Brazdeikis A, Decuzzi P. Design maps for the hyperthermic treatment of tumors with superparamagnetic nanoparticles. PLOS One. 2013 Feb 25;8(2):e57332.
203. Anselmo AC, Mitragotri S. Nanoparticles in the clinic: An update post COVID-19 vaccines. Bioengineering & Translational Medicine. 2021 Sep;6(3):e10246.
204. Zhang R, Kiessling F, Lammers T, Pallares RM. Clinical Translation of Gold Nanoparticles. Springer. 2022 Aug 31.
205. Mocellin S, Rossi CR, Pilati P, Nitti D. Tumor necrosis factor, cancer and anticancer therapy. Cytokine & Growth Factor Reviews. 2005 Feb 1;16(1):35–53.
206. Paciotti GF, Myer L, Weinreich D, Goia D, Pavel N, McLaughlin RE, Tamarkin L. Colloidal gold: A novel nanoparticle vector for tumor directed drug delivery. Drug Delivery. 2004 Jan 1;11(3):169–183.
207. Libutti SK, Paciotti GF, Byrnes AA, Alexander HR, Gannon WE, Walker M, Seidel GD, Yuldasheva N, Tamarkin L. Phase I and pharmacokinetic studies of CYT-6091, a novel PEGylated colloidal gold-rhTNF nanomedicine PEGylated colloidal gold-rhTNF nanomedicine phase I trial. Clinical Cancer Research. 2010 Dec 15;16(24):6139–6149.
208. Girelli D, Ugolini S, Busti F, Marchi G, Castagna A. Modern iron replacement therapy: clinical and pathophysiological insights. International Journal of Hematology. 2018 Jan;107(1):16–30.
209. Ulbrich K, Hola K, Subr V, Bakandritsos A, Tucek J, Zboril R. Targeted drug delivery with polymers and magnetic nanoparticles: Covalent and noncovalent approaches, release control, and clinical studies. Chemical Reviews. 2016 May 11;116(9):5338–5431.
210. Vangijzegem T, Stanicki D, Laurent S. Magnetic iron oxide nanoparticles for drug delivery: Applications and characteristics. Expert Opinion on Drug Delivery. 2019 Jan 2;16(1):69–78.
211. Coyne DW. Ferumoxytol for treatment of iron deficiency anemia in patients with chronic kidney disease. Expert Opinion on Pharmacotherapy. 2009 Oct 1;10(15):2563–2568.
212. Landry R, Jacobs PM, Davis R, Shenouda M, Bolton WK. Pharmacokinetic study of ferumoxytol: A new iron replacement therapy in normal subjects and hemodialysis patients. American Journal of Nephrology. 2005;25(4):400–410.
213. Spinowitz BS, Kausz AT, Baptista J, Noble SD, Sothinathan R, Bernardo MV, Brenner L, Pereira BJ. Ferumoxytol for treating iron deficiency anemia in CKD. Journal of the American Society of Nephrology. 2008 Aug 1;19(8):1599–1605.
214. Chopra I. The increasing use of silver-based products as antimicrobial agents: A useful development or a cause for concern? Journal of Antimicrobial Chemotherapy. 2007 Apr 1;59(4):587–590.
215. Cardozo VF, Oliveira AG, Nishio EK, Perugini MR, Andrade CG, Silveira WD, Durán N, Andrade G, Kobayashi RK, Nakazato G. Antibacterial activity of extracellular compounds produced by a Pseudomonas strain against methicillin-resistant Staphylococcus aureus (MRSA) strains. Annals of Clinical Microbiology and Antimicrobials. 2013 Jan;12(1):1–8.
216. Franci G, Falanga A, Galdiero S, Palomba L, Rai M, Morelli G, Galdiero M. Silver nanoparticles as potential antibacterial agents. Molecules. 2015 May 18;20(5):8856–8874.
217. Drake PL, Hazelwood KJ. Exposure-related health effects of silver and silver compounds: A review. The Annals of Occupational Hygiene. 2005 Oct 1;49(7):575–585.
218. Van de Voorde K, Nijsten T, Schelfhout K, Moorkens G, Lambert J. Long term use of silver containing nose-drops resulting in systemic argyria. Acta Clinica Belgica. 2005 Feb 1;60(1):33–35.
219. Huang H, Feng W, Chen Y, Shi J. Inorganic nanoparticles in clinical trials and translations. Nano Today. 2020 Dec 1;35:100972.
220. Champlin RE, Kastenberg WE, Gale RP. Radiation accidents and nuclear energy: medical consequences and therapy. Annals of Internal Medicine. 1988 Nov 1;109(9):730–744.

221. Thompson DF, Church CO. Prussian blue for treatment of radiocesium poisoning. Pharmacotherapy: The Journal of Human Pharmacology and Drug Therapy. 2001 Nov;21(11):1364–1367.
222. Pearce J. Studies of any toxicological effects of Prussian blue compounds in mammals—a review. Food and Chemical Toxicology. 1994 Jun 1;32(6):577–582.
223. Mody VV, Siwale R, Singh A, Mody HR. Introduction to metallic nanoparticles. Journal of Pharmacy & Bioallied Sciences. 2010 Dec;2(4):282.
224. Shnoudeh AJ, Hamad I, Abdo RW, Qadumii L, Jaber AY, Surchi HS, Alkelany SZ. Synthesis, characterization, and applications of metal nanoparticles. In Biomaterials and Bionanotechnology (pp. 527–612). Academic Press. 2019 Jan 1.
225. Rai M, Ingle AP, Birla S, Yadav A, Santos CA. Strategic role of selected noble metal nanoparticles in medicine. Critical Reviews in Microbiology. 2016 Sep 2;42(5):696–719.
226. Iv M, Telischak N, Feng D, Holdsworth SJ, Yeom KW, Daldrup-Link HE. Clinical applications of iron oxide nanoparticles for magnetic resonance imaging of brain tumors. Nanomedicine. 2015 Mar;10(6):993–1018.
227. Ahamed M, AlSalhi MS, Siddiqui MKJ. Silver nanoparticle applications and human health. Clinica Chimica Acta. 2010 Dec 14;411(23):1841–1848.
228. Alphandéry E, Idbaih A, Adam C, Delattre JY, Schmitt C, Guyot F, Chebbi I. Development of non-pyrogenic magnetosome minerals coated with poly-l-lysine leading to full disappearance of intracranial U87-Luc glioblastoma in 100% of treated mice using magnetic hyperthermia. Biomaterials. 2017 Oct 1;141:210–222.
229. Akhtar S, Asiri SM, Khan FA, Gunday ST, Iqbal A, Alrushaid N, Labib OA, Deen GR, Henari FZ. Formulation of gold nanoparticles with hibiscus and curcumin extracts induced anti-cancer activity. Arabian Journal of Chemistry. 2022 Feb 1;15(2):103594.
230. Safwat MA, Soliman GM, Sayed D, Attia MA. Fluorouracil-loaded gold nanoparticles for the treatment of skin cancer: Development, in vitro characterization, and in vivo evaluation in a mouse skin cancer xenograft model. Molecular Pharmaceutics. 2018 Apr 27;15(6):2194–205.
231. Thambiraj S, Vijayalakshmi R, Ravi Shankaran D. An effective strategy for development of docetaxel encapsulated gold nanoformulations for treatment of prostate cancer. Scientific Reports. 2021 Feb 2;11(1):1–7.
232. Shaker MA, Shaaban MI. Formulation of carbapenems loaded gold nanoparticles to combat multi-antibiotic bacterial resistance: In vitro antibacterial study. International Journal of Pharmaceutics. 2017 Jun 15;525(1):71–84.
233. Nazaripour E, Mousazadeh F, Moghadam MD, Najafi K, Borhani F, Sarani M, Ghasemi M, Rahdar A, Iravani S, Khatami M. Biosynthesis of lead oxide and cerium oxide nanoparticles and their cytotoxic activities against colon cancer cell line. Inorganic Chemistry Communications. 2021 Sep 1;131:108800.
234. Alangari A, Alqahtani MS, Mateen A, Kalam MA, Alshememry A, Ali R, Kazi M, AlGhamdi KM, Syed R. Iron oxide nanoparticles: Preparation, characterization, and assessment of antimicrobial and anticancer activity. Adsorption Science & Technology. 2022 Mar 11;9(9).
235. Sales DA, Marques TM, Ghosh A, Gusmão SB, Vasconcelos TL, Luz-Lima C, Ferreira OP, Hollanda LM, Lima IS, Silva-Filho EC, Dittz D. Synthesis of silver-cerium titanate nanotubes and their surface properties and antibacterial applications. Materials Science and Engineering: C. 2020 Oct 1;115:111051.
236. Chandrasekaran R, Gnanasekar S, Seetharaman P, Keppanan R, Arockiaswamy W, Sivaperumal S. Formulation of Carica papaya latex-functionalized silver nanoparticles for its improved antibacterial and anticancer applications. Journal of Molecular Liquids. 2016 Jul 1;219:232–238.
237. Yasir M, Singh J, Tripathi MK, Singh P, Shrivastava R. Green synthesis of silver nanoparticles using leaf extract of common arrowhead houseplant and its anticandidal activity. Pharmacognosy Magazine. 2017;13(Suppl 4):S840–S844.

238. Thapa R, Bhagat C, Shrestha P, Awal S, Dudhagara P. Enzyme-mediated formulation of stable elliptical silver nanoparticles tested against clinical pathogens and MDR bacteria and development of antimicrobial surgical thread. Annals of Clinical Microbiology. 2017 May 16;16(1):39.
239. Abdelnaby MA, Shoueir KR, Ghazy AA, Abdelhamid SM, El Kemary MA, Mahmoud HE, Baraka K, Abozahra RR. Synthesis and evaluation of metallic nanoparticles-based vaccines against Candida albicans infections. Journal of Drug Delivery Science and Technology. 2022 Feb 1;68:102862.
240. Namasivayam SKR, Venkatachalam G, Bharani RSA, Kumar JA, Sivasubramanian S. Molecular intervention of colon cancer and inflammation manifestation by tannin capped biocompatible controlled sized gold nanoparticles from Terminalia bellirica: A green strategy for pharmacological drug formulation based on nanotechnology principles. 3 Biotech. 2021 Sep;11(9):401.

7 In Vivo and In Vitro Toxicity Study of Metallic Nanoparticles

Biswakanth Kar, Anindya Bose, Sudipta Roy, Pranabesh Chakraborty, Soumalya Chakraborty, Sanjoy Kumar Das, Gautam Pal, Farheen Waziri, and Md Sabir Alam

7.1 INTRODUCTION

A pathogen is an invader that tries to live and propagate in our bodies. They have a specialized mechanism system for crossing our cellular layer and then producing a specific response in the host, like colonization, drawing nutrition from the host body, circumventing host immune response, replicating by using host machinery, and exiting and transmitting another host. During this process, many pathogens cause infection or disease in the host. The common types of pathogens are bacteria, viruses, and infectious eukaryotic organisms like fungi and protozoa [1]. Each type of pathogen has a distinct virulence power through which it causes infection differently. Sometimes these infectious pathogens are carried and transmitted from one living organism to another by a disease vector; the agents are considered intermediate parasites or microbes [2, 3]. Common disease vectors like mosquitoes, lice, and ticks transmit a huge amount of infectious pathogens in different ways [4–6]. For example, the Anopheles mosquito acts as a vector for malaria and filariasis by using its mouthpart for feeding and drawing host blood, which is transmitted to another living organism.

This chapter will provide an outline of the bacterial pathogen, pathogenicity, virulence, and virulence factor, as well as the interaction of the pathogen with the host and its detection. A bacterial pathogen is any bacterium that can cause infection. Pathogenicity refers to the mechanism of infection or the ability to cause disease [1]. Virulence is defined as a quantitative measure of pathogenicity. A virulence factor is defined as gene products that are produced by the pathogens and are capable of establishing themselves within a host and augmenting their potential to cause infections. There are different types of virulence factors, like (a) adherence factors, (b) invasion factors, (c) capsules, (d) endotoxins, (e) exotoxins, and (f) siderophores. Some examples of virulence factors are toxins, cell surface proteins that prevent phagocytosis and protect the bacterium, and a surface receptor that helps to bind host cells [7]. The bacterial toxin possesses the ability to produce disease. Based on

 DOI: 10.1201/9781003317319-7

the chemical level, bacterial toxin is mainly of two types: lipopolysaccharides and proteins. Lipopolysaccharides are cell wall-associated substances of gram-negative bacteria, which are known as endotoxins and are parts of the cell envelope. The gram-negative pathogens, in particular *Pseudomonas, Neisseria, Salmonella, Bordetella pertussis, Haemophilus influenzae, Shigella, Escherichia coli, and Vibrio cholera*, are associated with the outer membrane of the lipopolysaccharide complex [8, 9]. Proteins are secreted from bacterial cells and are known as exotoxin. During exponential growth of live bacteria, exotoxin is secreted, and production of the toxin is specific to a particular type of bacterial strain; for example, tetanus toxin is produced by only *Clostridium tetani* and diphtheria toxin is produced by only *Corynebacterium diphtheria*. These exotoxins cause damage at a different target site, indicate the location of activity such as enterotoxin, a neurotoxin, and leukocidin [10]. It was found that most of the bacterial pathogens have developed specific gene products that allow them to multiply in their host without any interference from body defense. However, some virulence products are evolved in a definite strain of pathogens; for example, *E. coli* produces diarrhea-causing enterotoxins, which are named enterotoxigenic *E. coli*; these toxins cause mucosal cell damage and cytotoxic effects; the second group of the strains of *E. coli* produce shigella-like enterotoxins that are associated with tissue destruction and inflammation; the third group of enteropathogenic *E. coli* is associated with neonatal diarrhea [11]. Like *E. coli*, another pathogen called *S. aureus* produces enterotoxins, a major cause of food poisoning, and symptoms include nausea, vomiting, and occasional diarrhea [12].

Bacterial exotoxins produced by pathogens and acting on the host target cell, especially macrophages and neutrophils, result in suppression of the natural immune response, providing the pathogen with an appropriate environment for replication. For example, adenylate cyclase and pertussis toxins are produced by *Bordetella pertussis*, which infects the ciliated respiratory epithelium of humans and modulates the host's innate immune and anti-inflammatory responses [13]. Several recent studies have established that the virulence factor of the pertussis toxin is quite effective in promoting the disease and respiratory infections. Carbonetti and team reported the role of purified pertussis toxin in the respiratory tract using a mutant strain of mouse intranasal infection model. Results show that 14 days before administration of bacterial inoculation, toxins act as a soluble factor for increasing respiratory infection [14]. A recent report cited the association of *Bordetella pertussis* infection with cytokine response in a wild-type strain of mouse model. The experimental result showed increased neutrophil count in the airway due to up-regulation of cytokine IL-17, IFN-γ, and TNF-α, indicating that the higher bacterial number in the airways promotes transmission [15].

Anthrax toxin produced from *Bacillus anthracis* acts on various tissues at the same time and sometimes causes lethal effects due to the actions of anthrax lethal toxin and edema toxin and its ubiquitous expression in host tissue. Liu et al. cited the lethal effect of anthrax toxin using two receptors, namely capillary morphogenesis protein-2 of null mice and cell-type-specific expressing mice. The result found that lethal effect occurred by the lethal effect toxin via cell type receptor, edema toxin-induced lethality occurs due to expression of hepatocytes, suggesting anthrax toxins induce lethal effect in the host due to damaging effect on the vital organs [16]. The previous studies revealed that both lethal toxin and edema toxin can interfere with host immunity by inhibiting the activation of macrophages along with the stimulation of neutrophils, which favors bacterial replication and contributes to severe bacteremia. The study discovered that both in vitro and in vivo models exhibit macrophage inhibition of pro-inflammatory cytokine production [17] as well as RAW2647 cell stimulation of in vitro apoptosis [18]. Tournier et al. revealed the effect of edema toxin and lethal toxin in murine bone marrow-derived dendritic cells. The results indicated there was inhibition of IL-12p70 and tumor necrosis factor-alpha secretion by edema toxin and lethal toxin, which suggests both toxins are linked with each other and impair dendritic cell response, which leads to suppression of immune functions in the host and promotes bacterial escape and multiplication [19].

Mycotoxins are fungal metabolites that are mainly found in agricultural products and humid or water-damaged environments. Human or animal exposure to mycotoxin causes dysfunction of the immune response, which increases the susceptibility of infection in the mucosal epithelia of the gastrointestinal and respiratory tracts of the host [20]. For example, aflatoxin, a toxin released by fungal pathogens, *Aspergillus flavus* and *Aspergillus parasiticus* causes infections in humans and suppresses cell-mediated immune reactions as well as disturbs T-cell and B-cell, which increase the risk of liver cancer [21]. Long-term exposure to low doses of aflatoxins causes many types of cancer, such as liver, lung, and skin cancer, as well as other diseases, including mutagenicity, teratogenicity, and immunosuppression in humans. Besides, aflatoxins are also associated with nutritional disorders known as kwashiorkor by interfering with the absorption of vitamins, minerals, and protein synthesis [22]. The basic mechanism of toxicity of aflatoxins is ascribed mostly to the intermediate metabolite AFB1-exo-8,9 epoxide, which reacts with nucleic acid and proteins to induce various cell disturbance and promote mutagenicity [23]. Ochratoxin A is another naturally occurring foodborne mycotoxin, produced by fungal pathogens including *Aspergillus ochraceus, Aspergillus carbonarius*, and a potent renal carcinogen in many animals with immunotoxicity [24]. Similarly, citrinin is also associated with nephrotoxicity; besides, it was found that citrinin is involved in the induction of apoptosis through oxidative stress and increases the possibility of carcinogenesis in humans [25].

Globally, food safety is a major concern for all human beings. There are many foodborne pathogens, which include, fungus, bacteria, and viruses, that are held responsible for the contamination and poisoning of the food [26]. The pathogenic foodborne bacteria in particular *Salmonella enterica, Campylobacter jejuni, Listeria monocytogenes, Escherichia coli, Staphylococcus aureus,* and *Bacillus cereus* are the leading causes of food poisoning mortality all over the world [27]. Most of the gram-negative bacteria are responsible for food poisoning due to the production of the toxin, for example, *Staphylococcus aureus* produces enterotoxin A and B, which is the main reason for food poisoning. Similarly, *E. coli* produces shiga-like toxins 1 and 2, responsible for dysentery, while *Listeria monocytogenes* secretes a toxin called listeriolysin O, which causes listeriosis [28].

Botulinum neurotoxins are the most potent naturally occurring toxins produced from the gram-positive anaerobic bacteria *Clostridium botulinum*, which causes severe food poisoning and bolulinum illness. This zinc-dependent botulinum neurotoxin inhibited the release of a parasympathetic neurotransmitter called acetylcholine at neuromuscular junctions, resulting in flaccid paralysis [29]. Food poisoning is brought on by a dangerous and extremely irritating toxin called lipopolysaccharides that is produced by the gram-negative bacterium cyanobacteria [30]. While the gram-positive bacteria *Bacillus cereus* produces different toxins, such as one emetic toxin and three enterotoxins or hemolysins, to be involved in food poisoning.

By monitoring the effects of chemical substances in living animal beings (in vivo) or in animal and human cell lines, toxicity that affects humans is explored (in vitro). In standardized experimental designs using various model systems, toxicogenomic research gathers gene expression profiles and histopathological evaluation data for hundreds of medicines and contaminants. These data are a priceless resource for studying how drugs affect biological systems across the entire genome.

One of the important factors is the conditions under which the toxins are able to express their virulence factors. This is a major concern and needs to be explored and monitored minutely with all significance. An insight study is also desirable in this particular field. It is a challenging task for the whole fraternity to assess pathogen-induced diseases around the globe. People who suffer from pathogens and foodborne toxins show various diseases that vary from acute to chronic, along with immunomodulation, indicating the seriousness of the contamination. Therefore, it is a real-life problem, and the scientists have accepted the challenge of detecting all the responsible pathogens toxin in all respects so that the level of infection or foodborne diseases can be controlled and monitored properly.

7.2 TYPES OF CONVENTIONAL MODEL AVAILABLE FOR TOXIN DETECTION BOTH IN VIVO AND IN VITRO

There are some conventional models available for the detection of pathogen toxins, which depend on how the agar plates are being used for the culturing of the microorganisms. Most of the conventional models are arduous, and it takes at least two days to get the initial results. In some cases, the time limit is even higher, and it roughly takes a week to detect the symptoms of the particular pathogen microorganism. Consequently, scientists have developed slightly modified models so that the pathogen can be detected more precisely, as there is a huge demand in the fields of biotechnology and the food industry. The main purpose of such a modification is to further confront the spread of bacterial diseases that occur due to food poisoning. Such detection methods for these particular diseases are categorized into (i) sensor-based and (ii) conventional methods. The objective of this chapter is to highlight and focus on the recent development of rapid methods of identification and proper detection of the bacterial toxins, as well as their advantages/disadvantages and their characteristics and behavior for the respective models. We have focused mainly on three types of conventional methods that are efficient enough for the detection of bacterial pathogens, in particular (i) the count method of culturing and colony, (ii) immunology-based, and (iii) polymerase chain reaction.

7.2.1 Immunological Detection Method

The detection of the pathogen toxin by the method based on immunology is mainly based on antibody–antigen interaction. In such cases, a specific antibody will be allowed to bind to a peculiar antigen, which may strengthen the binding and can be utilized for the detection of pathogens from gram-negative bacteria, in particular, *Salmonella* and *E. coli* [31]. Immunological detection techniques such as fluorescent assay with enzyme-linked (ELFA), enzyme immunoassay (EIA), immunosorbent assay with enzyme-linked (ELISA), and flow injection immunoassay use polyclonal and monoclonal antibodies for the identification of bacterial and foodborne pathogens. The major merit of this procedure is that the time is easily controlled for the preparation of the assay as compared to culturing techniques. However, the disadvantage is that a real-time pathogen detection technique is quite impossible [32]. EIA is used for the detection of various pathogenic antigens in body fluids, such as rotavirus antigen, Haemophilus influenza type B antigen, and hepatitis B virus antigen. The benefit of this method is a high degree of sensitivity, which is ensured by the enzyme-substrate reaction's intrinsic amplification and the use of objective endpoints without the need for radioactivity. However, the disadvantage is that it is not adequately sensitive to the detection of antigens from other viruses, bacteria, and parasites [33]. The detection of the shiga toxin-producing O157 nonmotile strain of *E. coli* was done by using a phage-derived ligand in ELFA before antibody detection [34]. The main benefit of the ELFA test is the detection of many types of phenotypic variants, no cross-reactivity to other serotypes of *E. coli*.

ELISA is the most common type of immunoassay used for quantitative measurement of the toxin of bacteria and foodborne pathogens; for example, determination of Clostridium difficile toxins A and B in fecal matter by using anti-toxin A and B antibodies, which are in two antibodies (sandwich). In the next set of anti-toxin antibodies, A and B antibodies are allowed to be sandwiched between the captured antigen. Then an anti-second antibody is added in conjugation with peroxidase enzyme and tetramethylbenzidine dye; the development of a yellow color specifies the presence of toxins A and B [35]. Similarly, sandwich ELISA is also used for the detection of pathogenic *Vibrio parahaemolyticus* in seafood, and the detection of toxins such as enterotoxin and botulinum toxin in foods such as *Clostridium perfringens* [36]. The advantage is high sensitivity and specificity, more accuracy than another immunoassay. The disadvantage of ELISA is that it requires specialized equipment and trained personnel. To overcome this disadvantage, a fast, inexpensive, simple, and dependable immunoassay called Lateral Flow includes dipsticks and immunochromatographic strips that have been well established for spotting foodborne pathogens. The lateral flow immunoassay device consists of four sections in a unique arrangement so that color can be visualized within 2–10 minutes upon addition of the desired sample [36]. This technique is useful for the detection of foodborne pathogens, which include *Campylobacter jejuni, Staphylococcus aureus, Escherichia coli* O157, *Listeria* spp., *Salmonella*, brevetoxins, and *Staphylococcal enterotoxin* B [37].

7.2.2 Count Method of Culturing and Colony

The plate counting method is basically supported by the growth of bacterial colonies on the nutrient agar medium, with serial dilution in an appropriate ratio such that the number of colonies on the agar plate can be well counted. For example, plate count agar or MacConkey agar is used to count *E. coli*. In general, the plates are incubated for various intervals of time. In our case, we have considered an incubation temperature of 22 °C for a duration of 24 hours in the first case and 37°C for a duration of 24 hours in the second case. Recently, a fluorescent agent was used for the automatic counting of the colonies. The total number of colonies is called the total viable count, expressed in terms of cfu/ml. The cell culture concentration is measured using turbidimetry, whereas the turbidity of cultures is detected using turbidometry. The demerit of this method is the requirement of several days to reveal the presence of a pathogen.

7.2.3 Polymerase Chain Reaction (PCR)

PCR is the most common detection model for the detection of the toxins of pathogens. There are many PCR methods available for the detection of pathogens, including reverse transcript PCR, real-time PCR or qPCR or digital PCR, and multiplex PCR [38]. Recently, qPCR has become a well-developed method for quantification and detection of the toxin of bacterial and foodborne pathogens. In the first part of this technique, DNA is extracted from the bacteria using the standard method. Then specific fragments of DNA are amplified with selected primers, followed by quantification of the DNA. Guilbault et al. reported the detection technique of *C. difficile* toxin B gene from stool samples using a PCR assay by comparing it with a standard cytotoxicity assay [39]. Besides bacterial pathogens, qPCR plays a vital role in the quantification and typing of viral pathogens. qPCR methods play an important role in detecting and quantifying foodborne pathogens [40]. The main advantages of such systems are that we get rapid and high-throughput detection and the quantification of specified DNA sequences in various matrices. Apart from these, simultaneous amplification and visualization of newly formed DNA amplicons in lower-time amplification are also facilitated in terms of safety to avoid cross-contaminations, as well as multiplexing of the amplification for various targets into an individual reaction [41]. The limitation of PCR requires that we have expert and trained personnel so that the complete procedure is operated in a systematic order. In the whole scenario, identification of antimicrobial resistance, typing and detection of bacterial strains, and most possibly quantification in non-processed and raw food, are considered with the utmost precaution.

7.2.4 Detection of Bacterial Endotoxin

Although there are several conventional model approaches, out of them, two are most appropriate and user-friendly and are mainly used for the detection of bacterial pathogens, that is, limulus amoebocyte lysate (LAL) and rabbit pyrogen test (RPT). The LAL test is used to specifically quantify the gram-negative bacteria that are mixed with the endotoxin-contaminated food, or in other cases, they are mixed in water with horseshoe crab blood to deduce the amoebocyte extraction that is observed from endotoxins. If gelation occurs in the sample, then it indicates the presence of endotoxin [42]. The advantage of the LAL technique is that it is highly sensitive as compared to all the other detection techniques. The disadvantage with the LAL technique is that it requires skilled personnel to complete all the complex steps so that external interference can be avoided. Apart from these, the testing kits used are a bit expensive. In the rabbit pyrogen test, an experimental solution is injected into a rabbit's body, and it is observed for some time for a change in body temperature. A rise in body temperature indicates endotoxin present in the sample. The drawback of this model is that it is time-consuming and expensive.

7.2.5 Botulism Bioassay for Neurotoxin Detection

The mouse botulism bioassay (MBA), also known as the lethal assay, is the standard protocol for analyzing and detecting neurotoxin. Initially, the sample is administered intraperitoneally to mice; after four days, LD_{50} determines the detection of neurotoxin in the sample. We have observed that by adding type-specific antitoxins to the aliquots of the as-prepared diluted sample, the type of the toxin can be easily determined. In yet another case, mice injected with a particular toxin-containing sample of antitoxin did not exhibit clear symptoms of botulism.

The main benefit of this method is that it is quick and sensitive, but a drawback is that a large sample is needed, which makes it challenging to collect it from baby stools or food scraps; the assay takes longer to produce clear results; it is also labor-intensive, expensive, and necessitates the euthanasia of lab animals. Scientists at the Centers for Disease Control and Prevention recently created matrix-assisted laser activity of ionization-time of flight mass spectrometry (MALDI-TOF MS) as a potential tool for the identification of a poison to get around these restrictions. The advantage of this technique is that it is rapid, highly sensitive, and economical in terms of cost and labor. MALDI-TOF-MS with advanced techniques like Bruker MALDI-TOF Biotyper is excellent for performing microbiological testing, strain typing, detection of foodborne pathogens, and identifying microorganisms from isolates [43]. We have also observed that with the use of these technologies, we can easily diagnose diseases caused by fungi, bacteria, and viruses. These tools are also used to identify strains of harmful bacteria that are resistant to certain antibiotics. For example, Bacteroides fragilis, where carbapenem resistance is mostly attributed to the presence of the cfiA gene, is resistant to this antibiotic [44].

7.3 RECENT TECHNIQUE DEVELOPED FOR DETECTION BOTH IN VIVO AND IN VITRO

Biosensor-based methods for detection of pathogen: A biosensor is based on the recognition of living organisms or antigens present in pathogens, which is completed by the determination of the molecular species that is used to create a binding with the target pathogen for the sensing application. Various receptors have been introduced in the biosensor applications so as to increase the overall efficiency of the system. There are different types of receptors that are used in the detection of such systems, which mainly include, antigen/antibody, biomimetic, bacteriophage, cellular structures/cells, nucleic acids/DNA, and enzymes [28]. Based on receptors, different biosensing techniques are established to detect pathogens, and the most sensitive and accurate methods are (a) optical, (b) electrochemical, (c) mass-sensitive, and (d) nanomaterial-based biosensors.

7.3.1 Optical-Biosensor Method for Detection of Pathogen

This method is very effective and consistent with both pathogens and endotoxins [45]. The basic principle of an optical biosensor was the development of an optical-based sensor that appropriately propagates the laser light through the tapered optical fiber with a complete internal reflection on the detection surface, and then the emitted light after total internal reflection (TIR) is properly detected. The propagated light is allowed to be incident on the core of the fiber or waveguide, and it detects pathogens and all other endotoxins. The pathogens detected by this method are *E. coli*, *Salmonella*, *Listeria*, and other endotoxins [46, 47].

7.3.2 Electrochemical-Biosensor Method for Detection of Pathogen

As discussed in the optical method, an electrochemical biosensor also uses a sensing electrode and interacts with the living sample; accordingly, we can observe the change in potential and current [48]. There are various biosensors used in these methods, including amperometric, conductometric, impedimetric, and potentiometric [49]. This approach has the advantages of being inexpensive, compact, and robust for liquid samples, but it also has the disadvantage of having poorer selectivity and sensitivity. Ion-selective electrodes or an ion-sensitive field-effect transistor are employed in potentiometric biosensors to provide a potential signal. There is no electricity flowing through the electrode. Pathogens cause potential to build up between the electrodes. This idea has been used to detect pathogens like *E. coli* [50]. The amperometric principle is used to measure a current flow that is connected to the pathogen concentration. Pathogens such as *E. coli, Salmonella, Mycobacterium smegmatis*, and *Bacillus cereus* have been found using it [51, 52]. In the impedimetric biosensor method, the impedance of the sensor is measured using electrochemical impedance spectroscopy that is affected by the biological reaction. Between the two electrode terminals, a small alternating voltage is applied with a wide range of frequency; the impedance is measured and is then analyzed using this method. The detection of pathogens like *E. coli*, *S. typhimurium*, marine biotoxins in seafood, and endotoxins has been reported [53–55].

7.3.3 Mass-Sensitive Biosensors

The transduction mechanism, which involves slight variations in the mass of the biosensor, is the foundation of the mass-sensitive biosensor's operating principle. It is perfectly able to detect even the smallest changes in mass, and scientists have given it the name piezoelectric biosensors. It primarily comprises two types of equipment: surface acoustic wave devices, also known as bulk wave devices, and quartz crystal microbalance devices [56, 57].

7.3.4 Surface Acoustic Wave (SAW) Device

The principle lying behind the SAW device is based on an interdigital transducer that produces acoustic waves at the surface due to a piezoelectric substrate [28], which is mainly used for the detection of the toxin of pathogens as well as endotoxins such as lipopolysaccharide from *E. coli and P. mirabilis* [58, 59].

7.3.5 Quartz Crystal Microbalance (QCM) Device

The QCM device is based on the piezoelectric principle. There are two gold plates on which an electric signal is applied that produces vibration, which is measured, and the resonance frequency of the crystal changes accordingly. This device has been used for the detection of *E. coli* pathogens and endotoxins [60, 61], and it is also used in the clinical field, biochemistry, and food industry. Lian et al. reported that *S. aureus* pathogens were detected in the culture medium and milk in the range of 4.1×10^1 and 4.1×10^5CFU/mL using a piezoelectric biosensor device, suggesting QCM can be used specifically for clinical diagnosis and food testing [62]. Further, a team of scientists uses a piezoelectric biosensor to detect *L. monocytogenes* in the range of 10^2 CFU/mL in milk [63] and *S. aureus* in a culture with QCM dissipation tracking [64]. The advantages of the QCM biosensor are real-time monitoring, easy handling, unlabeled detection, and electrode for ligand immobilization. The requirement for improvement in repeatability and stability can be regarded as a limitation of QRM.

7.3.6 Nanomaterial-Based Biosensors

By varying the morphology of the nanomaterials, there is a drastic change in the features and applications of the nano-based products. Recently, different types of biosensors have been developed for the detection of various diseases, like antibacterial, antiviral, neurological, and anticancer, all based on nanotechnology [65–90]. When we discuss nanomaterial-based biosensors, they generally consist of a sensing electrode that is modified by varying the nanomaterials and ultimately leads to a fast reaction process [28]. Such kinds of biosensors are used to detect pathogens like *E. coli, Salmonella,* and *L. monocytogenes* apart from foodborne pathogens such as respiratory syncytial virus, Giardia parasites, Cryptosporidium parasites [91, 92]. Researchers established the use of nanomaterial-based biosensors for the detection of bacterial pathogens and foodborne toxins in the food industry, biochemistry, and biotechnology, as shown in Figure 7.1.

Yang and co-workers discovered a nano-based sensor for the detection of *Salmonella* spp. using AuNPs consisting of glassy carbon electrodes [93]. Similarly, Yang et al. reported the use of TiO_2 nanowire microelectrodes to detect *Listeria monocytogenes* rapidly and sensitively, without interference from other foodborne pathogens, and aluminum anodized oxide nanopore membranes to detect *E. coli* [93]. Wan et al. cited the reduced graphene sheets-based sensor as having excellent conductivity and being used to detect sulfate-reducing bacterial pathogens [94]. The advantage of this biosensor is that it is cost-effective; measurement can be done in real-time, whereas the demerit is feared due to the toxicity of the non-material. The nanomaterial-based in vitro and in vivo activities are discussed in Table 7.1.

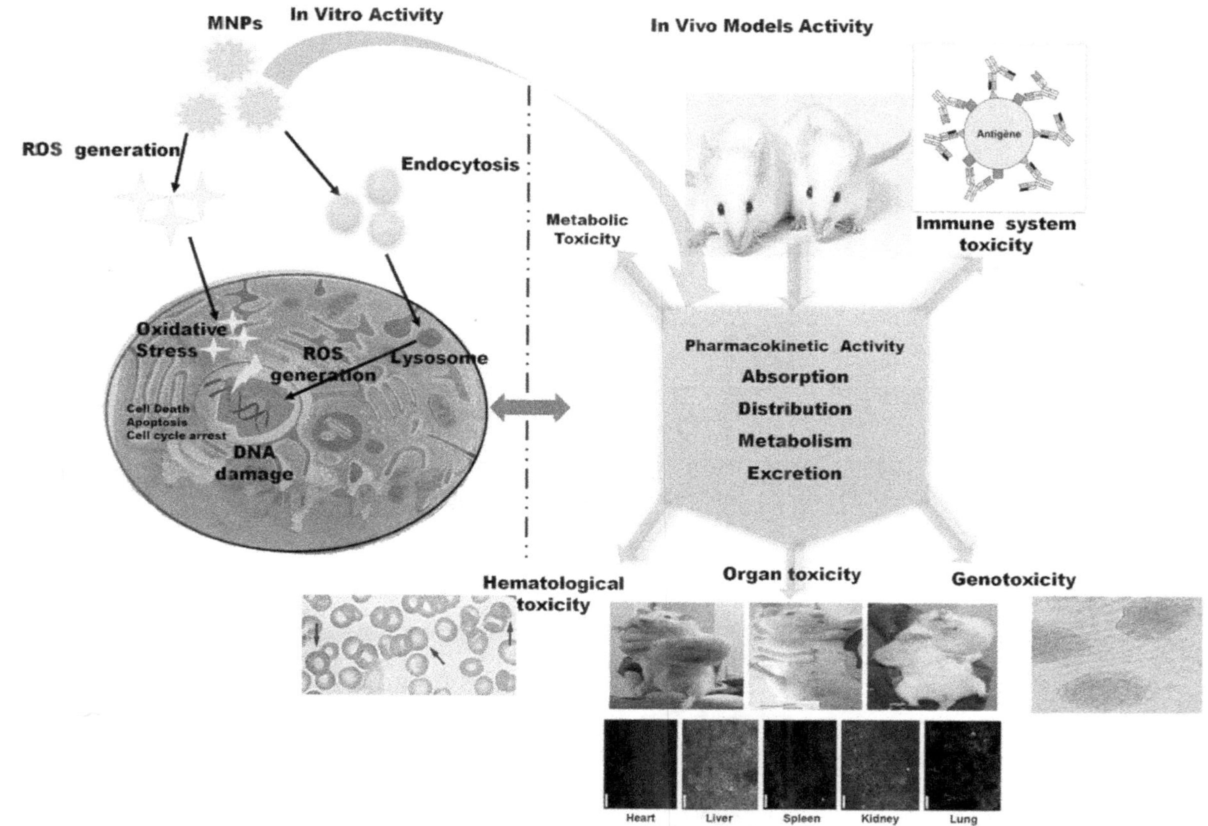

FIGURE 7.1 Schematic illustration for in vitro and in vivo toxicity study of MNPs.

TABLE 7.1
In Vitro and In Vivo Toxicity Activities of Various Nanocarriers

Nanocarrier with particle size	Assay technique	Species/cell culture	Mechanism	Reference
Gold (10, 50, 100, 250 nm)	ICP-MS	**Male Wister rate**	The liver and spleen were the organs with the smallest AuNPs, and the particles were distributed throughout the body in the blood, heart, lungs, kidney, thymus, brain, and reproductive organs.	[95]
Gold (10–35 nm)	ICP-MS	**In vivo**	The biodistribution of AuNPs was largely concentrated in the liver and spleen, but no evidence of toxicity to the liver or kidneys was discovered. The safety of AuNPs was further supported by the animals' normal behavior and the lack of acute and chronic toxicities in important organs.	[96]
Silver (15–100 nm)	MTT Glutathione DCFH-DA	BRL 3A	Cell viability ↓ LDH ↑ ROS ↑	[97]
Silver (30–50)	MTT DCFH-DA	Human alveolar cell line	Cell viability ↓ ROS ↑	[98]
Zinc oxide (50–70)	ELISA Flow-cytometry	Human colon carcinoma cells	Oxidative stress↑ Cell viability↓ Inflammatory biomarkers	[99]
Zinc oxide (307–419)	Comet micronucleus test MTT	Human cervix carcinoma cell line (HEp-2)	DNA damage Cell viability ↓	[100]
Iron oxide (100–150)	MTS	Human macrophages	Cell viability ↓	[101]
Iron oxide (30)	MTT	Murine macrophage cells	Cell viability ↓	[102]
Titanium oxide (<100)	ELISA Trypan blue DCFH-DA	Human lung cells	Oxidative stress ↑ DNA adduct formation Cytotoxicity ↑	[103]
Titanium oxide (160)	Comet micronucleus test	in vivo	DNA damage Genotoxicity	[104]
Fullerenes (178)	Micronucleus test	CHO HELA HEK293	DNA strand breakage Chromosomal damage	[105, 106]
pristine C60 fullerene	MCHC, MCV, phase contrast and fluorescent microscopy	human embryonic kidney HEK293 cells and in vitro activity of adult male mice	After 24 hours of incubation, C60 fullerenes displayed negligible toxicity toward HEK293 cells, with an IC50 value of 383.4 lg/mL and no harmful effect in the range of 75–150 mg/kg.	[107]
Silica (15–46)	DCFH-DA Commercial kit	Human bronchoalveolar carcinoma cells	ROS ↑ LDH ↑ Malondialdehyde ↑	[108]

TABLE 7.1 *(Continued)*

In Vitro and In Vivo Toxicity Activities of Various Nanocarriers

Nanocarrier with particle size	Assay technique	Species/cell culture	Mechanism	Reference
Silica (43)	DCFH-DA 5,5,6,6-tetraethyl-benzimidazo-lylcarbo-cyanide iodine	Hepatocellular carcinoma cells (HepG2)	ROS ↑ Mitochondrial damage Oxidative stress ↑	[109]
SWCNTs (10–30)	Commercial kits	in vivo	LDH ↑ AST ↑ ALT ↑	[110]
MWCNTs (20)	MTT	Lung cancer cells	Cell viability ↓	[111]
Copper oxide (50)	MTT LDH	Human lung epithelial cells	Cell viability ↓ LDH ↑ Lipid peroxidation ↑	[112]
CuO NPs (< 50 nm)	Glutathione levels (GSH), Catalase (CAT), superoxide dismutase SOD, lipid peroxidation MDA	Male wester albino rats	CuO nanoparticles administered orally to rats result in severe liver damage, which may be brought about by oxidative stress.	[113]
Aluminum oxide (8–12)	MTT DHE	HBMVECs	Mitochondrial function ↓ Oxidative stress ↑ Alter proteins expression of the BBB	[114]
Aluminum oxide (160)	MTT	HMSC	Cell viability ↓	[115]
CoO NPs	NADPH oxidase activity, superoxide anion generation, lipid peroxidation (MDA), lactose dehydrogenase (LDH)	In vitro homolysis assay and in vivo activity of Swiss mice	Reactive oxygen species, which generated TNF-a, and cell death were both considerably elevated by CoO NPs. Through the activation of caspase-8, p38 MAPK, and caspase-3, this TNF-a contributes significantly to cell death. Regarding CoNPs' in vitro carcinogenic potential in long-term environments, there is insufficient proof.	[116]
Co_3O_4 NPs 47.0 + 20.3) nm and (84.6+ 30.05) nm	ICP-MS	A549 lung carcinoma cells; gastrointestinal, Caco-2 colorectal adenocarcinoma cells and SH-SY5Y neuroblastoma cells	A549 was the cells that was most susceptible to Co3O4 NPs. Concerns regarding the hazards of airborne exposure to the pulmonary system at work and at home are raised by the DNA damage, apoptosis, and oxidative stress potentials of Co_3O_4 NPs on A549 cells, as well as the very low toxicity shown on the liver, gastrointestinal, and nervous systems.	[117]

7.4 TOXICOLOGICAL STUDIES OF METALLIC NANOPARTICLES

7.4.1 In Vitro Toxicity of MNPs

These insights into how the chemical stability of metallic nanoparticles affects their toxicity suggest that a sensible strategy for studying nanoparticle toxicity should start with an evaluation of the physico-chemical characteristics of the suspensions both at the start of the toxicity study and at its conclusion. While conducting these experiments, it is essential to keep an eye out for the possibility that metallic nanoparticles could lead to oxidative stress in cellular targets as a response to the nanoparticles [95]. On the basis of in vitro evidence, "the smaller the more toxic" cannot, however, be generalized because NPs of some materials may become more toxic as their size decreases, while in other instances, the toxicity may become more poisonous as their size increases. For instance, 100-nm titanium dioxide (TiO_2) is claimed to trigger programmed cell death compared to 15-nm particles, in contrast to gold nanoparticles, where larger particles are thought to be harmless. The results from in vitro toxicity studies of MNPs are difficult to compare because different researchers used different cell lines, different NP sizes, distinctive surface modification, varying amounts and periods of NP application onto the cells. Therefore, before any conclusions are reached for therapeutic uses, it is crucial to collect substantial data from in vitro tests of exposure. For instance, mice were used. Chen et al. studied various-sized citrate-capped gold nanospheres with diameters of 3, 5, 8, 12, 17, 37, 50, and 100 nm. The smallest and largest sizes 3 and 5 nm and 50 and 100 nm were found to be harmless at the doses tested, while mice exposed to the intermediate size range from 8 to 37 nm exhibited severe toxicity, including weight loss, loss of appetite, changes in the color of their fur, and shorter average lifespans. It was thought that the buildup of intermediate-sized particles caused damage to the liver, spleen, and lungs [96].

7.4.2 In Vivo Toxicity in Mammals

Numerous mammalian investigations have demonstrated the neurotoxicity of metallic NPs. AlNPs administered systemically raised the expression of genes associated with autophagy activities in the brain, which decreased the expression of tight-junction proteins and simultaneously improved the blood–brain barrier (BBB) permeability [97]. AuNPs detected in the brains of intraperitoneally administered mice were further shown to be able to cross the BBB; nevertheless, of all the organs that were examined, the brain's gold concentration was the lowest. It's interesting to note that neither obvious CNS damage nor behavioral alterations in mice were seen [98]. Due to acute exposure in the male rats to AuNPs, it reduced the amounts of carbonyl protein and thiol barbituric acid reactive compounds in the rats brains. In addition, the hippocampal, striatal, and cerebral cortex showed decreased catalase activity and decreased energy metabolism. Intriguingly, prolonged exposure to AuNPs only caused the brain's catalase to be inhibited and the cerebral cortex's energy metabolism to be repressed [99].

Furthermore, mice exposed repeatedly to citrate-stabilized AgNPs experienced severe synaptic degeneration, mainly in the hippocampal region of the brain, which may have a negative impact on nerve and cognitive function [100]. Accordingly, after

7 days of treatment, adult female rats fed AgNPs intragastrically experienced modest hippocampus and neuron shrinkage, apart from astrocyte swelling. The study also revealed a notable rise in interleukin-4 (IL-4) levels in the blood. Researchers hypothesized that neurodegeneration resulted from inflammatory responses after exposure to AgNPs [101]. After intraperitoneal administration, neurotoxicity and impaired BBB functions were also noted. These effects were linked to altered trace element concentrations in the serum and brain, decreased antioxidant enzyme activity, promotion of apoptosis and inflammatory processes, and down-regulated expression of tight junction proteins [102]. Similar to humans, rats exposed chronically and intragastrically to low doses of silver nanoparticles developed silver toxicity and had their hippocampal-dependent memory and cognitive coordination functions impaired [103]. Contrarily, Dbrowska-Bouta et al. showed in their studies on male Wistar rats that administration of AgNPs at low concentrations did not cause neurotoxicity in behavioral tests; however, more thorough examinations revealed anomalies in the myelin sheaths and a changed expression of myelin proteins [104].

7.4.3 Mechanisms of Metallic Nanoparticles Toxicity

Cellular toxicity is caused by a variety of mechanisms, including physical degeneration, the release of toxic ions, and the production of reactive oxygen species. Acute toxic reactions to the MNPs can be produced either directly at the cell membrane or by cellular absorption, which causes membrane damage. NPs have a variety of characteristics, including a wide surface area and different morphologies [105]. The fundamental reason for the non-specific damage to cell membranes is the abrasive nature of MNPs, such as magnesium oxide NPs, which damaged bacterial cell membranes and led to leakage of cell contents and ultimately cell death. In other investigations, it was discovered that using zinc oxide and silver in high amounts can harm bacterial cell membranes. The cell membrane destruction is more obvious in unicellular organisms [106].

MNPs have a strong potential to form reactive oxygen species (ROS), such as peroxides, oxygen ions, and free radicals, due to their physicochemical characteristics, which include a high chemical reactivity and a large surface area to volume ratio [107]. Different processes through which ROS interact with cells might result in oxidative stress. Several intracellular organelles may be damages. The primary bodily parts that are susceptible to oxidation, such as lipids, proteins, nucleic acids, lose their functionality as a result of oxidative stress. A few MNPs have a propensity to cause oxidative stress. The main contributors to the development of oxidative stress are particle cell interactions, active redox cycling on the surface of MNPs caused by transition metal-based NPs, and reactive surfaces of MNPs containing prooxidant functional groups [108].

7.4.4 Membrane Damage

Acute toxic reactions to the MNPs can be produced either directly at the cell membrane or by cellular absorption. NPs have a variety of characteristics, including a large surface area and a wide range of morphologies [109]. The main reason for the non-specific damage to cell membranes is the abrasive nature of MNPs, including magnesium oxide NPs, which damaged bacterial cell membranes and led to leakage of cell contents and finally cell death. In other investigations, it was discovered that using zinc oxide and silver in high amounts can harm bacterial cell membranes [110]. In unicellular organisms, cell membrane damage is more pronounced. It is unclear how NPs' antimicrobial activity works mechanistically. It is believed that NPs gather close to the microbial cell membrane and enter the cell through pit formation or membrane disruption. After crossing the bacterial cell membrane, NPs either generate free radicals or engage in protein–protein interactions that render the enzymes inactive. By altering the internal environment of microbial cells, these events seem to be the cause of cellular death [111].

7.4.5 Release of Toxic Dissolved Species

Ion release is a characteristic of MNPs. Due to the release of metal ions or even their interaction with one another, they contribute to the toxicity of nanometallic monomers and metallic oxide. It has been demonstrated that exposure to solubilized metallic ions is the primary cause of the biological effects of MNPs in cells. The put forth notion connects metallic nanoparticles chemical activity to their potential toxicity toward prokaryotic and eukaryotic cells. It is well known that the chemical stability of metallic particles or the speciation of dissolving metallic compounds influences the toxicity of dissolved and micron-sized metallic nanoparticles. We investigate the possibility of identifying which metallic nanoparticles, based on their redox characteristics and their capacity to be reduced, oxidized, or dissolved in biological environments. It is widely known that the toxicity of metallic compounds in non-nanoforms is correlated with their speciation [112]. The oxidation states of metals (As, Cr, Mn, Fe, etc.), their connection with particular ligands, and their concentration in solution can all significantly alter how these metals (and others) affect living things.

7.4.6 Reactive Oxygen Species Generation

Molecules possessing at least one oxygen atom and one or more unpaired electrons that are capable of surviving alone are referred to as reactive oxygen species (ROS). Free oxygen radicals such as the superoxide anion radical, hydroperoxyl radical, hydroxyl radical, and singlet oxygen, as well as free nitrogen radicals, are all present in this group. Small amounts of ROS are created during physiologically relevant cell processes, such as inflammatory or aerobic respiration processes, which are mostly found in macrophages and hepatocytes. Signaling molecules make up the majority of reactive oxygen species. Using the dye 2–7′-dichloro-dihydro-fluorescein diacetate, intracellular reactive oxygen species (ROS) were detected in serum-free media (DCFH-DA). Intracellular esterases convert the luminous 2–7′-dichlorodi hydrofluorescein (DCFH) from the stable, non-fluorescent molecule DCFH-DA, which is then swiftly oxidized into the highly fluorescent compound (DCF) in the presence of hydroxyl radicals. For IP15 and HK-2, respectively, sub-confluent cells cultured in 60-mm Petri dishes were treated for 15 min with 10 or 20 M of DCFH-DA [113]. After being treated with various particle concentrations for 4 hours, cells were rinsed with PBS. The induction of ROS in cells was monitored using tert-butyl hydroperoxide (TBHP) solution. Cells were removed after exposure, sonicated to lyse them, and then centrifuged. Using a fluorimeter, ROS levels in the supernatants were measured at 520 nm for emission and 480 nm for excitation [114].

7.4.7 Toxicological Assessments of NPs

Several factors must be considered in the critical examination of a chosen NP in order to acquire reliable results and reduce the over-emphasized toxicity or safety conclusions inherent in a non-rigorous inquiry [115]. Since NPs are roughly the same size as viruses, non-molecular processes are used to absorb and transport them through tissues. The effectiveness of present risk assessment techniques would be compromised because traditional toxicological tests may not be appropriate with regard to NPs. The development of trustworthy in vitro and in vivo protocols is required for the systematic evaluation of the toxicity of any given NP based on the mode of uptake (inhalation, injection, ingestion, and permeation), chemical characteristics, shape and size, composition, concentration, solubility, charge, and other characteristics. Additional study is required to determine how they move through the body and what biological and toxicological consequences they have on people and other living things. The evaluation of NPs' potential toxicity and an understanding of the underlying biological interaction mechanisms are crucial steps in the creation of biocompatible (safe-by-design) NPs, which will help prevent serious future negative consequences [116].

7.4.8 Effect of MNPs Toxicity in Different Organs

Metallic nanoparticles (MNPs) have great promise as medication delivery systems and therapeutic agents for a range of conditions. The evaluation of toxicity is a crucial step in the creation of nanoformulations and receives a lot of attention as a result. For targeted distribution and treatment, formulations with individual or combined nanoparticle suspensions may be employed. In light of potential future treatments based on MNPs, this may be further assessed for safety-related concerns. The surface properties of the body determine how nanoparticles are distributed there. Future tests must take into account nanoparticle size, dosage, and entry pathways. MNPs taken intravenously stand as key targets for the liver and spleen, but other organs and organ systems need to be tested for toxicity [117]. As a result, the current chapter focuses on the toxic effects of MNPs on healthy cells and organs, as shown in Figure 7.2.

When discussing tissue engineering and regenerative medicine, toxicity concerns are crucial and of great concern. As previously indicated, in order to use MNPs in regenerative medicine, cells (the therapeutic agents) must first be labeled with MNPs before being implanted into the body. The therapeutic effectiveness of cell-based therapy can be considerably reduced by using hazardous particles over an extended period of time [118]. It is accurate to say at this time that toxicology is the study of harmful effects on humans, animals, and the environment caused by chemical, physical, and biological agents. Impaired mitochondrial activity, membrane leakage, and morphological alterations are the results of toxic cellular impacts. This can reduce the effectiveness of the therapy and have negative effects on cell survival, proliferation,

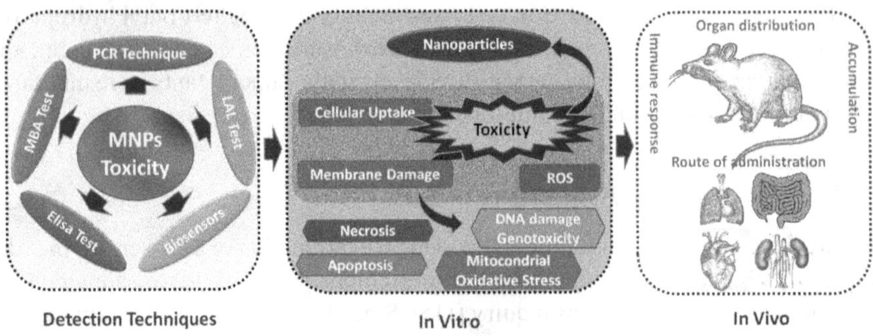

FIGURE 7.2 Techniques for detection of in vitro and in vivo toxicological parameters of MNPs.

and metabolic activity [119]. The possibility of MNPs migrating through the body, entering, and accumulating within organs is a persistent worry in situations where the MNPs are included in the therapy and transplanted into the body. The structural characteristics, dosage, and intended use of MNPs are among the key determinants of their toxicity to biological entities. Other important determinants include their intended use. The particles themselves may have a naturally hazardous chemical makeup. It's interesting to note that some substances, like gold, which are generally considered hazardous, are toxic at the nanoscale [120].

7.5 CONCLUSIONS

Bacterial pathogens may secrete a bench of diverse toxins that affect different aspects and cells of the host's innate immune response. In particular, it becomes apparent that pathogens change the immune cell function of macrophages and neutrophils and cause disease. Various in vitro and in vivo studies of the pathogens were conducted to explore the molecular functions of bacterial and foodborne toxins on host cells and how these pathogenic microorganisms promote infection in their hosts. More detailed knowledge of host responses toward infecting microbes needs to be analyzed. In the context of bacterial infections in different animal species, including humans, the detection of pathogens and toxins at such an early stage is a bit difficult and critical to preventing bacterial infection and food poisoning. By monitoring the effects of chemical substances in living animal beings (in vivo) or in animal and human cell lines, toxicity that affects humans is explored (in vitro). In standardized experimental designs using various model systems, toxicogenomic research gathers gene expression profiles and histopathological evaluation data for hundreds of medicines and contaminants. These data are a priceless resource for studying how drugs affect biological systems across the entire genome. There are various methods available for the detection of pathogens and toxins. Each developed protocol has its own features and corresponding advantages/disadvantages in terms of sensitivity, cost-effectiveness, and user friendliness. This chapter has been reviewed, considering various methods along with sensor-based applications for detection and availability to the scientific community.

ACKNOWLEDGEMENTS

The authors gratefully acknowledge the contributions of their collaborators and co-workers mentioned in the cited references. The authors are also thankful to **Dr. Jamilur R. Ansari**, Dronacharya College of Engineering, Gurgaon, India, for the needful editing in the Chapter. **MSA** is thankful to SGT College of Pharmacy, SGT University, Gurgaon-Badli Road, Chandu, Budhera, Gurugram, Haryana-122505 for their support. **BK** and **AS** are thankful to the School of Pharmaceutical Sciences (SPS), Siksha O Anusandhan University, Kalinganagar, Bhubaneswar, Odisha, and also **SKD** is thankful to Institute of Pharmacy, Kalyani, West Bengal, India.

CONFLICT OF INTEREST

The authors declare no conflict of interest.

REFERENCES

1. Casadevall, A., L. Pirofski. "Host-Pathogen Interactions: Basic Concepts of Microbial Commensalism, Colonization, Infection, and Disease". Infection and Immunity 68 (2000): 6511–6518.
2. Lounibos, L.P. "Invasions by Insect Vectors of Human Disease". Annual Review of Entomology 47 (2002): 233–266.
3. Ramalho-Ortigao, M., D.J. Gubler. 147—Human Diseases Associated with Vectors (Arthropods in Disease Transmission): Hunter's Tropical Medicine and Emerging Infectious Diseases (10th Edition, pp. 1063–1069). Elsevier (2020).
4. Farajollahi, A., D.M. Fonseca, L.D. Kramer, A.M. Kilpatrick. " 'Bird Biting' Mosquitoes and Human Disease: A Review of the Role of Culex Pipiens Complex Mosquitoes in Epidemiology". Infection, Genetics and Evolution 11 (2011): 1577–1585.
5. Badiaga, S., P. Brouqui. "Human Louse-Transmitted Infectious Diseases". Clinical Microbiology and Infection 18 (2012): 332–337.
6. Fuente, J., A. Estrda-Pena, J. Venzal, K. Kocan, D. Sonenshine. "Overview: Ticks as Vectors of Pathogens That Cause Disease in Humans and Animals". Frontiers in Bioscience 13 (2008): 6938–6946.
7. Al-Mebairik, N.F., T.A. El-Kersh, Y.A. Al-Sheikh, M. Ali, M. Marie. "A Review of Virulence Factors, Pathogenesis, and Antibiotic Resistance in Staphylococcus Aureus". Reviews in Medical Microbiology 27 (2016): 50–56.
8. Rudkin, J.K., R.M. Mcloughlin, A. Preston, R.C. Massey. "Bacterial Toxins: Offensive, Defensive, or Something Else Altogether?" PLOS Pathogens (2017): 1–12.
9. Edae, C., E.K. Wabalo. "Bacterial Toxins and Their Modes of Action: A Review Article Bacterial Toxins and Their Modes of Action : A Review Article". Journal of Medicine, Physiology and Biophysics 55 (2019): 10–16.
10. Balfanz, J., P. Rautenberg, U.W.E. Ullmann. "Molecular Mechanisms of Action of Bacterial Exotoxins". Zentralblatt für Bakteriologie 284 (1996): 170–206.
11. Evans, D.J., D.G. Evans. "Escherichia Coli in Diarrheal Disease". In Medical Microbiology (4th Edition). University of Texas Medical Branch (1996).
12. Argudín, M.Á., M.C. Mendoza, M.R. Rodicio. "Food Poisoning and Staphylococcus Aureus Enterotoxins". Toxins 2, (2010): 1751–1773.
13. Melvin, J.A., E.V. Scheller, J.F. Miller, P.A. Cotter. "Bordetella Pertussis Pathogenesis: Current and Future Challenges". Nature Reviews Microbiology 12, no. 4 (2014): 274–288.

14. Carbonetti, N.H., G.V. Artamonova, R.M. Mays, Z.E.V. Worthington. "Pertussis Toxin Plays an Early Role in Respiratory Tract Colonization by Bordetella Pertussis". Infection and Immunity 71, no. 11 (2003): 6358–6366.
15. Andreasen, C., D.A. Powell, N.H. Carbonetti. "Pertussis Toxin Stimulates IL-17 Production in Response to Bordetella Pertussis Infection in Mice". PLOS One 4, no. 9 (2009): e7079.
16. Liu, S., Y. Zhang, M. Moayeri, J. Liu, D. Crown, R.J. Fattah, A.N. Wein, Z.-X. Yu, T. Finkel, S.H. Leppla. "Key Tissue Targets Responsible for Anthrax-Toxin-Induced Lethality". Nature 501, no. 7465 (2013): 63–68.
17. Ribot, W.J., R.G. Panchal, K.C. Brittingham, G. Ruthel, T.A. Kenny, D. Lane, B. Curry, T.A. Hoover, A.M. Friedlander, S. Bavari. "Anthrax Lethal Toxin Impairs Innate Immune Functions of Alveolar Macrophages and Facilitates Bacillus Anthracis Survival". Infection and Immunity 74, no. 9 (2006): 5029–5034.
18. Popov, S.G., R. Villasmil, J. Bernardi, E. Grene, J. Cardwell, A. Wu, D. Alibek, C. Bailey, K. Alibek. "Lethal Toxin of Bacillus Anthracis Causes Apoptosis of Macrophages". Biochemical and Biophysical Research Communications 293, no. 1 (2002): 349–355.
19. Tournier, J.-N., A. Quesnel-Hellmann, J. Mathieu, C. Montecucco, W.-J. Tang, M. Mock, D.R. Vidal, P.L. Goossens. "Anthrax Edema Toxin Cooperates with Lethal Toxin to Impair Cytokine Secretion During Infection of Dendritic Cells". The Journal of Immunology 174, no. 8 (2005): 4934–4941.
20. Lee, J., H. Kim, J.-J. Jeon, H.-S. Kim, K.A. Zeller, L.L.A. Carter, J.F. Leslie, Y.-W. Lee. "Population Structure of and Mycotoxin Production by Fusarium Graminearum from Maize in South Korea". Applied and Environmental Microbiology 78, no. 7 (2012): 2161–2167.
21. Park, S.-H., D. Kim, J. Kim, Y. Moon. "Effects of Mycotoxins on Mucosal Microbial Infection and Related Pathogenesis". Toxins 7, no. 11 (2015): 4484–4502.
22. Turner, P.C. "The Molecular Epidemiology of Chronic Aflatoxin Driven Impaired Child Growth". Scientifica 2013 (2013): 152879.
23. Rushing, B.R., M.I. Selim. "Structure and Oxidation of Pyrrole Adducts Formed Between Aflatoxin B2a and Biological Amines". Chemical Research in Toxicology 30, no. 6 (2017): 1275–1285.
24. Bui-Klimke, T.R., F. Wu. "Ochratoxin A and Human Health Risk: A Review of the Evidence". Critical Reviews in Food Science and Nutrition 55, no. 13 (2015): 1860–1869.
25. Čulig, B., M. Bevardi, J. Bošnir, S. Serdar, D. Lasić, A. Racz, A. Galić, Ž. Kuharić. "Presence of Citrinin in Grains and Its Possible Health Effects". African Journal of Traditional, Complementary and Alternative Medicines 14, no. 3 (2017): 22–30.
26. Dwivedi, H.P., L.-A. Jaykus. "Detection of Pathogens in Foods: The Current State-of-the-Art and Future Directions". Critical Reviews in Microbiology 37, no. 1 (2011): 40–63.
27. Oliver, S.P., B.M. Jayarao, R.A. Almeida. "Foodborne Pathogens in Milk and the Dairy Farm Environment: Food Safety and Public Health Implications". Foodborne Pathogens & Disease 2, no. 2 (2005): 115–129.
28. Alahi, M.E.E., S.C. Mukhopadhyay. "Detection Methodologies for Pathogen and Toxins: A Review". Sensors 17, no. 8 (2017): 1885.
29. Raphael, B.H., M.J. Choudoir, C. Lúquez, R. Fernández, S.E. Maslanka. "Sequence Diversity of Genes Encoding Botulinum Neurotoxin Type F". Applied and Environmental Microbiology 76, no. 14 (2010): 4805–4812.
30. Seydel, U., A.B. Schromm, R. Blunck, K. Brandenburg. "Chemical Structure, Molecular Conformation, and Bioactivity of Endotoxins". Chemical Immunology 74 (2000): 5–24.
31. Chen, C.-S., R.A. Durst. "Simultaneous Detection of Escherichia Coli O157: H7, Salmonella spp. and Listeria Monocytogenes with an Array-Based Immunosorbent Assay Using Universal Protein G-Liposomal Nanovesicles". Talanta 69, no. 1 (2006): 232–238.

32. Meng, J., M.P. Doyle. "Introduction: Microbiological Food Safety". Microbes and Infection 4, no. 4 (2002): 395–397.
33. Yolken, R.H. "Enzyme Immunoassays for the Detection of Infectious Antigens in Body Fluids: Current Limitations and Future Prospects". Reviews of Infectious Diseases 4, no. 1 (1982): 35–68.
34. Rozand, C., P.C.H. Feng. "Specificity Analysis of a Novel Phage-Derived Ligand in an Enzyme-Linked Fluorescent Assay for the Detection of Escherichia Coli O157: H7". Journal of Food Protection 72, no. 5 (2009): 1078–1081.
35. Lall, S., G. Nataraj, P. Mehta. "Use of Culture-and ELISA-Based Toxin Assay for Detecting Clostridium Difficile, a Neglected Pathogen: A Single-Center Study from a Tertiary Care Setting". Journal of Laboratory Physicians 9, no. 4 (2017): 254.
36. Zhao, X., C.-W. Lin, J. Wang, D. Hwan Oh. "Advances in Rapid Detection Methods for Foodborne Pathogens". Journal of Microbiology and Biotechnology 24, no. 3 (2014): 297–312.
37. Zhou, Y., F.-G. Pan, Y.-S. Li, Y.-Y. Zhang, J.-H. Zhang, S.-Y. Lu, H.-L. Ren, Z.-S. Liu. "Colloidal Gold Probe-Based Immunochromatographic Assay for the Rapid Detection of Brevetoxins in Fishery Product Samples". Biosensors and Bioelectronics 24, no. 8 (2009): 2744–2747.
38. Mukhopadhyay, A., U.K. Mukhopadhyay. "Novel Multiplex PCR Approaches for the Simultaneous Detection of Human Pathogens: Escherichia Coli 0157: H7 and Listeria Monocytogenes". Journal of Microbiological Methods 68, no. 1 (2007): 193–200.
39. Guilbault, C., A.-C. Labbé, L. Poirier, L. Busque, C. Beliveau, M. Laverdiere. "Development and Evaluation of a PCR Method for Detection of the Clostridium Difficile Toxin B Gene in Stool Specimens". Journal of Clinical Microbiology 40, no. 6 (2002): 2288–2290.
40. Kralik, P., M. Ricchi. "A Basic Guide to Real Time PCR in Microbial Diagnostics: Definitions, Parameters, and Everything". Frontiers in Microbiology 8 (2017): 108.
41. Klein, D. "Quantification Using Real-Time PCR Technology: Applications and Limitations". Trends in Molecular Medicine 8, no. 6 (2002): 257–260.
42. Alhogail, S., G.A.R.Y. Suaifan, M. Zourob. "Rapid Colorimetric Sensing Platform for the Detection of Listeria Monocytogenes Foodborne Pathogen". Biosensors and Bioelectronics 86 (2016): 1061–1066.
43. Singhal, N., M. Kumar, P.K. Kanaujia, J.S. Virdi. "MALDI-TOF Mass Spectrometry: An Emerging Technology for Microbial Identification and Diagnosis". Frontiers in Microbiology 6 (2015): 791.
44. Drawz, S.M., R.A. Bonomo. "Three Decades of β-Lactamase Inhibitors". Clinical Microbiology Reviews 23, no. 1 (2010): 160–201.
45. Narsaiah, K., S.N. Jha, R. Bhardwaj, R. Sharma, R. Kumar. "Optical Biosensors for Food Quality and Safety Assurance: A Review". Journal of Food Science and Technology 49, no. 4 (2012): 383–406.
46. Kuo, T. Y. "Localized Surface Plasmon Coupled Fluorescence Fiber Optic Based Biosensing". Metal-Enhanced Fluorescence (2010): 183.
47. Villalobos, P., M.I. Chávez, Y. Olguín, E. Sánchez, E. Valdés, R. Galindo, M.E. Young. "The Application of Polymerized Lipid Vesicles as Colorimetric Biosensors for Real-Time Detection of Pathogens in Drinking Water". Electronic Journal of Biotechnology 15, no. 1 (2012): 4.
48. Wang, D., Z. Wang, J. Chen, A.J. Kinchla, S.R. Nugen. "Rapid Detection of Salmonella Using a Redox Cycling-Based Electrochemical Method". Food Control 62 (2016): 81–88.
49. Nag, A., A. Iqbal Zia, X. Li, S.C. Mukhopadhyay, J. Kosel. "Novel Sensing Approach for LPG Leakage Detection—Part II: Effects of Particle Size, Composition, and Coating Layer Thickness". IEEE Sensors Journal 16, no. 4 (2015): 1088–1094.

50. Ercole, C., M. Del Gallo, L. Mosiello, S. Baccella, A. Lepidi. "Escherichia Coli Detection in Vegetable Food by a Potentiometric Biosensor". Sensors and Actuators B: Chemical 91, no. 1–3 (2003): 163–168.
51. Pohanka, M., P. Skládal. "Electrochemical Biosensors-Principles and Applications". Journal of Applied Biomedicine 6, no. 2 (2008).
52. Gau, J., E.H. Lan, B. Dunn, C.-M. Ho, J.C.S. Woo. "A MEMS Based Amperometric Detector for E. Coli Bacteria Using Self-Assembled Monolayers". Biosensors and Bioelectronics 16, no. 9–12 (2001): 745–755.
53. Yang, L., R. Bashir. "Electrical/Electrochemical Impedance for Rapid Detection of Foodborne Pathogenic Bacteria". Biotechnology Advances 26, no. 2 (2008): 135–150.
54. Yang, L., Y. Li, C.L. Griffis, M.G. Johnson. "Interdigitated Microelectrode (IME) Impedance Sensor for the Detection of Viable Salmonella Typhimurium". Biosensors and Bioelectronics 19, no. 10 (2004): 1139–1147.
55. Syaifudin, A.R.M., K.P. Jayasundera, S.C. Mukhopadhyay. "A Low Cost Novel Sensing System for Detection of Dangerous Marine Biotoxins in Seafood". Sensors and Actuators B: Chemical 137, no. 1 (2009): 67–75.
56. Senturk, E., S. Aktop, P. Sanlibaba, B.U. Tezel. "Biosensors: A Novel Approach to Detect Food-Borne Pathogens". Applied Microbiology: Open Access 4, no. 3 (2018).
57. Yang, X., J. Kirsch, A. Simonian. "Campylobacter spp. Detection in the 21st Century: A Review of the Recent Achievements in Biosensor Development". Journal of Microbiological Methods 95, no. 1 (2013): 48–56.
58. Rocha-Gaso, M.-I., C. March-Iborra, Á. Montoya-Baides, A. Arnau-Vives. "Surface Generated Acoustic Wave Biosensors for the Detection of Pathogens: A Review". Sensors 9, no. 7 (2009): 5740–5769.
59. Hammer, M.U., A. Brauser, C. Olak, G. Brezesinski, T. Goldmann, T. Gutsmann, J. Andrä. "Lipopolysaccharide Interaction Is Decisive for the Activity of the Antimicrobial Peptide NK-2 Against Escherichia Coli and Proteus Mirabilis". Biochemical Journal 427, no. 3 (2010): 477–488.
60. Carmon, K.S., R.E. Baltus, L.A. Luck. "A Piezoelectric Quartz Crystal Biosensor: The Use of Two Single Cysteine Mutants of the Periplasmic Escherichia Coli Glucose/Galactose Receptor as Target Proteins for the Detection of Glucose". Biochemistry 43, no. 44 (2004): 14249–14256.
61. Shen, Z., M. Huang, C. Xiao, Y. Zhang, X. Zeng, P.G. Wang. "Nonlabeled Quartz Crystal Microbalance Biosensor for Bacterial Detection Using Carbohydrate and Lectin Recognitions". Analytical Chemistry 79, no. 6 (2007): 2312–2319.
62. Lian, Y., F. He, H. Wang, F. Tong. "A New Aptamer/Graphene Interdigitated Gold Electrode Piezoelectric Sensor for Rapid and Specific Detection of Staphylococcus Aureus". Biosensors and Bioelectronics 65 (2015): 314–319.
63. Sharma, H., R. Mutharasan. "Rapid and Sensitive Immunodetection of Listeria Monocytogenes in Milk Using a Novel Piezoelectric Cantilever Sensor". Biosensors and Bioelectronics 45 (2013): 158–162.
64. Guntupalli, R., I. Sorokulova, E. Olsen, L. Globa, O. Pustovyy, V. Vodyanoy. "Biosensor for Detection of Antibiotic Resistant Staphylococcus Bacteria". JoVE: Journal of Visualized Experiments 75 (2013): e50474.
65. Javed, M.N., E.S. Dahiya, A.M. Ibrahim, M. Alam, F.A. Khan, F.H. Pottoo. "Recent Advancement in Clinical Application of Nanotechnological Approached Targeted Delivery of Herbal Drugs". In Nanophytomedicine (pp. 151–172). Springer (2020).
66. Mishra, S., S. Sharma, M.N. Javed, F.H. Pottoo, M.A. Barkat, M.S. Alam, M. Amir, M. Sarafroz. "Bioinspired Nanocomposites: Applications in Disease Diagnosis and Treatment". Pharmaceutical Nanotechnology 7, no. 3 (2019 Jun 1): 206–219.

67. Javed, M.N., E.S. Dahiya, A.M. Ibrahim, M.S. Alam, F.A. Khan, F.H. Pottoo. "Recent Advancement in Clinical Application of Nanotechnological Approached Targeted Delivery of Herbal Drugs". In Nanophytomedicine (pp. 151–172). Springer (2020).
68. Aslam, M., M.N. Javed, H.H. Deeb, M.K. Nicola, M. Mirza, M.S. Alam, M. Akhtar, A. Waziri. "Lipid Nanocarriers for Neurotherapeutics: Introduction, Challenges, Blood-Brain Barrier, and Promises of Delivery Approaches". CNS & Neurological Disorders-Drug Targets (Formerly Current Drug Targets-CNS & Neurological Disorders) 21 (2022).
69. Javed, M.N., F.H. Pottoo, A. Shamim, M.S. Hasnain, M.S. Alam. "Design of Experiments for the Development of Nanoparticles, Nanomaterials, and Nanocomposites". In Design of Experiments for Pharmaceutical Product Development (pp. 151–169). Springer (2021).
70. Javed, M.N., M.S. Alam, A. Waziri, F.H. Pottoo, A.K. Yadav, M.S. Hasnain, F.A. Almalki. "QbD Applications for the Development of Nanopharmaceutical Products". In Pharmaceutical Quality by Design (pp. 229–253). Academic Press (2019 Jan 1).
71. Kumar, R., G. Dhamija, J.R. Ansari, M.N. Javed, M.S. Alam. "C-Dot Nanoparticulated Devices for Biomedical Applications". In Nanotechnology (pp. 271–299). CRC Press (2022).
72. Bharti, C., M.S. Alam, M.N. Javed, M. Khalid, F.A. Saifullah, R. Manchanda. "Silica Based Nanomaterial for Drug Delivery". Nanomaterials: Evolution and Advancement towards Therapeutic Drug Delivery (Part II) (2021 Jun 2): 57.
73. Mall, S.K., T. Yadav, A. Waziri, M.S. Alam. "Treatment Opportunities with *Fernandoa adenophylla* and Recent Novel Approaches for Natural Medicinal Phytochemicals as a Drug Delivery System". Exploration of Medicine 3 (2022): 516–539.
74. Naseh, M.F., J.R. Ansari, M.S. Alam, M.N. Javed. "Sustainable Nanotorus for Biosensing and Therapeutical Applications". In Handbook of Green and Sustainable Nanotechnology: Fundamentals, Developments and Applications (pp. 1–21). Springer International Publishing (2022 Aug 19).
75. Sunilbhai, C.A., M.S. Alam, K.K. Sadasivuni, J.R. Ansari. "SPR Assisted Diabetes Detection". In Advanced Bioscience and Biosystems for Detection and Management of Diabetes (pp. 91–131). Springer (2022).
76. Singhal, S., M. Gupta, M.S. Alam, M.N. Javed, J.R. Ansari. "Carbon Allotropes-Based Nanodevices: Graphene in Biomedical Applications". In Nanotechnology (pp. 241–269). CRC Press (2022).
77. Pottoo, F.H., N. Tabassum, M.N. Javed, S. Nigar, R. Rasheed, A. Khan, M. Barkat, M.S. Alam, A. Maqbool, M.A. Ansari, G.E. Barreto. "The Synergistic Effect of Raloxifene, Fluoxetine, and Bromocriptine Protects Against Pilocarpine-Induced Status Epilepticus and Temporal Lobe Epilepsy". Molecular Neurobiology 56, no. 2 (2019 Feb): 1233 1247.
78. Pottoo, F.H., S. Sharma, M.N. Javed, M.A. Barkat, Harshita, M.S. Alam, M.J. Naim, O. Alam, M.A. Ansari, G.E. Barreto, G.M. Ashraf. "Lipid-Based Nanoformulations in the Treatment of Neurological Disorders". Drug Metabolism Reviews 52, no. 1 (2020 Jan 2): 185–204.
79. Pottoo, F.H., M.N. Javed, M. Barkat, M.S. Alam, J.A. Nowshehri, D.M. Alshayban, M.A. Ansari. "Estrogen and Serotonin: Complexity of Interactions and Implications for Epileptic Seizures and Epileptogenesis". Current Neuropharmacology 17, no. 3 (2019 Mar 1): 214–231.
80. Pottoo, F.H., N. Tabassum, M.N. Javed, S. Nigar, S. Sharma, M.A. Barkat, M.S. Alam, M.A. Ansari, G.E. Barreto, G.M. Ashraf. "Raloxifene Potentiates the Effect of Fluoxetine Against Maximal Electroshock Induced Seizures in Mice". European Journal of Pharmaceutical Sciences 146 (2020 Apr 15): 105261.

81. Waziri, A., C. Bharti, M. Aslam, P. Jamil, M. Mirza, M.N. Javed, U. Pottoo, A. Ahmadi, M.S. Alam. "Probiotics for the Chemoprotective Role Against the Toxic Effect of Cancer Chemotherapy". Anti-Cancer Agents in Medicinal Chemistry (Formerly Current Medicinal Chemistry-Anti-Cancer Agents) 22, no. 4 (2022 Feb 1): 654–667.
82. Javed, M.N., M.H. Akhter, M. Taleuzzaman, M. Faiyazudin, M.S. Alam. "Cationic Nanoparticles for Treatment of Neurological Diseases". In Fundamentals of Bionanomaterials (pp. 273–292). Elsevier (2022 Jan 1).
83. Kumari, N., N. Daram, M.S. Alam, A.K. Verma. "Rationalizing the Use of Polyphenol Nano-Formulations in the Therapy of Neurodegenerative Diseases". CNS & Neurological Disorders-Drug Targets (Formerly Current Drug Targets-CNS & Neurological Disorders) 21, no. 10 (2022 Dec 1): 966–976.
84. Raj, S., R. Manchanda, M. Bhandari, M.S. Alam. "Review on Natural Bioactive Products as Radioprotective Therapeutics: Present and Past Perspective". Current Pharmaceutical Biotechnology 23 (2022).
85. Ibrahim, A.M., L. Chauhan, A. Bhardwaj, A. Sharma, F. Fayaz, B. Kumar, M. Alhashmi, N. AlHajri, M.S. Alam, F.H. Pottoo. "Brain-Derived Neurotropic Factor in Neurodegenerative Disorders". Biomedicines 10, no. 5 (2022 May): 1143.
86. Alam, M.S., A. Garg, F.H. Pottoo, M.K. Saifullah, A.I. Tareq, O. Manzoor, M. Mohsin, M.N. Javed. "Gum Ghatti Mediated, One Pot Green Synthesis of Optimized Gold Nanoparticles: Investigation of Process-Variables Impact Using Box-Behnken Based Statistical Design". International Journal of Biological Macromolecules. 104 (2017 Nov 1): 758–767.
87. Alam, M.S., M.N. Javed, F.H. Pottoo, A. Waziri, F.A. Almalki, M.S. Hasnain, A. Garg, M.K. Saifullah. "QbD Approached Comparison of Reaction Mechanism in Microwave Synthesized Gold Nanoparticles and Their Superior Catalytic Role Against Hazardous Nirto-Dye". Applied Organometallic Chemistry 33, no. 9 (2019 Sep): e5071.
88. Pandit, J., M.S. Alam, J.R. Ansari, M. Singhal, N. Gupta, A. Waziri, K. Sharma, F.H. Potto. "Multifaced Applications of Nanoparticles in Biological Science". In Nanomaterials in the Battle Against Pathogens and Disease Vectors (pp. 17–50). CRC Press (2022).
89. Alam, M.S., M.F. Naseh, J.R. Ansari, A. Waziri, M.N. Javed, A. Ahmadi, M.K. Saifullah, A. Garg. "Synthesis Approaches for Higher Yields of Nanoparticles". In Nanomaterials in the Battle Against Pathogens and Disease Vectors (pp. 51–82). CRC Press (2022).
90. Javed, M.N., F.H. Pottoo, M.S. Alam. "Metallic Nanoparticle Alone and/or in Combination as Novel Agent for the Treatment of Uncontrolled Electric Conductance Related Disorders and/or Seizure, Epilepsy & Convulsions". Patent Acquired on October 10 (2016): 40.
91. Agrawal, A., R.A. Tripp, L.J. Anderson, S. Nie. "Real-Time Detection of Virus Particles and Viral Protein Expression with Two-Color Nanoparticle Probes". Journal of Virology 79, no. 13 (2005): 8625–8628.
92. Zhu, L., S. Ang, W.-T. Liu. "Quantum Dots as a Novel Immunofluorescent Detection System for Cryptosporidium Parvum and Giardia Lamblia". Applied and Environmental Microbiology 70, no. 1 (2004): 597–598.
93. Yang, G.-J., J.-L. Huang, W.-J. Meng, M. Shen, X.-A. Jiao. "A Reusable Capacitive Immunosensor for Detection of Salmonella spp. Based on Grafted Ethylene Diamine and Self-Assembled Gold Nanoparticle Monolayers". Analytica Chimica Acta 647, no. 2 (2009): 159–166.
94. Wan, Y., Z. Lin, D. Zhang, Y. Wang, B. Hou. "Impedimetric Immunosensor Doped with Reduced Graphene Sheets Fabricated by Controllable Electrodeposition for the Non-Labelled Detection of Bacteria". Biosensors and Bioelectronics 26, no. 5 (2011): 1959–1964.

95. Xia, T., M. Kovochich, J. Brant, M. Hotze, J. Sempf, T. Oberley, C. Sioutas, J.I. Yeh, M.R. Wiesner, A.E. Nel. "Comparison of the Abilities of Ambient and Manufactured Nanoparticles to Induce Cellular Toxicity According to an Oxidative Stress Paradigm". Nano Letters 6, no. 8 (2006): 1794–1807.
96. Chen, Y.S., Y.C. Hung, I. Liau, G.S. Huang. "Assessment of the In Vivo Toxicity of Gold Nanoparticles". Nanoscale Research Letters 4 (2009): 858–864.
97. Chen L., B. Zhang, M. Toborek. "Autophagy Is Involved in Nanoalumina-Induced Cerebrovascular Toxicity". Nanomedicine 9, no. 2 (2013): 212–221.
98. Lasagna-Reeves C., D. Gonzalez-Romero, M.A. Barria, I. Olmedo, A. Clos, V.M. Sadagopa Ramanujam, A. Urayama., L. Vergara, M.J. Kogan, C. Soto. "Bioaccumulation and Toxicity of Gold Nanoparticles After Repeated Administration in Mice". Biochemical and Biophysical Research Communications 393, no. 4 (2010): 649–655.
99. Ferreira, G.K., E. Cardoso, F.S. Vuolo, L.S. Galant, M. Michels, C.L. Gonçalves, G.T. Rezin, F. Dal-Pizzol, R. Benavides, G. AlonsoNúñez, V.M. Andrade, E.L. Streck, M.M. da Silva Paula. "E of Acute and Long-Term Administration of Gold Nanoparticles on Biochemical Parameters in Rat Brain". Materials Science & Engineering C-Materials for Biological Applications 79 (2017): 748–755.
100. Skalska, J., M. Frontczak-Baniewicz, L. Strużyńska. "Synaptic Degeneration in Rat Brain After Prolonged Oral Exposure to Silver Nanoparticles". Neurotoxicology 46 (2015, Jan 1), 145–154. https://doi.org/10.1016/j.neuro.2014.11.002.
101. Xu, L., A. Shao, Y. Zhao, Z. Wang, C. Zhang, Y. Sun, J. Deng, L.L. Chou. "Neurotoxicity of Silver Nanoparticles in Rat Brain After Intragastric Exposure". Journal of Nanoscience and Nanotechnology 15, no. 6 (2015): 4.
102. Lebda, M.A., K.M. Sadek, H.G. Tohamy, T.K. Abouzed, M. Shukry, M. Umezawa, Y.S. El-Sayed. "Potential Role of α-Lipoic Acid and Ginkgo Biloba Against Silver Nanoparticles-Induced Neuronal Apoptosis and Blood-Brain Barrier Impairments in Rats". Life Sciences 212 (2018): 251–260.
103. Jeng, H.A., J. Swanson. "Toxicity of Metal Oxide Nanoparticles in Mammalian Cells". Journal of Environmental Science and Health Part A 41, no. 12 (2006): 2699–2711.
104. Węsierska, M., K. Dziendzikowska, J. Gromadzka-Ostrowska, J. Dudek, H. Polkowska-Motrenko, J.N. Audinot, A.C. Gutleb, A. Lankoff, M. Kruszewski. "Silver Ions Are Responsible for Memory Impairment Induced by Oral Administration of Silver Nanoparticles". Toxicology Letters 290 (2018): 133–144.
105. Dąbrowska-Bouta, B., M. Zięba, J. Orzelska-Górka, J. Skalska, G. Sulkowski, M. Frontczak-Baniewicz, S. Talarek, J. Listos, L. Strużyńska. "Influence of a Low Dose of Silver Nanoparticles on Cerebral Myelin and Behavior of Adult Rats". Toxicology (2016): 363–364, 29–36.
106. Klabunde, K.J., J. Stark, O. Koper, C. Mohs, D.G. Park, S. Decker et al. "Nanocrystals as Stoichiometric Reagents with Unique Surface Chemistry". The Journal of Physical Chemistry 100 (1996): 12142–12153.
107. Wiesner, M.R., G.V. Lowry, P. Alvarez, D. Dionysiou, P. Biswas. Assessing the Risks of Manufactured Nanomaterials. ACS Publications (2006).
108. Oberdörster, G., E. Oberdörster, J. Oberdörster. "Nanotoxicology: An Emerging Discipline Evolving from Studies of Ultrafine Particles". Environmental Health Perspectives 113 (2005): 823.
109. Manke, A., L. Wang, Y. Rojanasakul. "Mechanisms of Nanoparticle-Induced Oxidative Stress and Toxicity". BioMed Research International 2013 (2013): 942916.
110. Klabunde, K.J., J. Stark, O. Koper, C. Mohs, D.G. Park, S. Decker, Y. Jiang, I. Lagadic, D. Zhang. "Nanocrystals as Stoichiometric Reagents with Unique Surface Chemistry". The Journal of Physical Chemistry A 100 (1996): 1214212153.

111. Morones, J.R., J.L. Elechiguerra, A. Camacho, K. Holt, J.B. Kouri, J.T. Ramírez, M.J. Yacaman. "The Bactericidal Effect of Silver Nanoparticles". Nanotechnology 16 (2005): 2346.
112. Prabhu, R.R., M.A. Khadar. "Characterization of Chemically Synthesized CdS Nanoparticles". Journal of Physics 65, no. 5 (2005): 801–807.
113. Gupta, A.K., M. Gupta. "Synthesis and Surface Engineering of Iron Oxide Nanoparticles for Biomedical Applications". Biomaterials 26, no. 18 (2005): 3995.
114. Zhang, Z., S. Weichenthal, J.C. Kwong, R.T. Burnett, M. Hatzopoulou, M. Jerrett, A. van Donkelaar, L. Bai, R.V. Martin, R. Copes, H. Lu, P. Lakey, M. Shiraiwa, H. Chen. "A Population-Based Cohort Study of Respiratory Disease and Long-Term Exposure to Iron and Copper in Fine Particulate Air Pollution and Their Combined Impact on Reactive Oxygen Species Generation in Human Lungs". Environmental Science & Technology 55, no. 6 (2021): 3807–3818, https://doi.org/10.1021/acs.est.0c05931.
115. Johnston, L.J., N. Gonzalez-Rojano, K.J. Wilkinson, B. Xing. "Key Challenges for Evaluation of the Safety of Engineered Nanomaterials". NanoImpact 18 (2020): 100219, https://doi.org/10.1016/j.impact.2020.100219.
116. Frohlich, E., S. Salar-Behzadi. "Toxicological Assessment of Inhaled Nanoparticles: Role of In Vivo, Ex Vivo, In Vitro, and In Silico Studies". International Journal of Molecular Sciences 15, no. 3 (2014): 4795–4822, https://doi.org/10.3390/ijms15034795.
117. Awaad, A. "Histopathological and Immunological Changes Induced by Magnetite Nanoparticles in the Spleen, Liver and Genital Tract of Mice Following Intravaginal Instillation". The Journal of Basic & Applied Zoology 71 (2015): 32–47.
118. Huang, D.M., T.H. Chung, Y. Hung, F. Lu, S.-H. Wu, C.-Y. Mou, M. Yao, Y.-C. Chen. "Internalization of Mesoporous Silica Nanoparticles Induces Transient but Not Sufficient Osteogenic Signals in Human Mesenchymal Stem Cells". Toxicology and Applied Pharmacology 231, no. 2 (2008): 208–215.
119. C.-Y. Yang, J.-K. Hsiao, M.-F. Tai, S.-T. Chen, H.-Y. Cheng, J.-L. Wang, H.-M. Liu. "Direct Labeling of hMSC with SPIO: The Long-Term Influence on Toxicity, Chondrogenic Differentiation Capacity, and Intracellular Distribution". Molecular Imaging and Biology 13, no. 3 (2011): 443–451.
120. Chen, L., J.P. Giesy, P. Xie. "The Dose Makes the Poison". Nature Nanotechnology 6, no. 6 (2011): 329.

8 Metallic Nanoparticles for Skins and Photothermal Therapy

Manish Kumar, Navneet Mehan, Shailendra Bhatt, Md Sabir Alam, and Rupesh K. Gautam

8.1 INTRODUCTION

8.1.1 Metallic Nanoparticles

In nanotechnology, we can study the science and engineering of nanosized particles, that is, 100 nm or less than 100 nm. Despite the fact that the phrase is new, in the development of the most effective technologies nanomaterials are most commonly used. Nanotechnology has gained popularity in current years as a result of its applications in "targeted drug delivery, electronic storage systems, magnetic separation and pre-concentration of target analytes, biotechnology, and gene and drug delivery vehicles" [1–12]. These nanoparticles have the potential to have a substantial influence on society due to their wide variety of possible uses. Although nanomaterials are relatively young, "their history extends back to 1959, when a physicist at Cal Tech, Richard P. Feynman, predicted their arrival". "There is lots of room at the bottom", he said in one of his classes, signifying that scaling down to the nanoscale and the secret to future technology and advancement were to start from the bottom. As nanotechnology advanced, new nanostructures evolved with characteristics that set them apart from their bigger counterparts. This variance in physiochemical parameters is due to nanoparticles' high surface-to-volume ratio. Because of their unique features, they are most commonly used in biomedical applications because a wide range of biological functions take place on the nanoscale [13].

Generally, nanoparticles used in biotechnology have particle sizes ranging from 10 to 500 nm, rarely reaching 700 nm. Cell surfaces and inside cells can be evaluated and assigned various aspects of the cell because of the tiny size of these particles. Correspondingly, its prospective use in the delivery of drugs and noninvasive imaging provided a number of benefits over traditional pharmaceuticals. In order to fully exploit nanoparticles, systemic distribution, stability, biocompatibility, and the ability to target specific regions of the body are all requirements for nanoparticulate systems. Targeting systems that are more particular are meant to recognize certain cells, like cancer cells. Using a nanoparticle and ligand with specific binding activity for the target cells, this can be accomplished. It is also possible to connect several copies of therapeutic materials to nanoparticles, which increases the amount

DOI: 10.1201/9781003317319-8

of these molecules at the sick spot. By controlling the nanoparticle size (>3–5 nm), it is possible to influence the concentration and kinetics of the active chemical. This particle size regulation, along with a stealth ligand surface coating, allows them to hide from the body's natural immune system, extending the time during which they are able to circulate in the blood. These advancements in biotechnology have created a plethora of new possibilities for molecular diagnostics and therapy. Using a number of approaches, these nanocarriers may be constructed to operate as imaging probes, like "ultrasound (US), X-ray, computed tomography (CT), positron emission tomography (PET), magnetic resonance imaging (MRI), optical imaging, and surface-enhanced Raman imaging" once they have been targeted (active or passive) (SERS). As a result, these so-called "molecular imaging probes" might provide some evidence on identifying irregularities in various body parts and body organs to find out the severity of illness and evaluate treatment effectiveness. It is possible to see cell activity and monitor molecular processes in living organisms while inflicting no harm to them by utilizing short-term molecular imaging techniques. Nanoparticles like "gold and silver nanoparticles, magnetic nanoparticles (iron oxide), nano-shells, and nano-cages" have been employed and modified over the course of the year to allow their usage as a therapeutic agent and diagnostic agent for various disease targeting applications like anticancer, antibacterial, and neurological [14–23].

8.1.1.1 Iron Oxide Nanoparticles

Iron (III) oxide (Fe_2O_3) is a paramagnetic inorganic compound with a reddish brown color. It's one of three primary iron oxides, with FeO and Fe_3O_4 being the other two. The mineral magnetite, which is made up of Fe_3O_4, is also super-paramagnetic in nature. SPION nanoparticles ("super-paramagnetic iron oxide nanoparticles") have emerged as intriguing prospects for a wide range of biological applications, with "enhanced resolution contrast agents for MRI, targeted drug delivery and imaging, hyperthermia, gene therapy, stem cell tracking, molecular/cellular tracking, magnetic separation technologies (e.g., rapid DNA sequencing), and early detection of inflammatory, cancer, and diabetes diseases due to their ultrafine size, magnetic properties, and biocompatibility". To produce high-resolution MR images, high-magnetization nanoparticles are required for each of these biological applications. Super-paramagnetic nanoparticles, in general, resemble suitable imaging probes for use as contrast agents for MRI since the MR signal intensity is considerably modified without sacrificing in vivo stability. T1 as well as T2 relaxation times of water protons are reduced by all contrast agents, which affects the signal intensity of the imaging tissue [24].

Restricted size distribution between 10 and 250 nm in diameter for homogenous and targeted imaging has been developed in response to increasing knowledge of the molecular biology of a wide range of disorders. It's tough to create magnetic nanoparticles with this particular diameter, and several chemical approaches have been proposed for their synthesis. "Microemulsions, sol-gel syntheses, sono-chemical reactions, hydrothermal reactions, hydrolysis and hemolysis of precursors, flow injection syntheses, and electrospray syntheses" are examples of these approaches. Co-precipitation of iron salts using chemical means is, nevertheless, the most prevalent approach for producing magnetite nanoparticles. The key benefit of the co-precipitation method is that it may produce a high number of nanoparticles by controlling the size distribution. Because of this fact, the crystal's development is controlled by kinetic variables. Ultra-small particles of iron oxide (USPIO) (10–40 nm) and tiny particles of iron oxide (SPIO) (60–150 nm) are some of the magnetic contrast agents generated using these approaches. CLIO (10–30 nm) is also known as CLION (10–30 nm) because of the cross-linking of the iron oxide nanoparticles (MIONs) with dextran. Because of the carboxylation of the dextran coating, the blood clearance half-life is reduced. As a result, "ferumoxytol (AMAG Pharmaceuticals), a carboxyalkylated polysaccharide-coated iron oxide nanoparticle", has previously been approved as a first-pass contrast agent, macrophage uptake is unspecific and too rapid to improve uptake in macrophage-rich plaques [25].

A specific surface coating may be applied to these particles, allowing them to be readily linked to other particles like "medicines, proteins, enzymes, antibodies, or nucleotides and guided to an organ, tissue, or tumor", improving cellular uptake. These nanoparticles have been used to produce targeted molecular imaging probes that directly target bodily tissue or cells, whereas standard contrast agents disperse somewhat nonspecifically. Conroy and colleagues, for example, "created (chlorotoxin) (CTX), a biocompatible iron oxide nano-probe coated with poly (ethylene glycol) (PEG)" that may specifically target glioma tumors using a surface-bound targeting peptide. Furthermore, MRI investigations revealed that the nanoprobe accumulated preferentially within gliomas. Apopa et al. developed iron oxide nanoparticles, which are used to improve cell permeability by producing reactive oxygen species (ROS) and stabilizing microtubules, according to another study. Iron oxide nanoparticles have only a few applications in biological imaging [26].

These findings shed light on the bio reactivity of tailored iron nanoparticles, which may have uses in "medical imaging and drug delivery". The synthesis and alteration of iron oxide complexes, as well as "dendrimers, polymeric nanoparticles, liposomes, and solid lipid nanoparticles", is a most inventive topic of research. On specific types of neuronal cells, these magnetic nanoparticles cause some toxic effects; however, this remains a source of worry [27].

8.1.1.2 Gold Nanoparticles

Gold nanoparticles are suspensions (or colloids) of gold particles ranging in size from 1 to 100 nanometers. These colloidal solutions have an extensive history dating back to the Roman era, when they were employed to decorate glass. However, "it wasn't until Michael Faraday's study in the 1850s" that the contemporary scientific evaluation of colloidal gold began, when he realized that colloidal gold solutions are unique from bulk gold in terms of their properties. As a result, the colloidal solution has either a vivid red ("smaller than 100 nm") or a dirty yellowish tint ("for particles larger than 100 nm") (for larger particles). These gold nanoparticles' intriguing optical features are due to their one-of-a-kind collaboration with light. When there is a light source nearby, the surface plasmon decays either radioactively or by converting absorbed light to heat in a non-radioactive manner. Aqueous solution absorption peaks around 520 nm for gold nanospheres with a particle size of roughly 10 nm due to their LSPR. Larger particles suffer from electromagnetic retardation, so these nanospheres show a Stokes shift as the size of the nanospheres increases.

Furthermore, the structure of colloidal gold nanoparticles influences their properties and uses. Rod-shaped nanoparticles, for example, have two resonances: Because plasmon oscillation occurs along the short axis of each nanorods and long axis plasmon oscillation occurs in response to the nanorods aspect ratio (or length-to-width ratio), both of these phenomena may be explained. The red wavelength lies on the long-axis LSPR transitions from the visible to the near-infrared spectrum. When the nanorods aspect ratio is raised and the oscillator strength steadily rises. Rod-like particles, for example, exhibit both "transverse and longitudinal absorption peaks, and shape anisotropy influences their self-assembly". Gold nanoparticles are the topic of extensive investigation due to their remarkable optical properties, with several applications in "biological imaging, electronics, and materials science". Over the years, dependable and high-yielding technologies for developing gold nanoparticles for specific purposes have been developed; these include both spherical and non-spherical forms [28].

Turkevich et al. pioneered the most widely used technique for the manufacture of mono-disperse round gold nanoparticles in 1951, which was further developed by "Frens et al. in 1973". This approach uses citrate as a reducing agent to chemically reduce gold compounds like hydrogen tetra-chloro-aurate ($HAuCl_4$). This approach yields monodisperse spherical gold nanoparticles with a size range of 10–20 nm. Brown and Natan, the other way around, reported the creation of larger gold nanoparticles with a size range between 30 and 100 nm by seeding Au^{3+} with hydroxylamine. Following that, the shape of these gold nanoparticles was altered, resulting in "rod, triangular, polygonal rods, and spherical particles". Gold nanoparticles results have different characteristics that allow for a large surface area to volume ratio. On the surface of gold nanoparticles the ligands can be linked together, like "oligonucleotides, proteins, and antibodies that include functional groups like thiols, mercaptans, phosphines, and amines", all of these things like to stick to the surface of gold. The development of gold nanoconjugates with much better LSPR gold nanoparticles has led to imaging methods that are easier to use but more effective, such as "dark-field imaging, SERS, and optical imaging for the diagnosis of various disease states" [29–34].

El Sayed et al., for example, were the first to use gold nanoparticles for imaging cancer by putting AuNPs selectively into the nucleus of cancer cells. They used PEG to attach an arginine-glycine-aspartic acid peptide (RGD) and a nuclear localization signal peptide (NLS) to a 30-nm gold nanoparticle (AuNP) so that the AuNPs would preferentially go into the nucleus of cancer cells [35].

According to Qian et al., the production of gold nanoparticles, such as those used to target tumour sites, was also reported as an in vivo Raman scattering probe. A Raman reporter was used to encode these gold nanoparticles, which were then enclosed in a thiol-modified PEG shell. Once the gold nanoparticles had been pegylated, they were linked to an antibody against the epidermal growth factor receptor, which tends to be overproduced in cancer cells, to specifically target tumor cells. SERS was used to observe the Raman amplification from these customized particles at 633 or 785 nm by electronic transitions [36].

SERS might be utilized to make exceedingly precise identifications and can identify and actively target EGFR-positive tumor xenografts in animal models as well as human cancer cells, according to Qian and colleagues' results. Furthermore, gold nanorods' employment as photothermal agents distinguishes them from all other nanoparticles. Photothermal therapy (PTT) is a treatment that involves exposing a photosensitizer to a specific band of light (mainly IR). The sensitizer is thrilled after being activated, and it gives out heat as a kind of vibrational energy. The actual therapeutic method that destroys the targeted cells is heat. One of the most recent advances is the use of gold nanoparticles in photothermal therapy. Spherical gold nanoparticles' absorption qualities haven't been discovered to be appropriate for in vivo applications yet, according to previous research. This is because the peak absorptions for 10 nm diameter have been restricted at 520 nm. Furthermore, the transmission window for skin, tissues, and hemoglobin ranges from 650 to 900 nm. This was overcome by Murphy and coworkers' recent development of gold nanorods; they were able to change the absorption peak of these nanoparticles from 550 nm to 1 m merely by altering the aspect ratios of the nanorods. When laser light (in the IR region) is used to preferentially collect rod-shaped gold nanoparticles in tumors, the tissue around them is just mildly warmed, and the nanorods transform light into heat, destroying the cancerous cells. Gold nanorods are distinguished from other nanoprobes by this potential use. A lack of compatibility with existing high-resolution imaging technologies like MRI led to the creation of nano-cages and nano-shells [37].

8.1.1.3 Nano-shells and Nano-cages

According to "Neeves and Birnboim", LSPR modes with wavelengths adjustable over the electromagnetic spectrum might be produced by a combined round particle made up of a dielectric core and metallic shell. The near-infrared resonance of a silica–gold nano-shell particle was later established by "Naomi Halas and Peter Nordlander of Rice University". By recently generating amine-terminated silica spheres, they created silica–gold nanospheres. After that, gold colloidal suspension (1–2 nm in size) was added to the amine-end silica spheres. To protect the silica core and amine terminal, AuNPs was plated using chemical reduction. Although this development is frequently used, the difficulty of controlling the width and softness of these shells renders it unsuitable for the routine manufacture of particle-sized nano-shells. They also demonstrated that when these nano-shells were targeted onto tumor cells using photothermal ablation in vitro and in vivo, they were able to completely remove tumor tissue from the body. Nano-shells with near-infrared resonant properties can be used for whole blood immunoassays in a different study. Additionally, they proved that the nano-shells, when used in conjunction with antibodies, serve as specific analyte recognition sites. The analyte encourages dimmer formation, which alters the LSPR. As a continuation of previous research in this field, Jaeyun et al. developed multifunctional magnetic gold nano-shells (Mag-GNS) using Fe_3O_4 nanoparticles as the magnetic core. Photothermal treatment is possible thanks to the gold nano-shells, while the Fe_3O_4 nanoparticles allow for MRI diagnostics. An antibody can be used to specifically target cancer cells by attaching the Mag-GNS to the PEG linker. These particles can be used to identify malignancy using MRI once they have been localized, as well as to remove cancer cells, utilizing photothermal therapy [38]. The hollow, porous gold nanoparticles known as gold nano-cages absorb light in the near-infrared region, similar to gold nano-shells. In boiling water, Xia and his colleagues were the ones who made the initial discovery of the interaction between silver nanoparticles and chloroauric acid ($HAuCI_4$). LSPR peaks may also be tailored to the near-infrared region by changing the wall thickness and porosity. They've been used in the delivery and/or controlled release of drugs in a similar way as nano-shells. Furthermore, tiny things such as magnetic nanoparticles can be housed within the hollow interiors to create multifunctional hybrid nanostructures for diagnostic imaging and therapy [39].

8.1.1.4 Silver Nanoparticles

Silver nanoparticles are silver particles that are 1–100 nm in size. While many are categorized as "silver", some silver alloys include a significant amount of silver oxide due to the high surface-to-bulk silver atom ratio. The history of ionic silver, like that of gold nanoparticles, is extensive, and it was first used to make yellow glass. Bone cement, surgical instruments, surgical masks, and other products now all include silver nanoparticles. Furthermore, it has been established that the right quantity of ionic silver may be used to heal wounds. It has been shown that silver nanoparticles are more effective than silver sulfadiazine in the treatment of chronic wounds. Silver Nano, a material that combines silver nanoparticles on the surfaces of home appliances, was also created and commercialized by Samsung. Furthermore, because of their intriguing physiochemical features, these nanomaterials have aroused a lot of interest in SERS-based biological imaging. Because of their high surface plasmon resonance and broad effective scattering cross-section, individual silver nanoparticles are excellent molecular labeling candidates. As a result, several silver oxide nanoprobes with specific targets are now being developed. They're usually made by reducing a silver salt in the presence of a colloidal stabilizer using a reducing agent like "sodium borohydride. Polyvinyl alcohol, poly (vinyl pyrrolidone), bovine serum albumin (BSA), citrate, and cellulose" are the most commonly used colloidal stabilizers. Recent breakthroughs in the development of silver nanoparticles include the employment of fresh, innovative methods, such as the reduction of glucose using starch as a stabilizer and the implantation of ions. It's also worth noting that not all nanoparticles are created equal. Its efficacy has been demonstrated to be influenced by its size and form [40].

Elechiguerra et al. found that silver nanoparticles interact with HIV-1 in a size-dependent manner, with particles in the 1–10 nm range attaching predominantly to the virus. According to the researchers, when silver nanoparticles come into contact with HIV-1, they selectively target the virus' gp120 glycoprotein knobs. In a similar vein, Furno and colleagues created biomaterials by encasing silver oxide nanoparticles in silicone and then soaking them in CO_2. The development of these cutting-edge biomaterials was motivated by the need to reduce the spread of antibiotic-resistant bacteria. Silver impregnation (instead of coating) of medical polymers may lead to an antibacterial biomaterial, despite inconsistent results [41].

Despite their lack of popularity compared to gold nanoparticles and nano-shells, the current medical field owes a great deal to the discovery of these microscopic particles. The amazing thing about noble metals is that they may be utilized indefinitely, even when new uses and protocols are created [42].

8.1.2 Applications of MNPs [43]

1. Metallic nanoparticles are used in the delivery of drugs.
2. Metallic nanoparticles are used in the delivery of proteins and peptides.
3. Metallic nanoparticles play an important role in gene delivery.
4. Metallic nanoparticles are also used in tissue engineering.
5. They find application in enzymology.
6. It is also used in the form of a surface coating of nanoparticles.
7. They find application in biosensing devices.
8. Metallic nanoparticles are used in diagnostics and theranostics.
9. It is also used in cosmetics and wound dressings.

8.1.3 Types of Metallic Nanoparticles for Dermatological Disorders

8.1.3.1 Anti-fungal Activity of MNPs

Fifteen to thirty percent of the populations are found to be affected by fungal infection at this time. The prevalence is as high as 70% in some common groups, for example, athletes and miners. Approximately two thousand kinds of fungi cause these extremely contagious diseases. They are characterized as civilization diseases due to their high frequency of incidence. The main causes of the rise in fungal infections are frequently used antibiotics and immunosuppressive drugs, which suppress the immune system of the human body. As a result, scientists are now putting a lot of work into finding compounds/molecules that have a broad spectrum of antifungal activity. Metal NPs, such as "copper NPs (CuNPs) and silver NPs", are examples of compounds that have biological activity (AgNPs). Copper was used to maintain personal hygiene throughout the middle ages. As a result, research into the antifungal properties of nanosized copper was undertaken. CuNPs have been incorporated into medically relevant synthetic materials and were found to limit the development of germs on these materials. The fungi "*Aspergillus flavus* and *Penicillium chrysogenum*" were found to be sensitive to AgNPs and CuNPs. Its antifungal effectiveness against Candida albicans has also been demonstrated by Usman et al. According to the researchers, CuNPs inhibited the development of "methicillin-resistant *Staphylococcus aureus*, *Bacillus subtilis*, *Pseudomonas aeruginosa*, and *Salmonella choleraesuis*". CuNPs are now used as additives in "medical textiles such as socks, t-shirts, and workwear made of cotton fabric" as a result of this research. AgNPs are also effective antifungal agents. Monteiro et al. discovered that AgNPs were effective against "*Candida albicans* and *Candida glabrata*". Adding AgNPs to acrylic resins has also been proven to inhibit fungal adhesion to this material. On the other hand, there is still a scarcity of data on AgNPs "cytotoxic and gene-toxic effects" on human cells [44].

Because of their low cost and high biocidal activity, AgNPs are widely used in different dermatological goods and cosmetics ("spray, cream, colloidal NPs, ointment, gel") for the inhibition and management of fungal infection. At low inhibitory doses, a mixture of AgNPs and magnetic iron oxide nanoparticles has also been shown to have significant antibacterial and antifungal properties without harming mouse embryonic cells. When compared to other magnetic silver nano-composites identified

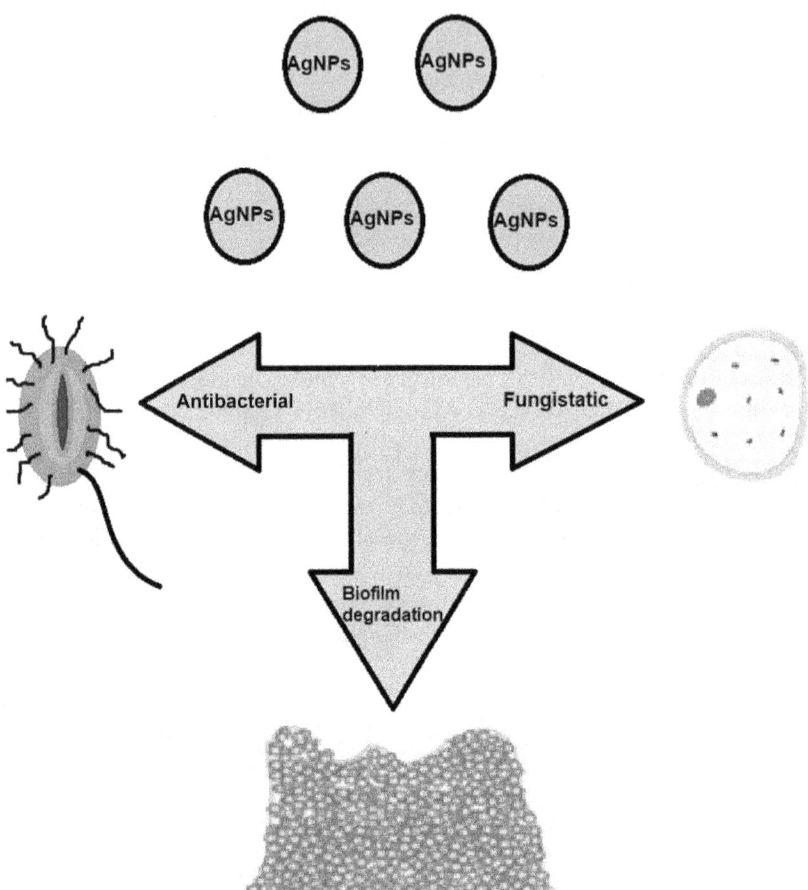

FIGURE 8.1 Antimicrobial activity of AgNPs.

so far, this form of nano-composite appears to have the most potent antibacterial properties. The antimicrobial activity of nanoparticles is shown in Figure 8.1 [45].

The mechanism behind metal nanoparticles' lethal impact on fungal cells is still unknown. Nasrollahi et al., on the other hand, observed that AgNPs have been found to bind to and permeate fungal membranes. Intracellular NPs may hinder cell division and cause cell death by targeting the respiratory chain [46].

8.1.3.2 Antibacterial Activity of MNPs

The most prevalent bacterial infections are those caused by *Streptococci* and *Staphylococci*. Because of antibiotic resistance, these opportunistic microbes produce infections that are difficult to treat. The most common symptoms are furuncles, carbuncles, and purulent irritation of the hair follicle. 7–10% of patients admitted to the hospital have skin or soft tissue infections, at a rate of 24.6 per 1000 people each year. Alternative therapies are being researched in order to save money on the $64,000 per person expense of treating a *Staphylococcus* infection. Folliculitis, furunculosis, and infectious impetigo are all caused by *Staphylococcus aureus*, among other bacteria. Furthermore, this bacterium is frequently linked to the development of atopic dermatitis (AD) [47].

Metal NPs have recently become popular due to advances in research. In plant dermatological remedies, silver and zinc NPs are used as microbicidal chemicals. AgNPs in the form of a colloidal solution were also shown to have a bactericidal effect against "*Streptococcus mutans, Streptococcus salivarius, Streptococcus salivarius*, and *Streptococcus mitis* strains". Furthermore, nano-metal oxides such as "titanium oxide nanoparticles (TiO_2 NPs) and zinc oxide nanoparticles (ZnO NPs)" have been demonstrated to suppress Staphylococcus strain proliferation [48].

8.1.3.3 Antimicrobial Activity of MNPs

The mechanism of NPs' antibacterial activity has yet to be fully understood. Nanosilver has been shown to influence single bacterial cells. In recent years, nanometals have been shown to react with thiol groups (–SH) on bacterial peptidoglycan. Murein is a protein that helps microorganisms form their cell walls. The bacterial cell is rendered inactive as a result of this procedure, as shown in Figure 8.2 [48].

Murein is a tri-peptide made up of "D-alanine, D-glutamic acid, and mezodiaminopimelic acid, N-acetylmuramic acid, and N-acetylglucosamine", which make up the peptidoglycan biopolymer. Murein is made up of polysaccharide chains that are

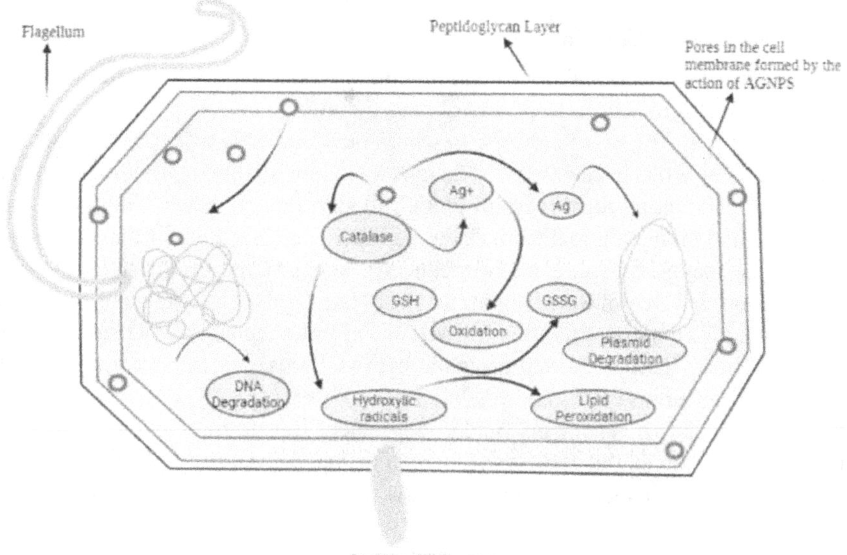

FIGURE 8.2 Mechanisms of nanoparticles antimicrobial activity.

joined together by peptide bridges. Chains are made up of "N-acetylglucosamine and N-acetylmuramic acid" disaccharides connected by a 1,4-glycosidic bond. A peptide bridge connects N-acetylmuramic molecules to the tetra-peptide. In terms of the cell wall's rigidity, or the polymer's crosslinking, it is determined by the number of peptide bridges. The peptidoglycan of gram-positive cells is more cross-linked than that of gram-negative cells. As a result, the more cross-linked the bacteria's cell membrane is, the greater their antibiotic resistance. This is due to the antibiotic's difficulty penetrating within the cell and, consequently, its destruction [49].

Methionine and cysteine, two building blocks of the cell wall, have an SH group on their side chains. The capacity to absorb oxygen and catalyze the process of oxidation is a distinguishing attribute of atomic silver. The oxygen deposited on the nanosurface silver combines with the bacterial cell wall's -SH, forming sulfur atom bonds. AgNPs cause respiratory chain enzymes to be inhibited, or by separating oxidative phosphorylation from respiratory electron transport, membrane permeability to protons and phosphates may be changed. As a result, the electron transport channels located across cell membranes (i.e., the respiratory chain) are damaged. Cell death arises as a result of this process [50].

AgNPs' reductive characteristics cause the tertiary protein structure to be destroyed, enzymatic activity to be lost, and protein denaturation to occur. AgNPs are absorbed by bacteria that have a broken or missing cell wall. AgNPs' catalytic properties cause the bacteria's genetic material to be oxidized. AgNPs have been discovered to link to bacterial DNA without breaking hydrogen bonds. Preventing the growth of DNA, this is necessary for bacterial multiplication. The mechanism underlying this procedure is still a mystery. Importantly, because eukaryotic species' cell walls are not formed of peptidoglycan, they are immune to the mechanism described earlier [51].

8.1.3.4 MNPs on Skin Cancer

Skin cancer is the most established malignant neoplasm in Caucasians in the United States and other countries, despite not being the most fatal form of malignancy. It accounts for 8 to 10% of all cancer cases in Poland, and new research suggests that this percentage will continue to rise. The most common malignant tumor is basal cell carcinoma, accounting for three-quarters of all skin malignancies. "Squamous cell carcinoma and malignant melanoma", on the other hand, account for the majority of mortality. A major risk factor for skin cancer is sunburn and excessive exposure to UV radiation. It is the most common cause of "basal cell carcinoma and squamous cell carcinoma". Ionizing radiation, as well as various chemical agents such as arsenic, insecticides, herbicides, and aromatic hydrocarbons, can cause normal cells to change into cancerous cells [52].

The molecular mechanisms underlying such transformations are currently unknown. It is thought that the slow change of melanocytes into dysplastic cells leads to the emergence of cancer cells in melanoma. The oxidative DNA damage UV radiation causes in human skin cells have been linked to the development of melanoma and non-melanoma skin cancers. As a result, never-ending hunt for effective chemical and physical filters, that operates as micro-mirrors and shields the skin from both UVA and UVB radiation in order to prevent skin cancer growth. Metal and metal oxide nanoparticles, such as ZnO NPs and TiO_2 NPs, can be used to protect the skin. These NPs, however, may have harmful consequences for the skin. Metal NP interactions with UV light, for example, can amplify their phototoxic and photoallergic properties. Reactive oxygen species (ROS) play a major role in photo-cancerogenicity and skin aging, and ZnO NPs and TiO_2 NPs have been shown to contribute to their generation. For example, NPs that include pro-oxidant functional groups on their reactive surfaces and transition metals in their structure can cause ROS production, as can particle–cell interactions. Nanostructured ZnO/TiO_2 is photo-stimulated by UV radiation, resulting in electron excitation. Surface-bound molecules (e.g., oxygen or water) can react with the developing exactions or electron-hole pairs to produce radicals. Photo-stimulated ZnO NPs, for example, may produce the hydroxyl radical (OH•) and the superoxide anion (O•), as well as singlet oxygen ($1O_2$). TiO_2NPs produce electron-hole pairs, where ROS such as O•, OH•, and hydrogen peroxide (H_2O_2) can be produced. In addition, the mitochondrial respiratory chain is disrupted, and NADPH-like enzyme systems are activated. NPs can cause intracellular ROS production. As a result, oxidative stress caused by NPs causes DNA cleavage, lipid peroxidation, and the formation of endogenous DNA adducts (8-OHdG). Consequently, more research into the effects of NPs on skin cells exposed to UV light is needed [53].

8.1.3.5 MNPs as UV Filters

UV radiation is one of the most prominent variables impacting the texture and appearance of the skin. Exposure to the sun's rays can have a variety of negative effects, from increased redness and the look of sunburn to more serious consequences including skin cancer and premature aging. In addition, disorders induced by light hypersensitivity are becoming more common. As a result, skin disease prevention relies heavily on the use of preventative cosmetics. UV filters are a major component in such compositions. TiO_2 NPs are one of the materials with important UV filter characteristics. These NPs can absorb light with a wavelength of 365 nm and are very stable in the presence of UV radiation. TiO_2 NPs have also been found to absorb light with wavelengths ranging from 200 to 400 nm. Furthermore, to boost their UVA (320–400 nm) and UVB (400–600 nm) safety, these nanoparticles have now been found to be added to a variety of cosmetic products (280–320 nm). Furthermore, TiO_2 NPs are non-irritating to the skin and have anti-allergic effects. Furthermore, unlike TiO_2 NPs, they do not penetrate the skin but rather concentrate in the stratum corneum, where they may be detected. ZnO NPs can be used successfully as sunscreen filters. Ultraviolet a (UVA) radiation may be absorbed by ZnO nanoparticles that have a wavelength of up to 380 nm in the same way that TiO_2 NPs do. However, their appeal may be attributed to their lack of durability and the fact that they do not leave white marks on the skin. Due to their physical qualities, TiO_2 NP and ZnO NPs both protect the skin. On the skin, they create agglomerates that operate as a micromirror, reflecting damaging radiation [54].

8.1.4 Skin Penetration of Metallic Nanoparticles

In addition to protecting the body from mechanical and physical injury, skin also acts as a barrier against invasive viruses, parasites, and hazardous chemicals entering the body. Additionally, skin aids in the digestion of proteins, lipids, carbs, and vitamins and helps keep fluids within the body from evaporating. Langerhans cells in the skin: the body's ability to regulate temperature and respond to antigens is heavily reliant on cells that can recognize and digest certain antigens, causing an inflammatory response. When it comes to the body's most vital chemicals, the skin is where they are synthesized. Neurosensory functions, including heat, pressure, and pain, are also controlled by the skin [55].

The skin consists of three layers: "the epidermis, dermis, and subcutaneous tissue", which are all located under the surface. Squamous keratinizing epithelium is the skin's thinnest and outermost layer and is composed of primary cells that form the epidermis. The most important cells in the epidermis are the keratinocytes, which account for 90–95% of all cells in the epidermis. As previously mentioned, in addition to the basal stratum basal and the prickly stratum spinosum, the epidermis also contains the granular strata of strata, the lamellar strata, the lamellar strata, and the lamellar lucidum, as well as the horny strata, the stratum corneum. The stratum basal also contains melanocytes, which are responsible for the production of melanin, as well as Merkel and Langerhans cells, which are sensory receptors. When the dermis is paired with connective tissue, it forms the basement membrane. Nerves and hair follicles are located in the dermis. The dermis also includes the sebaceous and sweat glands, blood vessels, and lymphatic vessels. Reticular fibers and collagen are two of the main components, along with fibroblasts, histiocytes, and mast cells, lymphocytes, and plasma cells. Energy is stored in the loose connective and adipose tissue in this layer of skin [56].

Due to the complexity of this structure, it is not surprising that nanometric structures penetrate human skin. Information may be passively carried by particles having a mass of less than 600 Da, as long as they are smaller than this. A recent study, on the other hand, demonstrates that the epidermis' stratum corneum is impermeable to nanoparticles (SC). For example, iron nanostructures have been demonstrated to penetrate the skin's lipid layer and hair follicle openings, and finally reach the granular layer of skin. Those creatures were found in the basal and spinous layers, where they had previously been thought to be absent. "Size, shape, chemical composition, stability, surface area, and charge" are only a few of the NPs' physicochemical characteristics that have an impact on their penetration, as shown in Figure 8.3. The skin's ability to absorb NPs is also influenced by its health. Intact skin can be permeated and penetrated by nanoparticles (NPs) less than 4 nm; damaged skin can only be permeated and penetrated by NPs 21 to 45 nm; and intact skin cannot be permeated or penetrated by NPs greater than 45 nm [57].

8.1.5 Commonly Used Metals and Metal Oxides Nanoparticles for Dermatological Disorders

8.1.5.1 Titanium Dioxide Nanoparticles (TiO_2 NPs)

Cosmetics, dazzling paints, meals, fabrics, self-cleaning surfaces, and computing equipment all use TiO_2 NPs. Sunscreens are increasingly including TiO_2 NPs as an active ingredient, acting as a UV filter, due to their excellent photo stability and ability to minimize photoallergy. TiO_2 NPs absorb light with wavelengths shorter than

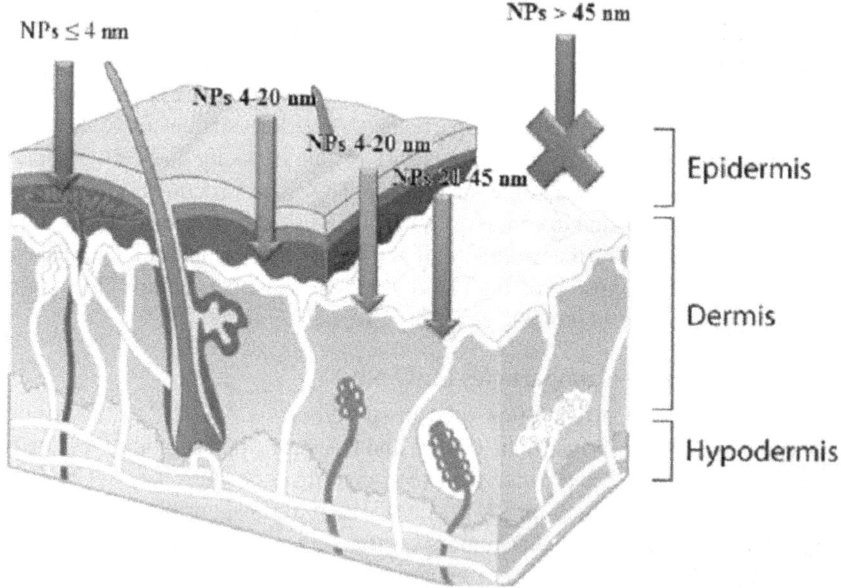

FIGURE 8.3 Human skin penetrations of nanoparticles.

365 nm, but their capacity to form aggregates and agglomerates decreases their protective capabilities. Agglomerates (100–200 nm) have been shown to have reduced UVB protection effectiveness and shift the absorption band to the longer UVA wavelength. TiO_2 NPs photocatalytic activity causes unfavorable free radical production in the skin. As a result, SiO_2 or Al_2O_3 is commonly used to coat these NPs [58].

TiO_2 NPs have been linked to toxicity in several studies. In an in vitro investigation, human keratinocytes (HaCaT cells) accumulated TiO_2 NPs, which were then ingested through endocytosis. In human keratinocytes, as a result, ROS were produced, mitochondrial DNA was damaged, and genotoxic effects were seen. TiO_2 NPs, on the other hand, can trigger autophagy and protect primary human keratinocytes from injury and death. In keratinocytes, autophagy maintains intracellular homeostasis while also degrading insoluble cytoplasmic toxicants like TiO_2 NPs. Autophagy, on the other hand, is unable to cease detrimental effects at high (cytotoxic) quantities of NPs and may even cause cell death. There is currently inadequate information to explain how skin cells and TiO_2 NPs come into contact. Consequently, additional study is needed to better understand the therapeutic features of these NPs in the field of dermatology (Gilbert et al., 2004). TiO_2 NPs are employed in pharmaceutics as a cream because they speed up wound healing. As a result, the effects of rutile and anatase, two of the most frequent crystalline forms of TiO_2 NPs, on human dermal fibroblasts were investigated. This sort of cell is important in the healing of wounds. Both forms of TiO_2 nanostructures were found to affect cell function, resulting in a reduction in fibroblast area, cell proliferation, motility, and the capacity to compress collagen. As a consequence, an anionic and hydrophobic polymer brush was grafted onto the NPs to protect them from oxidation. Functionalized NPs, on the other hand, did not adhere to the fibroblast membrane or penetrate the cells, therefore safeguarding the cells. The toxicity of NPs under UV light is an essential factor. It is possible that exposure to ultraviolet light may injure human keratinocyte cells that are protected by nanoparticles (NPs). The amount of ROS created during these reactions determines the severity of phototoxicity and cell membrane damage. It was discovered that the smaller the TiO_2 nanostructures, the more cell damage was generated, and in comparison to TiO_2 NPs, the rutile form demonstrated reduced phototoxicity. Furthermore, titanium dioxide nanoparticles (TiO_2 NPs) have the potential to decompose proteins in keratinocyte cells, particularly the tyrosine residues, increasing the phototoxic effect of UVA radiation [59].

8.1.5.2 Zinc Oxide Nanoparticles (ZnO NPs)

Another kind of nanoparticle often used in sunscreens is ZnO NP, which absorbs light with wavelengths shorter than 380 nm and has strong photostability. Compared to TiO_2NPs, ZnO NPs are a better sunscreen ingredient because they are transparent and almost completely absorb UVA rays, which are responsible for skin aging, allergic reactions, and malignant tumors on the skin. In the textile business, antibacterial ZnO NPs (20–30 nm) are employed in the production of textiles. They also have anti-inflammatory qualities, which is a bonus, as well as the ability to speed up wound healing by reducing skin infections. To understand why they are used in medicinal ointments, consider these qualities [60]. However, human skin may be adversely affected by these NPs. In human epidermal keratinocyte HaCaT cells, ZnO NPs have been shown to disturb mitochondrial function and cause lactate dehydrogenase (LDH) leakage into the medium. As well as producing reactive oxygen species (ROS), ZnO NPs kill HaCaT cells by damaging their organelles through oxidative stress and lipid peroxidation in the membrane. To put it another way, ZnO NPs have a significant impact on the expression of genes linked to the death of cells, included in this group are heme oxygenase 1 (HMOX 1), superoxide dismutation and glutathione peroxidation (GPx), and heat shock 70 kDa protein (HSP70). SOD (superoxide dismutase) is an enzyme that transforms the O• radical into molecular oxygen and peroxide. ZnO NPs increased the activity of SOD (H_2O_2). Glutathione levels, which protect cells from oxidative damage, were also much lower in these patients [79]. ZnO NPs interfered with mitochondrial activity, according to research, and provoked cell death in human dermal fibroblasts by activating p38 and p53 and inhibiting p53. This might be connected to their carcinogenic potential, as ZnO NPs have been shown to cause DNA damage in primary human keratinocyte skin cells. ZnO NPs are a common form of photocatalyst. Nanodermatology and nanotoxicology rely heavily on investigations of skin cell phototoxicity. We found that hydroxyl radicals were formed in keratinocyte HaCaT cells due to LDH leakage and ZnO NPs photo-genotoxic effects, which resulted in the generation of 8-OHdG, a prominent marker of oxidative DNA damage and carcinogenesis [61].

8.1.5.3 Gold Nanoparticles (AuNPs)

Because of their ease of "synthesis, chemical stability, and distinctive optical features", AuNPs are an appealing material for biological research and development.

AuNPs with a diameter of 14 nm have been discovered to simply pass over the cell membrane and concentrate in the vacuole. AuNPs disrupted the synthesis of intracellular matrix proteins, Deficiencies in cell proliferation, adhesion, and motility were attributed to aberrant actin filaments and extracellular matrix construction. Another study found that AuNPs' wide surface increased keratinocyte cell adhesion and proliferation, hinting that they may be used in skin tissue engineering and biomedical materials. 34 nm AuNPs were tested in vitro on the development of primary mouse keratinocytes. The findings showed that a modest quantity of AuNPs (5 ppm) enhanced cell proliferation, whereas a high dosage caused toxicity. As a result, the therapeutic window of AuNPs is still unknown [62].

8.1.5.4 Silver Nanoparticles (AgNPs)

Because of the growth in bacterial resistance to commonly used antibiotics, infectious illnesses caused by dangerous bacteria are becoming more common, and novel antibacterial medicines are urgently needed. Since the First World War, silver nitrate has been used to clean wounds. Widely available antibiotics have restricted the use of silver as a systemic treatment, and silver-containing treatments are now only used as a topical cream to treat burns. Nanosilver was made possible by the advancement of nanotechnology. AgNPs have a bigger surface area and a broader range of biological activities than silver nitrate. AgNPs, for example, slowed wound healing by lowering inflammation and suppressing bacterial activity. As part of the multi-step re-epithelialization process necessary for wound healing, keratinocyte cells migrate and proliferate in the epidermal layer of the skin. As a result, a lot of work has been put into figuring out the chemistry behind the interaction between AgNP and skin cells. The survival, metabolic activity, and proliferative and migratory abilities of primary "normal human epidermal keratinocytes" (NHEK) were decreased when they were exposed to 15 nm AgNPs. In addition, after a lengthy period of exposure, the activation of 3/7 caspase and DNA damage occurred, demonstrating that the chemicals were cytotoxic and genotoxic in nature. UVB-induced DNA damage, apoptosis, and mutation can be prevented by AgNPs (10–50 nm) [63].

8.1.6 Metallic Nanoparticles for Photothermal Therapy

Phototherapy, like other medicine delivery techniques, depends on PS molecule transit to the target area, as previously indicated. Active chemicals can only be released over a specified length of time if nanoparticle size control is maintained. For the manufacture of nanoparticles, it's also crucial to evaluate the raw materials used and their biodegradability. Organic and inorganic NPs, including silica and magnetic NPs, are among the most commonly employed in PDT. The following sections will highlight some of the most impressive feats made with nanocarriers [64].

Organic nanoparticles: The most common techniques for encapsulating compounds that can be employed in PDT are organic nanoparticles. There are numerous types based on the materials used and how they are organized [64].

Solid lipid nanoparticles: Since their development in the 1990s, solid lipid nanoparticles (SLNs) have been safe, occlusive nanoparticles with a protective effect that can also boost medication penetration in the skin. The surfactant layer and lipidic nucleus make up these NPs, and they may be generated using either the "Müller and Lucks" method of high-pressure homogenization (1996) or the Gasco microemulsion approach (1998).

In any case, certain lipids, such as triglycerides or glycerides combinations, are necessary for the synthesis of SLNs at body temperature. It is because of their composition that SLNs are safe, biodegradable, and inexpensive to produce on a large scale. SLNs, on the other hand, have several drawbacks, including limited encapsulation efficiency and the possibility of drug leakage during storage. A second generation of SLNs, known as nanostructured lipid carriers (NLCs), was developed. They're made up of SLNs with a lipid-based solid matrix that's less organized. Imperfect, amorphous, and numerous NLCs are the three types of NLCs. The imperfect NLCs are made up of a mix of solid lipids with varying chain lengths and lipid saturation levels, all of which contribute to the formation of an imperfect solid matrix. Amorphous lipids are made up of specific solid and liquid lipids that combine to form a solid particle that does not crystallize. The multiple type is created when solid lipids are combined with increased levels of liquid lipids, resulting in the formation of oil nano-compartments within the solid matrix [65]. In any case, for SLN to be made, lipids that are solid at body temperature, like triglycerides and glycerides, must be present. It is because of their composition that SLNs are safe, biodegradable, and cheap to make in large quantities [66]. SLNs, on the other hand, have several drawbacks, such as poor encapsulation and the chance that the drug will leak out during storage. To deal with these problems, SLNs, now in their second generation, known as nanostructured lipid carriers (NLCs), were developed. They're made up of SLNs with a lipid-based solid matrix that's less organized. Imperfect, amorphous, and numerous NLCs are the three types of NLCs. The imperfect NLCs are made up of a mix of solid lipids with varying chain lengths and lipid saturation levels, all of which contribute to the formation of an imperfect solid matrix. Amorphous lipids are made up of specific solid and liquid lipids that combine to form a solid particle that does not crystallize. The multiple type is created when solid lipids are combined with increased levels of liquid lipids, resulting in the formation of oil nano-compartments within the solid matrix [66].

Liposomes: Liposomes are formed when phospholipid bilayers fold in on themselves and form vesicles in an aqueous medium, allowing them to self-organize. Because lipids are amphiphilic, they can store drugs that are both water-loving and water-hating in different compartments. As a model for cell membranes, liposomes are extensively employed, but they may also be used to provide medicine. In addition to being biocompatible and biodegradable, the lipid composition has no harmful effects on the human body. The addition of polyethylene glycol (PEG) to these particles can result in the development of stealth liposomes that can avoid the immune system and improve blood circulation. Antibodies, for example, can be utilized as ligands in the functionalization process, which, in turn, is responsible for a reliable targeted drug delivery system. Liposomes are excellent candidates for photodynamic treatment because of their versatility [67].

Micelles: Even though amphiphilic molecules organize themselves spontaneously to form micelles, which are similar to liposomes, because the packing parameters are different, their final structures are not the same as those of vesicles. To make stable micelles with a hydrophobic interior that is separate from watery fluids, the concentration of amphiphilic molecules must be higher than the critical micellar concentration (CMC). Micelles can also be made from polymers that have both water-repelling and water-attracting parts. Because the CMC values are different, it is important to choose the right amphiphilic molecule [68]. Zhang et al. looked into making micelles from block copolymers so that the anticancer drugs doxorubicin and pheophorbide A could be delivered together to treat melanoma. Melanoma cells were able to take up molecules with these compounds both in the lab and in the wild. This process was driven by light-induced ROS production. More importantly, micelles with irradiation slowed tumor development far more than micelles without irradiation [68].

Nanoemulsion: It takes a lot of energy to form minuscule droplets in a nanoemulsion, which is a combination of oil and surfactant. Using nanoemulsions to enhance the bioavailability of lipophilic medications is an effective strategy. Researchers who looked at nanoemulsions with the natural chemical curcumin as a photosensitizer drug found that the curcumin-nanoemulsion was especially harmful to breast cancer cells and made a lot of ROS. Natural raw materials were also used by Mongue-Fuentes et al. for developing nanoemulsions for PDT. The nanoemulsions were made from acai oil, which, when combined with light irradiation, killed melanoma cells by 85%, according to their findings, which were also substantiated in animal testing in mice, where tumor volumes dropped by 85%.

8.1.7 Metallic Nanoparticles in Targeted Therapy

Targeted therapy is a new way of treating cancer. It is based on the use of "intelligent" medications that modify the proteins involved in the tumor formation process specifically. The study of the molecular pathways of tumor cell proliferation is necessary for the creation of this therapy technique. Targeted therapies provide a higher level of specificity, which can lessen therapy side effects (Table 8.1) [74].

TABLE 8.1
Role of Metallic Nanoparticles in Photothermal Therapy

Metallic nanoparticles	Technique	Inference	Reference
Iron-oxide nanoparticles	Franz diffusion	Increase the temperature of tumor cells and cell was illuminated by laser by which the energy of radiation transformed into heat.	[69]
Silver-carbon nanoparticles	Franz diffusion	Increase the penetration ability of metallic nanoparticles and helps in reduction of melanoma tumor size.	[70]
Gold Nanoparticles	Electro-thermal atomic absorption spectroscopy	Increase the topical absorption by decreasing size of particles.	[71]
Silver nanoparticles	Infrared absorption Plasmonic Therapy	Due to increased intracellular silver content, it acts beneficially on breast cancer cells and act as effective photothermal agent.	[72]
Gold Nanoparticles	Infrared absorption Plasmonic Therapy	Produce localized heat sufficient to damage tumor cells	[73]

TABLE 8.2
Metallic Nanoparticles for the Treatment of Different Disease

Metallic nanoparticles	Technique	Treatment of disease	Targeting	References
Silver nanoparticles	Electro-thermal atomic absorption spectroscopy	Damaged skin	Passive targeting	[76]
Gold nanoparticle	Franz diffusion	Acne treatment	Active targeting	[77]
Cobalt nanoparticle	Franz diffusion	Damaged skin	Passive targeting	[78]
Doxorubicin loaded nanoparticles	Franz diffusion	–	Active targeting	[79]
5-flourouracil loaded gold nanoparticles	Electro-thermal atomic absorption spectroscopy	Colorectal cancer	Active targeting	[80]
Temozolomide loaded nanoparticles	Franz diffusion	Lung cancer	Active targeting	[81]

FeONPs can also be used as a contrast agent in magnetic resonance imaging (MRI) to find cancers of the skin. Even though they are small, they are a high-contrast unit, lowering the amount of material to be supplied and thereby lowering the risk of toxicity. The aggregation of magnetic NPs in vivo, on the other hand, restricts their therapeutic and diagnostic capabilities. The aggregation of magnetic NPs may reduce biodistribution, heating capability, and MRI imaging performance. Bharath et al., for example, showed that adding a mesoporous silica shell to the surface of FesONPs can stop them from sticking together and improve clinical planning and use (Table 8.2) [75].

8.2 CONCLUSIONS

Nanotechnology is one of the fields of science that is changing the most quickly. Much more dermoactive medicines have appeared as medicine and pharmaceutical sciences have progressed. Because of their intriguing features, including significant antibacterial and antifungal activity, NPs are now used in a variety of dermatological and cosmetic treatments. In the near future, nanomaterials and nanoproducts are likely to play a big role in cleaning, pharmacy, medicine, and beauty. In terms of how nanotechnology is used in medicine, targeted cancer therapy using NPs to carry active chemicals is said to be at its peak. Melanoma is already treated with targeted therapy using NPs, and research into its usage in the treatment of other skin malignancies is ongoing. Nanotechnology is also helping to develop tailored anticancer skin therapies. Nanotechnology has also revolutionized the pharmacodynamics and pharmacokinetics of medications, instrumentation, molecular and noninvasive imaging, biomarker identification for particular diseases, and the miniaturization of dermatology equipment. Aside from the benefits of improved medication absorption and therapeutic usage of current drugs, there is a risk that nanoproducts would damage humans. Because NPs are commonly utilized in cosmetics, such as sunscreens and personal care items, they come into close contact with skin. As a result, determining the toxicity of exposure via the cutaneous route is crucial. Physical and chemical features of NPs, such as "size, shape, solubility, zeta potential, functionalization, concentration, and exposure period", all influence their pharmacological efficacy and toxicity. Because of their enormous surface area, the reactivity of NPs is hampered by their higher ability to bind other molecules, such as proteins. For their application in cosmetology and dermatology, there must be an understanding of the nanostructures' molecular mechanism of action on human cells and a determination of the safe concentration. Dermatological infections and skin problems can be treated in unique ways thanks to nanotechnology. Biocompatibility and dispersion of particular combinations can be improved by modifying traditional chemicals to the nanoscale. Cosmetic and dermatological therapies may now be developed that are more successful and have fewer side effects thanks to this new technology.

ACKNOWLEDGMENTS

The authors are thankful to their respective institutes for providing the facilities and encouragement. The authors are also thankful to Dr. Jamilur R. Ansari, Dronacharya College of Engineering, Gurgaon, India, for the needful editing in the chapter. MSA is thankful to SGT University for their support and cooperation.

REFERENCES

1. Javed MN, Dahiya ES, Ibrahim AM, Alam M, Khan FA, Pottoo FH. Recent advancement in clinical application of nanotechnological approached targeted delivery of herbal drugs. In Nanophytomedicine 2020 (pp. 151–72). Springer, Singapore.
2. Mishra S, Sharma S, Javed MN, Pottoo FH, Barkat MA, Alam MS, Amir M, Sarafroz M. Bioinspired nanocomposites: Applications in disease diagnosis and treatment. Pharmaceutical Nanotechnology. 2019 Jun 1;7(3):206–19.
3. Javed MN, Dahiya ES, Ibrahim AM, Alam M, Khan FA, Pottoo FH. Recent advancement in clinical application of nanotechnological approached targeted delivery of herbal drugs. In Nanophytomedicine 2020 (pp. 151–72). Springer, Singapore.
4. Aslam M, Javed M, Deeb HH, Nicola MK, Mirza M, Alam M, Akhtar M, Waziri A. Lipid nanocarriers for neurotherapeutics: Introduction, challenges, blood-brain barrier, and promises of delivery approaches. CNS & Neurological Disorders-Drug Targets (Formerly Current Drug Targets-CNS & Neurological Disorders). 2022; 21(10):952–965. doi.org/10.2174/1871527320666210706104240.
5. Javed MN, Pottoo FH, Shamim A, Hasnain MS, Alam MS. Design of experiments for the development of nanoparticles, nanomaterials, and nanocomposites. In Design of Experiments for Pharmaceutical Product Development 2021 (pp. 151–69). Springer, Singapore.
6. Javed MN, Alam MS, Waziri A, Pottoo FH, Yadav AK, Hasnain MS, Almalki FA. QbD applications for the development of nanopharmaceutical products. In Pharmaceutical Quality by Design 2019 Jan 1 (pp. 229–53). Academic Press. https://doi.org/10.1016/B978-0-12-815799-2.00013-7.
7. Kumar R, Dhamija G, Ansari JR, Javed MN, Alam MS. C-dot nanoparticulated devices for biomedical applications. In Nanotechnology 2022 (pp. 271–99). CRC Press, Boca Raton. https://doi.org/10.1201/9781003220350-15.

8. Bharti C, Alam MS, Javed MN, Khalid M, Saifullah FA, Manchanda R. Silica based nanomaterial for drug delivery. Nanomaterials: Evolution and Advancement Towards Therapeutic Drug Delivery (Part II). 2021 Jun 2:57.
9. Sangeet Kumar Mall SK, Yadav T, Waziri A, Alam MS. Treatment opportunities with *Fernandoa adenophylla* and recent novel approaches for natural medicinal phytochemicals as a drug delivery system; Exploration of Medicine. 2022;3:516–39.
10. Naseh MF, Ansari JR, Alam MS, Javed MN. Sustainable Nanotorus for Biosensing and Therapeutical Applications. In Handbook of Green and Sustainable Nanotechnology: Fundamentals, Developments and Applications 2022 Aug 19 (pp. 1–21). Springer International Publishing, Cham.
11. Sunilbhai CA, Alam M, Sadasivuni KK, Ansari JR. SPR assisted diabetes detection. In Advanced Bioscience and Biosystems for Detection and Management of Diabetes 2022 (pp. 91–131). Springer, Cham.
12. Singhal S, Gupta M, Alam MS, Javed MN, Ansari JR. Carbon allotropes-based nanodevices: Graphene in biomedical applications. In Nanotechnology 2022 (pp. 241–69). CRC Press. https://doi.org/10.1201/9781003220350-14.
13. Ghaseminezhad SM, Hamedi S, Shojaosadati SA. Green synthesis of silver nanoparticles by a novel method: Comparative study of their properties. Carbohydrate Polymers. 2012 Jun 20;89(2):467–72.
14. Abbai R, Mathiyalagan R, Markus J, Kim YJ, Wang C, Singh P, Ahn S, Farh ME, Yang DC. Green synthesis of multifunctional silver and gold nanoparticles from the oriental herbal adaptogen: Siberian ginseng. International Journal of Nanomedicine. 2016;11:3131.
15. Pottoo FH, Tabassum N, Javed M, Nigar S, Rasheed R, Khan A, Barkat M, Alam M, Maqbool A, Ansari MA, Barreto GE. The synergistic effect of raloxifene, fluoxetine, and bromocriptine protects against pilocarpine-induced status epilepticus and temporal lobe epilepsy. Molecular Neurobiology. 2019 Feb;56(2):1233–47.
16. Pottoo FH, Sharma S, Javed MN, Barkat MA, Harshita, Alam MS, Naim MJ, Alam O, Ansari MA, Barreto GE, Ashraf GM. Lipid-based nanoformulations in the treatment of neurological disorders. Drug Metabolism Reviews. 2020 Jan 2;52(1):185–204.
17. Pottoo FH, Javed M, Barkat M, Alam M, Nowshehri JA, Alshayban DM, Ansari MA. Estrogen and serotonin: Complexity of interactions and implications for epileptic seizures and epileptogenesis. Current Neuropharmacology. 2019 Mar 1; 17(3):214–31.
18. Pottoo FH, Tabassum N, Javed MN, Nigar S, Sharma S, Barkat MA, Alam MS, Ansari MA, Barreto GE, Ashraf GM. Raloxifene potentiates the effect of fluoxetine against maximal electroshock induced seizures in mice. European Journal of Pharmaceutical Sciences. 2020 Apr 15;146:105261.
19. Waziri A, Bharti C, Aslam M, Jamil P, Mirza M, Javed MN, Pottoo U, Ahmadi A, Alam MS. Probiotics for the chemoprotective role against the toxic effect of cancer chemotherapy. Anti-Cancer Agents in Medicinal Chemistry (Formerly Current Medicinal Chemistry-Anti-Cancer Agents). 2022 Feb 1;22(4):654–67.
20. Javed MN, Akhter MH, Taleuzzaman M, Faiyazudin M, Alam MS. Cationic nanoparticles for treatment of neurological diseases. In Fundamentals of Bionanomaterials 2022 Jan 1 (pp. 273–92). Elsevier. https://doi.org/10.1016/B978-0-12-824147-9.00010-8.
21. Kumari N, Daram N, Alam MS, Verma AK. Rationalizing the use of polyphenol nano-formulations in the therapy of neurodegenerative diseases. CNS & Neurological Disorders-Drug Targets (Formerly Current Drug Targets-CNS & Neurological Disorders). 2022 Dec 1;21(10):966–76.
22. Raj S, Manchanda R, Bhandari M, Alam M. Review on natural bioactive products as radioprotective therapeutics: Present and past perspective. Current Pharmaceutical Biotechnology. 2022. 1721–1738. https://doi.org/10.2174/1389201023666220110104645.

23. Ibrahim AM, Chauhan L, Bhardwaj A, Sharma A, Fayaz F, Kumar B, Alhashmi M, AlHajri N, Alam MS, Pottoo FH. Brain-derived neurotropic factor in neurodegenerative disorders. Biomedicines. 2022 May;10(5):1143.
24. Safwat MA, Soliman GM, Sayed D, Attia MA. Gold nanoparticles enhance 5-fluorouracil anticancer efficacy against colorectal cancer cells. International Journal of Pharmaceutics. 2016 Nov 20;513(1–2):648–58.
25. Abdelrasoul GN, Farkas B, Romano I, Diaspro A, Beke S. Nanocomposite scaffold fabrication by incorporating gold nanoparticles into biodegradable polymer matrix: Synthesis, characterization, and photothermal effect. Materials Science and Engineering: C. 2015 Nov 1;56:305–10.
26. Filon FL, Crosera M, Timeus E, Adami G, Bovenzi M, Ponti J, Maina G. Human skin penetration of cobalt nanoparticles through intact and damaged skin. Toxicology in Vitro. 2013 Feb 1;27(1):121–7.
27. Ahamed M, AlSalhi MS, Siddiqui MK. Silver nanoparticle applications and human health. Clinica Chimica Acta. 2010 Dec 14;411(23–24):1841–8.
28. Abe S, Maity B, Ueno T. Functionalization of protein crystals with metal ions, complexes and nanoparticles. Current Opinion in Chemical Biology. 2018 Apr 1;43:68–76.
29. Carvalho MD, Henriques F, Ferreira LP, Godinho M, Cruz MM. Iron oxide nanoparticles: The influence of synthesis method and size on composition and magnetic properties. Journal of Solid State Chemistry. 2013 May 1;201:144–52.
30. Alam MS, Garg A, Pottoo FH, Saifullah MK, Tareq AI, Manzoor O, Mohsin M, Javed MN. Gum ghatti mediated, one pot green synthesis of optimized gold nanoparticles: Investigation of process-variables impact using Box-Behnken based statistical design. International Journal of Biological Macromolecules. 2017 Nov 1;104:758–67.
31. Alam MS, Javed MN, Pottoo FH, Waziri A, Almalki FA, Hasnain MS, Garg A, Saifullah MK. QbD approached comparison of reaction mechanism in microwave synthesized gold nanoparticles and their superior catalytic role against hazardous nirto-dye. Applied Organometallic Chemistry. 2019 Sep;33(9):e5071.
32. Pandit J, Alam MS, Ansari JR, Singhal M, Gupta N, Waziri A, Sharma K, Potto FH. Multifaced applications of nanoparticles in biological science. In Nanomaterials in the Battle Against Pathogens and Disease Vectors 2022 (pp. 17–50). CRC Press.
33. Alam MS, Naseh MF, Ansari JR, Waziri A, Javed MN, Ahmadi A, Saifullah MK, Garg A. Synthesis approaches for higher yields of nanoparticles. In Nanomaterials in the Battle Against Pathogens and Disease Vectors 2022 (pp. 51–82). CRC Press.
34. Javed MN, Pottoo FH, Alam MS. Metallic nanoparticle alone and/or in combination as novel agent for the treatment of uncontrolled electric conductance related disorders and/or seizure, epilepsy & convulsions. Patent Acquired on October. 2016;10:40.
35. Bansal V, Rautaray D, Bharde A, Ahire K, Sanyal A, Ahmad A. Sastry M. Fungus-mediated biosynthesis of silica and titania particles. Journal of Materials Chemistry. 2005;15:2583–9.
36. Bhatia S. Nanoparticles types, classification, characterization, fabrication methods and drug delivery applications. In Natural Polymer Drug Delivery Systems 2016 (pp. 33–93). Springer, Cham.
37. Agarwal A, Huang SW, O'donnell M, Day KC, Day M, Kotov N, Ashkenazi S. Targeted gold nanorod contrast agent for prostate cancer detection by photoacoustic imaging. Journal of Applied Physics. 2007 Sep 15;102(6):064701.
38. Van Der Bruggen P, Zhang Y, Chaux P, Stroobant V, Panichelli C, Schultz ES, Chapiro J, Van den Eynde BJ, Brasseur F, Boon T. Tumor-specific shared antigenic peptides recognized by human T cells. Immunological Reviews. 2002 Oct;188(1):51–64.
39. Safwat MA, Soliman GM, Sayed D, Attia MA. Gold nanoparticles enhance 5-fluorouracil anticancer efficacy against colorectal cancer cells. International Journal of Pharmaceutics. 2016 Nov 20;513(1–2):648–58.

40. Hasnain MS, Javed MN, Alam MS, Rishishwar P, Rishishwar S, Ali S, Nayak AK, Beg S. Purple heart plant leaves extract-mediated silver nanoparticle synthesis: Optimization by Box-Behnken design. Materials Science and Engineering: C. 2019 Jun 1;99:1105–14.
41. Kaur G, Dogra V, Kumar R, Kumar S, Bhanjana G, Dilbaghi N, Singhal NK. DNA interaction, anti-proliferative effect of copper oxide nanocolloids prepared from metallosurfactant based microemulsions acting as precursor, template and reducing agent. International Journal of Pharmaceutics. 2018 Jan 15;535(1–2):95–105.
42. Ahmed S, Kaur G, Sharma P, Singh S, Ikram S. Fruit waste (peel) as bio-reductant to synthesize silver nanoparticles with antimicrobial, antioxidant and cytotoxic activities. Journal of Applied Biomedicine. 2018 Aug 1;16(3):221–31.
43. Khan I, Saeed K, Khan I. Nanoparticles: Properties, applications and toxicities. Arabian Journal of Chemistry. 2019;12(7):908–931. https://doi.org/10.1016/j.arabjc.2017.05.011.
44. Kalaiselvam S, Parameshwaran R, Harikrishnan S. Analytical and experimental investigations of nanoparticles embedded phase change materials for cooling application in modern buildings. Renewable Energy. 2012 Mar 1;39(1):375–87.
45. Iravani S, Korbekandi H, Mirmohammadi SV, Zolfaghari B. Synthesis of silver nanoparticles: Chemical, physical and biological methods. Research in Pharmaceutical Sciences. 2014 Nov;9(6):385.
46. Kitching M, Ramani M, Marsili E. Fungal biosynthesis of gold nanoparticles: Mechanism and scale up. Microbial Biotechnology. 2015 Nov;8(6):904–17.
47. Kim KB, Kim YW, Lim SK, Roh TH, Bang DY, Choi SM, Lim DS, Kim YJ, Baek SH, Kim MK, Seo HS. Risk assessment of zinc oxide, a cosmetic ingredient used as a UV filter of sunscreens. Journal of Toxicology and Environmental Health, Part B. 2017 Apr 3;20(3):155–82.
48. Kharisov BI, Dias HR, Kharissova OV. Mini-review: Ferrite nanoparticles in the catalysis. Arabian Journal of Chemistry. 2019 Nov 1;12(7):1234–46.
49. Jurašin DD, Ćurlin M, Capjak I, Crnković T, Lovrić M, Babič M, Horák D, Vrček IV, Gajović S. Surface coating affects behavior of metallic nanoparticles in a biological environment. Beilstein Journal of Nanotechnology. 2016 Feb 15;7(1):246–62.
50. Jana NR, Wang ZL, Sau TK, Pal T. Seed-mediated growth method to prepare cubic copper nanoparticles. Current Science-Bangalore. 2000 Nov 10;79(9):1367–9.
51. Jain KK. Applications of nanobiotechnology in clinical diagnostics. Clinical Chemistry. 2007 Nov 1;53(11):2002–9.
52. Jain KK. Nanomedicine: Application of nanobiotechnology in medical practice. Medical Principles and Practice. 2008;17(2):89–101.
53. Iravani S, Zolfaghari B. Green synthesis of silver nanoparticles using Pinus Eldarica bark extract. BioMed Research International. 2013 Oct;2013.
54. Hamzawy MA, Abo-Youssef AM, Salem HF, Mohammed SA. Antitumor activity of intratracheal inhalation of temozolomide (TMZ) loaded into gold nanoparticles and/or liposomes against urethane-induced lung cancer in BALB/c mice. Drug Delivery. 2017 Jan 1;24(1):599–607.
55. He W, Wamer W, Fu PP, Yin JJ. Enzyme-mimetic effects of gold@ platinum nanorods on the antioxidant activity. Journal of Environmental Science and Health, Part C: Environmental Carcinogenesis and Ecotoxicology Reviews. 2014;32:186–211.
56. Hu H, Nie L, Feng S, Suo J. Preparation, characterization and in vitro release study of gallic acid loaded silica nanoparticles for controlled release. Die Pharmazie-An International Journal of Pharmaceutical Sciences. 2013 Jun 1;68(6):401–5.
57. Guo Z, Ganawi AA, Liu Q, He L. Nanomaterials in mass spectrometry ionization and prospects for biological application. Analytical and Bioanalytical Chemistry. 2006 Feb;384(3):584–92.
58. Filon FL, Crosera M, Timeus E, Adami G, Bovenzi M, Ponti J, Maina G. Human skin penetration of cobalt nanoparticles through intact and damaged skin. Toxicology in Vitro. 2013 Feb 1;27(1):121–7.

59. Larese FF, D'Agostin F, Crosera M, Adami G, Renzi N, Bovenzi M, Maina G. Human skin penetration of silver nanoparticles through intact and damaged skin. Toxicology. 2009 Jan 8;255(1–2):33–7.
60. Selvaraj S, Thangam R, Fathima NN. Electrospinning of casein nanofibers with silver nanoparticles for potential biomedical applications. International Journal of Biological Macromolecules. 2018 Dec 1;120:1674–81
61. Anandan S, Grieser F, Ashokkumar M. Sonochemical synthesis of Au– Ag core–shell bimetallic nanoparticles. The Journal of Physical Chemistry C. 2008 Oct 2; 112(39):15102–5.
62. Anandalakshmi K, Venugobal J, Ramasamy V. Characterization of silver nanoparticles by green synthesis method using Pedalium murex leaf extract and their antibacterial activity. Applied Nanoscience. 2016 Mar;6(3):399–408.
63. Alaqad K, Saleh TA. Gold and silver nanoparticles: Synthesis methods, characterization routes and applications towards drugs. Journal of Environmental and Analytical Toxicology. 2016;6(4):525–2161.
64. Agarwal H, Kumar SV, Rajeshkumar S. A review on green synthesis of zinc oxide nanoparticles–An eco-friendly approach. Resource-Efficient Technologies. 2017 Dec 1;3(4):406–13.
65. Ahamed M, AlSalhi MS, Siddiqui MK. Silver nanoparticle applications and human health. Clinica Chimica Acta. 2010 Dec 14;411(23–24):1841–8.
66. Agarwal A, Huang SW, O'donnell M, Day KC, Day M, Kotov N, Ashkenazi S. Targeted gold nanorod contrast agent for prostate cancer detection by photoacoustic imaging. Journal of Applied Physics. 2007 Sep 15;102(6):064701.
67. Guo Z, Ganawi AA, Liu Q, He L. Nanomaterials in mass spectrometry ionization and prospects for biological application. Analytical and Bioanalytical Chemistry. 2006 Feb;384(3):584–92.
68. Jana NR, Wang ZL, Sau TK, Pal T. Seed-mediated growth method to prepare cubic copper nanoparticles. Current Science-Bangalore. 2000 Nov 10;79(9):1367–9.
69. Estelrich J, Busquets MA. Iron oxide nanoparticles in photothermal therapy. Molecules. 2018 Jun 28;23(7):1567.
70. Behnam MA, Emami F, Sobhani Z, Koohi-Hosseinabadi O, Dehghanian AR, Zebarjad SM, Moghim MH, Oryan A. Novel combination of silver nanoparticles and carbon nanotubes for plasmonic photo thermal therapy in melanoma cancer model. Advanced Pharmaceutical Bulletin. 2018 Mar;8(1):49.
71. Liu X, Shan G, Yu J, Yang W, Ren Z, Wang X, Xie X, Chen HJ, Chen X. Laser heating of metallic nanoparticles for photothermal ablation applications. AIP Advances. 2017 Feb 24;7(2):025308.
72. Thompson EA, Graham E, MacNeill CM, Young M, Donati G, Wailes EM, Jones BT, Levi-Polyachenko NH. Differential response of MCF7, MDA-MB-231, and MCF 10A cells to hyperthermia, silver nanoparticles and silver nanoparticle-induced photothermal therapy. International Journal of Hyperthermia. 2014 Aug 1;30(5):312–23.
73. Pattani VP, Tunnell JW. Nanoparticle-mediated photothermal therapy: A comparative study of heating for different particle types. Lasers in Surgery and Medicine. 2012 Oct;44(8):675–84.
74. Boris I, Kharisova H, Dias VR, Oxana B, Kharissova V. Mini-review: Ferrite nanoparticles in the catalysis. Arabian Journal of Chemistry. 2014, 1234–1246.
75. Kharisov BI, Dias HR, Kharissova OV. Mini-review: Ferrite nanoparticles in the catalysis. Arabian Journal of Chemistry. 2019 Nov 1;12(7):1234–46.
76. Larese FF, D'Agostin F, Crosera M, Adami G, Renzi N, Bovenzi M, Maina G. Human skin penetration of silver nanoparticles through intact and damaged skin. Toxicology. 2009 Jan 8;255(1–2):33–7.
77. Larese Filon F, Crosera M, Adami G, Bovenzi M, Rossi F, Maina G. Human skin penetration of gold nanoparticles through intact and damaged skin. Nanotoxicology. 2011 Dec 1;5(4):493–501.

78. Filon FL, Crosera M, Timeus E, Adami G, Bovenzi M, Ponti J, Maina G. Human skin penetration of cobalt nanoparticles through intact and damaged skin. Toxicology in Vitro. 2013 Feb 1;27(1):121–7
79. Elbialy NS, Fathy MM, Khalil WM. Doxorubicin loaded magnetic gold nanoparticles for in vivo targeted drug delivery. International Journal of Pharmaceutics. 2015 Jul 25;490(1–2):190–9.
80. Safwat MA, Soliman GM, Sayed D, Attia MA. Gold nanoparticles enhance 5-fluorouracil anticancer efficacy against colorectal cancer cells. International Journal of Pharmaceutics. 2016 Nov 20;513(1–2):648–58.
81. Hamzawy MA, Abo-Youssef AM, Salem HF, Mohammed SA. Antitumor activity of intratracheal inhalation of temozolomide (TMZ) loaded into gold nanoparticles and/or liposomes against urethane-induced lung cancer in BALB/c mice. Drug Delivery. 2017 Jan 1;24(1):599–607.

9 Advancement towards Brain-Targeted MNPs in Neurological Disorders

Sainu Gopika, Girish Kumar, Reshu Virmani, Ashwani Sharma, Anjali Sharma, and Tarun Virmani

9.1 INTRODUCTION

The prevalence of cerebrospinal nervous system disease is progressively rising, posing serious threats to human survival and driving up socioeconomic expenses associated with providing medical care. Neurological disorders continued to be a considerable element of disability-adjusted life years (DALYs) worldwide in 2019, accounting for 45–11.6% of DALYs and 16.5% of deaths. Acute brain injury and chronic neurodegenerative diseases are two categories of CNS circumstances that damage the structure and functionalities of the brain's neurons (NDs) [1]. The primary causes of acute neurological damage are stroke, cerebral ischemia, brain injury, and epilepsy, whereas a chronic condition comprises Alzheimer's disease, Parkinson's disease, and Huntington's disease. Numerous therapeutic drug moieties have been discovered to combat neurological disorders, but their effectiveness is restricted due to the existence of various barriers in the CNS, among which the BBB is the main impediment in drug transport to the brain [2]. The BBB is a penetrable barrier that is extremely stable. This convoluted neurochemical blocker demarcates circulating brain matter from brain extra-cellular fluid and prevents propagating residues from the blood from freely dissipating into the brain. The BBB is regarded as a non-endothelium attached by adherent synapses, two basement membranes, and astrocytic end-feet. Gap junctions among adjoining segmental endothelial lipid bilayers nearly utterly enclose the BBB, limiting the entry of pervasively propagating intrinsic and extrinsic residues further into the CNS, like infinitesimal artifacts (e.g., bacteria) and humongous or water-soluble particles [3]. As a direct outcome, although the BBB is indeed an instinctual protective framework, it also poses a considerable impairment toward the procurement of numerous medications to the CNS through the systemic route. Even just a relatively small proportion of prescription medications, that is, modest, lipid-soluble substances (400–500 Da) or water-insoluble substances (e.g., O_2, CO_2, hormones), could traverse the BBB, making medication distribution toward the CNS complicated [4]. Then, as consequence, the simplified BBB renders medication distribution toward the brain a complicated mechanism that

DOI: 10.1201/9781003317319-9

constitutes even further exploration. As an outcome, unprecedented research has been directed toward strengthening drug transfer throughout the BBB and then into the CNS. It has been difficult to optimize the prognostic value of neurological conditions like Alzheimer's disease (AD), Parkinson's disease (PD), brain tumors [5, 6], Huntington's disease (HD), multiple sclerosis (MS), Friedrich ataxia (FRDA), amyotrophic lateral sclerosis (ALS), ischemic stroke, and epilepsy in the CNS, especially in contrast to other parts of the body. A tremendous amount of research has been put into the exploration and design of therapeutic strategies of the CNS that can lead to novel and efficient therapies; even so, over 90% of recently developed prescription medications have already endeavored to secure interventional authorization from the US Food and Drug Administration (FDA) [7]. The inclusion of a stringent BBB minimizing drug intrusion toward the CNS area has become one of the most significant issues for the progression of neuronal therapeutic interventions [8]. Besides that, the non-targeted distribution of analytic reactants or active compounds could have an adverse influence on neural stem cells as well as glial cells, which become highly fragile neuronal constituents across the CNS that sustain brain processes as well as equilibrium. AD, PD, HD, MS, ALS, FRDA, ischemic stroke, and epilepsy each have protracted pathophysiological impact conditions that cause incalculable detriment to individuals and family members [9]. NPs had already incrementally joined the range of perspectives for investigators researching neurological disorders. Numerous different NPs are often used in studies, along with lipid-based NPs (liposomes and solid lipid NPs (SLN)), polymeric NPs (micelles, dendrimers, nanocapsules, and nanospheres), and inorganic NPs [10–12]. This contextual concern examines NP implementations for brain disorders, and also how a diverse set of NPs are being used in the clinical field. Aptamers, nanostructured lipid carriers, micelles, nanoemulsions, nanocapsules, nanocrystals, nanotubes, and inorganic and organic NPs are indications of all these. Currently, the field of study of NPs in neurological disorders is centered on therapeutics, drug permeation maintenance, and the precise initial assessment and therapy of ailments premised on NP attributes. Liposome NPs and polymeric NPs, for instance, demonstrated excellent therapeutic efficacy and lipid bilayer perforating strength [13, 14]. They have distinct benefits over conventional prescription medications, such as optimally traversing the BBB as well as the medication's slow and regulated discharge [15]. The idea of integrating, encompassing, or conjugating a wide range of medications into NPs for delivery to particular cell populations is one of the possible benefits of using NPs over traditional distribution systems. NPs can be orchestrated to have properties such as biocompatibility, non-toxicity, the ability to undertake innumerable projectiles of targeted therapies, the ability to safeguard treatment strategies from in vivo deterioration, the potential to regulate drug distribution for extended durations, and the capability to transverse the BBB. Encapsulation or complexation of minuscule particles (e.g., drugs, siRNA) into NPs has been seen in research to augment brain penetrability. Because of their advantageous properties, nanoparticles (NPs) or metallic nanoparticles (MNPs) offer several advantages in varied clinical domains for delivering imaging agents or therapies for age-related disorders [16–19]. The magnitude of NPs in the 10–200 nm wavelength spectrum not only results in greater physiochemical characteristics such as excitability, potency, size distribution, responsiveness, and consistency [20], but also augmented permeation characteristics throughout a BBB and depth into disease-ridden brain tissues [21]. The surface modulation of NPs improves BBB infiltration as well as the disease-targeting effectiveness of the system. Furthermore, the BBB, a preferential obstacle constituted primarily via neuronal epithelium cells, is the biggest impediment to eliminating brain diseases. BBB precludes nearly all drugs from entering the brain, impeding drug discovery. Significant advancements have been made in traversing the BBB and treating neurological disorders over the last few years. In this particular instance, one exemplar of MNPs is gold nanoparticles (AuNPs), which exhibit the attributes of extendable size, interesting optical characteristics, flexible surface alteration, and excellent biocompatibility, all of which account for AuNPs being one of the most efficacious development candidates in the biomedical fields [22-24]. In this context, the configuration, implementation, and present studies of the advancement of various MNPs are discussed as promising implementation systems containing analytic chemicals or restorative medications for neurological conditions [25–29]. Various types of MNPs along with their drug distribution efficiency in NDs, benefits, and applications have been depicted in Figure 9.1. Currently, advanced proficient MNPs scan represent both analytical and multidisciplinary therapeutic intervention processes at a relatively similar moment and have demonstrated exemplary prospects for approaching CNS serious illnesses; however, a multitude of functionality is still needed for relevant therapeutic use [30–35]. This chapter presents a concise characterization of neurological diseases such as AD, PD, HD, MS, ALS, FRDA, ischemic stroke, and epilepsy with trademark indicators extensively used in the MNPs field, a summary of advanced and powerful MNPs used for therapeutics and detection, as well as encouraging initiatives to enhance BBB infiltration and/or disease approaching the efficiency of MNPs. Henceforth, the incidence of MNPs for managing CNS conditions, their significant advances and emphasis on development as compared to the existing interventions, as well as the potential of NPs and forthcoming research directions for interventional research accomplishment, are studied [36, 37].

9.1.1 Metallic NP in Alzheimer's Disease (AD)

AD is a chronic, prolonged, and unmitigated neurological disease that affects the elderly, particularly those over 65 years of age, and therefore is distinguished via irrevocable cognition deterioration, behavioral functional limitations, and dementia as the condition worsens [21]. This is the leading consequence of dementia, culminating in cognitive problems, psychological and speech impairment, and behavioral problems [38] and influencing a broad area of the cerebral hippocampal [39], afflicting an estimated 30–35 million individuals globally. It is revealed that now almost 47 million people worldwide struggle with cognitive decline, and almost 36 million of them have Alzheimer's disease, with the amount expected to rise by 131 million by 2050 [40, 41]. The causation of AD is still unknown, but many assumptions, including genetic, amyloid, and tau protein hypothesized relationships, are helping to shed light on the disease's etiologies. Moreover, no conclusive curative treatment for AD exists due to a lack of knowledge of its biochemical as well as intracellular methodologies. Depositions of amyloid-beta (Aβ) peptides, neurofibrillary tangle buildup of

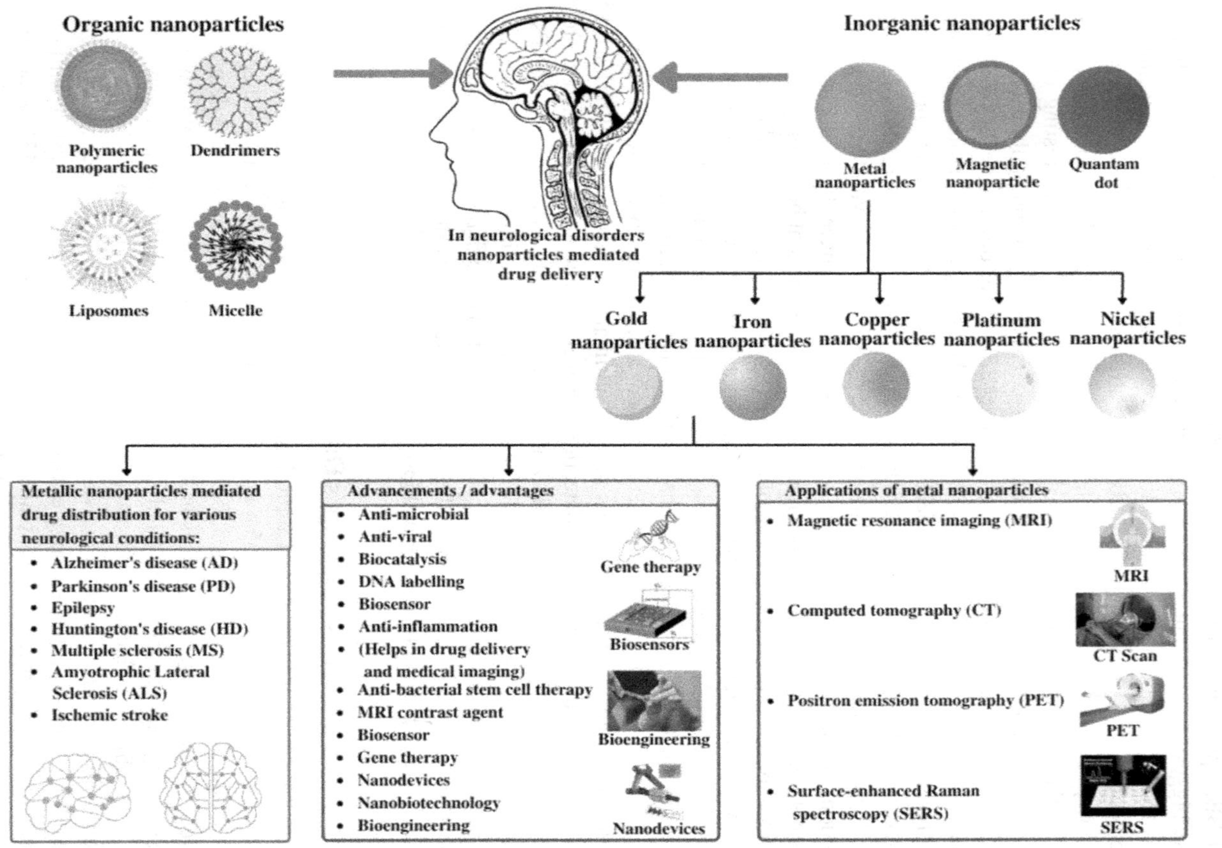

FIGURE 9.1 Metallic nanoparticles and their advantage and applications.

phosphoric tau proteins, and destructive neuro-inflammation throughout the human brain contributing to synaptic deficits, demyelination, and neuronal death are pivotal markers of AD advancement [42]. The anomalies of Aβ and tau proteins are indeed substantial defining characteristics of AD, which somewhat aggravate neuronal cell death through induction of apoptosis or autophagy but also contribute toward the instability of glial cells, astrocytes, and mitochondria, as well as the generation of ROS, facilitating the emergence of AD [43, 44]. The involvement of Aβ markers all over AD brains solubilized both proteins in cerebrospinal fluids (CSFs), which are key physiological signifiers that have been focused on for prognosis. Prevailing treatment modalities aim to prevent the emergence of Aβ plaques as well as neurofibrillary tangles and neutralize their agglomerations across nerve cells [45, 46]. Some medically authorized medications could only relieve signs or slow the development of AD by procuring neurotransmitters that also enhance interconnections among nerve cells in Alzheimer's patients' brains [47]. In the meantime, there is no proper intervention for AD because the etiopathogenesis of the disease is unspecified. Medical interventions are targeted at possible sign control; they're incapable of slowing or counteracting the disease's development. Seeing as therapeutically accessible prognostic strategies are incapable of detecting AD in its beginning phases, which is critical for precluding irrevocable neurotoxicity, attempts have been made to recognize consistent biochemical indicators to measure the disease's growth to achieve correct identification of AD just before the disease attains epidemiological seriousness. As a result, the prevailing assessment of AD is based primarily on interventional assessment and perceptual screening, with pertinent investigations including diagnostic imaging and blood examinations to exclude other conceivable serious illnesses [48]. It is broadly acknowledged that the abnormal emergence of intracerebral beta-amyloid (Aβ) accumulation particles is a quintessential example of AD pathogenesis [49]. As a possible consequence, precise and reliable measurement of Aβ and its intermediate products is critical not just for additional acknowledgment of etiology but also for initial rapid screening of AD development. Considerable initiatives had already been taken to establish NPs-derived image processing substances for the Aβ target identification in vitro or in vivo. Neuroimaging methodologies have been used to discover Aβ inside the brains of Alzheimer patient populations, which include fluorescence imaging [50], colorimetry sensor assay [51, 52], tip-enhanced Raman spectroscopy, surface-enhanced Raman scattering (SERS) [53], immuno-infrared sensor assay [54], electrochemical aptasensor assay [55], MRI [56], or a combination of fluorescence imaging/MRI [57]. A wide range of NPs, including gold NPs [58, 59], silver NPs [60], nanodiamonds [61], silica NPs [62], and nanoliposomes [63], with or without Aβ antibodies or numerous different substituents, were used to create ultrasensitive NP-based sensing devices to monitor Aβ levels in vitro. For instance, an Au/AgNPs-based colorimetric sensor array was built to diagnose the levels of amyloid protein 40 as well as 42 in biological fluid specimens at distinctive intensities by interpreting their transmittance and color modifications [64]. Furthermore, fluorescent probes as well as NPs have been configured to identify Aβ proteins in the presence of AD in vivo. Moreover, the aforementioned nanoagents could be cross-linked with targeted therapies to create a novel restorative framework for initial detection, treatments, and advancement surveillance in AD. Equivalent dynamic-modality probes have been evaluated [43], and they have been presumed to provide a potent and selective method of detecting Aβ. In simple terms, it can become a good potential platform to develop a high sensitivity and specificity nanosystem to accomplish incipient and efficient accurate identification of Aβ in vivo/vitro, which will be required for quick as well as efficient assessment and therapy of AD. However, apart from Aβ, the tau molecules are regarded as a fundamental diagnostic candidate for AD because protein hyperphosphorylation may indeed be embroiled in the formation of neuritic plaques, which serves as a significant influence in the pathogenicity of AD [65]. In this perspective, fluorescent probes based on pyrazine and quinoxaline have been applied for the determination of cortical neurofibrillary tangles [66]. Until one of all such probes, Probe 3, was tested in vitro along with tau accumulation particles, which showed maximum fluorescence as well as an enforceable predilection for tau aggregate particles. The probe's fluorophore was overlapping with transmissions from tau particles. Furthermore, stem cell implantation for regrowing nerve cells in AD tissues has been regarded as promising, and precursors of stem cells in experimental animals could drastically enhance AD-related abnormalities [67]. In two distinct ways, NPs could help with stem cell transplants for AD. On the one hand, NP-based treatments may make stem cell therapy easier. NPs, on either side, may be capable of producing greater in vivo imaging of stem cells to track the outcome of transplanted cells. The use of nanotechnology in tissue regeneration is estimated to increase the recuperative and pharmacological benefits of stem cell transplants as well as the identification of AD. As a result, from this perspective, NPs have a promising possibility for clinical identification as soon as possible and then delivering effective approaches. Throughout this due consideration, the exploration of novel AD identifiers and the advancement of NPs encompassing the AD identifiers have been high on the priority list [68–71].

9.1.2 Metallic NP in Parkinson's Disease (PD)

PD, a motor neurological condition, is among the second most prevalent neuropathological as well as neurodevelopmental disorders of the aging population, with numerous factors [72]. With the deterioration of dopaminergic nerve cells inside the SNpc area of the mid-brain, it is highly correlated to Lewy bodies (LBs) dystrophy as well as the destruction of dopamine pathways positioned in the SNpc [73], which causes decline concentrations of dopamine in the striatal region, and there will be further disruption in the dynamic equilibrium of neurochemicals [74], resulting in the incidence of motor impairments. It is distinguished by motor abnormalities such as dyskinesia (inability to move), loss of coordination, high stiffness, and spasticity. PD is also distinguished by the deposition of neurofibrillary tangles, proteinaceous Lewy bodies (LBs), or Lewy neurites made up of the synaptic protein α-syn as well as ubiquitin moieties [75]. Mitochondrial instability, OS, and microglial instability are all linked to PD, as are many other neurological abnormalities [76]. The primary histopathologic principle is the gradual decline of cells in the basal ganglia, particularly dopaminergic neural cells in the SNpc, culminating in the apparent lack of dopamine and the availability of LBs caused by α-syn accumulation in several residual nerve cells [77]. However, the exact pathophysiology of the induction of apoptosis has somehow remained largely unexplored. At the moment, the prognosis of PD is attributed to the existence of quintessential therapeutic outgrowths and the ostracization of potentially corresponding incurable diseases through alternative studies [78]. Patients suffering from Parkinson's, like those suffering from Alzheimer's, have no restorative interference, and the preponderance of prevailing therapies are preventative [79]. Earlier research discovered that intracellularly misfolded α-syn can agglomerate and form noxious residues in the LBs, exacerbating dopaminergic neuron impairment [80]. Furthermore, the overabundance of α-syn aggregate particles in dopaminergic neuronal cells cytosol provokes neuronal mitochondrial disturbance and anomalous dopamine metabolic activity, massively continuing to increase ROS levels and accelerating α-syn agglomeration. Intriguingly, α-syn consolidates are also responsible for expanded metal ion thresholds (including copper ions+, calcium ions+, and ferrous ion (Fe^{2+}) [81], which results in the instability of brain lymphocytes like T helper (Th) 17 cells [82] and therefore leads to systemic inflammation in the nerve cells. The majority of prevailing PDNP investigations have concentrated on genetic engineering. For example, it was discovered that nerve growth factors can be used to reactivate resting neurons to improve the release of dopamine. For this objective, NPs comprising compressed DNA have been tested [83], and a substantial change from the base point was identified in both interventional manifestations and aberrant brain metabolic activity as determined by tomography. While the possibility of incorporating nanotechnology to cure and prevent neurological diseases is very intriguing, the studies that have been investigated thus far are still preliminary. Nonetheless, these experiments offer useful insights into how functionally defective brain tissues respond to NPs. Initial competent assessment of Parkinsonism permits the administration of therapeutic applications before the degradation of dopaminergic nerve cells potentially delays the development of the disease. The identification of significantly reduced dopamine and an enhanced Ach threshold in neural tissue, blood, and urine specimens is the primary strategy for diagnosing PD [84]. The latest research used dopamine-binding aptamer-coated gold NPs mounted on an electrode to enhance specificity up to 1×10^{-8} M, allowing for higher precision detection of significantly lowered dopamine concentrations in urine specimens [85]. Moreover, prevailing detection technologies predicated on the molecular and electrical properties of dopamine binding to receptor sites/aptamers necessitate numerous stages and take longer than imaging modalities like MRI, CT, and so forth. The preferred approach for delivering dopamine or even another medicinal fragments/DNA to enhance neural potency in the Parkinsonism area is to improve BBB infiltration through rearranging NPs with transferrin or lactoferrin, culminating in medication accretion in the neuronal core area [86]. Chemical signals that entangle the dopaminergic, that is, D2 and D3 receptors, have been postulated as effective ligands for approaching PD brains. Because the dopamine concentration in specimens is absurdly minimal, NPs-based biosensors must have a very increased susceptibility to evaluate Parkinsonism in particular individuals using dopamine [87, 88]. For instance, by reconfiguring GO with tyrosine and Fe^{2+} ions, an immensely preferential fluorescent biosensor (GO-Y-Fe) for the identification of dopamine secreted from living neurons was generated [89]. Recently, a nanoplatform (NPoM) proficient in quantitatively identifying dopamine in the CSF utilizing SERS with a concentration range of 0.03

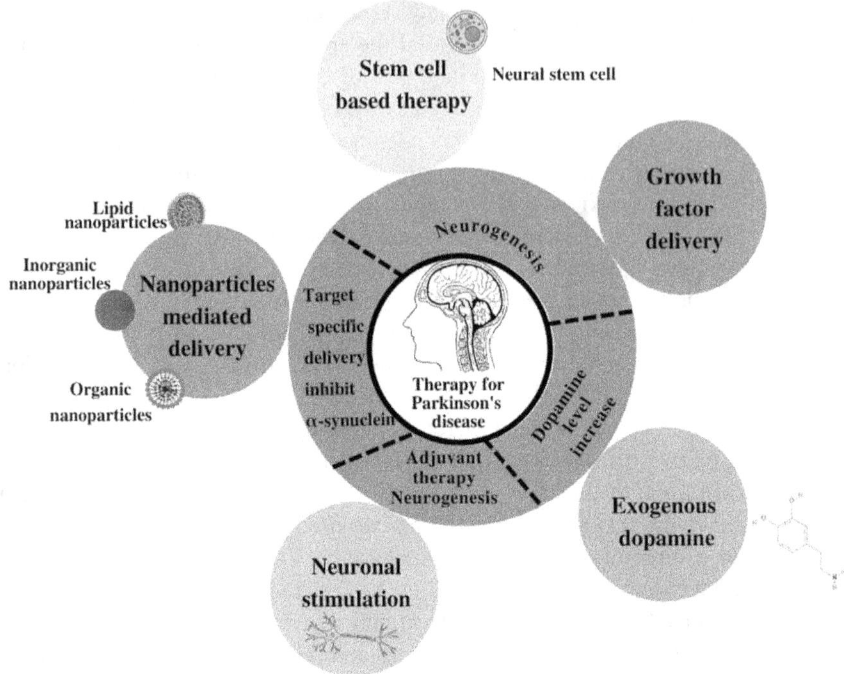

FIGURE 9.2 Mechanistical approach to treat Parkinson's disease using metallic nanoparticles.

pM was developed. On the surface of gold thin films, 3,3′-dithiodipropionic acid di (N-hydroxysuccinimide ester), a dopamine catcher, and 3-mercaptophenylboronic acid-decorated silver NPs (3-MPBA-SNPs) were sequentially retained. The band frequency at 998 cm^{-1} increased as the dopamine amount increased. The concentrations of dopamine in two human CSF specimens were 42.8 and 55.3 pM, respectively. NPoM demonstrated high specificity for the identification of dopamine without the involvement of several other chemical compounds [90]. The aforementioned optimistic findings suggest that using NPs to create detective biosensors with greater precision and susceptibility could retain tremendous potential in cellular examination and the clinical prognosis of PD. In brief, recent advancements in NPs have also influenced the development of a wide variety of neuro-biosensor nanosystems with good accuracy, selectivity, bioactivity, reliability, and versatility, which are supposed to develop an interpretive and concise introspective framework for initial assessment of PD, appraisal of pharmacotherapy clinical efficacy against Parkinsonism, and follow-up surveillance of PD advancements. A mechanistic approach for the management of Parkinson's disease using MNPs is depicted in Figure 9.2.

9.2 METALLIC NP IN HUNTINGTON'S DISEASE (HD)

HD, also referred to as Huntington's chorea, is a life-threatening neurological state caused by a mutation in the Huntington gene [91] that is characterized by gradual motor, intellectual, and behavioral problems [92]. The Huntington (HTT) gene's augmented cytosine-adenine-guanine (CAG) repeated series provokes gene mutation in the HTT protein (mHTT) by constituting an exceptionally elongated polyglutamine repeat series [93]. The genetic variation causes the development of abnormal Huntingtin protein, which itself is neurotoxic, but the underlying process causing HD is unknown. The mHTT is splintered into noxious portions, which could then cluster into noxious solubilized lifeforms such as Ab, designed to induce neuronal mitochondrial damage, overstimulation of microglia (proinflammatory secretion), a decrease of synaptic plasticity, as well as debilitated astrocyte-mediated glutamate uptake (link to dopamine neurotransmission variation), inevitably resulting in neuronal damage [94]. The indications of HD differ and are regulated by an individual as well as phase, but they include lack of coordination, incremental and unintended body motions, and learning impairment, including gradual mental impairment and even memory loss. Prevailing alternative treatments are centered on symptomatic treatment and nursing to delay expansion and minimize several associated abnormalities. Current findings have shown that NPs have great potential for use in biomaterials for identifying biochemical landmarks for the biochemical assessment of HD [95]. Conventional therapeutic imaging techniques, like CT and MRI, have a challenging time detecting HD in its early stages. The identification of critical biochemical parameters is critical for the interventional assessment of this disorder. Comparable to PD, the instability of the DA framework is linked to HD [96], and DA is becoming a primary indicator for HD. As a consequence, detecting DA with elevated sensitivity is critical for HD assessment and management [97]. Nonetheless, contemporary DA measurement techniques, including chromatography, electrochemistry, and SERS, have shortcomings in sensitivity and consistency [98]. Nanotechnology's scientific advances may strengthen these approaches. A notable study used synthesized AgNP dimers for rapid identification of DA via SERS. A new analysis promulgated graphene quantum dots (GQD) concocted by oxidative cutting of natural graphite for identification of monomeric and fibrillar A concentrations and surveillance of the fibrillization of A monomers [99]. The GQD was sheet-like and had a canonical graphene lattice of 3.4 with a larger range of almost 5.1±0.3 nm and a length of around 2±0.2 nm, according to TEM images. The highest emission amplitude was discovered at 500 nm under amplification at 400 nm. The photoluminescence (PL) frequency of GQDs steadily declined as the multitude of A1–42 monomers increased, whereas higher levels of A1–42 fibrils in a combination of A1–42 fibrils and monomers could augment the Pl spectra, illustrating their capability of detecting A1–42 fibrils and monomers. Besides that, when A1–42 monomers were added to pristine GQDs, there was a significant decline in PL intensity, while A1–42 fibrils caused only a minor decline. GQDs that were equivalent to ThT could have been used as an intermediary to regulate the method of A1–42 fibrillization. The fibrillization operation attained stability after 11 hours, as indicated by a plateau in the data [99]. For instance, by modifying the surface with a functional substituent, those nanosystems could intervene as a design to target restorative treatment for the expulsion of amyloidogenic proteins in neurological disorders. Moreover, comparable to AD, HD has been linked to debilitating AChE features, which could be a prospective diagnostic for HD assessment [100, 101].

To assess AChE action, a novel colorimetric nanoplatform (MnO_2-TMB) relying on manganese dioxide (MnO_2) nanosheets capable of oxidizing colorless TMB to blue oxidized TMB was developed. The occurrence of AChE could stimulate acetylthiocholine to produce thiocholine, resulting in Mn^{2+} generation without catalytic potential due to the significant decrease of MnO_2 nanosheets [102]. The ultrathin MnO_2 nanosheets formed by the redox process of potassium permanganate as well as sodium dodecyl benzene sulfonate had a maximum lateral measurement and a typical absorption peak for about 380 nm, creating a significant adequate surface area for the response with TMB. The amount of MnO_2 nanosheets increased, resulting in an increment in absorbance value at 652 nm and a strong blue color. A significant decline in the absorbance magnitude of the MnO_2-TMB process was observed at the AChE level marking from 0 to 17.5 mU mL^{-1}, along with a substantial color transformation from blue to colorless. Furthermore, a colorimetric framework with high amplitude demonstrated significant outcomes ranging from 0.1 to 15 mU mL^{-1}, with a concentration range of 0.035 mU mL^{-1}. Meanwhile, it was highly selective for AChE and had a high resistance to involvement by many other bioactive molecules. Assorted sugars as well as proteins had an insignificant influence on the absorptivity of the MnO_2-TMB system, and the MnO_2-TMB system could identify AChE actions even with 0.5 g mL^{-1} conflicting agents [102]. As a result, significant recent advancements in nanotechnology may initiate the progression of additional screening NPs to distinguish and ascertain a set of consistent biological markers for the specific prognosis of HD [70, 71, 103, 104].

9.3 METALLIC NP IN MULTIPLE SCLEROSIS (MS)

MS is another very prevalent autoimmune neurological disorder as well as a demyelinated brain disease, determined by the deterioration of the thermal barrier cover of neurones present inside the cerebrospinal region [105]. Since 2013, the global population with MS has ranged between 2 million and 2.5 million, with frequency and amplitude greatly varying across geographic areas and population groups [106]. The very first individual to characterize MS was Jean-Martin Charcot. Well, it is defined as the formation of innumerable lesions on the white matter of the cerebrospinal region [107]. The condition usually shows up between the ages of 20 and 50, and it affects more women than men [106]. The evaluation of this disorder is compared with experimental research results like magnetic resonance imaging (MRI) as well as 18-fluorodeoxyglucose positron emission tomography, among other things [108]. There are several findings on the antimicrobial effects of MNPs toward human pathogens, which expressed notable antimicrobial properties toward the evaluated human pathogenic organisms like bacteria [109, 110], fungi [111], viruses [112], and some other parasites. As a result, MNPs with antimicrobial properties could also be used to cure these CNS disease-causing prerequisites. Aside from that, nanotechnology is utilized in the prognosis and drug delivery for neurological disorders. This chapter covers many attributes of MNPs related to neurological disorders, such as prevailing therapeutic approaches and their constraints, nanoneuroparticles and their importance, the involvement of nanotechnology in the management and therapy of CNS afflictions, and cytotoxicity concerns.

9.4 METALLIC NP IN AMYOTROPHIC LATERAL SCLEROSIS (ALS)

ALS, another neurogenic emerging neurological condition, is not a rare condition, with approximately 2.6 per 100,000 people in Europe suffering from it each year [113], and it is strongly related to damage to efferent nerve cells in the cerebrospinal region. Almost 15–20% of patient populations have neurocognitive deformities characterized by behavioral problems, which inevitably lead to memory loss [114]. Even if some gene mutations are linked to patient populations with family history conditions, the fundamental processes for the majority of patient populations remain unknown [115]. ALS is distinguished by the deterioration of nerve cells that control muscle contractions, but the exact pathophysiology is unknown, though hereditary and ecological considerations are thought to be implicated in this neurological disease [116]. For people with ALS, the disintegration of nerve cells causes several conditions that are pertinent, with the primary focus being on mobility impairment at diverse scales, including muscle spasms, speech problems, dysphagia, and walking, and reduced ability to begin and regulate all motor muscle movement [117]. Emerging research suggests that TDP-43, SOD1, and ubiquitin-2, three amyloids that resemble Ab and accumulate into hazardous entities in the brains of ALS patients, cause cytoskeleton degradation, microglial and astrocyte instability, and abnormalities in axonal transport. Presently, diagnostic features are the primary factor used to determine an ALS prognosis. Considering that there is no curative treatment for ALS, the conventional health system mostly focuses on symptomatic relief, improving individuals' well-being, and reducing their fatality [118]. The ALS development is typically too modest to notice. Although MRI has been used to find elevated indications in the internal posterior capsule, ALS cannot be characterized specifically based on these symptoms. Up to 80% of anterior horn cells in the affected peripheral nerve pool have been eliminated whenever evident muscle weakness and loss of function manifest [119]. To implement the preceding therapies, it is necessary to create a comprehensive and focused technique to identify ALS as early as possible. Nanotechnology appears to be especially appealing for the identification of ALS because of this prerequisite. Even though the precise pathway is unknown, it is hypothesized that several cell-based and molecular mechanisms, including the modifications of the superoxide dismutase 1 (SOD1) gene, are important in the advancement of ALS [120]. The critical feature of ALS disease is intracellular agglomeration along with misfolding of the mutated SOD1 proteins, which can sometimes result in the similar functioning of wild-type SOD1 proteins in nerve cells [117]. The sensitive screening and categorization of such an anomalously clustered protein may help with the identification of ALS. In this specific instance, a colorimetric identification process for SOD1 protein aggregate particles was established. The SOD1 subunit was mounted on the interface of AuNPs (SOD1-AuNP) to create an immensely sensitive sensing probe for monitoring the systemic progression of SOD1 aggregate particles at a very minimal concentration. Furthermore, the sequential developmental state of SOD1 aggregate particles could be distinguished [121]. In many other phrases, a biomedical probe predicated on metallic NPs allows for the sensitive and quantifiable differentiation of the evolved state of aberrant protein aggregate particles in ALS, which has tremendous possibilities for perceiving the pathogenic mechanisms, providing convincing diagnostic tests, and directing target intervention for ALS. The prospective neuroinflammatory mechanisms may be elucidated in animal studies as a result of NP labeling to understand the physiological and pathological modalities of ALS and initiate commensurate medical approaches for ALS. Moreover, notable research found that sensitive phononics predicated on graphene accumulated on Si/SiO_2 substrates have immense prospects for discerning ALS by utilizing shift patterns in the Raman peak positions of graphene [122]. When graphene interacted with CSF from ALS patient populations, there was a clear n-doping effect and a red shift of ~3–3.5 cm^{-1} in the 2D value of the Raman spectral when contrasted to graphene in the involvement of a negative control CSF specimen. The cumulative carrier intensity in human CSF specimens from ALS in the SOD1G93A rat experimental model was substantially distinctive in comparison to many other chronic conditions, indicating their high efficacy in differentiating ALS from several other neurological disorders like MS. Following the irrevocable and transient functionalities of ALS advancement, an initial concise and differing assessment of ALS is critical for ALS patients' interventional consequences. This study illustrated that by using metallic nanoparticles, it is prudent and viable to distinguish ALS from other equivalent and conveniently misidentified disorders, allowing for more appropriate and expeditious therapies.

9.5 METALLIC NP IN ISCHEMIC STROKE

Ischemic stroke, chronic state in which there is inefficacious blood flow because of interference of cerebral vasculatures like thromboembolism [123]. It is proclaimed that ischemic stroke, along with increased morbidity, infirmity, and mortality, is the principal element of death and impairment globally [124] and also a major root for severe, prolonged impairment in the United States [125] and the third highest prevalent origin of fatality in emerging nations [126]. In the United States, it is predicted that around 800,000 individuals undergo stroke every year, of which ischemic stroke accounts for 87%. More than a quarter of patients with stroke survivors were older than 65 years and were physically challenged after six months [127]. The preponderance of acute ischemic strokes is thrombotic. Prompt, efficacious therapy for acute ischemic strokes is crucial for rescuing lives and decreasing the expansion of afterward defects. Stroke is another ND with a greater number of fatalities that are driven primarily by thrombosis (Moskowitz et al., 2010). Stroke is a major threat for people over the age of 65, and ischemic stroke accounts for about 87% of all strokes [129], which is closely linked to cerebral arterial thrombus, neuronal inflammation, ROS increased formation, as well as poor post-stroke rehabilitation [124]. The average individual tends to lose 1.9 million brain cells, 13.8 billion junctions, and 12 km of axonal fibers with every minute of therapeutic postponement, and the brain keeps losing numerous more neuronal cells as it does throughout nearly 3.6 years of baseline aging in every hour of therapeutic time lag. Prevailing ischemic stroke imaging techniques depend heavily on diagnostic manifestations, along with CT and MRI brain scans. The goal of restorative medication is to save the penumbra, a region in the brain where tissues do not obtain adequate blood but still obtain adequate oxygen supply to avoid cell death for a limited amount of time [130, 131], and to reduce ischemia-reperfusion concussion. Although the consistency of the BBB could be arbitrated from an ischemic perspective, macromolecular agents are unable to reach the brain proficiently [132]. As a result, the contemporary approach to treating acute ischemic stroke is to reconstruct the blood supply to the ischemic brain as promptly as possible. Besides that, the probability of hemorrhage and the late regimen frequently restrict application of clot-dissolving substances like tissue plasminogen activator (tPA). In actuality, less than 3% of patients with ischemic stroke started to receive tPA intervention within 3 h of the commencement of the stroke [127]. Furthermore, the production of ROS associated with cerebral ischemia and reperfusion worsens restorative efficacy. Because of their exceptional characteristics, NPs have already been extensively researched in terms of a definitive diagnosis of ischemic stroke. Recent ischemia-based clinical spectrums are not particularly selective. The current medication options for ischemic stroke are far from acceptable. Conversely, unlike hemorrhagic stroke, which is frequently preceded by blood splitting, edema, and pressure around the lesions, ischemic stroke has low optimistic indications within 6 to 24 hours of the onset [133]. Although MRI's diffusion-weighted imaging (DWI) and perfusion-weighted imaging (PWI) can identify ischemic discrepancies previously, the imaging period is too long for MRI to be an optimal technique for diagnosing ischemic stroke in an urgent situation, and the consequences for the preliminary phase of ischemic stroke may be unreliable [134]. NPs are being investigated to acquire a concise identification of ischemia occurrence as well as alterations in cerebrospinal tissues following a cerebral infarction. In rats with perpetual mid-cerebral artery obstruction, PEG-modified liposomes labeled with 1-[18F] fluoro-3,6-dioxatetracosane were utilized (p-MCAO). Within 3 hours of the onset of the ischemia, a real-time demonstration of ischemic lesions was realized using PET illustrations [135]. According to the findings of the preceding research, USPIO NPs can be utilized as a better intensity substance for MRI brain imaging after the occurrence of an ischemic stroke [136]. Thus, nanotechnology is expected to transform stroke therapeutic perspectives by facilitating targeted drug transfer and controlled drug release locally. Perfluorocarbon NPs with fibrin targeting have been evolved for the specialized delivery of plasminogen activator streptokinase to human plasma clots [137]. The researchers demonstrated in vitro that targeted streptokinase NPs reduced clot amount by 30% in 1 h at comparatively much lower concentrations than the free drug. SOD is a free radical scavenger with a brief in vivo half-life (six minutes) and low permeation beyond the BBB. SOD NPs reduced infarct intensity by 65%, increased subsistence, and restored functional ability in rats. Platinum nanoparticles, a novel and potent ROS scavenger, were tested for their implication on ischemic strokes in a mid-cerebral artery-occluded mouse laboratory model [138]. The use of platinum NPs improved motor activities and lessened brain infarct amplitude substantially. By integrating NPs with increased permeability to raise vascular permeability, such a nanosystem could serve as a susceptible screening aid in addition to an unconventional treatment possibility for ischemic stroke. These encouraging findings indicate that nanotechnology has an enormous possibility of strengthening the intervention in ischemic strokes.

9.6 METALLIC NP IN EPILEPSY

Epilepsy is a widespread and highly prevalent neurological disease that influences 50–70 million individuals globally [139]. It affects 1% of individuals over the age of 20 and up to 3% of individuals over the age of 75 [140]. Epilepsy is a collection of diseases or neuropathies, not a single disorder [141]. Moreover, epilepsy is defined as a neurological diagnostic and therapeutic disorder characterized by impulsive and repetitive epileptic fits caused by an aberrant paroxysmal neural release that ranges from a nearly imperceptible duration to a significant duration of intense shaking [142], as well as a rapid neuronal exoneration and disproportionate outflow of specialized neurons, and it influences 50 million individuals internationally [141]. Focal epilepsy is estimated to account for 60% of all epilepsy and is directly linked to gene alterations (such as genes encode K^+ ion channels as well as genes that govern cellular proliferation) [143] and neuroinflammation [144]. Present methods of diagnosing epilepsy depend primarily on seizure occurrence and assessments. EEG can discern the sequence of brain wave patterns, whereas MRI and CT can show unusual systemic lesions. Existing diagnostic methodologies, notwithstanding, aren't particularly selective. For instance, several brain signal patterns may not be caused by epilepsy, whereas others may be ordinary by exempting epilepsy disorder. Anti-epileptic drugs (AEDs) and surgery are used to cure epilepsy when drug resistance develops as a consequence of prolonged chemoradiation therapy. Despite significant advancements in AED pharmacology, up to 35–40% of patients with epilepsy have been resilient to pharmacotherapy and do not achieve remission, a condition known as refractory epilepsy [145]. In this perspective, NPs are predicted to accomplish key functions in epilepsy assessment as well as pharmacotherapy surveillance. A new micropillar electrode array (PEA) with extremely good flexibility and bioactivity depending on the lotus leaf was recently developed for multichannel sensing of epileptiform action in a rat experimental model [146]. A two-step molding procedure was used to create hierarchies of micropillars of PEA. The PEA was created by immobilizing a chromium/gold electrode layer and an insulating layer of SU-8 onto micropillars. The μPEA was made up of nine recording sites with a diameter of about 120 μm on average. It was 18 μm dense, with micropillars ranging in length from 6 to 17 μm. Penicillin, a GABA receptor blocker, was used to prompt epileptiform action in the rat cortex. The presumptuous micropillars enveloped by neuronal tissues were competent in detecting three epileptiform durations: basal, latent, and epileptiform action. The spectroscopic intensity was increased in the recurrence spectrum of 5–25 Hz, and the outflow was at its peak 30 minutes after penicillin infusion. Moreover, they accomplished multichannel data of the epileptiform outflow, and the channel region was directly linked with spike-time setback ranging from 9 to 13 ms. Immunohistochemical examination revealed that the PEAs had prolonged bioactivity. The average noise and signal-to-noise proportions of the micropillar electrodes were 22 μV and 234, respectively, which were preferable to those of a planar electrode array (55 μV and 149), indicating that PEAs have a lot of potential in neural recording [146]. A nanoscale instrument for observing increased EEG from laboratory mice was also configured to analyze the intricate EEG procedure and probe into the neurophysiologic framework of EEG signals. They created a polyimide-based microelectrode array on the mouse skull as well as measured large-scale activity in the brain in experimental mice to determine the neuronal association among both EEG rhythmic patterns and behaviors [147]. In the coming years, advancements in nanoscale devices may aid in the discovery of the underlying element of EEG rhythm generation for improved epilepsy prognosis. Table 9.1 summarizes the multiple metallic nanoparticles for NDs, and Table 9.2 depicts different types of nanoparticle-mediated delivery systems.

TABLE 9.1
Various Metallic Nanoparticles for NDs

S. No.	Type of nanoparticles	Uses	Reference
1	MnO-PEG-Cy5.5	Identification of gliomas	[148]
2	Iron oxide NPs	MRI-monitored magnetic glioma-targeted delivery of drugs	[149]
3	Iron oxide NPs coated with amphiphilic polymers	Targeting many stages of viral replication in the CNS	[150]
4	Iron oxide NPs with DDNP-SPIO	determine the β-amyloid plaques (Aβ)	[151]
5	Iron oxide NPs	Magnetic resonance imaging of brain tumors	[152]
6	Gold NPs	Prion disorder treatment and brain tumor image	[153]
7	Gold NPs	Helps in the identification of tumors	[154]
8	Platinum NPs	Treatment of PD	[155]
9	Zinc oxide NPs	Reduce oxidative stress in brain	[156]
10	Silver NPs PNPs radiolabeled with ^{99m}Tc	Intervention of glioblastoma	[157]
11	Gold NPs	BBB and lung targeting for tuberculosis and other lung illness	[158]
12	Cerium oxide NPs	Decreased oxidative and nitrosative damage after a stroke	[159]
13	Gadolinium Metallofullerene NPs	Brain tumor delivery of drug and longitudinal imaging	[160]

TABLE 9.2
Different Types of Nanoparticle-Mediated Delivery System

Lipid-based nanoparticle	Polymeric nanoparticle	Inorganic nanoparticle	References
High bioavailability, Formulation simplicity with a proportions of physicochemical effects, low encapsulation efficiency	Precise control of particle characteristics, easy surface modulations, possibility for accumulation and toxicity	Increased uptake due to ionic interaction with BBB. Unique electrical, magnetic and optical effects, small size	[161, 162]

9.7 MECHANISTIC APPROACHES OF MNPs FOR NEUROLOGICAL DISEASES

Multiple categories of MNPs (metallic nanoparticles) have been reviewed for the treatment/diagnosis of assorted neurological disorders such as PD, AD, HD, MS, ALS, ischemic stroke, and epilepsy. In the therapies as well as management of PD, a variety of interventions involving MNPs and many other biomaterials are used, such as nanoparticle-mediated delivery mechanisms, exogenous dopamine, neuronal stimulation, growth factor delivery, and stem cell-based therapy. The application of growth factors with neuroprotective and neurogenesis attributes has been investigated as a reliable therapy for the management of PD. Deep brain stimulation, on the other hand, is a protruding method that employs impenetrable electrodes to implement electric pulses to inhibit neural signals in a precise area and reduce dyskinesia and tremors [163]. It is also linked with negative consequences and is commonly used as an alternative therapy, particularly in instances where increased chemoradiation therapy causes complications. Aside from the intensity and length of the applied potential, as well as the orientation of the electrode prospects, all of which impact the results of DBS, miniaturization of the electrodes with improved signal/noise proportions, rapid conduction, biological inertness, reliability, and consistency are all crucial criteria from a biomaterial standpoint. Graphene, nitrogen-doped ultra-nanocrystalline diamond, and hybrid diamond-coated carbon fiber microelectrodes have all been revealed as promising electrode materials for highly precise neuronal stimulation [164]. Henceforth, in nanoparticle-mediated delivery systems, lipid nanoparticles (e.g., liposomes), inorganic and organic NPs explicitly concentrate on the target, as when given intravenously, liposomes efficaciously penetrate the BBB and steadily increase concentrations of dopamine in the striatal region while inhibiting a-syn. MNP-based treatment options have been discovered to be quite efficient in the particular circumstance of stem cell therapy, as they are competent in developing wider in vivo imaging of stem cells to measure all consequences of transplanted cells. The application of nanotechnologies in tissue regrowth is expected to raise the restorative and medicinal functionalities of stem cell transplants, as well as the detection of AD. As a result, advances have been made in the fields of HD, MS, ALS, ischemic stroke, and epilepsy. Figure 9.3 shows the graphical representation of various metallic nanoparticles and their roles in Neurological disorders.

Several more subsequent aspects and scientific advances (e.g., PET, MRI, gene therapy, neuronal stimulation, nanodevices, bioengineering) have been made in the field of MNPs, as well as those attributable to their diverse characteristics such as biocompatibility, antiviral, antimicrobial, and anti-inflammatory action, which are preferable for better assessment and treatment. Table 9.3 represents various metallic nanoparticles and their mechanistic approaches for brain-related diseases.

9.8 CONCLUSION AND FUTURE PROSPECTIVE

To summarize, nanotechnology has the potential to focus on providing exhilarating possibilities for future pharmacotherapy monitoring of neurodegenerative disorders. Nanotechnology implementations in brain research are critical for providing

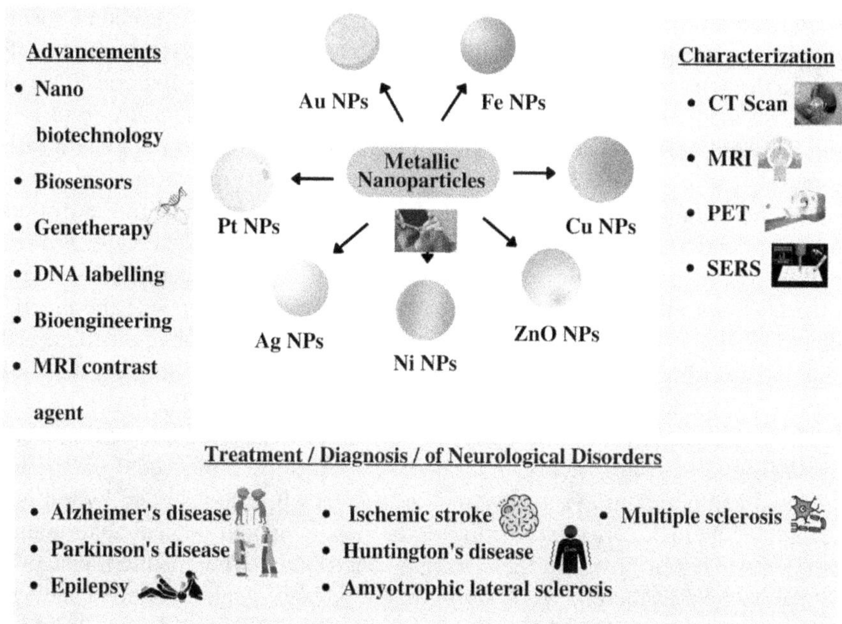

FIGURE 9.3 Graphical representations of various metallic nanoparticles roles in neurological disorders.

TABLE 9.3
Various Metallic Nanoparticles for NDs

Metallic nanoparticles	Mechanistic approaches	Size	Disease	References
Gold	Improve selectivity to the brain, readily cross BBB, decreasing inflammation and oxidative stress	1–50 nm	AD	[165, 166]
Silver	Optical properties are good, drug delivery is efficient	3–200 nm	Brain tumors	[167]
Carbon	Stem cell therapy, platelet aggregation	10–45 nm	Stroke, AD, PD, brain cancer	[168, 169]
Zinc oxide	Efficient drug delivery, reduce oxidative stress	30–150 nm	Brain tumors	[167]

optimism to patient populations struggling with neurological diseases. Due to their astonishing physiological, physical, and chemical characteristics, as well as exceptional spectral functionalities due to their nanoscale level, NPs have always shown considerable possibilities as an evolving agent for the medical assessment of neurological diseases. The establishment of a novel creation of MNPs with minimal toxicity can aid in the regulation of long-term medication and efficacious, specifically targeted drug distribution. Multiple NPs, like magnetic NPs and organic NPs, assist in the assessment of neurological diseases. Furthermore, polymeric NPs, liposomes, SLNPs, and other drug-loaded NPs can be used efficaciously in targeted drug delivery, which subsequently aids in the monitoring of neurological disease. Moreover, the modern processing nanoplatforms for diagnostic requirements as well as the suitable approaches for the targeted delivery of drugs to the dysfunctional neuronal areas are efficacious for AD, PD, HD, MS, ALS, FRDA, ischemic stroke, and epilepsy. A series of studies have constructed productive advancement in the last handful of years to conquer various barriers that impede the practical implementation of CNS-targeting prescription medications, attributable to the challenge of exploring biochemical markers and establishing NPs that target specific biochemical markers. Furthermore, newly suggested hybrid NPs, which integrate several or additional NPs, have performed a variety of activities, including double or numerous targeting systems, multimodal therapeutics, multimodal diagnostics, or related combinations. Such research has completely demonstrated hybrid NPs' exceptional characteristics for identifying or managing neurological abnormalities both in vitro and in vivo, and therefore may contribute to therapeutic translational accomplishment in the end. As a result, major advancements in nanotechnologies for the biosynthesis and functionalization of adaptable NPs will lead to the development of new indicators and NPs combined with diagnostics in the foreseeable future, bringing potential ways to overcome CNS illnesses. In conclusion, with the insightful utilization of MNPs, the intervention of neurological disorders outbreaks has experienced a significant evolution. Because prevailing targeted therapies only identify one target at a time, such alternative techniques are anticipated to enhance the efficacy of conventional restorative regimens. From this perspective, it is an enormous challenge that, if approaches including multimodal mediators are aggressively applied, could be readily conquered with nanotechnology. Nonetheless, the progress offers hope for the field of NPs to be fully utilized for prospective enhanced treatment techniques.

ACKNOWLEDGMENTS

The authors are grateful to MVN University in Palwal, Haryana, India, for their cooperative assistance and to Canva for imparting an interface for creating proficient scientific figures. The authors are also thankful to Dr. Jamilur R. Ansari, Dronacharya College of Engineering, Gurgaon, India, for the needful editing in the chapter.

REFERENCES

[1] T.T. Nguyen, T.T. Dung Nguyen, T.K. Vo, N.-M.-A. Tran, M.K. Nguyen, T. Van Vo, G. Van Vo, Nanotechnology-based drug delivery for central nervous system disorders, Biomed Pharmacother. 143 (2021) 112117. https://doi.org/10.1016/j.biopha.2021.112117.

[2] V. Vio, M.J. Marchant, E. Araya, M.J. Kogan, Metal nanoparticles for the treatment and diagnosis of neurodegenerative brain diseases, Curr Pharm Des. 23 (2017) 1916–1926. https://doi.org/10.2174/1381612823666170105152948.

[3] Y. Serlin, I. Shelef, B. Knyazer, A. Friedman, Anatomy and physiology of the blood-brain barrier, Semin Cell Dev Biol. 38 (2015) 2–6. https://doi.org/10.1016/j.semcdb.2015.01.002.
[4] N.J. Abbott, Prediction of blood–brain barrier permeation in drug discovery from in vivo, in vitro and in silico models, Drug Discov Today: Technol. 1 (2004) 407–416. https://doi.org/10.1016/j.ddtec.2004.11.014.
[5] M. Koshy, J.L. Villano, T.A. Dolecek, A. Howard, U. Mahmood, S.J. Chmura, R.R. Weichselbaum, B.J. McCarthy, Improved survival time trends for glioblastoma using the SEER 17 population-based registries, J Neurooncol. 107 (2012) 207–212. https://doi.org/10.1007/s11060-011-0738-7.
[6] W. Poewe, K. Seppi, C.M. Tanner, G.M. Halliday, P. Brundin, J. Volkmann, A.-E. Schrag, A.E. Lang, Parkinson disease, Nat Rev Dis Primers. 3 (2017) 17013. https://doi.org/10.1038/nrdp.2017.13.
[7] J. Cummings, J. Mintzer, H. Brodaty, M. Sano, S. Banerjee, D.P. Devanand, S. Gauthier, R. Howard, K. Lanctôt, C.G. Lyketsos, E. Peskind, A.P. Porsteinsson, E. Reich, C. Sampaio, D. Steffens, M. Wortmann, K. Zhong, Agitation in cognitive disorders: International Psychogeriatric Association provisional consensus clinical and research definition, Int Psychogeriatr. 27 (2015) 7–17. https://doi.org/10.1017/S1041610214001963.
[8] N.J. Abbott, L. Rönnbäck, E. Hansson, Astrocyte-endothelial interactions at the blood-brain barrier, Nat Rev Neurosci. 7 (2006) 41–53. https://doi.org/10.1038/nrn1824.
[9] D.E. Bredesen, R.V. Rao, P. Mehlen, Cell death in the nervous system, Nature. 443 (2006) 796–802. https://doi.org/10.1038/nature05293.
[10] M.N. Javed, F.H. Pottoo, A. Shamim, M.S. Hasnain, M.S. Alam, Design of experiments for the development of nanoparticles, nanomaterials, and nanocomposites, in: S. Beg (Ed.), Design of Experiments for Pharmaceutical Product Development, Springer Singapore, Singapore, 2021: pp. 151–169. https://doi.org/10.1007/978-981-33-4351-1_9.
[11] M.S. Alam, M.N. Javed, F.H. Pottoo, A. Waziri, F.A. Almalki, M.S. Hasnain, A. Garg, M.K. Saifullah, QbD approached comparison of reaction mechanism in microwave synthesized gold nanoparticles and their superior catalytic role against hazardous nirto-dye, Appl Organometal Chem. (2019). https://doi.org/10.1002/aoc.5071.
[12] M.N. Javed, E.S. Dahiya, A.M. Ibrahim, Md.S. Alam, F.A. Khan, F.H. Pottoo, Recent advancement in clinical application of nanotechnological approached targeted delivery of herbal drugs, in: S. Beg, M.A. Barkat, F.J. Ahmad (Eds.), Nanophytomedicine, Springer Singapore, Singapore, 2020: pp. 151–172. https://doi.org/10.1007/978-981-15-4909-0_9.
[13] A. Di Stefano, M. Carafa, P. Sozio, F. Pinnen, D. Braghiroli, G. Orlando, G. Cannazza, M. Ricciutelli, C. Marianecci, E. Santucci, Evaluation of rat striatal l-dopa and DA concentration after intraperitoneal administration of l-dopa prodrugs in liposomal formulations, J Control Release. 99 (2004) 293–300. https://doi.org/10.1016/j.jconrel.2004.07.010.
[14] C.-Y. Lin, Y.-C. Lin, C.-Y. Huang, S.-R. Wu, C.-M. Chen, H.-L. Liu, Ultrasound-responsive neurotrophic factor-loaded microbubble-liposome complex: Preclinical investigation for Parkinson's disease treatment, J Control Release. 321 (2020) 519–528. https://doi.org/10.1016/j.jconrel.2020.02.044.
[15] A. McRae, A. Dahlström, Transmitter-loaded polymeric microspheres induce regrowth of dopaminergic nerve terminals in striata of rats with 6-OH-DA induced parkinsonism, Neurochem Int. 25 (1994) 27–33. https://doi.org/10.1016/0197-0186(94)90049-3.
[16] R. Kumar, G. Dhamija, J.R. Ansari, Md.N. Javed, Md.S. Alam, C-dot nanoparticulated devices for biomedical applications, in: Nanotechnology, 1st ed., CRC Press, Boca Raton, 2022: pp. 271–299. https://doi.org/10.1201/9781003220350-15.
[17] S. Singhal, M. Gupta, Md.S. Alam, Md.N. Javed, J.R. Ansari, Carbon allotropes-based nanodevices, in: Nanotechnology, 1st ed., CRC Press, Boca Raton, 2022: pp. 241–269. https://doi.org/10.1201/9781003220350-14.

[18] J.R. Ansari, N. Singh, R. Ahmad, D. Chattopadhyay, A. Datta, Controlling self-assembly of ultra-small silver nanoparticles: Surface enhancement of Raman and fluorescent spectra, Opt Mater. 94 (2019) 138–147. https://doi.org/10.1016/j.optmat.2019.05.023.

[19] M. Kumar, M. Na, J.R Ansari, Analysis of the heating ability by varying the size of Fe_3O_4 magnetic nanoparticles for hyperthermia, Nano Sci Technol Int J. (2022). https://doi.org/10.1615/NanoSciTechnolIntJ.2022040075.

[20] de Jong, Drug delivery and nanoparticles: Applications and hazards, IJN. (2008) 133. https://doi.org/10.2147/IJN.S596.

[21] A. Kumar, J. Sidhu, A. Goyal, J.W. Tsao, Alzheimer disease, in: StatPearls, StatPearls Publishing, Treasure Island (FL), 2022. www.ncbi.nlm.nih.gov/books/NBK499922/ (accessed May 10, 2022).

[22] Applications of gold nanopaticles in brain diseases . . . Google Scholar. (n.d.). https://scholar.google.com/scholar?hl=en&as_sdt=0%2C5&q=applications+of+gold+nanopaticles+in+brain+diseases+across+the+blood+brain+barrier&btnG= (accessed November 13, 2022).

[23] S. Mishra, S. Sharma, M.N. Javed, F.H. Pottoo, M.A. Barkat, Harshita, M.S. Alam, M. Amir, M. Sarafroz, Bioinspired nanocomposites: Applications in disease diagnosis and treatment, PNT. 7 (2019) 206–219. https://doi.org/10.2174/2211738507666190425121509.

[24] M.N. Javed, F.H. Pottoo, A. Shamim, M.S. Hasnain, M.S. Alam, Design of experiments for the development of nanoparticles, nanomaterials, and nanocomposites, in: S. Beg (Ed.), Design of Experiments for Pharmaceutical Product Development, Springer Singapore, Singapore, 2021: pp. 151–169. https://doi.org/10.1007/978-981-33-4351-1_9.

[25] A.M. Ibrahim, L. Chauhan, A. Bhardwaj, A. Sharma, F. Fayaz, B. Kumar, M. Alhashmi, N. AlHajri, M.S. Alam, F.H. Pottoo, Brain-derived neurotropic factor in neurodegenerative disorders, Biomedicines. 10 (2022) 1143. https://doi.org/10.3390/biomedicines10051143.

[26] F.H. Pottoo, Md.N. Javed, Md.A. Barkat, Md.S. Alam, J.A. Nowshehri, D.M. Alshayban, M.A. Ansari, Estrogen and serotonin: Complexity of interactions and implications for epileptic seizures and epileptogenesis, CN. 17 (2019) 214–231. https://doi.org/10.2174/1570159X16666180628164432.

[27] M. Aslam, Md.N. Javed, H.H. Deeb, M.K. Nicola, Mohd.A. Mirza, Md.S. Alam, Md.H. Akhtar, A. Waziri, Lipid nanocarriers for neurotherapeutics: Introduction, challenges, blood-brain barrier, and promises of delivery approaches, CNSNDDT. 21 (2022) 952–965. https://doi.org/10.2174/1871527320666210706104240.

[28] F.H. Pottoo, S. Sharma, Md.N. Javed, Md.A. Barkat, Harshita, Md.S. Alam, Mohd. J. Naim, O. Alam, M.A. Ansari, G.E. Barreto, G.Md. Ashraf, Lipid-based nanoformulations in the treatment of neurological disorders, Drug Metabolism Reviews. 52 (2020) 185–204. https://doi.org/10.1080/03602532.2020.1726942.

[29] A. Waziri, C. Bharti, M. Aslam, P. Jamil, Mohd. A. Mirza, M.N. Javed, U. Pottoo, A. Ahmadi, M.S. Alam, Probiotics for the chemoprotective role against the toxic effect of cancer chemotherapy, ACAMC. 22 (2022) 654–667. https://doi.org/10.2174/1871520621666210514000615.

[30] Md.S. Alam, Md.F. Naseh, J.R. Ansari, A. Waziri, Md.N. Javed, A. Ahmadi, M.K. Saifullah, A. Garg, Synthesis approaches for higher yields of nanoparticles, in: Nanomaterials in the Battle Against Pathogens and Disease Vectors, 1st ed., CRC Press, Boca Raton, 2022: pp. 51–82. https://doi.org/10.1201/9781003126256-3.

[31] N. Singh, J.R. Ansari, M. Pal, N.T.K. Thanh, T. Le, A. Datta, Synthesis and magnetic properties of stable cobalt nanoparticles decorated reduced graphene oxide sheets in the aqueous medium, J Mater Sci: Mater Electron. 31 (2020) 15108–15117. https://doi.org/10.1007/s10854-020-04075-2.

[32] Md.F. Naseh, J.R. Ansari, Md.S. Alam, Md.N. Javed, Sustainable nanotorus for biosensing and therapeutical applications, in: U. Shanker, C.M. Hussain, M. Rani (Eds.), Handbook of Green and Sustainable Nanotechnology, Springer International Publishing, Cham, 2022: pp. 1–21. https://doi.org/10.1007/978-3-030-69023-6_47-1.

[33] M. Kumar, Madhavi, J.R. Ansari, Studies on the heating ability by varying the size of Fe_3O_4 magnetic nanoparticles for hyperthermia, Nano Sci Technol Int J. 13 (2022) 33–45. https://doi.org/10.1615/NanoSciTechnolIntJ.2022040075.

[34] M. Kumar, J.R. Ansari, M.T. Beig, Studies on superparamagnetic behaviour of Ni100-xCux alloy films deposited by DC magnetron sputtering, Mater Res Innov. (2021) 1–6. https://doi.org/10.1080/14328917.2021.1987069.

[35] C.A. Sunilbhai, Md.S. Alam, K.K. Sadasivuni, J.R. Ansari, SPR Assisted diabetes detection, in: K.K. Sadasivuni, J.-J. Cabibihan, A.K. A M Al-Ali, R.A. Malik (Eds.), Advanced Bioscience and Biosystems for Detection and Management of Diabetes, Springer International Publishing, Cham, 2022: pp. 91–131. https://doi.org/10.1007/978-3-030-99728-1_6.

[36] N. Singh, J.R. Ansari, M. Pal, A. Das, D. Sen, D. Chattopadhyay, A. Datta, Enhanced blue photoluminescence of cobalt-reduced graphene oxide hybrid material and observation of rare plasmonic response by tailoring morphology, Appl Phys A. 127 (2021) 568. https://doi.org/10.1007/s00339-021-04697-1.

[37] J.R. Ansari, N. Singh, S. Anwar, S. Mohapatra, A. Datta, Silver nanoparticles decorated two dimensional MoS2 nanosheets for enhanced photocatalytic activity, Colloids and Surf A: Physicochem Eng Asp. 635 (2022) 128102. https://doi.org/10.1016/j.colsurfa.2021.128102.

[38] A. Kumar, A. Singh, Ekavali, A review on Alzheimer's disease pathophysiology and its management: An update, Pharmacol Rep. 67 (2015) 195–203. https://doi.org/10.1016/j.pharep.2014.09.004.

[39] C.L. Masters, R. Bateman, K. Blennow, C.C. Rowe, R.A. Sperling, J.L. Cummings, Alzheimer's disease, Nat Rev Dis Primers. 1 (2015) 15056. https://doi.org/10.1038/nrdp.2015.56.

[40] I. van Dijken, M. van der Vlag, R. Flores Hernández, A. Ross, Perspectives on treatment of Alzheimer's disease: A closer look into EphB2 depletion, J Neurosci. 37 (2017) 11296–11297. https://doi.org/10.1523/JNEUROSCI.0214-17.2017.

[41] G. Leavey, A. Abbott, M. Watson, S. Todd, V. Coates, S. McIlfactrick, B. McCormack, B. Waterhouse-Bradley, E. Curran, The evaluation of a healthcare passport to improve quality of care and communication for people living with dementia (EQuIP): A protocol paper for a qualitative, longitudinal study, BMC Health Serv Res. 16 (2016) 363. https://doi.org/10.1186/s12913-016-1617-x.

[42] H. Akiyama, Inflammation and Alzheimer's disease, Neurobiol Aging. 21 (2000) 383–421. https://doi.org/10.1016/S0197-4580(00)00124-X.

[43] L. Cassidy, F. Fernandez, J.B. Johnson, M. Naiker, A.G. Owoola, D.A. Broszczak, Oxidative stress in Alzheimer's disease: A review on emergent natural polyphenolic therapeutics, Complement Ther Med. 49 (2020) 102294. https://doi.org/10.1016/j.ctim.2019.102294.

[44] M.T. Heneka, M.J. Carson, J.E. Khoury, G.E. Landreth, F. Brosseron, D.L. Feinstein, A.H. Jacobs, T. Wyss-Coray, J. Vitorica, R.M. Ransohoff, K. Herrup, S.A. Frautschy, B. Finsen, G.C. Brown, A. Verkhratsky, K. Yamanaka, J. Koistinaho, E. Latz, A. Halle, G.C. Petzold, T. Town, D. Morgan, M.L. Shinohara, V.H. Perry, C. Holmes, N.G. Bazan, D.J. Brooks, S. Hunot, B. Joseph, N. Deigendesch, O. Garaschuk, E. Boddeke, C.A. Dinarello, J.C. Breitner, G.M. Cole, D.T. Golenbock, M.P. Kummer, Neuroinflammation in Alzheimer's disease, The Lancet Neurol. 14 (2015) 388–405. https://doi.org/10.1016/S1474-4422(15)70016-5.

[45] P.S. Aisen, S. Gauthier, S.H. Ferris, D. Saumier, D. Haine, D. Garceau et al., Tramiprosate in mild-to-moderate Alzheimer's disease - a randomized, double-blind, placebo-controlled, multi-centre study (the Alphase Study), Arch Med Sci 7 (2011), pp. 102–111. DOI: 10.5114/aoms.2011.20612.

[46] M. Janusz and A. Zabłocka, Colostral proline-rich polypeptides—immunoregulatory properties and prospects of therapeutic use in Alzheimer's disease, Curr Alzheimer Res 7 (2010), pp. 323–333. https://doi.org/10.2174/156720510791162377.

[47] R. Kandimalla, P.H. Reddy, Therapeutics of neurotransmitters in Alzheimer's disease, J Alzheimers Dis. 57 (2017) 1049–1069. https://doi.org/10.3233/JAD-161118.

[48] M.F. Mendez, The accurate diagnosis of early-onset dementia, Int J Psychiatry Med. 36 (2006) 401–412. https://doi.org/10.2190/Q6J4-R143-P630-KW41.

[49] J. Hardy, D. Allsop, Amyloid deposition as the central event in the aetiology of Alzheimer's disease, Trends Pharmacol Sci. 12 (1991) 383–388. https://doi.org/10.1016/0165-6147(91)90609-v.

[50] S.-C. Lee, H.-H. Park, S.-H. Kim, S.-H. Koh, S.-H. Han, M.-Y. Yoon, Ultrasensitive fluorescence detection of Alzheimer's disease based on polyvalent directed peptide polymer coupled to a nanoporous ZnO nanoplatform, Anal Chem. 91 (2019) 5573–5581. https://doi.org/10.1021/acs.analchem.8b03735.

[51] M.P. Dorsey, B.M. Nguelifack and E.A. Yates, Colorimetric Detection of Mutant β-Amyloid(1–40) Membrane-Active Aggregation with Biosensing Vesicles, ACS Appl. Bio Mater. 2 (2019), pp. 4966–4977. https://doi.org/10.1021/acsabm.9b00694.

[52] Ghasemi, F., Hormozi-Nezhad, M.R. and Mahmoudi, M., 2015. A colorimetric sensor array for detection and discrimination of biothiols based on aggregation of gold nanoparticles. Analytica chimica acta, 882, pp. 58–67. https://doi.org/10.1016/j.aca.2015.04.011.

[53] C.M. Jin, J.B. Joo, I. Choi, Facile amplification of solution-state surface-enhanced raman scattering of small molecules using spontaneously formed 3D nanoplasmonic wells, Anal Chem. 90 (2018) 5023–5031. https://doi.org/10.1021/acs.analchem.7b04674.

[54] B. Budde, J. Schartner, L. Tönges, C. Kötting, A. Nabers, K. Gerwert, Reversible immuno-infrared sensor for the detection of Alzheimer's disease related biomarkers, ACS Sens. 4 (2019) 1851–1856. https://doi.org/10.1021/acssensors.9b00631.

[55] Y. Zhou, C. Li, X. Li, X. Zhu, B. Ye, M. Xu, A sensitive aptasensor for the detection of β-amyloid oligomers based on metal–organic frameworks as electrochemical signal probes, Anal Methods. 10 (2018) 4430–4437. https://doi.org/10.1039/C8AY00736E.

[56] S.H. Nasr, H. Kouyoumdjian, C. Mallett, S. Ramadan, D.C. Zhu, E.M. Shapiro, X. Huang, Detection of β-amyloid by sialic acid coated bovine serum albumin magnetic nanoparticles in a mouse model of Alzheimer's disease, Small. 14 (2018). https://doi.org/10.1002/smll.201701828.

[57] Z. Du, N. Gao, X. Wang, J. Ren, X. Qu, Near-infrared switchable fullerene-based synergy therapy for Alzheimer's disease, Small. (2018) e1801852. https://doi.org/10.1002/smll.201801852.

[58] D.-Y. Kang, J.-H. Lee, B.-K. Oh, J.-W. Choi, Ultra-sensitive immunosensor for beta-amyloid (1–42) using scanning tunneling microscopy-based electrical detection, Biosens Bioelectron. 24 (2009) 1431–1436. https://doi.org/10.1016/j.bios.2008.08.018.

[59] W.A. El-Said, T.-H. Kim, C.-H. Yea, H. Kim, J.-W. Choi, Fabrication of gold nanoparticle modified ITO substrate to detect β -amyloid using surface-enhanced Raman scattering, J Nanosci Nanotechnol. 11 (2011) 768–772. https://doi.org/10.1166/jnn.2011.3268.

[60] Y. Xing, X.-Z. Feng, L. Zhang, J. Hou, G.-C. Han, Z. Chen, A sensitive and selective electrochemical biosensor for the determination of beta-amyloid oligomer by inhibiting the peptide-triggered in situ assembly of silver nanoparticles, IJN. 12 (2017) 3171–3179. https://doi.org/10.2147/IJN.S132776.

[61] F. Morales-Zavala, N. Casanova-Morales, R.B. Gonzalez, A. Chandía-Cristi, L.D. Estrada, I. Alvizú, V. Waselowski, F. Guzman, S. Guerrero, M. Oyarzún-Olave, C. Rebolledo, E. Rodriguez, J. Armijo, H. Bhuyan, M. Favre, A.R. Alvarez, M.J. Kogan, J.R. Maze, Functionalization of stable fluorescent nanodiamonds towards reliable detection of biomarkers for Alzheimer's disease, J Nanobiotechnol. 16 (2018) 60. https://doi.org/10.1186/s12951-018-0385-7.
[62] M. Hülsemann, C. Zafiu, K. Kühbach, N. Lühmann, Y. Herrmann, L. Peters, C. Linnartz, J. Willbold, K. Kravchenko, A. Kulawik, S. Willbold, O. Bannach, D. Willbold, Biofunctionalized silica nanoparticles: Standards in amyloid-β oligomer-based diagnosis of Alzheimer's disease, JAD. 54 (2016) 79–88. https://doi.org/10.3233/JAD-160253.
[63] S. Mourtas, A.N. Lazar, E. Markoutsa, C. Duyckaerts, S.G. Antimisiaris, Multifunctional nanoliposomes with curcumin-lipid derivative and brain targeting functionality with potential applications for Alzheimer disease, Eur J Med Chem. 80 (2014) 175–183. https://doi.org/10.1016/j.ejmech.2014.04.050.
[64] F. Ghasemi, M.R. Hormozi-Nezhad, M. Mahmoudi, Label-free detection of β-amyloid peptides (Aβ40 and Aβ42): A colorimetric sensor array for plasma monitoring of Alzheimer's disease, Nanoscale. 10 (2018) 6361–6368. https://doi.org/10.1039/C8NR00195B.
[65] K. Buerger, M. Ewers, T. Pirttilä, R. Zinkowski, I. Alafuzoff, S.J. Teipel, J. DeBernardis, D. Kerkman, C. McCulloch, H. Soininen, H. Hampel, CSF phosphorylated tau protein correlates with neocortical neurofibrillary pathology in Alzheimer's disease, Brain. 129 (2006) 3035–3041. https://doi.org/10.1093/brain/awl269.
[66] B. Zhu, T. Zhang, Q. Jiang, Y. Li, Y. Fu, J. Dai, G. Li, Q. Qi, Y. Cheng, Synthesis and evaluation of pyrazine and quinoxaline fluorophores for in vivo detection of cerebral tau tangles in Alzheimer's models, Chem. Commun. 54 (2018) 11558–11561. https://doi.org/10.1039/C8CC06897F.
[67] A Mesenchymal Stem Cell Line Transplantation Improves Neurological Function and Angiogenesis in Intraventricular Amyloid β-Infused Rats—PubMed. (n.d.). https://pubmed.ncbi.nlm.nih.gov/30207232/ (accessed November 14, 2022).
[68] M.S. Alam, M.N. Javed, F.H. Pottoo, A. Waziri, F.A. Almalki, M.S. Hasnain, A. Garg, M.K. Saifullah, QbD approached comparison of reaction mechanism in microwave synthesized gold nanoparticles and their superior catalytic role against hazardous nirto-dye, Appl Organometal Chem. (2019). https://doi.org/10.1002/aoc.5071.
[69] M.S. Hasnain, Md.N. Javed, Md.S. Alam, P. Rishishwar, S. Rishishwar, S. Ali, A.K. Nayak, S. Beg, Purple heart plant leaves extract-mediated silver nanoparticle synthesis: Optimization by Box-Behnken design, Mater Sci Eng: C. 99 (2019) 1105–1114. https://doi.org/10.1016/j.msec.2019.02.061.
[70] F.H. Pottoo, N. Tabassum, Md.N. Javed, S. Nigar, S. Sharma, Md.A. Barkat, Harshita, Md.S. Alam, M.A. Ansari, G.E. Barreto, G.M. Ashraf, Raloxifene potentiates the effect of fluoxetine against maximal electroshock induced seizures in mice, Eur J Pharm Sci. 146 (2020) 105261. https://doi.org/10.1016/j.ejps.2020.105261.
[71] N. Kumari, N. Daram, Md.S. Alam, A.K. Verma, Rationalizing the use of polyphenol nano-formulations in the therapy of neurodegenerative diseases, CNSNDDT. 21 (2022) 966–976. https://doi.org/10.2174/1871527321666220512153854.
[72] A. Siderowf, M. Stern, Update on Parkinson disease, Ann Intern Med. 138 (2003) 651–658. https://doi.org/10.7326/0003-4819-138-8-200304150-00013.
[73] L.V. Kalia, A.E. Lang, Parkinson's disease, Lancet. 386 (2015) 896–912. https://doi.org/10.1016/S0140-6736(14)61393-3.
[74] M. Bordoni, E. Scarian, F. Rey, S. Gagliardi, S. Carelli, O. Pansarasa, C. Cereda, Biomaterials in neurodegenerative disorders: A promising therapeutic approach, Int J Mol Sci. 21 (2020) E3243. https://doi.org/10.3390/ijms21093243.
[75] K. Ebanks, P.A. Lewis, R. Bandopadhyay, Vesicular dysfunction and the pathogenesis of Parkinson's disease: Clues from genetic studies, Front Neurosci. 13 (2019) 1381. https://doi.org/10.3389/fnins.2019.01381.

[76] T.Y. Kim, E. Leem, J.M. Lee, S.R. Kim, Control of reactive oxygen species for the prevention of Parkinson's disease: The possible application of flavonoids, Antioxidants. 9 (2020) 583. https://doi.org/10.3390/antiox9070583.

[77] N. Zhang, F. Yan, X. Liang, M. Wu, Y. Shen, M. Chen, Y. Xu, G. Zou, P. Jiang, C. Tang, H. Zheng, Z. Dai, Localized delivery of curcumin into brain with polysorbate 80-modified cerasomes by ultrasound-targeted microbubble destruction for improved Parkinson's disease therapy, Theranostics. 8 (2018) 2264–2277. https://doi.org/10.7150/thno.23734.

[78] J. Jankovic, Parkinson's disease: Clinical features and diagnosis, J Neurol Neurosurg Psychiatry. 79 (2008) 368–376. https://doi.org/10.1136/jnnp.2007.131045.

[79] A. Samii, J.G. Nutt, B.R. Ransom, Parkinson's disease, Lancet. 363 (2004) 1783–1793. https://doi.org/10.1016/S0140-6736(04)16305-8.

[80] J.H. Kordower, C.W. Olanow, H.B. Dodiya, Y. Chu, T.G. Beach, C.H. Adler, G.M. Halliday, R.T. Bartus, Disease duration and the integrity of the nigrostriatal system in Parkinson's disease, Brain. 136 (2013) 2419–2431. https://doi.org/10.1093/brain/awt192.

[81] K. Jomova, D. Vondrakova, M. Lawson, M. Valko, Metals, oxidative stress and neurodegenerative disorders, Mol Cell Biochem. 345 (2010) 91–104. https://doi.org/10.1007/s11010-010-0563-x.

[82] H. Kebir, K. Kreymborg, I. Ifergan, A. Dodelet-Devillers, R. Cayrol, M. Bernard, F. Giuliani, N. Arbour, B. Becher, A. Prat, Human TH17 lymphocytes promote blood-brain barrier disruption and central nervous system inflammation, Nat Med. 13 (2007) 1173–1175. https://doi.org/10.1038/nm1651.

[83] J. Baskin, J.E. Jeon, S.J.G. Lewis, Nanoparticles for drug delivery in Parkinson's disease, J Neurol. 268 (2021) 1981–1994. https://doi.org/10.1007/s00415-020-10291-x.

[84] J.G. Nutt, J.H. Carter, G.J. Sexton, The dopamine transporter: Importance in Parkinson's disease, Ann Neurol. 55 (2004) 766–773. https://doi.org/10.1002/ana.20089.

[85] Y. Xu, X. Hun, F. Liu, X. Wen, X. Luo, Aptamer biosensor for dopamine based on a gold electrode modified with carbon nanoparticles and thionine labeled gold nanoparticles as probe, Microchimica Acta (Online). 182 (2015) 1797–1802.

[86] R. Huang, W. Ke, Y. Liu, D. Wu, L. Feng, C. Jiang, Y. Pei, Gene therapy using lactoferrin-modified nanoparticles in a rotenone-induced chronic Parkinson model, J Neurol Sci. 290 (2010) 123–130. https://doi.org/10.1016/j.jns.2009.09.032.

[87] D.-J. Park, J.-H. Choi, W.-J. Lee, S. Um and B.-K. Oh, Selective Electrochemical Detection of Dopamine Using Reduced Graphene Oxide Sheets-Gold Nanoparticles Modified Electrode, Journal of Nanoscience and Nanotechnology 17 (2017), pp. 8012–8018. https://doi.org/10.1166/jnn.2017.15073.

[88] J.H. An, T.-H. Kim, B.-K. Oh and J.W. Choi, Detection of dopamine in dopaminergic cell using nanoparticles-based barcode DNA analysis, J Nanosci Nanotechnol 12 (2012), pp. 764–768. https://doi.org/10.1166/jnn.2012.5403.

[89] S.-J. Jeon, C. Choi, J.-M. Ju, S. Lee, J.H. Park, J.-H. Kim, Tuning the response selectivity of graphene oxide fluorescence by organometallic complexation for neurotransmitter detection, Nanoscale. 11 (2019) 5254–5264. https://doi.org/10.1039/C9NR00643E.

[90] K. Zhang, Y. Liu, Y. Wang, R. Zhang, J. Liu, J. Wei, H. Qian, K. Qian, R. Chen, B. Liu, Quantitative SERS detection of dopamine in cerebrospinal fluid by dual-recognition-induced hot spot generation, ACS Appl Mater Interfaces. 10 (2018) 15388–15394. https://doi.org/10.1021/acsami.8b01063.

[91] C. Zuccato, M. Valenza, E. Cattaneo, Molecular mechanisms and potential therapeutical targets in Huntington's disease, Physiol Rev. 90 (2010) 905–981. https://doi.org/10.1152/physrev.00041.2009.

[92] G.P. Bates, R. Dorsey, J.F. Gusella, M.R. Hayden, C. Kay, B.R. Leavitt, M. Nance, C.A. Ross, R.I. Scahill, R. Wetzel, E.J. Wild, S.J. Tabrizi, Huntington disease, Nat Rev Dis Primers. 1 (2015) 15005. https://doi.org/10.1038/nrdp.2015.5.

[93] A. Yamamoto, J.J. Lucas, R. Hen, Reversal of neuropathology and motor dysfunction in a conditional model of Huntington's disease, Cell. 101 (2000) 57–66. https://doi.org/10.1016/S0092-8674(00)80623-6.
[94] M. Jimenez-Sanchez, F. Licitra, B.R. Underwood, D.C. Rubinsztein, Huntington's disease: Mechanisms of pathogenesis and therapeutic strategies, Cold Spring Harb Perspect Med. 7 (2017) a024240. https://doi.org/10.1101/cshperspect.a024240.
[95] W. Cong, R. Bai, Y.-F. Li, L. Wang, C. Chen, Selenium nanoparticles as an efficient nanomedicine for the therapy of Huntington's disease, ACS Appl Mater Interfaces. 11 (2019) 34725–34735. https://doi.org/10.1021/acsami.9b12319.
[96] J.M. Gil, A.C. Rego, Mechanisms of neurodegeneration in Huntington's disease, Eur J Neurosci. 27 (2008) 2803–2820. https://doi.org/10.1111/j.1460-9568.2008.06310.x.
[97] T. Lee, L.X. Cai, V.S. Lelyveld, A. Hai, A. Jasanoff, Molecular-level functional magnetic resonance imaging of dopaminergic signaling, Science. 344 (2014) 533–535. https://doi.org/10.1126/science.1249380.
[98] L. Zhang, Y. Cheng, J. Lei, Y. Liu, Q. Hao, H. Ju, Stepwise chemical reaction strategy for highly sensitive electrochemiluminescent detection of dopamine, Anal Chem. 85 (2013) 8001–8007. https://doi.org/10.1021/ac401894w.
[99] H. Huang, P. Li, M. Zhang, Y. Yu, Y. Huang, H. Gu, C. Wang, Y. Yang, Graphene quantum dots for detecting monomeric amyloid peptides, Nanoscale. 9 (2017) 5044–5048. https://doi.org/10.1039/C6NR10017A.
[100] E. Cubo, K.M. Shannon, D. Tracy, J.A. Jaglin, B.A. Bernard, J. Wuu et al., Effect of donepezil on motor and cognitive function in Huntington disease, Neurology 67 (2006), pp. 1268–1271. https://doi.org/10.1212/01.wnl.0000238106.10423.00.
[101] Y. Zhang, T. Hei, Y. Cai, Q. Gao and Q. Zhang, Affinity binding-guided fluorescent nanobiosensor for acetylcholinesterase inhibitors via distance modulation between the fluorophore and metallic nanoparticle, Anal Chem 84 (2012), pp. 2830–2836. https://doi.org/10.1021/ac300436m.
[102] X. Yan, Y. Song, X. Wu, C. Zhu, X. Su, D. Du et al., Oxidase-mimicking activity of ultrathin MnO2 nanosheets in colorimetric assay of acetylcholinesterase activity, Nanoscale 9 (2017), pp. 2317–2323. https://doi.org/10.1039/C6NR08473G.
[103] M.N. Javed, E.S. Dahiya, A.M. Ibrahim, Md.S. Alam, F.A. Khan, F.H. Pottoo, Recent advancement in clinical application of nanotechnological approached targeted delivery of herbal drugs, in: S. Beg, M.A. Barkat, F.J. Ahmad (Eds.), Nanophytomedicine, Springer Singapore, Singapore, 2020: pp. 151–172. https://doi.org/10.1007/978-981-15-4909-0_9.
[104] S. Raj, R. Manchanda, M. Bhandari, Md.S. Alam, Review on natural bioactive products as radioprotective therapeutics: Present and past perspective, CPB. 23 (2022) 1721–1738. https://doi.org/10.2174/1389201023666220110104645.
[105] K. Berer, G. Krishnamoorthy, Microbial view of central nervous system autoimmunity, FEBS Lett. 588 (2014) 4207–4213. https://doi.org/10.1016/j.febslet.2014.04.007.
[106] R. Milo, E. Kahana, Multiple sclerosis: Geoepidemiology, genetics and the environment, Autoimmun Rev. 9 (2010) A387–394. https://doi.org/10.1016/j.autrev.2009.11.010.
[107] D.R. Kumar, F. Aslinia, S.H. Yale, J.J. Mazza, Jean-Martin Charcot: The father of neurology, Clin Med Res. 9 (2011) 46–49. https://doi.org/10.3121/cmr.2009.883.
[108] P. Berlit, Review: Diagnosis and treatment of cerebral vasculitis, Ther Adv Neurol Disord. 3 (2010) 29–42. https://doi.org/10.1177/1756285609347123.
[109] R. Thomas, A. Janardhanan, R.T. Varghese, E.V. Soniya, J. Mathew and E.K. Radhakrishnan, Antibacterial properties of silver nanoparticles synthesized by marine Ochrobactrum sp, Braz J Microbiol 45 (2015), pp. 1221–1227. https://doi.org/10.1590/S1517-83822014000400012.
[110] B. Perito, E. Giorgetti, P. Marsili and M. Muniz-Miranda, Antibacterial activity of silver nanoparticles obtained by pulsed laser ablation in pure water and in chloride solution, Beilstein J Nanotechnol 7 (2016), pp. 465–473. https://doi.org/10.3762/bjnano.7.40.

[111] M. Gajbhiye, J. Kesharwani, A. Ingle, A. Gade, M. Rai, Fungus-mediated synthesis of silver nanoparticles and their activity against pathogenic fungi in combination with fluconazole, Nanomed: Nanotechnol Biol Med. 5 (2009) 382–386. https://doi.org/10.1016/j.nano.2009.06.005.

[112] S. Galdiero, A. Falanga, M. Vitiello, M. Cantisani, V. Marra, M. Galdiero, Silver nanoparticles as potential antiviral agents, Molecules. 16 (2011) 8894–8918. https://doi.org/10.3390/molecules16108894.

[113] O. Hardiman, A. Al-Chalabi, C. Brayne, E. Beghi, L.H. van den Berg, A. Chio, S. Martin, G. Logroscino, J. Rooney, The changing picture of amyotrophic lateral sclerosis: Lessons from European registers, J Neurol Neurosurg Psychiatry. 88 (2017) 557–563. https://doi.org/10.1136/jnnp-2016-314495.

[114] R.H. Brown, A. Al-Chalabi, Amyotrophic lateral sclerosis, N Engl J Med. 377 (2017) 162–172. https://doi.org/10.1056/NEJMra1603471.

[115] J.P. Taylor, R.H. Brown, D.W. Cleveland, Decoding ALS: From genes to mechanism, Nature. 539 (2016) 197–206. https://doi.org/10.1038/nature20413.

[116] T.S. Wingo, D.J. Cutler, N. Yarab, C.M. Kelly, J.D. Glass, The heritability of amyotrophic lateral sclerosis in a clinically ascertained United States research registry, PLoS One. 6 (2011) e27985. https://doi.org/10.1371/journal.pone.0027985.

[117] O. Hardiman, A. Al-Chalabi, A. Chio, E.M. Corr, G. Logroscino, W. Robberecht, P.J. Shaw, Z. Simmons, L.H. van den Berg, Amyotrophic lateral sclerosis, Nat Rev Dis Primers. 3 (2017) 17071. https://doi.org/10.1038/nrdp.2017.71.

[118] M.A. van Es, O. Hardiman, A. Chio, A. Al-Chalabi, R.J. Pasterkamp, J.H. Veldink, L.H. van den Berg, Amyotrophic lateral sclerosis, The Lancet. 390 (2017) 2084–2098. https://doi.org/10.1016/S0140-6736(17)31287-4.

[119] A. Eisen, Amyotrophic lateral sclerosis: A 40-year personal perspective, J Clin Neurosci. 16 (2009) 505–512. https://doi.org/10.1016/j.jocn.2008.07.072.

[120] S. Martin, A. Al Khleifat, A. Al-Chalabi, What causes amyotrophic lateral sclerosis? F1000Res. 6 (2017) 371. https://doi.org/10.12688/f1000research.10476.1.

[121] S. Hong, I. Choi, S. Lee, Y.I. Yang, T. Kang, J. Yi, Sensitive and colorimetric detection of the structural evolution of superoxide dismutase with gold nanoparticles, Anal Chem. 81 (2009) 1378–1382. https://doi.org/10.1021/ac802099c.

[122] B. Keisham, A. Seksenyan, S. Denyer, P. Kheirkhah, G.D. Arnone, P. Avalos, A.D. Bhimani, C. Svendsen, V. Berry, A.I. Mehta, Quantum capacitance based amplified graphene phononics for studying neurodegenerative diseases, ACS Appl Mater Interfaces. 11 (2019) 169–175. https://doi.org/10.1021/acsami.8b15893.

[123] X. J, Z. Y, X. J, L. G, D. C, Z. X et al., Engineered Nanoplatelets for Targeted Delivery of Plasminogen Activators to Reverse Thrombus in Multiple Mouse Thrombosis Models, Advanced materials (Deerfield Beach, Fla.) (2020). 32(4):1905145. https://doi.org/10.1002/adma.201905145.

[124] E.J. Benjamin, M.J. Blaha, S.E. Chiuve, M. Cushman, S.R. Das, R. Deo, S.D. de Ferranti, J. Floyd, M. Fornage, C. Gillespie, C.R. Isasi, M.C. Jiménez, L.C. Jordan, S.E. Judd, D. Lackland, J.H. Lichtman, L. Lisabeth, S. Liu, C.T. Longenecker, R.H. Mackey, K. Matsushita, D. Mozaffarian, M.E. Mussolino, K. Nasir, R.W. Neumar, L. Palaniappan, D.K. Pandey, R.R. Thiagarajan, M.J. Reeves, M. Ritchey, C.J. Rodriguez, G.A. Roth, W.D. Rosamond, C. Sasson, A. Towfighi, C.W. Tsao, M.B. Turner, S.S. Virani, J.H. Voeks, J.Z. Willey, J.T. Wilkins, J.HY. Wu, H.M. Alger, S.S. Wong, P. Muntner, Heart disease and stroke statistics—2017 update: A report from the American heart association, Circulation. 135 (2017). https://doi.org/10.1161/CIR.0000000000000485.

[125] V.L. Roger, A.S. Go, D.M. Lloyd-Jones, R.J. Adams, J.D. Berry, T.M. Brown, M.R. Carnethon, S. Dai, G. de Simone, E.S. Ford, C.S. Fox, H.J. Fullerton, C. Gillespie, K.J. Greenlund, S.M. Hailpern, J.A. Heit, P.M. Ho, V.J. Howard, B.M. Kissela, S.J. Kittner, D.T. Lackland, J.H. Lichtman, L.D. Lisabeth, D.M. Makuc, G.M. Marcus, A. Marelli,

D.B. Matchar, M.M. McDermott, J.B. Meigs, C.S. Moy, D. Mozaffarian, M.E. Mussolino, G. Nichol, N.P. Paynter, W.D. Rosamond, P.D. Sorlie, R.S. Stafford, T.N. Turan, M.B. Turner, N.D. Wong, J. Wylie-Rosett, American heart association statistics committee and stroke statistics subcommittee, Heart disease and stroke statistics-2011 update: A report from the American heart association, Circulation. 123 (2011) e18–e209. https://doi.org/10.1161/CIR.0b013e3182009701.

[126] J.L. Saver, Time is brain-quantified, Stroke. 37 (2006) 263–266. https://doi.org/10.1161/01.STR.0000196957.55928.ab.

[127] L.B. Goldstein, Acute ischemic stroke treatment in 2007, Circulation. 116 (2007) 1504–1514. https://doi.org/10.1161/CIRCULATIONAHA.106.670885.

[128] M.A. Moskowitz, E.H. Lo, C. Iadecola, The science of stroke: Mechanisms in search of treatments, Neuron. 67 (2010) 181–198. https://doi.org/10.1016/j.neuron.2010.07.002.

[129] A.G. Thrift, H.M. Dewey, R.A.L. Macdonell, J.J. McNeil, G.A. Donnan, incidence of the major stroke subtypes: Initial findings from the north east Melbourne stroke incidence study (NEMESIS), Stroke. 32 (2001) 1732–1738. https://doi.org/10.1161/01.STR.32.8.1732.

[130] A.M. Hakim, Ischemic penumbra: the therapeutic window, Neurology 51 (1998), pp. S44–46. https://doi.org/10.1212/WNL.51.3_Suppl_3.S44.

[131] M.D. Ginsberg and W.A. Pulsinelli, The ischemic penumbra, injury thresholds, and the therapeutic window for acute stroke, Ann Neurol 36 (1994), pp. 553–554. DOI: 10.1002/ana.410360402.

[132] V.N. Bharadwaj, D.T. Nguyen, V.D. Kodibagkar, S.E. Stabenfeldt, Nanoparticle-based therapeutics for brain injury, Adv Healthcare Mater. 7 (2018) 1700668. https://doi.org/10.1002/adhm.201700668.

[133] J.B. Fiebach, P.D. Schellinger, O. Jansen, M. Meyer, P. Wilde, J. Bender, P. Schramm, E. Jüttler, J. Oehler, M. Hartmann, S. Hähnel, M. Knauth, W. Hacke, K. Sartor, CT and diffusion-weighted MR imaging in randomized order: Diffusion-weighted imaging results in higher accuracy and lower interrater variability in the diagnosis of hyperacute ischemic stroke, Stroke. 33 (2002) 2206–2210. https://doi.org/10.1161/01.str.0000026864.20339.cb.

[134] T. Fukuta, T. Ishii, T. Asai, G. Nakamura, Y. Takeuchi, A. Sato et al., Real-time trafficking of PEGylated liposomes in the rodent focal brain ischemia analyzed by positron emission tomography, Artif Organs 2014 Aug;38(8):662–6. https://doi.org/10.1111/aor.12350.

[135] T. Fukuta, T. Ishii, T. Asai, G. Nakamura, Y. Takeuchi, A. Sato, Y. Agato, K. Shimizu, S. Akai, D. Fukumoto, N. Harada, H. Tsukada, A.T. Kawaguchi, N. Oku, Real-Time Trafficking of PEGylated liposomes in the rodent focal brain ischemia analyzed by positron emission tomography: Real-time trafficking of PEGylated liposomes, Artificial Organs. 38 (2014) 662–666. https://doi.org/10.1111/aor.12350.

[136] K.P. Doyle, L.N. Quach, H.E.D. Arceuil, M.S. Buckwalter, Ferumoxytol administration does not alter infarct volume or the inflammatory response to stroke in mice, Neuroscience Letters. 584 (2015) 236–240. https://doi.org/10.1016/j.neulet.2014.10.041.

[137] J. Marsh, A. Senpan, G. Hu, M. Scott, P. Gaffney, S. Wickline, G. Lanza, Fibrin-targeted perfluorocarbon nanoparticles for targeted thrombolysis, Nanomedicine. 2 (2007) 533–543. https://doi.org/10.2217/17435889.2.4.533.

[138] M. Takamiya, Y. Miyamoto, T. Yamashita, K. Deguchi, Y. Ohta, Y. Ikeda, T. Matsuura, K. Abe, Neurological and pathological improvements of cerebral infarction in mice with platinum nanoparticles, J Neurosci Res. 89 (2011) 1125–1133. https://doi.org/10.1002/jnr.22622.

[139] S.-Y. Kwan, Y.-C. Chuang, C.-W. Huang, T.-C. Chen, S.-B. Jou, A. Dash, Zonisamide: Review of recent clinical evidence for treatment of epilepsy, CNS Neurosci Ther. 21 (2015) 683–691. https://doi.org/10.1111/cns.12418.

[140] D. Hirtz, D.J. Thurman, K. Gwinn-Hardy, M. Mohamed, A.R. Chaudhuri, R. Zalutsky, How common are the "common" neurologic disorders? Neurology. 68 (2007) 326–337. https://doi.org/10.1212/01.wnl.0000252807.38124.a3.
[141] B.S. Chang, D.H. Lowenstein, Epilepsy, N Engl J Med. 349 (2003) 1257–1266. https://doi.org/10.1056/NEJMra022308.
[142] R.S. Fisher, C. Acevedo, A. Arzimanoglou, A. Bogacz, J.H. Cross, C.E. Elger, J. Engel, L. Forsgren, J.A. French, M. Glynn, D.C. Hesdorffer, B.I. Lee, G.W. Mathern, S.L. Moshé, E. Perucca, I.E. Scheffer, T. Tomson, M. Watanabe, S. Wiebe, ILAE Official Report: A practical clinical definition of epilepsy, Epilepsia. 55 (2014) 475–482. https://doi.org/10.1111/epi.12550.
[143] O. Devinsky, A. Vezzani, T.J. O'Brien, N. Jette, I.E. Scheffer, M. de Curtis, P. Perucca, Epilepsy, Nat Rev Dis Primers. 4 (2018) 18024. https://doi.org/10.1038/nrdp.2018.24.
[144] A. Vezzani, J. French, T. Bartfai, T.Z. Baram, The role of inflammation in epilepsy, Nat Rev Neurol. 7 (2011) 31–40. https://doi.org/10.1038/nrneurol.2010.178.
[145] A. Rosillo-de la Torre, Pharmacoresistant epilepsy and nanotechnology, Front Biosci. E6 (2014) 329. https://doi.org/10.2741/709.
[146] M. Du, S. Guan, L. Gao, S. Lv, S. Yang, J. Shi, J. Wang, H. Li, Y. Fang, Flexible micropillar electrode arrays for in vivo neural activity recordings, Small. 15 (2019) 1900582. https://doi.org/10.1002/smll.201900582.
[147] M. Lee, D. Kim, H.-S. Shin, H.-G. Sung and J.H. Choi, High-density EEG recordings of the freely moving mice using polyimide-based microelectrode, J Vis Exp (2011), pp. 2562. doi: 10.3791/2562.
[148] N. Chen, C. Shao, S. Li, Z. Wang, Y. Qu, W. Gu, C. Yu, L. Ye, Cy5.5 conjugated MnO nanoparticles for magnetic resonance/near-infrared fluorescence dual-modal imaging of brain gliomas, J Colloid Interface Sci. 457 (2015) 27–34. https://doi.org/10.1016/j.jcis.2015.06.046.
[149] B. Chertok, B.A. Moffat, A.E. David, F. Yu, C. Bergemann, B.D. Ross, V.C. Yang, Iron oxide nanoparticles as a drug delivery vehicle for MRI monitored magnetic targeting of brain tumors, Biomaterials. 29 (2008) 487–496. https://doi.org/10.1016/j.biomaterials.2007.08.050.
[150] L. Fiandra, M. Colombo, S. Mazzucchelli, M. Truffi, B. Santini, R. Allevi, M. Nebuloni, A. Capetti, G. Rizzardini, D. Prosperi, F. Corsi, Nanoformulation of antiretroviral drugs enhances their penetration across the blood brain barrier in mice, Nanomed: Nanotechnol Biol Med. 11 (2015) 1387–1397. https://doi.org/10.1016/j.nano.2015.03.009.
[151] D. Zhang, H.-B. Fa, J.-T. Zhou, S. Li, X.-W. Diao, W. Yin, The detection of β-amyloid plaques in an Alzheimer's disease rat model with DDNP-SPIO, Clin Radiol. 70 (2015) 74–80. https://doi.org/10.1016/j.crad.2014.09.019.
[152] M. Iv, N. Telischak, D. Feng, S.J. Holdsworth, K.W. Yeom, H.E. Daldrup-Link, Clinical applications of iron oxide nanoparticles for magnetic resonance imaging of brain tumors, Nanomedicine (Lond). 10 (2015) 993–1018. https://doi.org/10.2217/nnm.14.203.
[153] E.S. Day, L. Zhang, P.A. Thompson, J.A. Zawaski, C.C. Kaffes, M.W. Gaber, S.M. Blaney, J.L. West, Vascular-targeted photothermal therapy of an orthotopic murine glioma model, Nanomedicine. 7 (2012) 1133–1148. https://doi.org/10.2217/nnm.11.189.
[154] E. Schültke, R. Menk, B. Pinzer, A. Astolfo, M. Stampanoni, F. Arfelli, L.-A. Harsan, G. Nikkhah, Single-cell resolution in high-resolution synchrotron X-ray CT imaging with gold nanoparticles, J Synchrotron Radiat. 21 (2014) 242–250. https://doi.org/10.1107/S1600577513029007.
[155] J. Nellore, C. Pauline, K. Amarnath, Bacopa monnieri phytochemicals mediated synthesis of platinum nanoparticles and its neurorescue effect on 1-Methyl 4-Phenyl 1,2,3,6 tetrahydropyridine-induced experimental parkinsonism in Zebrafish, J Neurodegener Dis. 2013 (2013) 972391. https://doi.org/10.1155/2013/972391.

[156] M. Afifi, O.A. Almaghrabi, N.M. Kadasa, Ameliorative effect of zinc oxide nanoparticles on antioxidants and sperm characteristics in streptozotocin-induced diabetic rat testes, Biomed Res Int. 2015 (2015) 153573. https://doi.org/10.1155/2015/153573.

[157] E. Locatelli, M. Naddaka, C. Uboldi, G. Loudos, E. Fragogeorgi, V. Molinari, A. Pucci, T. Tsotakos, D. Psimadas, J. Ponti, M.C. Franchini, Targeted delivery of silver nanoparticles and alisertib: In vitro and in vivo synergistic effect against glioblastoma, Nanomedicine. 9 (2014) 839–849. https://doi.org/10.2217/nnm.14.1.

[158] M. Schäffler, F. Sousa, A. Wenk, L. Sitia, S. Hirn, C. Schleh, N. Haberl, M. Violatto, M. Canovi, P. Andreozzi, M. Salmona, P. Bigini, W.G. Kreyling, S. Krol, Blood protein coating of gold nanoparticles as potential tool for organ targeting, Biomaterials. 35 (2014) 3455–3466. https://doi.org/10.1016/j.biomaterials.2013.12.100.

[159] A.Y. Estevez, S. Pritchard, K. Harper, J.W. Aston, A. Lynch, J.J. Lucky, J.S. Ludington, P. Chatani, W.P. Mosenthal, J.C. Leiter, S. Andreescu, J.S. Erlichman, Neuroprotective mechanisms of cerium oxide nanoparticles in a mouse hippocampal brain slice model of ischemia, Free Radic Biol Med. 51 (2011) 1155–1163. https://doi.org/10.1016/j.freeradbiomed.2011.06.006.

[160] T. Li, S. Murphy, B. Kiselev, K.S. Bakshi, J. Zhang, A. Eltahir, Y. Zhang, Y. Chen, J. Zhu, R.M. Davis, L.A. Madsen, J.R. Morris, D.R. Karolyi, S.M. LaConte, Z. Sheng, H.C. Dorn, A new interleukin-13 amino-coated gadolinium metallofullerene nanoparticle for targeted MRI detection of glioblastoma tumor cells, J Am Chem Soc. 137 (2015) 7881–7888. https://doi.org/10.1021/jacs.5b03991.

[161] M.J. Mitchell, M.M. Billingsley, R.M. Haley, M.E. Wechsler, N.A. Peppas and R. Langer, Engineering precision nanoparticles for drug delivery, Nat Rev Drug Discov 20 (2021), 20, 101–124. https://doi.org/10.1038/s41573-020-0090-8.

[162] V. Kasina, R.J. Mownn, R. Bahal and G.C. Sartor, Nanoparticle delivery systems for substance use disorder, Neuropsychopharmacology (2022), 47, 1431–1439. https://doi.org/10.1038/s41386-022-01311-7.

[163] J. Volkmann, Deep brain stimulation for the treatment of Parkinson's disease, J Clin Neurophysiol (2004), 21(1):6–17.

[164] M.A. Hejazi, W. Tong, A. Stacey, A. Soto-Breceda, M.R. Ibbotson, M. Yunzab et al., Hybrid diamond/ carbon fiber microelectrodes enable multimodal electrical/chemical neural interfacing, Biomaterials 230 (2020), pp. 119648. https://doi.org/10.1016/j.biomaterials.2019.119648.

[165] G. de B. Silveira, A.P. Muller, R.A. Machado-de-Ávila and P.C.L. Silveira, Advance in the use of gold nanoparticles in the treatment of neurodegenerative diseases: new perspectives, Neural Regen Res (2021), 16(12):2425. doi: 10.4103/1673-5374.313040.

[166] Siddiqi, K.S., Husen, A., Sohrab, S.S. and Yassin, M.O., 2018. Recent status of nanomaterial fabrication and their potential applications in neurological disease management. Nanoscale research letters, 2018, 13, 1–17. https://doi.org/10.1186/s11671-018-2638-7.

[167] G. Caruso, G. Raudino and M. Caffo, Patented nanomedicines for the treatment of brain tumors, Pharm Pat Anal 2013 Nov;2(6):745–54. https://doi.org/10.4155/ppa.13.56.

[168] Mukhtar, M., Bilal, M., Rahdar, A., Barani, M., Arshad, R., Behl, T., Brisc, C., Banica, F. and Bungau, S. Nanomaterials for diagnosis and treatment of brain cancer: Recent updates. Chemosensors, 2020 Nov 20;8(4):117. https://doi.org/10.3390/chemosensors8040117.

[169] A. Estella-Hermoso de Mendoza, V. Préat, F. Mollinedo and M.J. Blanco-Prieto, In vitro and in vivo efficacy of edelfosine-loaded lipid nanoparticles against glioma, J Control Release 2011 Dec 20;156(3):421–6. https://doi.org/10.1016/j.jconrel.2011.07.030.

10 Catalyst Metallic Nanoparticles
Types, Mechanism, and Trends

Namita, Arti, Md Sabir Alam, Md Noushad Javed, Md Naushad Alam, and Jamilur R. Ansari

10.1 INTRODUCTION

Metallic nanoparticles have recently been utilized in the practical domains of nanofabrication, nano-biosensors and optoelectronic devices. Nanocatalysis is an essential branch of nanoscience, where nanoparticles (NPs) are used as catalysts. Owing to their high surface energy, excessive surface-to-volume ratio, tunable morphologies, energetic atoms available on the surface, and distinctive electronic constructions, metallic nanoparticles effectively participate in heterogeneous catalysis. Ultrafine metallic nanoparticles (sub-nanometer to ~3 nm), usually called nanoclusters, manifest remarkable catalytic properties. Metallic nanoparticles are, however, unstable, especially at high temperatures, on account of their high surface energy. Consequently, their migration and amalgamation are observed during the process of catalysis [1]. These features of metallic NPs [2] abruptly affect their catalytic properties [3] and behaviors. In order to overcome the stability [4] issue, metallic NPs are encapsulated [5] in nano-shells or nanopores. In order to investigate the catalytic properties of silver nanoparticle catalysts, methylene blue (MB) [6] underwent a reduction reaction with $NaBH_4$, and this reaction was chosen as a replica in the green synthesis process [7–9]. The morphology, uniformity, configuration, and functionality of the metallic NPs are not under control. Recent studies revealed that the capping molecules are effectively removed through thermal oxidation or UV/ozone treatment [10–14]. Metallic nanoparticle catalysts act as bridge builders between homogeneous and heterogeneous catalysts. The generation of hot electrons at the metal surface has been revealed by many studies done on metal oxide-covered metal. The factor of great concern is the toxicity [15] associated with metallic nanoparticle synthesis (based on metals like palladium, platinum, and rhodium) [16] synthesis. The level of toxicity [17] depends on the size of the particles, interrelated ligands, and some other related factors. Although non-noble-metal (iron, copper, nickel, and cobalt) based NP catalysts [18] are reported to be less harmful and hazardous than those of noble-metal-based NP catalysts (gold, platinum, and silver) [19]. To administer

DOI: 10.1201/9781003317319-10

the morphology of the metallic nanoparticles has become a big deal in the past few years [20–24]. When the solute has a low concentration and is incorporated with the polymeric monolayer so as to cling on to the subsequent growth region, mono-sized [25] metallic nanoparticles are evolved [26].

For ensuing industrial applications and commercialization based on metallic nanoparticles as prominent catalysts, there are still some barricades that need to be knocked down, such as inaccessible synthesis methods [27], toxicity, designing strategies, and maintenance of the catalytic property under extreme situations [28–31].

10.2 TYPES OF METALLIC NANOCATALYSTS

Metallic nanocatalysts are grouped into three sections: homogeneous catalysts, heterogeneous catalysts, and biocatalysts. Heterogeneous catalysts are comparatively preferred over homogeneous catalysts due to several factors such as the number of active sites, efficiency, selectivity [32], and rate of reaction. Heterogeneous catalysts proved to be an integral part of the chemical industry for several mechanisms like metamorphosis, renewal, or changeover. Metallic NPs are classified into two forms: noble-metal (gold, platinum, and silver)-based NPs and non-noble-metal (iron, copper, nickel, and cobalt)-based NPs. The latter one has been reported to have prominent catalytic properties, abundant availability, brilliant stability, and environmental sustainability. Non-noble metal-based nanoparticles broadly participate in various applications based on heterogeneous catalysis [33].

There are many metal-based nanoparticles that have been used as catalysts in many fields so far.

Gold-Based NPs

When the size is <5 nm, gold (Au) [34], being the least reactive metal, shows high catalytic activity toward the CO [35] oxidation reaction.

Copper-based NPs

These are the efficient catalysts for the production of methanol from the synthesis gas (also called syngas, $CO + H_2$) [36].

Silver-based NPs

These perform remarkably well as catalysts for photocatalysis-based water splitting [37], carbon dioxide reduction processes [38], and the deterioration of organic contaminants.

Platinum-based, iridium-based, and ruthenium-based NPs

These are used as catalysts in some important half-reactions like HER (hydrogen evolution reaction) [39], ORR (oxygen reduction reaction), and OER (oxygen evolution reaction), which are employed to convert and store renewable energy [40, 41].

10.3 SYNTHESIS MECHANISM

The techniques or mechanisms of synthesis can be classified under two heads: the thermodynamic equilibrium approach and the kinetic approach. Under the thermodynamic approach, the synthesis process constitutes the phenomena of supersaturation, nucleation, and subsequent growth. On the other hand, under the kinetic approach, metallic NPs are synthesized by confining the growth in the restricted region.

10.3.1 Mechanism of Thermodynamic Approach

In the thermodynamical approach, mono-sized silver nanoparticles are synthesized following polyol process, where ethylene glycol (EG) has been extensively used owing to its excellent reducing ability and remarkably increased breaking point. The evolution of metallic NPs requires anisotropic extension, which is controlled by stabilizers like PVP. This approach furnishes metallic nanoparticles with controlled shapes in the form of nanowires [42], nanocubes, nanospheres, and nanopolyhedra.

10.3.2 Mechanism of Polyol Process (AgNP Catalysts)

A chemical reaction between $AgNO_3$ and PVP was undertaken in the presence of ethylene glycol (EG) at 433 K. The formation of silver nanoparticles was reported on account of the gradual appearance of bright yellow color. Below 433 K, polyol process doesn't obtain AgNPs. In this process, EG played a significant role as a reducing agent, which reduces Ag^+ ions into Ag atoms. Gradually, the color of the silver solution turned out to be gray. The AgNPs were separated from EG by diluting them in acetone and isolated from PVP by centrifugation. For $AgNO_3$/PVP mole-to-mole ratios equal to 1.5, silver nanowires were generated. For $AgNO_3$/PVP mole-to-mole ratios ranging from 1.5 to 3, silver nanocubes were reported. When silver nanocubes were maintained in the reaction mixture for a longer duration, silver nanopolyhedra catalysts were obtained.

10.3.3 Mechanism of Kinetic Approach

Under the kinetic approach, preparation of gold nanoparticles was reported inside micelles or in microemulsions by hampering the growth reaction in a circumscribed space. In micelle construction, the polymers [43] in bulk or surfactant assemble themselves into hydrocarbons when dissolved in solvent. The reaction takes place inside the micelle, with the reactants present there. As soon as reactants are absorbed, the subsequent growth of particle is obstructed.

10.4 SYNTHESIS PROCESS

There are generally two ways through which nanoparticles can be synthesized: physical method [44] and chemical method.

10.4.1 Physical Method

It is also known as the top-down approach. All states, like solid, liquid, gas, fluid, or vacuum, can engage this technique [45]. An external force in the form of mechanical, chemical, etc., can be applied to heavy materials to split them down into finer, smaller components.

10.4.2 Chemical Method

It is also known as the bottom-up approach. This method can also be applied to all the physical states of matter. Chemical method is the reverse or opposite of physical method. Here, small atomic particles can be combined chemically to become finer, smaller components on which experiments can be done. This method is more reliable for the synthesis of nanoparticles.

10.4.3 Chemical Reduction

Most of the nanoparticles can be synthesized through the reduction of metallic ions in solution. This is the method through which metal ions can be reduced, producing nanoparticles. External energy and chemical resistance can be used in the reduction process [46]. For decomposing bulk material, this external energy is used. This extra energy can be mostly in the form of photoenergy, electricity, or thermal energy. Gold nanoparticles [47, 48] can be obtained by reducing gold hydrochlorate solution with sodium tricitrate or by reducing chloroauric acid.

10.4.4 Mechanism of Encapsulation

The synthetic strategies for the process of encapsulation should be carefully selected, keeping in mind the properties of the materials used for encapsulation [49] and also the requirements of the metallic nanoparticles. Based on this, there are three categories in which metallic nanoparticle encapsulation materials are classified:

- Inorganic oxide nano-shells—carbon, SiO_2, TiO_2, etc.
- Porous materials—zeolites, mesoporous materials, metal organic [50] frameworks (MOFs), and covalent organic frameworks (COFs)
- Organic capsules—porous organic cages (POCs) and dendrimers [51].

10.4.4.1 Encapsulation With Inorganic Nano-shells

When metallic nanoparticles are encapsulated in organic nano-shells, their migration and amalgamation are effectively suppressed. At the same time, their catalytic properties are enhanced by creating new active sites at the metal–metal oxide interface [52].

10.4.4.2 Encapsulation in Nanopores

During catalysis, the highly porous [53] materials expedite the enhanced transfer of molecules, which in turn enables the reactants to approach metallic nanoparticles in the external medium. The nanopores attain the properties of shape selectivity by permitting the relocation of molecules smaller than a nominated size [54]. The metallic nanoparticles become stable and catalytically active [55], as they are now closely bonded with the mesoporous materials [56]. The catalytic performance can be promoted by inducing strong metal-support interaction.

10.4.4.3 Encapsulation in Organic Capsules

One of the organic capsules, dendrimer [57], serves as a brilliant pattern and an efficient stabilizing [58] agent to encapsulate even extremely small metallic NPs. The PAMAM (polyamido-amine) dendrimers were accredited and established their significant role as nanoreactors to synthesize ultrasmall metallic nanoparticles for homogeneous catalysis [59]. The metallic nanoparticles encapsulated with dendrimers or any other organic capsules manifest a better turnover frequency and enhance stability by the steric effect or by changing some properties of dendrimers.

10.5 MECHANISM OF SIZE CONTROL

For efficient catalytic reactions to take place, a large number of active sites are expected in order to minimize the activation energy. High-surface-area nanoparticles can effectively fulfill this requirement. The dimensions of nanoparticles greatly influence the selectivity rate of catalytic reactions. When the size of the particles is reduced below 5 nm, the surface-to-volume ratio is altered sharply.

10.6 COLLOIDAL SYNTHESIS MECHANISM (MONODISPERSED METALLIC NANOPARTICLES)

The evolution of colloidal nanoparticles [60] can be well understood by the LaMer model, as shown in Figure 10.1. The evolution mechanism, with the thermodynamic approach, is carried out in three stages: generation of supersaturation, nucleation, and slow growth kinetics. Colloidal synthesis of metallic NPs requires four basic

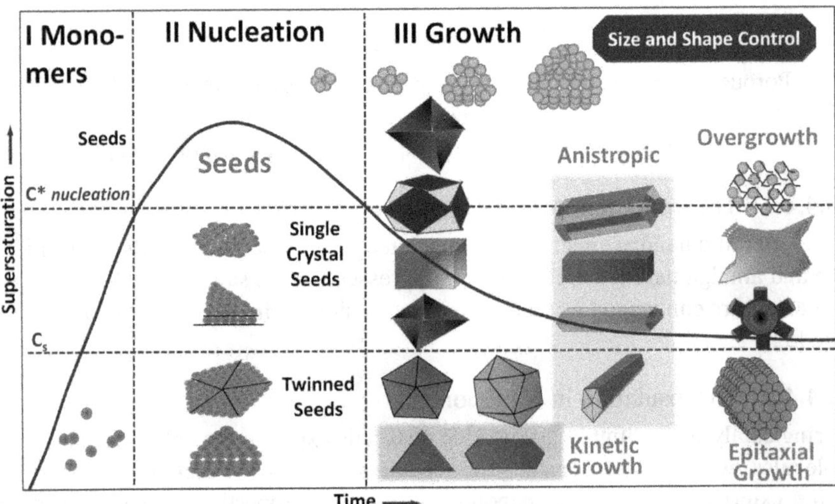

FIGURE 10.1 An illustration of general synthetic strategies for size and shape control of metal nanoparticles combining with the LaMer diagram.

ingredients: metallic precursor, surfactant, dissolvent, and eliminator. The final dimension of metallic NPs can be governed by the nimble amalgamation of those basic components. The size [61] can be controlled by synchronizing several variables such as precursor concentration, the precursor to surfactant ratio, and a selective reducing temperature depending on the dissolvent [62].

In order to promote colloidal stability, metallic nanoparticles are generated by incorporating different capping molecules at the end of the reaction. For particles of size less than 1 nm, the surface-to-volume ratio is extremely high. Owing to this, reductions in the closest neighbor separation, melting points, and ionization energy have been monitored.

By altering various experimental parameters during colloidal synthesis, metallic nanoparticles with varied shapes can be attained. In order to obtain bimetallic or multi-metallic structures, many precursors are introduced. Surfactants play a major role in maintaining stability. Surfactants from a diverse range are chosen depending on the target morphology of the metallic nanoparticles. While working on the size confinement, additional surfactants can be employed to administer the shape [63]. These surfactants are utilized as assistants for commanding the structure.

10.7 METALLIC NANOPARTICLES CATALYSTS

The 2D metal nanoparticle heterogeneous catalysts are fabricated by the monolayer [64, 65] deposit [66] of as-synthesized metallic nanoparticles on a two-dimensional substrate [67] or on the supporting oxide, which is then followed by the removal of the organic capping molecules by UV treatment.

On account of their systematized structure, large surface area, and increased pore volume, mesoporous oxides [68, 69] are used as excellent support materials for metallic nanoparticles for fabricating 3D metal nanoparticle heterogeneous catalysts. Such 3D metal NP catalysts with a controlled nanoparticle size are achieved by two approaches: capillary-induced inclusion and nanoparticle encapsulation, as shown in Figure 10.2.

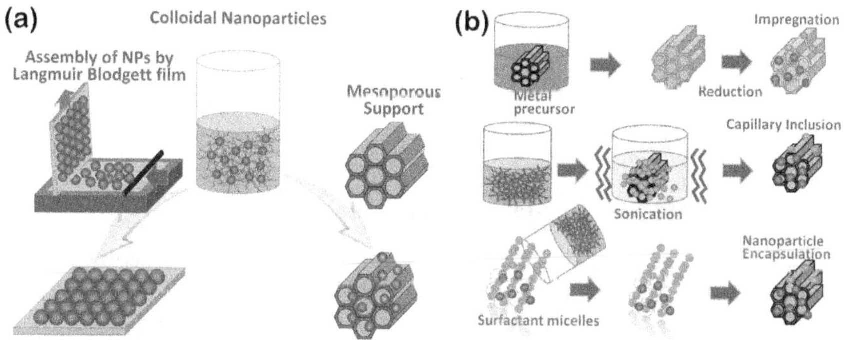

FIGURE 10.2 (a) The design of colloidal nanoparticle-based catalysts and (b) synthesis strategies for 3D catalysts.

10.8 MECHANISM OF SHAPE CONTROL

The process of epoxidation of olefins in the gaseous phase well explains the mechanism of shape control in metallic nanoparticles. Olefins are grouped into styrene and propene. Styrene comes under the no-allylic olefins category, whereas propene is constituted under the allylic olefins category. In order to synthesize metallic nanoparticles with different shapes, nanoparticle seeds, shaping-agents, or supplementary ions (Ag^+, Fe^{3+}, Co^+) are introduced. The colloidal synthesis method enables prompt heading toward the size and shape control techniques of the metallic nanoparticles. Diverse shapes like mono-metallic, bi-metallic, shell-shaped, core-shaped, and simple and complicated branched nanostructures are investigated. These can all be competently achieved by employing nano-metallic seeds and impurity atoms for the purpose of nucleation and substantial growth. It has been published recently that the catalytic reactions entail selectivity procedures that are based on the shape and size [70] of the nanoparticles.

The critical shapes of metallic nanoparticles can be determined explicitly by certain strategies, which include nucleation-generated nanoseeds, supplementary halogen ions, and shape-directing agents. Surface free energy of structure surfaces should be extremely low in order to achieve shape stability. The surface free energy, which is capable of creating new surface area, should be at a minimum. This can be carried out on account of the evolution of twinned defects, either singly or multiply twinned. It has been found that the metal surface of metallic nanoparticles and the capping material used for the shape-control strategies synergize so effectively that it becomes very important to select the capping materials wisely. Due to the interaction, the surface energy abruptly varies. In this shape-control phenomenon, polyvinyl pyrrolidone (PVP) [71], a polymer-based surfactant, has also emerged as an integral part of the mechanism. There are some structure-directing metallic ions such as Ag^+, Co^+, and Fe^{3+} and structure-directing reactive gases like H_2, O_2, NO, and CO that take part in the shape-determining approach during colloidal synthesis. These agents participate in the shape-control mechanism by promoting the growth rate of the crystal and by controlling the reduction [72] rate. Seed-mediated growth is one of the techniques frequently utilized for controlling the shape of metallic nanoparticles. For the subsequent crystal growth, as-developed metallic NPs are effectively utilized as seeds. Based on the nature of the metal atoms (similar or different), the shapes of metallic nanoparticles are decided. If the metal atoms (added) are the same, single metallic nanoparticles [73] are generated, and if they are different, bimetallic nanoparticles are generated.

10.9 MECHANISM OF SELECTIVITY

The selectivity rate can be promoted by several parameters, such as the size and nature of NPs, support materials, the oxidants being used, and the material used as a reaction enhancer. It has been reported that the epoxidation of propene [74] manifests smaller selectivity when compared with the epoxidation of ethylene. This has been attributed to the presence of allylic hydrogen, which shows higher reactivity in propene than the vinyl hydrogen present in ethylene. The oxidants for the selective oxidation [75] of propene are TiO_2, SiO_2, and N_2O. The oxidizing agent nitrous oxide (N_2O) is reported to favor the selective oxidation of C_6H_6 to C_6H_5OH. N_2O also helps free the oxygen atom more easily, which oxygen molecules cannot. It becomes the responsibility of the catalyst used for the propene epoxidation process to turn on the oxidant N_2O without deteriorating it.

10.9.1 Selectivity (Based on Size)

Metallic nanoparticles having sizes in the range of 1–10 nm demonstrate substantial changes in the reaction selectivity. The following variations depict how the selectivity leans on the size of platinum [76, 77] nanocatalysts. The decrease in the size of metallic nanoparticles up to 1 nm or less than that remarkably affects the reaction selectivity of the metallic nanoparticle catalysts. Similar studies were undertaken for the oxidation reaction of CO [78], which revealed that turnover rates were hiked eight times when the size of rhodium nanoparticles switched from 12 to 2 nm. Recent studies on the hydrogenation reaction [79] of *n-hexane* over platinum nanoparticles [80] have been reported to reveal the reliance of size on the selectivity of NPs.

10.9.2 Selectivity (Based on Shape)

The hydrogenation of benzene over platinum nanoparticles [81, 82] was performed to demonstrate the reliance of shape on the selectivity of NPs. Apart from this, a hydrogenolysis of MCP (methyl-cyclopentane) over platinum nanoparticles was carried out, which also revealed the influence of particle shape on selectivity. Several dramatic deviations were explored in incredible detail.

10.10 MECHANISM OF GREEN CATALYSIS

Metallic nanoparticles synthesized by green catalysis [83] have some exciting features, such as having more open active sites, thus leading to a faster reaction rate, lowering the activation energy, being eco-friendly, preventing waste disposal [84], indispensable and insoluble in the reaction mixture, which enables such metallic nanoparticles to get separated easily, reused, and harnessed. These nanoparticles prepared by green synthesis [85] promote excellent bonding, thereby permitting the reactants to tie up together at the metal sites, enhancing the rate of reaction.

10.10.1 Synthesis Mechanism

Green synthesis of metallic nanoparticles [86] employs two approaches: physical-based process and chemical-based process.

Physical-based process involves electrospray process, spray pyrolysis, flame spray pyrolysis (FSP), and unconventional machining process. In FSP technique, known for large-scale synthesis, a metal salt solution is sprayed over a flame using a narrow tube. Droplets are formed, and metal salts are converted into metal oxides.

Nanoparticles are formed after the mass collection of these oxides. *Chemical-based process* involves enzymatic biomaterials, fungus, and hot injection techniques [87].

10.10.2 Silver Nanoparticle Green Catalysts

An environmentally friendly biogenic synthesis [88, 89] of silver nanoparticles [90] uses fruit extracts of medicinal plants. Such green synthesis process [91] of metallic nanoparticles utilizes natural resources such as extracts of roots, fruits, leaves, bark, bud, stem, and latex. These natural extracts [92, 93] are capable of reducing metal ions and promoting stability. An aqueous fruit derivative of *Gmelina arborea*, the herbal plant [94] found in abundance in the north-east of India, was used as an efficient reducing and stabilizing agent in the green synthesis of silver nanoparticles. The aqueous solution of fruit extract was prepared using double-distilled water (ddH_2O). Silver nanoparticles (AgNPs) were manufactured by heating the fruit derivative sample on a hot plate magnetic [95, 96] stirrer. The change in color [97] was observed, and the formation of nanoparticles was reported. The as-synthesized AgNPs were investigated, described, and portrayed by UV-Vis spectroscopy analysis, transmission electron spectroscopy (TEM), selected area electron diffraction (SAED) pattern, and energy dispersive X-ray (EDX) spectrum.

When as-synthesized colloidal AgNPs were added to the methylene blue (MB) sample, the color of the sample changed from dark blue to pale blue, and thereby no color was reported. This MB degradation was investigated using a UV-Vis spectrophotometer. To determine the catalytic behavior of AgNPs catalysts, reduction reaction of methylene blue (MB) by $NaBH_4$ was chosen. The sample containing MB was first observed in the absence of as-synthesized AgNPs and then in the presence of the same catalyst but with a gradually increasing amount.

The graph shown in Figure 10.3 depicts the slower reduction rate of MB (30–40% reduction) as compared to another one (100% reduction). It was further investigated that with the increase in concentration of AgNP catalysts, the reduction rate increases. This finding infers that MB is efficiently degraded using AgNP catalysts, hence serving as an effective green catalyst [98]. In fact, as-synthesized colloidal AgNPs play the role of a potential intermediate that transfers electrons during the reduction reaction of MB by $NaBH_4$. In this model reduction reaction, $NaBH_4$ acts as a donor and the MB dye as an acceptor.

10.11 MECHANISM OF PHASE TRANSFER PROCESS (AUNP CATALYSTS)

When a mixture of liquefied gold metal ions and toluene suspension undergo vigorous stirring for about two days away from the light a colloidal suspension is obtained [99]. Two layers were observed, the top one with a reddish orange-colored organic phase and the bottom one with a dyed-orange liquified phase. Finally, a deep yellow color appeared, indicating the presence of $AuCl_4^-$ ions in the solution. This solution was then irradiated with UV light at 254 nm for about 40 h. The solvent of the resultant wine-red toluene solution was dispersed slowly and then calcined in air for 4 h at 623 K to eliminate amine, and gold nanoparticle catalysts were obtained.

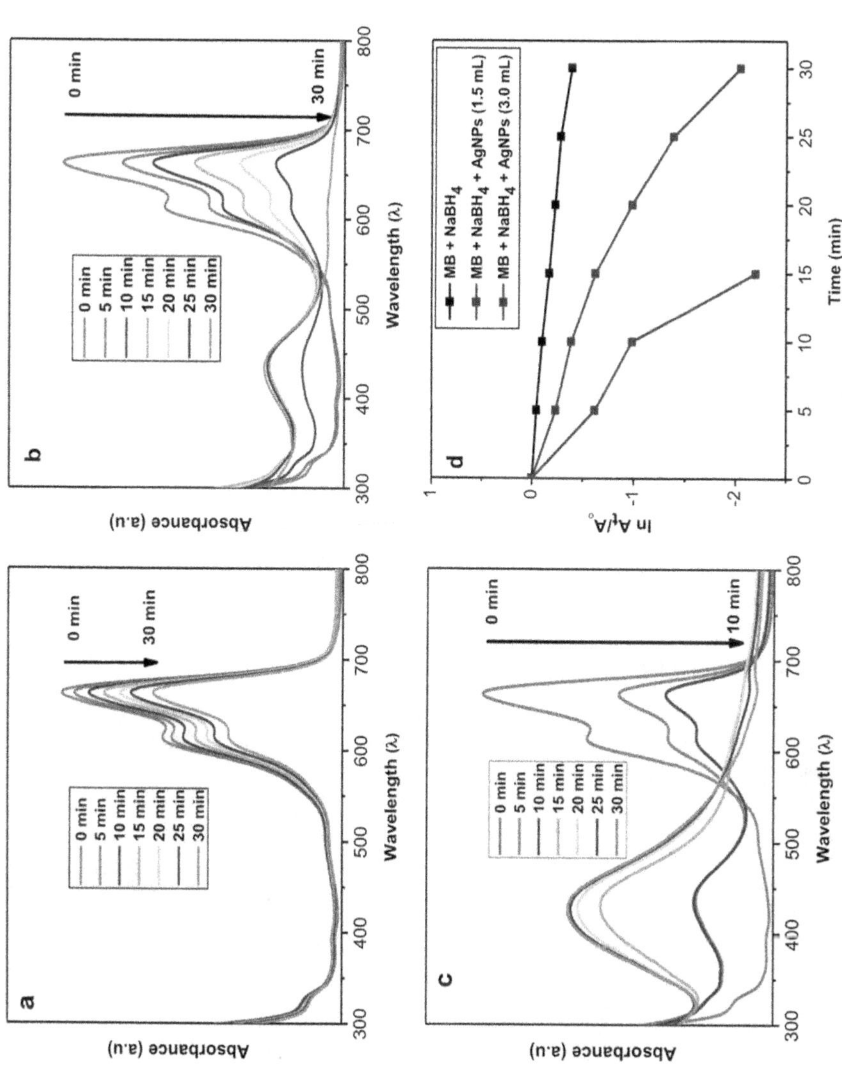

FIGURE 10.3 Successive UV-Vis absorbance spectra at 5 min intervals showing reduction of MB by $NaBH_4$ at ambient temperature; (a) reaction mixture without AgNPs (b) reaction mixture with 1.5 mL of AgNPs (c) reaction mixture with 3.0 mL of AgNPs (d) Smooth plots of ln A_t/A_0 against time for the degradation of MB at ambient temperature [6].

10.12 MECHANISM OF TESTING CATALYTIC BEHAVIOR

To test the catalytic behavior of nanoparticles in solution, we need some chemical reactions so that kinetic data can be obtained.

a) The chemical reaction must be of first order so that it can produce a single product.
b) UV-VIS spectroscopy technique is used to measure the kinetic behavior of the reaction.
c) The reaction should take place at room temperature.

To acquire these three conditions, the reaction should take place at surface level. The active sites on the surface of nanoparticles are responsible for the catalysis.

10.12.1 Reduction of 4-Nitrophenol by Borohydride Ions

To test the catalytic behavior of metal nanoparticles, a reaction including the reduction of 4-nitrophenol by borohydride ions in aqueous solution can be done. The steps involved are shown in Figure 10.4.

- 4-Nitrophenol is first reduced to nitrosophenol, which is quickly converted to 4-hydroxylaminophenol which is the first stable intermediate.
- Reduction to the final product, 4-aminophenol, takes place.
- There is an adsorption/desorption equilibrium for all compounds takes place in all steps. The reaction takes place at the surface of the particles.

10.12.2 Reduction of Hexacyanoferrate(III) Ions by Borohydride Ions

One more reaction to examine the catalytic activity of metallic nanoparticles is the reduction of hexacynoferrate (III) by borohydride ions. This reaction also takes place at surface level.

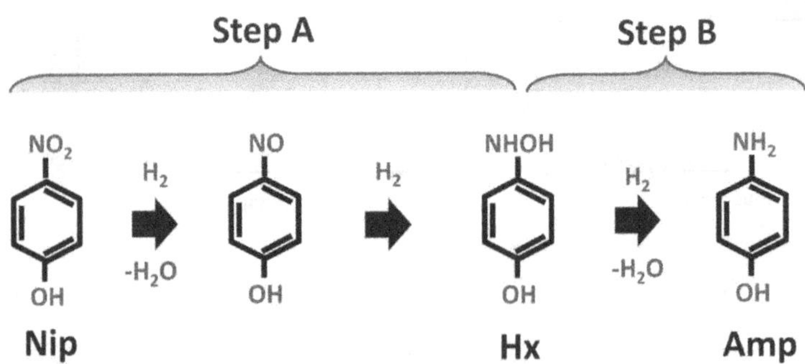

FIGURE 10.4 Reduction of 4-nitrophenol.

10.13 TRENDS

10.13.1 Encapsulated-Metallic Nanoparticles as Exceptional Catalysts

As Thermo-catalysts: Thermo-catalysis includes oxidation reactions (carbon monoxide-oxidation, WGS reaction, SEWGS reaction, combustion of methane, automobile three-way reaction, and alcohol oxidation) and hydrogenation reactions (CO_2 reduction, hydrogenation of unsaturated [100] organic group).

Oxidation—When metallic nanoparticles are encapsulated with metal oxide, the interface developed between metal and metal-oxides becomes catalytically active and intensifies many oxidation reactions. Before encapsulation, the lifetime of the metallic nanoparticle catalysts was greatly reduced by sintering, whereas on encapsulation, the confinement impact and the metal-oxide association build up the permanence of the metallic catalysts.

Hydrogenation—Encapsulated metallic nanoparticles have high resistance against sintering, even at the high temperature required for the hydrogenation reaction. Consequently, encapsulated metallic nanoparticles [101, 102] serve as better thermocatalysts which amplify the selectivity rate of catalysts and the durability and reliability of the metallic nanoparticles.

As Photocatalysts:

Encapsulated metallic nanoparticles, for plasmonic [103] metals Ag and Au, act as photocatalysts that advance the surface redox reaction, ameliorate sunlight consumption, reinforce the production of holes as well as electrons, and promote the transfer of those generated particles to the surface. Photocatalysis includes reduction reaction (photocatalytic CO_2) [104], degradation of organic [105] pollutants, and photocatalytic hydrogen production.

As Electrocatalysts:

Encapsulated metallic nanoparticles have a wide area of interest as electrocatalysts on account of their strong association with metal-encapsulating material. Transition metals such as iron, cobalt [106], nickel, and their alloys.

(i) With encapsulation of graphite [107] carbon shells, serve as excellent electrocatalysts in OER; and (ii) With encapsulation of nitrogen-doped carbon, serve as the most efficient electrocatalysts in ORR as well as HER (Pt-free catalysts).

Encapsulation of metallic nanoparticles with different substances such as silicon dioxide, zeolites, metallic oxides, carbon, dendrimers, MOF, COF, and organic cages provides a huge opportunity to synthesize different types of metallic nanoparticles with different compositions. It has been investigated that encapsulation enhances many properties of metal NPs, such as stability, recyclability, selectivity, and effective collaboration between the metal and the material of encapsulation. Not only these, but the opportunity for tandem catalysis is gained as well. One of the motives to manufacture the encapsulated metallic NPs is to inaugurate the integration between metallic

NPs and encapsulation material and to enhance the stability because, on encapsulation, the electron structure shows several modifications and the boundary developed between metal and oxide of metal serves as a diligent location for many chemical reactions.

10.13.2 Green Catalyst Silver Nanoparticles

The synthetic organic dyes [108], which are used in a huge number of industries like textile industries, cosmetic industries, food industries, paint industries, pharmacy, paper, and plastic industries, have remarkable effects on the health of humans as well as the environment. The degradation of synthetic dyes is attempted by several conventional treatments like reverse osmosis, adsorption, and coagulation, but the AgNPs prepared by the green synthesis method proved to be the most promising catalysts for the deterioration of hazardous MB dye. The as-prepared green catalyst, silver NPs, had been investigated to catalytically degrade methylene blue (MB) [109] dye almost within 10 minutes. This fast degradation of MB dye signifies the usefulness of metallic catalysts in effluent treatment in diverse fields.

Green synthesis is a convenient single-step method that doesn't demand high pressure, temperature, energy, toxic reagents, and hazardous chemicals during the synthesis process.

Green catalyst metallic nanoparticles are highly recommended, as these are cost effective, inexpensive, eco-friendly, and easy to handle.

10.13.3 Epoxidation Technique

The epoxidation of olefins (styrene and propene), often employed in chemical industries, best demonstrates the catalytic behavior of metal NPs. The catalytic activities of Ag, Au, Cu NPs can be manifested by styrene epoxidation and propene epoxidation.

Styrene Epoxidation (for AgNP catalysts)

The catalytic behavior of Ag nanopolyhedra and Ag nanocubes supported [110] on α-Al_2O_3 is similar to that of Ag nanowires. But when the support material was changed to $CaCO_3$, the performance and selectivity were both significantly enhanced. The catalytic performance of silver nanowire catalysts was remarkably enhanced when cesium, an electron-donating element, was added. Cesium, an electron donor, increases the electron density of silver and reduces the electrophilicity of adsorbed oxygen on the Ag facet. Thus, the activation energy of epoxidation decreases. The oxidation-reduction cycle is comparatively easier with Cs based silver nanowire catalysts.

Studies also revealed that the catalytic performance of AgNP catalysts is exceptional and remarkable over that of copper nanosphere [111] catalysts for the same reaction conditions. Further, XRD analysis manifests that silver existed in its metallic form, whereas copper existed in the form of CuO and Cu_2O during the reaction when performing as catalysts. This is due to the stronger bonding between oxygen and copper as compared to that of silver.

Propene Epoxidation (for Au and Cu NP catalysts)

The synthesis of AuNPs involves the Stober method with support from TiO_2 spheres. The catalytic performance in the epoxidation of propene is based on the size parameter of the metal particles. Owing to this, gold fragments up to 5 nm in diameter proved to be exquisite catalysts for propene epoxidation.

The 1% Au and Cu catalysts with support of TiO_2 under propene epoxidation by N_2O proved to be superior catalysts over Ag-nanowires and Cu-nanospheres [112] derived using the Polyol process.

10.14 CATALYSTS CHARACTERIZATION TECHNIQUE

The commonly used ex situ and in situ techniques for characterization of nanoparticles are shown in Table 10.1.

10.15 CONCLUSIONS AND FUTURE PROSPECTIVE

Metal particles in nanoscale technologies have been evolving and developing as excellent catalysts, especially when encapsulation techniques and green technologies are involved in the field of nanoscience. These are emerging as eco-friendly, resource-efficient, and fuel-efficient catalysts that limit the intake of external resources. The matter of the stability factor of nanosized catalysts has also been eradicated. With innumerable multiple electrifying features of nanocatalysts such as robustness, increased number of active sites, efficient conditioning, re-modelling, insolvable property, and helpful in peaceful separation from the resulting product,

TABLE 10.1
Commonly Used Ex Situ and In Situ Techniques for Characterizations of Nanoparticles

	Characterization Techniques
Ex situ characterization	Transmission electron microscopy (TEM) Energy-dispersive X-ray (EDX) Analysis scanning electron microscopy (SEM) Atomic force microscopy (AFM) X-ray diffraction (XRD) Small-angle X-ray scattering (SAXS) X-ray photoelectron spectroscopy (XPS) Physisorption and chemisorption by Brunauer-Emmett-Teller (BET) analysis UV/vis spectroscopy Thermogravimetric analysis (TGA) Inductively coupled plasma—optical emission spectroscopy (ICP-OES)
In situ characterization	Environmental transmission electron microscopy (TEM) Electron energy-loss spectroscopy (EELS) Time-resolved X-ray diffraction (XRD) Diffuse reflectance infrared Fourier-transform spectroscopy (DRIFTS) High-Pressure Scanning Tunneling Microscopy (HPSTM) Sum Frequency Generation Vibrational Spectroscopy (SFGVS) Ambient-Pressure X-ray Photoelectron Spectroscopy (APXPS) UV-Raman and Surface Enhanced Raman Spectroscopy (SERS) Near-Edge X-ray Absorption Fine Structure (NEXAFS)

nano-sized metallic catalysts proved to be the most demanding stimulants for the ensuing applications in the fields of optics, medicine, biomedicine, and technologies. Transition metal dichalcogenides (TMDs) such as molybdenum disulfide (MoS_2), tungsten disulfide (WS_2), molybdenum telluride ($MoTe_2$), molybdenum di-selenide ($MoSe_2$), and many more with a repeated number of layers have brilliant and exceptional electrical and optical properties with insignificant energy densities. Manifold shapes have been recently realized, including one- and two-dimensional shapes in the form of rods, discs, triangular or hexagonal plates, tetrapods, hyperbranched, and bipyramid nanostructures. Nanoparticle-selectivity enables the customization of active sites and the surface structures of metallic nanoparticles at the subatomic and microscopic levels. The morphologies of metallic nanoparticles play a leading role in the selectivity of catalytic reactions. Apart from morphology, reaction temperature dominates activity and selectivity as well. This is attributed to the spectroscopic inclination of the nanoparticle surfaces. Polymer-based surfactant PVP serves as a stabilizing agent for the platinum nanoparticle surface and is a less toxic capping material. These capping agents need to be eliminated at the final stage of the chemical reactions. This is done effectively by the thermal oxidation process or ultraviolet/ozone therapy. Several parameters of chemical reactions strongly influence the catalytic reaction selectivity of metallic nanoparticles undergoing far-field reactions. Those parameters include the nature of surface of metallic NPs, their oxidation states, reaction intermediates, relocation of charge, configuration, and mobility.

CONFLICT OF INTEREST

The authors declare that there is no conflict of interest

ACKNOWLEDGMENTS

Namita and Arti are thankful to the Department of Physics, L.N.M. University for the financial assistance in the form of a fellowship. MSA is thankful to SGT University for their cooperation and support. Md Naushad Alam thankfully acknowledges LNMU, Darbhanga for the support. JRA gratefully acknowledges the HoD, Applied Science and Director Sir, DCE for their cooperation and financial support.

REFERENCES

[1] G.J. Hutchings, M. Haruta, A golden age of catalysis: A perspective, Appl. Catal. Gen. 291 (2005) 2–5. https://doi.org/10.1016/j.apcata.2005.05.044.

[2] A. Moisala, A.G. Nasibulin, E.I. Kauppinen, The role of metal nanoparticles in the catalytic production of single-walled carbon nanotubes—a review, J. Phys. Condens. Matter. 15 (2003) S3011–S3035. https://doi.org/10.1088/0953-8984/15/42/003.

[3] C. Walkey, S. Das, S. Seal, J. Erlichman, K. Heckman, L. Ghibelli, E. Traversa, J.F. McGinnis, W.T. Self, Catalytic properties and biomedical applications of cerium oxide nanoparticles, Environ. Sci. Nano. 2 (2015) 33–53. https://doi.org/10.1039/C4EN00138A.

[4] A.L. Gould, S. Kadkhodazadeh, J.B. Wagner, C.R.A. Catlow, A.J. Logsdail, M. Di Vece, Understanding the thermal stability of silver nanoparticles embedded in a-Si, J. Phys. Chem. C. 119 (2015) 23767–23773. https://doi.org/10.1021/acs.jpcc.5b07324.

[5] C. Gao, F. Lyu, Y. Yin, Encapsulated metal nanoparticles for catalysis, Chem. Rev. 121 (2021) 834–881. https://doi.org/10.1021/acs.chemrev.0c00237.
[6] J. Saha, A. Begum, A. Mukherjee, S. Kumar, A novel green synthesis of silver nanoparticles and their catalytic action in reduction of methylene blue dye, Sustain. Environ. Res. 27 (2017) 245–250. https://doi.org/10.1016/j.serj.2017.04.003.
[7] R. Kumar, G. Dhamija, J.R. Ansari, Md.N. Javed, Md.S. Alam, C-dot nanoparticulated devices for biomedical applications, in: Nanotechnology, 1st ed., CRC Press, Boca Raton, 2022: pp. 271–299. https://doi.org/10.1201/9781003220350-15.
[8] S. Singhal, M. Gupta, Md.S. Alam, Md.N. Javed, J.R. Ansari, Carbon allotropes-based nanodevices, in: Nanotechnology, 1st ed., CRC Press, Boca Raton, 2022: pp. 241–269. https://doi.org/10.1201/9781003220350-14.
[9] J. Pandit, Md.S. Alam, J.R. Ansari, M. Singhal, N. Gupta, A. Waziri, K. Sharma, F. Hyder Potto, Multifaced applications of nanoparticles in biological science, in: Nanomaterials in the Battle Against Pathogens and Disease Vectors, 1st ed., CRC Press, Boca Raton, 2022: pp. 17–50. https://doi.org/10.1201/9781003126256-2.
[10] S. Mishra, S. Sharma, M.N. Javed, F.H. Pottoo, M.A. Barkat, Harshita, M.S. Alam, M. Amir, M. Sarafroz, Bioinspired nanocomposites: Applications in disease diagnosis and treatment, Pharm. Nanotechnol. 7 (2019) 206–219. https://doi.org/10.2174/2211738507666190425121509.
[11] A.M. Ibrahim, L. Chauhan, A. Bhardwaj, A. Sharma, F. Fayaz, B. Kumar, M. Alhashmi, N. AlHajri, M.S. Alam, F.H. Pottoo, Brain-derived neurotropic factor in neurodegenerative disorders, Biomedicines. 10 (2022) 1143. https://doi.org/10.3390/biomedicines10051143.
[12] M.N. Javed, F.H. Pottoo, A. Shamim, M.S. Hasnain, M.S. Alam, Design of experiments for the development of nanoparticles, nanomaterials, and nanocomposites, in: S. Beg (Ed.), Des Expositions Pharmaceutical Product Development, Springer Singapore, Singapore, 2021: pp. 151–169. https://doi.org/10.1007/978-981-33-4351-1_9.
[13] F.H. Pottoo, S. Sharma, Md.N. Javed, Md.A. Barkat, Harshita, Md.S. Alam, Mohd.J. Naim, O. Alam, M.A. Ansari, G.E. Barreto, G.Md. Ashraf, Lipid-based nanoformulations in the treatment of neurological disorders, Drug Metab. Rev. 52 (2020) 185–204. https://doi.org/10.1080/03602532.2020.1726942.
[14] M.S. Hasnain, Md.N. Javed, Md.S. Alam, P. Rishishwar, S. Rishishwar, S. Ali, A.K. Nayak, S. Beg, Purple heart plant leaves extract-mediated silver nanoparticle synthesis: Optimization by Box-Behnken design, Mater. Sci. Eng. C. 99 (2019) 1105–1114. https://doi.org/10.1016/j.msec.2019.02.061.
[15] P.V. Asharani, Y. lianwu, Z. Gong, S. Valiyaveettil, Comparison of the toxicity of silver, gold and platinum nanoparticles in developing zebrafish embryos, Nanotoxicology. 5 (2011) 43–54. https://doi.org/10.3109/17435390.2010.489207.
[16] L. Xu, X.-C. Wu, J.-J. Zhu, Green preparation and catalytic application of Pd nanoparticles, Nanotechnology. 19 (2008) 305603. https://doi.org/10.1088/0957-4484/19/30/305603.
[17] M. Mahmoudi, A. Simchi, A.S. Milani, P. Stroeve, Cell toxicity of superparamagnetic iron oxide nanoparticles, J. Colloid Interface Sci. 336 (2009) 510–518. https://doi.org/10.1016/j.jcis.2009.04.046.
[18] C.T. Kresge, M.E. Leonowicz, W.J. Roth, J.C. Vartuli, J.S. Beck, Ordered mesoporous molecular sieves synthesized by a liquid-crystal template mechanism, Nature. 359 (1992) 710–712. https://doi.org/10.1038/359710a0.
[19] J. Conde, G. Doria, P. Baptista, Noble metal nanoparticles applications in cancer, J. Drug Deliv. 2012 (2012) 1–12. https://doi.org/10.1155/2012/751075.
[20] M.S. Alam, M.N. Javed, F.H. Pottoo, A. Waziri, F.A. Almalki, M.S. Hasnain, A. Garg, M.K. Saifullah, QbD approached comparison of reaction mechanism in microwave synthesized gold nanoparticles and their superior catalytic role against hazardous nirto-dye, Appl. Organomet. Chem. (2019). https://doi.org/10.1002/aoc.5071.

[21] N. Kumari, N. Daram, Md.S. Alam, A.K. Verma, Rationalizing the use of polyphenol nano-formulations in the therapy of neurodegenerative diseases, CNS Neurol. Disord. Drug Targets. 21 (2022) 966–976. https://doi.org/10.2174/1871527321666220512153854.
[22] M.N. Javed, E.S. Dahiya, A.M. Ibrahim, Md.S. Alam, F.A. Khan, F.H. Pottoo, Recent advancement in clinical application of nanotechnological approached targeted delivery of herbal drugs, in: S. Beg, M.A. Barkat, F.J. Ahmad (Eds.), Nanophytomedicine, Springer Singapore, Singapore, 2020: pp. 151–172. https://doi.org/10.1007/978-981-15-4909-0_9.
[23] S. Raj, R. Manchanda, M. Bhandari, Md.S. Alam, Review on natural bioactive products as radioprotective therapeutics: Present and past perspective, Curr. Pharm. Biotechnol. 23 (2022) 1721–1738. https://doi.org/10.2174/1389201023666220110104645.
[24] F.H. Pottoo, N. Tabassum, Md.N. Javed, S. Nigar, R. Rasheed, A. Khan, Md.A. Barkat, Md.S. Alam, A. Maqbool, M.A. Ansari, G.E. Barreto, G.M. Ashraf, The synergistic effect of raloxifene, fluoxetine, and bromocriptine protects against pilocarpine-induced status epilepticus and temporal lobe epilepsy, Mol. Neurobiol. 56 (2019) 1233–1247. https://doi.org/10.1007/s12035-018-1121-x.
[25] J. Guzman, B.C. Gates, Structure and reactivity of a mononuclear gold-complex catalyst supported on magnesium oxide, Angew. Chem. Int. Ed. 42 (2003) 690–693. https://doi.org/10.1002/anie.200390191.
[26] J.R. Ansari, N. Singh, R. Ahmad, D. Chattopadhyay, A. Datta, Controlling self-assembly of ultra-small silver nanoparticles: Surface enhancement of Raman and fluorescent spectra, Opt. Mater. 94 (2019) 138–147. https://doi.org/10.1016/j.optmat.2019.05.023.
[27] D. Sarvamangala, K. Kondala, N. Sivakumar, M.S. Babu, S. Manga, Synthesis, characterization and anti microbial studies of Agnp's using probiotics, Int. Res. J. Pharm. 4 (2013) 240–243. https://doi.org/10.7897/2230-8407.04352.
[28] J.R. Ansari, S.M. Hegazy, M.T. Houkan, K. Kannan, A. Aly, K.K. Sadasivuni, Nanocellulose-based materials/composites for sensors, in: Nanocellulose Based Compos Electron., Elsevier, 2021: pp. 185–214. https://doi.org/10.1016/B978-0-12-822350-5.00008-4.
[29] C.A. Sunilbhai, Md.S. Alam, K.K. Sadasivuni, J.R. Ansari, SPR assisted diabetes detection, in: K.K. Sadasivuni, J.-J. Cabibihan, A.K. Al-Ali, R.A. Malik (Eds.), Advanced Bioscience and Biosystems for Detection Manag Diabetes, Springer International Publishing, Cham, 2022: pp. 91–131. https://doi.org/10.1007/978-3-030-99728-1_6.
[30] Md.F. Naseh, J.R. Ansari, Md.S. Alam, Md.N. Javed, sustainable nanotorus for biosensing and therapeutical applications, in: U. Shanker, C.M. Hussain, M. Rani (Eds.), Handbook of Green and Sustainable Nanotechnology, Springer International Publishing, Cham, 2022: pp. 1–21. https://doi.org/10.1007/978-3-030-69023-6_47-1.
[31] Md.S. Alam, Md.F. Naseh, J.R. Ansari, A. Waziri, Md.N. Javed, A. Ahmadi, M.K. Saifullah, A. Garg, Synthesis approaches for higher yields of nanoparticles, in: Nanomaterials in the Battle Against Pathogens and Disease Vectors, 1st ed., CRC Press, Boca Raton, 2022: pp. 51–82. https://doi.org/10.1201/9781003126256-3.
[32] K. An, G.A. Somorjai, size and shape control of metal nanoparticles for reaction selectivity in catalysis, ChemCatChem. 4 (2012) 1512–1524. https://doi.org/10.1002/cctc.201200229.
[33] X.-H. Li, M. Antonietti, Metal nanoparticles at mesoporous N-doped carbons and carbon nitrides: Functional Mott–Schottky heterojunctions for catalysis, Chem. Soc. Rev. 42 (2013) 6593. https://doi.org/10.1039/c3cs60067j.
[34] I. Hagarová, L. Nemček, M. Šebesta, O. Zvěřina, P. Kasak, M. Urík, Preconcentration and separation of gold nanoparticles from environmental waters using extraction techniques followed by spectrometric quantification, Int. J. Mol. Sci. 23 (2022) 11465. https://doi.org/10.3390/ijms231911465.
[35] A.A. Herzing, C.J. Kiely, A.F. Carley, P. Landon, G.J. Hutchings, Identification of active gold nanoclusters on iron oxide supports for CO Oxidation, Science. 321 (2008) 1331–1335. https://doi.org/10.1126/science.1159639.

[36] M.B. Gawande, A. Goswami, F.-X. Felpin, T. Asefa, X. Huang, R. Silva, X. Zou, R. Zboril, R.S. Varma, Cu and cu-based nanoparticles: Synthesis and applications in catalysis, Chem. Rev. 116 (2016) 3722–3811. https://doi.org/10.1021/acs.chemrev.5b00482.
[37] F. Wang, T.A. Shifa, X. Zhan, Y. Huang, K. Liu, Z. Cheng, C. Jiang, J. He, Recent advances in transition-metal dichalcogenide based nanomaterials for water splitting, Nanoscale. 7 (2015) 19764–19788. https://doi.org/10.1039/C5NR06718A.
[38] H. Hu, J.H. Xin, H. Hu, X. Wang, D. Miao, Y. Liu, Synthesis and stabilization of metal nanocatalysts for reduction reactions—a review, J. Mater. Chem. A. 3 (2015) 11157–11182. https://doi.org/10.1039/C5TA00753D.
[39] K. Hu, T. Ohto, L. Chen, J. Han, M. Wakisaka, Y. Nagata, J. Fujita, Y. Ito, Graphene layer encapsulation of non-noble metal nanoparticles as acid-stable hydrogen evolution catalysts, ACS Energy Lett. 3 (2018) 1539–1544. https://doi.org/10.1021/acsenergylett.8b00739.
[40] Y. Zhao, E.A. Hernandez-Pagan, N.M. Vargas-Barbosa, J.L. Dysart, T.E. Mallouk, A High yield synthesis of ligand-free iridium oxide nanoparticles with high electrocatalytic activity, J. Phys. Chem. Lett. 2 (2011) 402–406. https://doi.org/10.1021/jz200051c.
[41] S.H. Joo, J.Y. Park, C.-K. Tsung, Y. Yamada, P. Yang, G.A. Somorjai, Thermally stable Pt/mesoporous silica core–shell nanocatalysts for high-temperature reactions, Nat. Mater. 8 (2009) 126–131. https://doi.org/10.1038/nmat2329.
[42] Y.W. Wang, L.D. Zhang, G.Z. Wang, X.S. Peng, Z.Q. Chu, C.H. Liang, Catalytic growth of semiconducting zinc oxide nanowires and their photoluminescence properties, J. Cryst. Growth. 234 (2002) 171–175. https://doi.org/10.1016/S0022-0248(01)01661-X.
[43] J. Virkutyte, R.S. Varma, Green synthesis of metal nanoparticles: Biodegradable polymers and enzymes in stabilization and surface functionalization, Chem Sci. 2 (2011) 837–846. https://doi.org/10.1039/C0SC00338G.
[44] P.K. Singh, P. Kumar, A.K. Das, Unconventional physical methods for synthesis of metal and non-metal nanoparticles: A review, Proc. Natl. Acad. Sci. India Sect. Phys. Sci. 89 (2019) 199–221. https://doi.org/10.1007/s40010-017-0474-2.
[45] P. Migowski, J. Dupont, Catalytic applications of metal nanoparticles in imidazolium ionic liquids, Chem.—Eur. J. 13 (2007) 32–39. https://doi.org/10.1002/chem.200601438.
[46] A.O. Neto, R.R. Dias, M.M. Tusi, M. Linardi, E.V. Spinacé, Electro-oxidation of methanol and ethanol using PtRu/C, PtSn/C and PtSnRu/C electrocatalysts prepared by an alcohol-reduction process, J. Power Sources. 166 (2007) 87–91. https://doi.org/10.1016/j.jpowsour.2006.12.088.
[47] J. Kimling, M. Maier, B. Okenve, V. Kotaidis, H. Ballot, A. Plech, Turkevich method for gold nanoparticle synthesis revisited, J. Phys. Chem. B. 110 (2006) 15700–15707. https://doi.org/10.1021/jp061667w.
[48] F. Schulz, T. Homolka, N.G. Bastús, V. Puntes, H. Weller, T. Vossmeyer, Little adjustments significantly improve the turkevich synthesis of gold nanoparticles, Langmuir. 30 (2014) 10779–10784. https://doi.org/10.1021/la503209b.
[49] Y. Mei, Y. Lu, F. Polzer, M. Ballauff, M. Drechsler, Catalytic activity of palladium nanoparticles encapsulated in spherical polyelectrolyte brushes and core–shell microgels, Chem. Mater. 19 (2007) 1062–1069. https://doi.org/10.1021/cm062554s.
[50] Q. Yang, Q. Xu, H.-L. Jiang, Metal–organic frameworks meet metal nanoparticles: Synergistic effect for enhanced catalysis, Chem. Soc. Rev. 46 (2017) 4774–4808. https://doi.org/10.1039/C6CS00724D.
[51] R.M. Crooks, M. Zhao, L. Sun, V. Chechik, L.K. Yeung, Dendrimer-encapsulated metal nanoparticles: Synthesis, characterization, and applications to catalysis, Acc. Chem. Res. 34 (2001) 181–190. https://doi.org/10.1021/ar000110a.
[52] N. Li, Q. Zhang, J. Liu, J. Joo, A. Lee, Y. Gan, Y. Yin, Sol–gel coating of inorganic nanostructures with resorcinol–formaldehyde resin, Chem. Commun. 49 (2013) 5135. https://doi.org/10.1039/c3cc41456f.

[53] F. Farzaneh, S. Haghshenas, Facile synthesis and characterization of nanoporous NiO with folic acid as photodegredation catalyst for Congo red, Mater. Sci. Appl. 03 (2012) 697–703. https://doi.org/10.4236/msa.2012.310102.

[54] B.R. Cuenya, Synthesis and catalytic properties of metal nanoparticles: Size, shape, support, composition, and oxidation state effects, Thin Solid Films. 518 (2010) 3127–3150. https://doi.org/10.1016/j.tsf.2010.01.018.

[55] G. Zhang, G. Wang, Y. Liu, H. Liu, J. Qu, J. Li, Highly active and stable catalysts of phytic acid-derivative transition metal phosphides for full water splitting, J. Am. Chem. Soc. 138 (2016) 14686–14693. https://doi.org/10.1021/jacs.6b08491.

[56] Z. Kónya, V.F. Puntes, I. Kiricsi, J. Zhu, J.W. Ager, M.K. Ko, H. Frei, P. Alivisatos, G.A. Somorjai, Synthetic insertion of gold nanoparticles into mesoporous silica, Chem. Mater. 15 (2003) 1242–1248. https://doi.org/10.1021/cm020824a.

[57] A.K. Ilunga, R. Meijboom, Catalytic oxidation of methylene blue by dendrimer encapsulated silver and gold nanoparticles, J. Mol. Catal. Chem. 411 (2016) 48–60. https://doi.org/10.1016/j.molcata.2015.10.009.

[58] P. Raveendran, J. Fu, S.L. Wallen, Completely "green" synthesis and stabilization of metal nanoparticles, J. Am. Chem. Soc. 125 (2003) 13940–13941. https://doi.org/10.1021/ja029267j.

[59] S. Shylesh, V. Schünemann, W.R. Thiel, Magnetically separable nanocatalysts: Bridges between homogeneous and heterogeneous catalysis, Angew. Chem. Int. Ed. 49 (2010) 3428–3459. https://doi.org/10.1002/anie.200905684.

[60] S. Kumar, M. Singh, D. Halder, A. Mitra, Mechanistic study of antibacterial activity of biologically synthesized silver nanocolloids, Colloids Surf. Physicochem. Eng. Asp. 449 (2014) 82–86. https://doi.org/10.1016/j.colsurfa.2014.02.027.

[61] C.L. Haynes, R.P. Van Duyne, Nanosphere lithography: A versatile nanofabrication tool for studies of size-dependent nanoparticle optics, J. Phys. Chem. B. 105 (2001) 5599–5611. https://doi.org/10.1021/jp010657m.

[62] T. Yang, H. Ling, J.-F. Lamonier, M. Jaroniec, J. Huang, M.J. Monteiro, J. Liu, A synthetic strategy for carbon nanospheres impregnated with highly monodispersed metal nanoparticles, NPG Asia Mater. 8 (2016) e240–e240. https://doi.org/10.1038/am.2015.145.

[63] Y. Sun, Y. Xia, Shape-controlled synthesis of gold and silver nanoparticles, Science. 298 (2002) 2176–2179. https://doi.org/10.1126/science.1077229.

[64] S.T. Hunt, M. Milina, A.C. Alba-Rubio, C.H. Hendon, J.A. Dumesic, Y. Roman-Leshkov, Self-assembly of noble metal monolayers on transition metal carbide nanoparticle catalysts, Science. 352 (2016) 974–978. https://doi.org/10.1126/science.aad8471.

[65] X. Cui, P. Ren, D. Deng, J. Deng, X. Bao, Single layer graphene encapsulating nonprecious metals as high-performance electrocatalysts for water oxidation, Energy Environ. Sci. 9 (2016) 123–129. https://doi.org/10.1039/C5EE03316K.

[66] E. Ostuni, C.S. Chen, D.E. Ingber, G.M. Whitesides, Selective deposition of proteins and cells in arrays of microwells, Langmuir. 17 (2001) 2828–2834. https://doi.org/10.1021/la001372o.

[67] H.-H. Wang, C.-Y. Liu, S.-B. Wu, N.-W. Liu, C.-Y. Peng, T.-H. Chan, C.-F. Hsu, J.-K. Wang, Y.-L. Wang, Highly Raman-enhancing substrates based on silver nanoparticle arrays with tunable sub-10nm gaps, Adv. Mater. 18 (2006) 491–495. https://doi.org/10.1002/adma.200501875.

[68] C.Y. Ma, Z. Mu, J.J. Li, Y.G. Jin, J. Cheng, G.Q. Lu, Z.P. Hao, S.Z. Qiao, Mesoporous Co_3O_4 and Au/Co_3O_4 catalysts for low-temperature oxidation of trace ethylene, J. Am. Chem. Soc. 132 (2010) 2608–2613. https://doi.org/10.1021/ja906274t.

[69] W. Zhu, A. Noureddine, J.Y. Howe, J. Guo, C.J. Brinker, Conversion of metal–organic cage to ligand-free ultrasmall noble metal nanocluster catalysts confined within mesoporous silica nanoparticle supports, Nano Lett. 19 (2019) 1512–1519. https://doi.org/10.1021/acs.nanolett.8b04121.

[70] K.L. Kelly, E. Coronado, L.L. Zhao, G.C. Schatz, The optical properties of metal nanoparticles: The influence of size, shape, and dielectric environment, J. Phys. Chem. B. 107 (2003) 668–677. https://doi.org/10.1021/jp026731y.

[71] R. Narayanan, M.A. El-Sayed, Effect of catalysis on the stability of metallic nanoparticles: Suzuki reaction catalyzed by PVP-palladium nanoparticles, J. Am. Chem. Soc. 125 (2003) 8340–8347. https://doi.org/10.1021/ja035044x.

[72] Y. Lan, B. Deng, C. Kim, E.C. Thornton, H. Xu, Catalysis of elemental sulfur nanoparticles on chromium(VI) reduction by sulfide under anaerobic conditions, Environ. Sci. Technol. 39 (2005) 2087–2094. https://doi.org/10.1021/es048829r.

[73] P. Zijlstra, M. Orrit, Single metal nanoparticles: Optical detection, spectroscopy and applications, Rep. Prog. Phys. 74 (2011) 106401. https://doi.org/10.1088/0034-4885/74/10/106401.

[74] R. Nilsson, T. Lindblad, A. Andersson, Ammoxidation of propene over antimony-vanadium-oxide catalysts, Catal. Lett. 29 (1994) 409–420. https://doi.org/10.1007/BF00807120.

[75] Y. Hong, X. Yan, X. Liao, R. Li, S. Xu, L. Xiao, J. Fan, Platinum nanoparticles supported on Ca(Mg)-zeolites for efficient room-temperature alcohol oxidation under aqueous conditions, Chem. Commun. 50 (2014) 9679. https://doi.org/10.1039/C4CC02685C.

[76] S. Cheong, J.D. Watt, R.D. Tilley, Shape control of platinum and palladium nanoparticles for catalysis, Nanoscale. 2 (2010) 2045. https://doi.org/10.1039/c0nr00276c.

[77] R.M. Rioux, H. Song, J.D. Hoefelmeyer, P. Yang, G.A. Somorjai, High-surface-area catalyst design: Synthesis, characterization, and reaction studies of platinum nanoparticles in mesoporous SBA-15 silica, J. Phys. Chem. B. 109 (2005) 2192–2202. https://doi.org/10.1021/jp048867x.

[78] C. Yang, M. Kalwei, F. Schüth, K. Chao, Gold nanoparticles in SBA-15 showing catalytic activity in CO oxidation, Appl. Catal. Gen. 254 (2003) 289–296. https://doi.org/10.1016/S0926-860X(03)00490-3.

[79] J. Dupont, G.S. Fonseca, A.P. Umpierre, P.F.P. Fichtner, S.R. Teixeira, Transition-metal nanoparticles in imidazolium ionic liquids: Recycable catalysts for biphasic hydrogenation reactions, J. Am. Chem. Soc. 124 (2002) 4228–4229. https://doi.org/10.1021/ja025818u.

[80] S. Hrapovic, Y. Liu, K.B. Male, J.H.T. Luong, Electrochemical biosensing platforms using platinum nanoparticles and carbon nanotubes, Anal. Chem. 76 (2004) 1083–1088. https://doi.org/10.1021/ac035143t.

[81] Y. Ou, X. Cui, X. Zhang, Z. Jiang, Titanium carbide nanoparticles supported Pt catalysts for methanol electrooxidation in acidic media, J. Power Sources. 195 (2010) 1365–1369. https://doi.org/10.1016/j.jpowsour.2009.09.031.

[82] G.-H. Wang, J. Hilgert, F.H. Richter, F. Wang, H.-J. Bongard, B. Spliethoff, C. Weidenthaler, F. Schüth, Platinum–cobalt bimetallic nanoparticles in hollow carbon nanospheres for hydrogenolysis of 5-hydroxymethylfurfural, Nat. Mater. 13 (2014) 293–300. https://doi.org/10.1038/nmat3872.

[83] N. Narayan, A. Meiyazhagan, R. Vajtai, Metal nanoparticles as green catalysts, Materials. 12 (2019) 3602. https://doi.org/10.3390/ma12213602.

[84] Y. Yang, C. Zhang, Z. Hu, Impact of metallic and metal oxide nanoparticles on wastewater treatment and anaerobic digestion, Env. Sci Process. Impacts. 15 (2013) 39–48. https://doi.org/10.1039/C2EM30655G.

[85] F.T. Thema, E. Manikandan, M.S. Dhlamini, M. Maaza, Green synthesis of ZnO nanoparticles via Agathosma betulina natural extract, Mater. Lett. 161 (2015) 124–127. https://doi.org/10.1016/j.matlet.2015.08.052.

[86] T.N.J.I. Edison, Y.R. Lee, M.G. Sethuraman, Green synthesis of silver nanoparticles using Terminalia cuneata and its catalytic action in reduction of direct yellow-12 dye, Spectrochim. Acta. A. Mol. Biomol. Spectrosc. 161 (2016) 122–129. https://doi.org/10.1016/j.saa.2016.02.044.

[87] K. Pal, T. Zaheer, Nanomaterials in the Battle Against Pathogens and Disease Vectors, 1st ed., CRC Press, Boca Raton, 2022.

[88] N. Kulkarni, U. Muddapur, Biosynthesis of metal nanoparticles: A review, J. Nanotechnol. 2014 (2014) 1–8. https://doi.org/10.1155/2014/510246.

[89] L. Sintubin, B. De Gusseme, P. Van der Meeren, B.F.G. Pycke, W. Verstraete, N. Boon, The antibacterial activity of biogenic silver and its mode of action, Appl. Microbiol. Biotechnol. 91 (2011) 153–162. https://doi.org/10.1007/s00253-011-3225-3.

[90] S. Hamedi, S.A. Shojaosadati, A. Mohammadi, Evaluation of the catalytic, antibacterial and anti-biofilm activities of the Convolvulus arvensis extract functionalized silver nanoparticles, J. Photochem. Photobiol. B. 167 (2017) 36–44. https://doi.org/10.1016/j.jphotobiol.2016.12.025.

[91] V.K. Sharma, R.A. Yngard, Y. Lin, Silver nanoparticles: Green synthesis and their antimicrobial activities, Adv. Colloid Interface Sci. 145 (2009) 83–96. https://doi.org/10.1016/j.cis.2008.09.002.

[92] S.P. Chandran, M. Chaudhary, R. Pasricha, A. Ahmad, M. Sastry, Synthesis of gold nanotriangles and silver nanoparticles using aloe vera plant extract, Biotechnol. Prog. 22 (2006) 577–583. https://doi.org/10.1021/bp0501423.

[93] R. Mariselvam, A.J.A. Ranjitsingh, A. Usha Raja Nanthini, K. Kalirajan, C. Padmalatha, P. Mosae Selvakumar, Green synthesis of silver nanoparticles from the extract of the inflorescence of Cocos nucifera (Family: Arecaceae) for enhanced antibacterial activity, Spectrochim. Acta. A. Mol. Biomol. Spectrosc. 129 (2014) 537–541. https://doi.org/10.1016/j.saa.2014.03.066.

[94] Z.-R. Mashwani, T. Khan, M.A. Khan, A. Nadhman, Synthesis in plants and plant extracts of silver nanoparticles with potent antimicrobial properties: Current status and future prospects, Appl. Microbiol. Biotechnol. 99 (2015) 9923–9934. https://doi.org/10.1007/s00253-015-6987-1.

[95] R. Ramprasad, P. Zurcher, M. Petras, M. Miller, P. Renaud, Magnetic properties of metallic ferromagnetic nanoparticle composites, J. Appl. Phys. 96 (2004) 519–529. https://doi.org/10.1063/1.1759073.

[96] C. Di Paola, R. D'Agosta, F. Baletto, Geometrical effects on the magnetic properties of nanoparticles, Nano Lett. 16 (2016) 2885–2889. https://doi.org/10.1021/acs.nanolett.6b00916.

[97] I. Díez, M. Pusa, S. Kulmala, H. Jiang, A. Walther, A.S. Goldmann, A.H.E. Müller, O. Ikkala, R.H.A. Ras, Color tunability and electrochemiluminescence of silver nanoclusters, Angew. Chem. Int. Ed. 48 (2009) 2122–2125. https://doi.org/10.1002/anie.200806210.

[98] A. Fukuoka, P.L. Dhepe, Sustainable green catalysis by supported metal nanoparticles, Chem. Rec. 9 (2009) 224–235. https://doi.org/10.1002/tcr.200900004.

[99] R. Yousefi, F. Jamali-Sheini, M. Cheraghizade, S. Khosravi-Gandomani, A. Sáaedi, N.M. Huang, W.J. Basirun, M. Azarang, Enhanced visible-light photocatalytic activity of strontium-doped zinc oxide nanoparticles, Mater. Sci. Semicond. Process. 32 (2015) 152–159. https://doi.org/10.1016/j.mssp.2015.01.013.

[100] S. Martínez-Méndez, Y. Henríquez, O. Domínguez, L. D'Ornelas, H. Krentzien, Catalytic properties of silica supported titanium, vanadium and niobium oxide nanoparticles towards the oxidation of saturated and unsaturated hydrocarbons, J. Mol. Catal. Chem. 252 (2006) 226–234. https://doi.org/10.1016/j.molcata.2006.02.041.

[101] P. Yang, Y. Xu, L. Chen, X. Wang, B. Mao, Z. Xie, S.-D. Wang, F. Bao, Q. Zhang, encapsulated silver nanoparticles can be directly converted to silver nanoshell in the gas phase, Nano Lett. 15 (2015) 8397–8401. https://doi.org/10.1021/acs.nanolett.5b04328.

[102] M. Zhang, J. Guan, Y. Tu, S. Chen, Y. Wang, S. Wang, L. Yu, C. Ma, D. Deng, X. Bao, Highly efficient H_2 production from H_2S *via* a robust graphene-encapsulated metal catalyst, Energy Environ. Sci. 13 (2020) 119–126. https://doi.org/10.1039/C9EE03231B.

[103] P.K. Jain, X. Huang, I.H. El-Sayed, M.A. El-Sayed, Review of some interesting surface plasmon resonance-enhanced properties of noble metal nanoparticles and their applications to biosystems, Plasmonics. 2 (2007) 107–118. https://doi.org/10.1007/s11468-007-9031-1.

[104] S. Solomon, G.-K. Plattner, R. Knutti, P. Friedlingstein, Irreversible climate change due to carbon dioxide emissions, Proc. Natl. Acad. Sci. 106 (2009) 1704–1709. https://doi.org/10.1073/pnas.0812721106.

[105] V.K. Vidhu, D. Philip, Catalytic degradation of organic dyes using biosynthesized silver nanoparticles, Micron. 56 (2014) 54–62. https://doi.org/10.1016/j.micron.2013.10.006.

[106] R.V. Jagadeesh, K. Murugesan, A.S. Alshammari, H. Neumann, M.-M. Pohl, J. Radnik, M. Beller, MOF-derived cobalt nanoparticles catalyze a general synthesis of amines, Science. 358 (2017) 326–332. https://doi.org/10.1126/science.aan6245.

[107] G.M. Scheuermann, L. Rumi, P. Steurer, W. Bannwarth, R. Mülhaupt, Palladium nanoparticles on graphite oxide and its functionalized graphene derivatives as highly active catalysts for the Suzuki–Miyaura coupling reaction, J. Am. Chem. Soc. 131 (2009) 8262–8270. https://doi.org/10.1021/ja901105a.

[108] A. Rostami-Vartooni, M. Nasrollahzadeh, M. Alizadeh, Green synthesis of seashell supported silver nanoparticles using Bunium persicum seeds extract: Application of the particles for catalytic reduction of organic dyes, J. Colloid Interface Sci. 470 (2016) 268–275. https://doi.org/10.1016/j.jcis.2016.02.060.

[109] M. Oz, D.E. Lorke, M. Hasan, G.A. Petroianu, Cellular and molecular actions of Methylene Blue in the nervous system, Med. Res. Rev. 31 (2011) 93–117. https://doi.org/10.1002/med.20177.

[110] J.M. Campelo, D. Luna, R. Luque, J.M. Marinas, A.A. Romero, Sustainable preparation of supported metal nanoparticles and their applications in catalysis, ChemSusChem. 2 (2009) 18–45. https://doi.org/10.1002/cssc.200800227.

[111] F. Rispoli, A. Angelov, D. Badia, A. Kumar, S. Seal, V. Shah, Understanding the toxicity of aggregated zero valent copper nanoparticles against Escherichia coli, J. Hazard. Mater. 180 (2010) 212–216. https://doi.org/10.1016/j.jhazmat.2010.04.016.

[112] H. Zhao, F. Zhang, S. Zhang, S. He, F. Shen, X. Han, Y. Yin, C. Gao, Scalable synthesis of sub-100 nm hollow carbon nanospheres for energy storage applications, Nano Res. 11 (2018) 1822–1833. https://doi.org/10.1007/s12274-017-1800-3.

11 MNPs for Remediation of Toxicants and Wastewater Treatment

Emna Melliti, Alma Mejri, Md Sabir Alam, Jamilur R. Ansari, Hamza Elfil, and Abdelmoneim Mars

11.1 INTRODUCTION

Water is considered the most important life-giving source and indispensable for the endurance of all living beings. Interestingly, humans need about 0.03% of the planet's entire freshwater to meet their supply needs [1]. Nonetheless, in recent decades, water quality has been adversely affected by the rapid growth in the worldwide demographic, which has exhibited an uncontrollable increase in urbanization and industrialization and severe environmental and health consequences [2]. Indeed, the freshwater supplies have been contaminated by toxins, namely, organic and inorganic pollutants, pesticides, pharmaceutical composites, fertilizers, and heavy metals, posing a major risk to water resources despite the recognition of sustainable water management policies [3]. Wastewater is stemmed from numerous sources, such as hospitals, industrial and agricultural activities, and residential areas. Facing these threats, various classical methods of wastewater treatment have been developed, including precipitation, chlorination, ionization, ion exchange, electrochemical treatments, membrane filtration and reverse osmosis, evaporation, and floatation [4]. Also, the recent trends nanotechnology-based nanocarriers have vast applications in the fields of biomedical science, bio-nanotechnology. Various nanocarriers like metallic nanoparticles (Ag, Au, Zn, Cu, Co, Ti, etc.) have very broad applications in the field of drug delivery and various disease targeting (cancer, neurological, antibacterial, antifungal, and many more) [5–31]. However, these treatment techniques are either reasonably expensive as compared to new wastewater treatment nanotechnologies or have caused the formation of high levels of poisonous byproducts [32]. Thus, there is an urgent need for new, simple, efficient, and eco-friendly technologies. In this context, water remediation nanotechnologies prove their capability of being an efficient method for wastewater treatment without the aforementioned snags of traditional disinfection methodologies or the formation of toxic byproducts [33]. Furthermore, nanoparticles have recently been used for removing or degrading water pollutants, citing heavy metals, pesticides, dyes, toxins, viruses, and antibiotics [34]. Metallic nanoparticles, carbonaceous nanomaterials, zeolites, and dendrimers are the most commonly used nanomaterials in wastewater treatment [35]. The present chapter summarizes the advanced applications of metallic nanoparticles in the

DOI: 10.1201/9781003317319-11

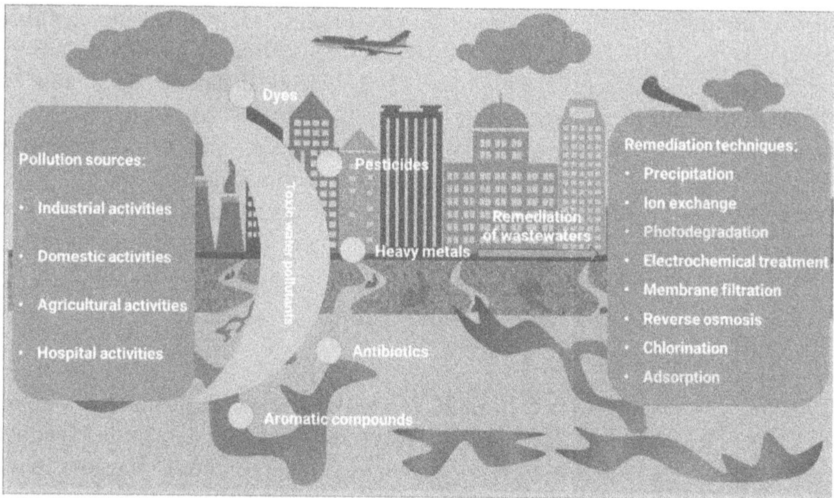

FIGURE 11.1 Representation of water pollution sources, common contaminants, and wastewater remediation techniques.

treatment of wastewater, including the removal and degradation of water contaminants (Figure 11.1). A roadmap of future trends is also presented to provide a perspective for improving treatment effectiveness.

11.2 METALLIC NANOPARTICLES FOR REMOVAL OF WATER POLLUTANTS

11.2.1 Key Parameters for the Removal of Water Pollutants

So far, a huge amount of research on metallic nanomaterials and their application in the removal of water contaminants has been done to determine the interaction mechanisms between metallic nanoparticles and contaminants and to define the key parameters of the adsorption processes [36]. As regards nanomaterial criteria, various crucial properties, including organization of the structure, filtration competence, size, diameter of the pore, the surface-volume ratio, and adsorption capacity, affect the removal efficiencies [37]. On the other hand, the influence of some experimental parameters such as pH, temperature, and content of organic matter is the main experimental parameter that is known to monitor the removal process from domestic, industrial, and biomedical wastewater effluents [38]. Interestingly, the miscibility of the contaminants in water and the viscosity of the wastewater have been shown to affect mainly the adsorption rate [39]. Indeed, several studies have demonstrated that less viscous solvents flow easily into the pores, contrary to very viscous liquids, which mainly diffuse slowly. Recognizing that viscosity depends on temperature, wastewater can be heated by various mechanisms through the joule effect, reducing viscosity, resulting in faster spread in pores, and promoting faster removal [40].

For the water-dispersible contaminations, electrostatic interactions, and chemistry between the surface of metallic nanoparticles and contaminants, monitor the adsorption affinity [41]. These can be improved by the chemical functionalization of the surface of metallic nanoparticles, which increases the number of dynamic sites. Furthermore, the choice of sorption approaches depends essentially on the chemical structure of the target pollutants, whether they have molecular or ionic structures. Commonly, the physicochemical characteristics of metallic nanoparticles, their physical state, and the chemical composition of the contaminants play major roles in the sorption process [42].

11.2.2 Removal of Dyes

Dyes are used in most aspects of daily life, citing the painting of textiles, paper, and leather [43]. In this context, the clothing sector is one of the main international generators of dyes. Indeed, this sector evacuates a large volume of wastewater into ecosystems and sewage systems without the proper pretreatment. These wastewaters are frequently laden with difficult-to-treat pollutants, including coloring and non-coloring compounds that, if discharged directly into the environment, cause a serious hazard to the ecosystem [44]. Moreover, dyes are mainly aromatic and unsaturated organic molecules. Their ability to color is mostly due to the existence of chromophores, which are unsaturated chemical groups that allow dyes to penetrate the material and stay fixed [45]. However, the specific chemical structure of dyes makes them exceedingly hazardous, undegradable, and carcinogenic, posing a serious threat to both the environment and human health [46]. In fact, various dyes are undegradable by sunlight, heating, or chemical oxidants, making their elimination from wastewater of enormous significance. Nowadays, there is a keen interest in applying metallic nanoparticles as dye removal tools due to their outstanding physicochemical properties, such as their enormous surface area, nanometric dimension, chemical structure, and particularly the possibility of modifying their surfaces by introducing various molecules and probes, which considerably enhance the adsorption process [47]. The adsorption mechanism between metallic nanoparticles and dye essentially includes four kinds, citing hydrogen bonding and electrostatic, π–π, and synergistic interactions [48]. Over the past few years, many researchers have studied the capacity of various metallic nanoparticles (e.g., Au, Ag, and magnetic nanoparticles) and their hybrid nanocomposites to remove synthetic dyes [49]. As of late, magnetic nanoparticle-based adsorbents have demonstrated significant demands and excellent efficacy in the elimination of dyes due to a variety of features, such as outstanding affinity for an external magnet, allowing quick and simple separation [50]. According to research, reported by Xia et al., magnetic nanoparticles with bare surfaces show a propensity to aggregate into larger nanoparticles when exposed to high surface energies, leading to a considerable reduction in surface area and reduced dye adsorption [51]. The functionalization of magnetic nanoparticles by simply coating them with different functional groups can be an effective solution to agglomeration behavior. Besides, coating magnetic nanoparticles with organic molecules makes biocompatible and nontoxic and increases their dispersion in aqueous media. Absalan et al. used magnetic nanoparticles tailored with an ionic liquid to remove reactive red-120 dye

[52]. According to the study, adsorption equilibrium was attained in about 2 minutes, and nearly 99.98% of the dye was removed from the water. One cycle of adsorption yielded a maximum absorption of 166.67 mg of reactive red-120 dye per 1 g of the adsorbent. Meanwhile, 98% of the adsorbed dye was desorbing from the adsorbent, and the regenerated adsorbent could be utilized three times without significantly losing its adsorption ability. More recently, Xu et al. have prepared polyacrylic acid-functionalized magnetite nanoparticles for use as an effective dye adsorbent [53]. The obtained nanoparticles were applied to remove rhodamine 6G dye from an aqueous solution with a removal rate of 90%. The maximum amount of adsorbed dye reached 55.8 mg per 1 g of the adsorbent in only 90 minutes. Interestingly, magnetic hierarchical hollow silica nanospheres have been prepared by Zhang et al. and have been used as adsorbents to extract methylene blue dye from water [54]. The outcomes showed that the proposed system had a significant proportion of 90% dye molecule adsorption in basic conditions. The greatest amount of adsorption for 1 g of the adsorbent was reported to be 71.45 mg, and equilibrium was reached in 120 minutes. They showed that the developed adsorbent could be recycled and used five times more. Furthermore, silver nanoparticles (AgNPs) were similarly used to remove various dyes, including carmoisine, tartrazine, methylene blue, and brilliant blue FCF. In this context, Z. Lockman and collaborators have used silver nanoparticles, which are prepared by reacting Kyllinga brevifolia (KBE) extract with silver nitrate [55]. The eco-friendly silver nanoparticles were applied to remove methylene blue (MB) dye with a removal rate of 93% and a rate of extraction of 0.2663 min^{-1}. More interestingly, M. Zahoor et al. designed an adsorbent based on palladium-nickel nanocomposite hosted on carbon and cerium oxide and utilized it to remove the azo dye acid orange-8 (AO-8) from aqueous solution [56]. They claimed that the removal rate was 90%, with maximal adsorption capacities of 769 mg/g at 333 K. The findings demonstrated that the adsorbent can be reused for the other 6 cycles with up to 80% removal efficiency. On the other hand, carbon-based nanomaterials have attracted significant importance in the removal of dyes due to their high adsorption capacity, established between the dyes and carbonated materials by electrostatic or van Der Waals-type interactions [57]. Within this framework, M. Siddik and coworkers described the preparation of graphene oxide–gold nanocomposite and its application as efficient nanoadsorbents for the extraction of malachite green and ethyl violet dyes from water samples [58]. The findings revealed that the synthesized nanocomposite showed effective adsorption features and recoveries. In a similar manner, an in situ prepared Au-reduced graphene oxide (Au-RGO) nanohybrid was applied to extract rhodamine 123 (Rh123) from aqueous media, and results showed excellent adsorption efficiency and great recyclability of the nanomaterial-based adsorption system. In addition, the authors reported that a prototype device for the removal of a wide range of dyes has been developed using Au-RGO nanohybrid on a syringe filter system, leading to the fabrication of a portable, reusable adsorbent for water purification.

11.2.3 Removal of Heavy Metals

Water pollution, especially with heavy metals, has emerged as a worldwide concern due to its numerous damaging effects on health and the environment [59].

Indeed, heavy metals can be released neatly into water mainly through diverse natural and industrial sources, commonly metallurgy, coating, mining, chemical plants, agriculture, domestic wastewater, etc. [60]. According to recent statistics, heavy metal contamination has affected 40% of the world's lakes and rivers [61]. For instance, the Coyuni river basin has been affected by significant mercury contamination induced by amalgam mining techniques employed in artisanal gold mining in Venezuela [62]. Moreover, arsenic-containing water is consumed daily by over 140 million people in 50 countries on different continents at levels above the recommended limits for drinking water (the WHO reference value of 10 μg/L) [63]. The hazard of heavy metals is mainly due to their adverse effects, notably their non-biodegradability by microorganisms once they are discharged into the environment [64]. Thus, they are tending to bioaccumulate in living organisms and food chains, causing multiple serious damages affecting various organs and neurodegenerative disorders including Alzheimer's and Parkinson's diseases [65]. Based on the aforementioned background, the urgent need to develop effective technologies for extracting heavy metals from aqueous solutions has become critical and has attracted enormous attention. Up to date, multiple technologies have demonstrated their ability to address the issue of heavy metals in wastewater, including electrochemical treatments, separation techniques based on precipitation, and membrane filtration [66]. Besides, adsorption was suggested as the most widely used technology because of its affordability and ease of use. Interestingly, metallic nanoparticles and their related nanocomposites are potential synergies that allow the combination of two or more nanomaterials to produce an adsorbent with superior qualities to those of its constituents. Several studies revealed that silver, gold, and magnetic nanoparticle-based adsorbents have shown outstanding performances in the extraction of heavy metals from wastewater [67]. In this framework, Lisha et al. explored the affinity of mercury (Hg) toward gold (Au), well recognized because they can form different amalgams, namely AuHg, $AuHg_3$, and Au_3Hg [68]. The authors applied gold nanoparticles (AuNPs) to remove Hg using aluminum support. The findings demonstrated that Au nanoparticles had a substantially better removal capacity for Hg(0) than conventional adsorbents, up to 4.065 g/g of adsorbent. They also claim that the developed system can be utilized to treat heavy metal-contaminated water effectively, showing that the costs of using an adsorbent based on AuNPs were minimal and that recovered Au nanoparticles were simple to obtain. Likewise, citrate-stabilized gold nanoparticles were used by Jimenez and collaborators to eliminate mercury (II) from water [69]. In the reported work, citrate ions were used once as stabilizer agents of gold nanoparticles and once as a reducing agent of mercury (II) to mercury (0). The results revealed that an elaborated system can remove mercury (II) at a concentration lower than 5 ppb. Regarding reusability, the isolated final mercury derivative was determined to be an Au_3Hg alloy that can be easily desorbed at a high temperature or pressure, recovering gold nanoparticles. On the other hand, mercaptosuccinic acid (MSA)-stabilized silver nanoparticles were recently used as mercury adsorbents by E. Sumesh et al. [70]. The investigation showed that, as compared to conventional adsorbents, they had a greater removal capacity for Hg (II) (800 mg/g) was found. The authors also suggested that the reported nanoadsorbent can be used as a promising alternative for mercury removal, noting that the cost of mercury removal (II) through the use of Ag@MSA was competitive.

As far as magnetic nanoparticles are concerned, these materials are potentially applied in environmental remediation and water purification [71]. Hence, nanosized zero-valent iron, or its oxides, have been considered the appropriate option for scavenging various kinds of pollutants like dyes, pesticides, and especially heavy metals [72]. It should be emphasized that the nanoscale zero iron consists of a core of Fe (0) that can reduce heavy metal ions and a shell of ferric oxide to assure electrostatic interaction with heavy metals through reactive sites displayed on nanoparticle surface [73]. The outstanding performance of zero iron in the removal of heavy metals from wastewater is significantly influenced by its high reduction capacity and wide specific surface area [74]. In this context, Huang et al. prepared sodium dodecyl sulfate (SDS)-modified iron (0) nanoparticles to remove chromium (VI) from wastewater [75]. The choice of SDS was based on its properties including its outstanding migration and high dispersion in water. The findings indicated a high removal capacity of 253.68 mg/g and a remarkable removal efficiency of 98.919%. Interestingly, Zarime et al. proposed a solution to the issue of aggregation of iron nanoparticles (0) by functionalizing their surface with bentonite [76]. The nanocomposite was applied to eliminate numerous heavy metal ions, including lead, copper, cadmium, cobalt, nickel, and zinc, in aqueous solutions. Compared to using simply bentonite as an adsorbent, they report that adding bentonite gives the nanocomposites more heavy metal adsorbing sites, displaying a greater removal capability for the heavy metals indicated earlier. As regards iron oxide-based nanomaterials, Khezami and collaborators reported the preparation and use of goethite (α-FeOOH) nanocrystalline powders for the remediation of contaminated wastewater with cadmium [77]. The experiments showed an impressive maximum adsorption capacity of 167 mg/g, at a pH 7 and a temperature of 328 K. The results revealed that the adsorption data and the Langmuir and Freundlich isotherms agreed well and that the pseudo-second-order model well described the adsorption kinetics. Moreover, the reported system displayed a good ability to eliminate several heavy metals, including manganese, cobalt, nickel, and zinc. Interestingly, many reports have discussed the potential of magnetic nanomaterials in wastewater treatment due to how easily they are separated from aqueous solutions when applying a magnetic field [78]. Recently, Giraldo et al. described the use of magnetite nanoparticles in the elimination of several heavy metals such as copper, lead, manganese, and zinc [79]. The findings showed that nanosized magnetite has a superior ability to bind Pb (II), with a maximum adsorption capacity of 0.180 mmol/g, while manganese (II) adsorption has the lowest adsorption capacity. The authors cited various electrostatic interactions between the heavy metal ions and the adsorbent sites as a possible explanation for the observed variation. Indeed, the adsorption capacity rises as the distance between the surface of the adsorbent and the heavy metal ion decreases, which is somewhat connected to the hydrated ionic rays. Given that Pb (II) had the smallest hydrated ionic radius, it was reasonable that it would have the highest adsorption capacity. Likewise, zinc oxide nanoparticles have received considerable interest as adsorbents due to their excellent adsorption capacity not only for heavy metals but also for a wide range of water pollutants [80]. In that respect, Somu et al. applied casein-modified zinc oxide nanoparticles to treat the simultaneous presence of three heavy metal ions, namely cadmium, lead, and cobalt [81]. The Langmuir model was used to fit the adsorption data, and the adsorption capacities of Cd (II), Pd (II), and Co (II) were determined to be 156.74, 194.93, and 67.93 mg/g, respectively. Given the excellent antimicrobial activity of the casein-modified ZnO nanoparticles, the results indicate that the reported system may be a promising adsorbent for practical wastewater.

11.2.4 Removal of Pesticides

Currently, around 2 million tons of pesticides are used annually to protect crops against weeds, insects, and pests [82]. Indeed, the protective role of pesticides makes them a crucial tool to meet urgent food needs and improve the standard of living of the world's population. Nevertheless, the unabated rise in pesticide usage serves as a reminder of the poisonous consequences these chemicals have on both the environment and human health. Indeed, pesticides are discharged into aquatic ecosystems through a variety of indirect methods, including sprinkling, soil erosion, and particularly careless chemical dumping, given that pesticides are not easily degraded [83]. The lethal effects are generally due to persistent and bioaccumulative behaviors that disrupt the sex hormone and interfere with reproduction due to endocrine deficiency [84]. In this perspective, pesticides are rated by WHO assessments as highly poisonous, highly hazardous, and extremely dangerous; hence, it is urgently important to eliminate them from water (The WHO, 2009). As was previously stated, adsorption, which is theoretically a surface phenomenon, is an efficient method for cleaning up water and reusing it. The number of reactive sites, porosity, and specific surface area of the adsorbent are among the crucial variables that influence adsorption [85]. A broad range of metal oxide nanoparticles was applied to remediate pesticide-contaminated waters, including oxides of ferric, manganese, zinc, titanium, magnesium, and cerium [86]. For instance, Z. Zongshane et al. have described the use of polystyrene-decorated magnetic nanospheres for the removal of organochlorine from water samples [87]. The adsorption findings showed that organochlorine pesticides can be readily adsorbed. The pseudo-second-order model of adsorption kinetics was also demonstrated by fitted data. Moreover, the obtained results demonstrated that the developed system displayed acceptable performance with a removal capacity of 93.3%. Recently, P. A. Azar and coworkers demonstrated the applicability of chitosan-modified zinc oxide nanoparticles for the removal of permethrin pesticide [88]. The authors indicated that 0.5 g of the nanocomposite can remove 99% of the pesticide under optimal conditions. They also studied the reuse of the adsorption system, and the results showed promising performance in water treatment with 56% regeneration after 3 cycles. Interestingly, A. B. Tesfamichael et al. described the preparation of cerium oxide nanoparticles modified with aluminum oxide and their utilization in the extraction of dimethyl methylphosphonate pesticide from water [89]. The results of the adsorption investigation showed that the highest adsorption capacity was 775 mol/g. The proposed adsorption platform, according to the authors, can be a great tool for getting rid of a variety of pesticides that contain organophosphorus. More recently, V. Vasic and coworkers described the use of both gold nanoparticles and nanorods for the elimination of the organophosphorus pesticide dimethoate [90]. The investigations demonstrated that the fitted experimental data obeyed the Langmuir adsorption isotherm for both kinds of gold nanomaterials. The adsorption capacities for nanoparticles and nanorods were determined to be 456 mg/g and 57.1 mg/g, respectively, demonstrating that nanoparticles are more effective than nanorods at removing the target pesticide. Furthermore, M. A. Khatani et al. detailed the elimination of imidacloprid pesticide from an aqueous solution using chitosan-decorated silver nanoparticles (AgNPs@chitosan) on a modified membrane [91]. The investigations demonstrated that imidacloprid may be removed by both chitosan and chitosan-modified silver nanoparticle membranes with adsorption rates of 40% and 85%, respectively. The authors declared that the obtained membranes are auspicious materials for the elimination of both imidacloprid and a broad range of pesticides, including organochlorine, from contaminated waters. More interestingly, T. Pradeep and collaborators investigated the mechanistic aspects of the interaction established between metallic nanoparticles (gold and silver NPs) and pesticides [92]. The authors chose chlorpyrifos (CP) as a model pesticide. The interaction between CP pesticide and nanoparticle surfaces was investigated by several microscopic and spectroscopic techniques, and the findings demonstrated that the adsorption of CP pesticide occurred through the creation of silver or gold–S complexes. Furthermore, the rate of CP degradation speeds up with the increase in temperature and pH, reaching its maximum level in 3 hours. Compared to gold NPs, silver NPs achieve a higher rate of CP adsorption and degradation. The authors proposed that silver and gold NPs can be utilized in environmentally friendly treatments, as they can be applied to water at ambient temperature without the need for additional stimulation, such as UV radiation.

11.2.5 Reuse and Risk of Used Metallic Nanoparticles

Nowadays, the use of nanomaterials, particularly metal nanoparticles, has seen considerable growth in a wide range of applications, such as the remediation and reuse of wastewater from industrial, medical, electronic, energy, and agricultural. Therefore, new challenges have emerged in terms of the recyclability of used nanoparticles, handling desorbed water pollutants, safety, and risks to human health and the environment [93]. Along with the application of metallic nanoparticles in wastewater treatment, special attention is given to recycling and recovering them from various matrices in order to provide sustainable solutions and to develop new ecological remediation approaches (Figure 11.2). As contaminants are adsorbed, post-treatment procedures must be developed, and effective separation processes must ensue. In this framework, several separation techniques are developed, such as magnetic separation, antisolvents, aqueous dispersion, liquid–liquid extraction, precipitation, centrifugation/solvent evaporation, and so on [94]. Interestingly, the two most popular ways to produce useful metallic nanoparticles and their oxides in significant amounts are hydrometallurgy and pyrometallurgy techniques, which are primarily based on the concepts of precipitation and heat treatments [95]. Nevertheless, metallic nanoparticles, which are regenerated from wastewater, must prove their fundamental and initial properties citing physico-chemical properties and stability, to show that recycling techniques are effective enough to produce the necessary results. Numerous

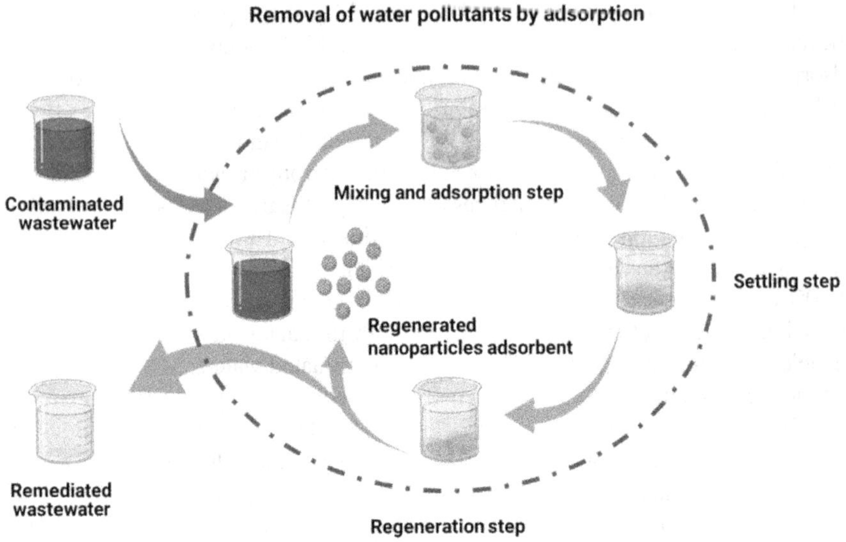

FIGURE 11.2 Illustration of wastewater treatment by adsorption and the regeneration cycle of used metallic nanoparticle based-adsorbents.

studies have demonstrated that factors such as particle size, surface chemistry, solubility, and shape are crucial in evaluating potential dangers and exposure from inhaled nanoparticles [94]. In this context, numerous regulatory authorities, such as the Registration, Evaluation, Authorization and Restriction of Chemicals (REACH) and the Toxic Substance Control Act (TSCA), regularly evaluate the risks of nanomaterials by assessing various criteria and providing essential guidelines regarding their use [96]. Based on the cited information, there is no doubt about the urgency of recycling and the recovery of used nanoparticles into valuable products. Nonetheless, because recycling designs are not cost-effective, there is a chance of developing a secondary supply chain that runs parallel to the production chain. Research in this field must thus continue, and the findings must be turned into suitable technology.

11.3 PHOTODEGRADATION OF WATER POLLUTANTS

The out-of-control growth of the world population leads to the continual growth of industrial, textile, pharmaceutical, and agricultural activities, resulting in the widespread use of contaminated wastewater. Nevertheless, these waters are often contaminated persistent hazardous pollutants with low biodegradability and a complex composition [97]. Although research into the development of promising chemical treatment methods is very advanced, the oxidants utilized during the chemical treatment of water are still inefficient in the degradation and mineralization of the complex structure of contaminants [98]. In this framework, photocatalysis is an advantageous advanced oxidation process (AOP) that has recently gained popularity [99]. According to numerous studies, heterogeneous photocatalysis has demonstrated excellent results in the degradation of persistent organic pollutants. To select an appropriate photocatalyst, certain criteria must be met. Certainly, the catalyst's effectiveness at removing the target pollutants, the material's stability, and particularly the band gap are important for providing high photocatalytic efficiency [100]. According to their physical and chemical characteristics, many element categories, including metallic nanoparticles (noble metals, transition metals, etc.), can be used to create photocatalysts. Mainly, the mechanism of the photocatalytic process contains four steps, as illustrated in Figure 11.3. Furthermore, the pH, the initial amount of the contaminant, and the catalyst charge are three key parameters that affect the pollutant's ability to adsorb on the catalyst surface [101]. Indeed, due to a decrease in hydroxyl radical generation on the catalyst's surface, it has been discovered that rises in pollutant concentration decrease the photodegradation capacity. In experiments, it has also been confirmed that adding more catalyst will result in more pollutants being adsorbed at the catalyst site.

11.3.1 Degradation of Organics Dyes

A dye is a colorful, poisonous material that, when applied to fabric, imparts a semi-permanent color. Its coloration is at the origin of light absorption in the entire visible range of the spectrum at a given wavelength [102]. However, the aggressive application of dyes in various industrial fields, particularly, printing, textiles, cosmetics, plastics, and so on, affects aquatic ecosystems as they reduce sunlight transmission through water. Indeed, the contamination prevents light from entering the bottom of the river and the sea, causing a disorder in photosynthesis activity and oxygen deficiency and leading to a disturbance of the ecosystem [103]. Therefore, the degradation of such toxic substances from waste effluents becomes an urgent need. Although silver nanoparticles (AgNPs) are well known for their unique electronic properties, which set them apart from other nanoparticles, their applications are not limited to optics and photoelectronic devices, but also water treatment. Its great capacity to disinfect is beneficial in treatment technology, where it is extensively researched as a powerful adsorbent, co-catalyst, and membrane nanofiller [104]. Recently, Zhu et al. have prepared a number of hybrid silver salt/collagen fiber photocatalysts for the degradation of methyl orange [105]. In this work, the precipitation procedure was used to produce a variety of silver salts on the surface of the collagen fiber, including silver chloride, silver acetate, silver molybdate, silver tungstate, and silver phosphate. According to the findings, among photocatalysts used under visible light, the AgCl/collagen hybrid displayed the greatest degrading efficacy (about 90%). Interestingly, platinum nanoparticles are the most efficient co-catalyst for extracting photogenerated electrons to reduce contaminants due to their rapid trapping of conductance bond electrons in semiconductors, which prevents load-bearing recombination between conductance and valence bonds and improves photocatalytic performance. Furthermore, numerous studies have shown that Pt is able to increase visible light absorption and transport it to the semiconductor photocatalyst, thereby increasing the photocatalytic performance. More recently, J. H. Kim et al. have combined gold nanoparticles as plasmonic nanoparticles with polydimethylsiloxane-TiO_2 semiconductor photocatalysts to photodegrade Rhodamine B dye [106].

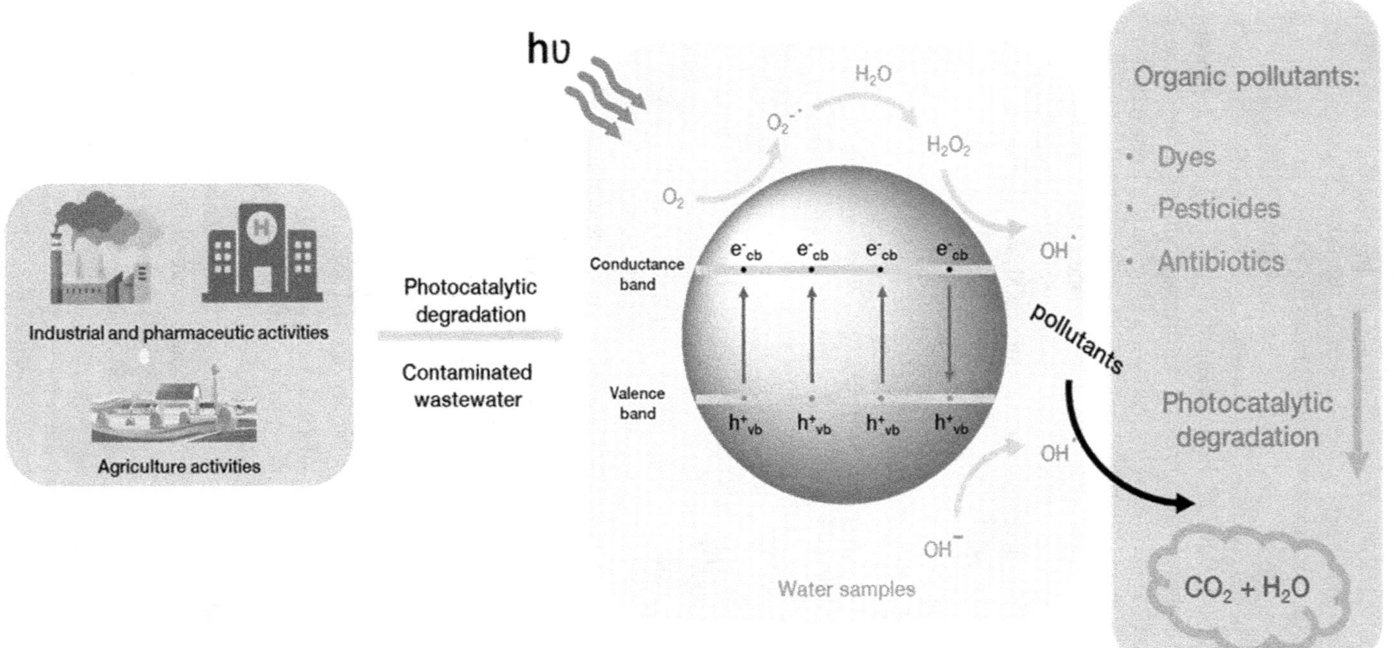

FIGURE 11.3 Schematic of the photocatalytic degradation mechanism of organic contaminants in water.

They suggested that due to their capacity to considerably increase overall photocatalytic efficiency, immobilized titanium dioxide-gold nanoparticles on the surface of the PDMS substrate can be a promising approach for highly effective photocatalysts using the spectrum of solar radiation. In comparison to the PDMS-TiO_2 and PDMS-Au sponges, the results demonstrated significantly improved photocatalytic efficiency under UV and visible light, as well as excellent reusability even over several cycles. According to the authors, PDMS-TiO_2-Au sponge's remarkable photocatalytic efficacy and straightforward manufacture make it potentially useful in environmental applications, particularly for the remediation of water pollution. On the other hand, zinc oxide nanoparticle-based photocatalysts were widely applied in the degradation of various dyes, including malachite green, azo, and methylene blue dyes [107]. For instance, the photocatalytic activity of ZnO nanoparticles synthesized using laser ablation was investigated by Nikša Krstulovic and collaborators [108]. The studies on the photodegradation of solutions containing various concentrations of methylene blue and rhodamine B, irradiated with UV, using various mass concentrations of ZnO catalyst, revealed remarkable photocatalytic efficiency. Therefore, they concluded that photocatalytic efficiency declines as dye concentration rises, while it can be increased by raising the concentration of the catalyst. Nonetheless, very few reports have described the application of magnetic nanoparticles for the photo-mineralization of harmful dyes.

Recently, E. Alzahrani reported the application of silver-magnetic nanoparticles (Fe_3O_4) for the photodegradation of Eosin Y dye [133]. According to the authors, the UV/Ag-Fe_3O_4 combination can be suggested as an efficient technique for the elimination of hazardous dyes, including azo and Eosin Y dyes. More interestingly, L. Hariani and coworkers reported the application of a tin oxide–iron oxide nanohybrid for the photodegradation of Congo red dye [134]. The findings revealed that the percentage of photodegradation was 50.76% under ideal conditions: a contact time of 90 min and a catalyst concentration of 18 mg/L. Moreover, the authors suggested that Congo red photodegradation kinetics using SnO_2–Fe_3O_4 nanocomposite obeyed the pseudo-second-order equation. Finally, various metal nanoparticle-based photocatalysts for the degradation of several dyes used so far under different light sources are summarized in Table 11.1. The summary of studies on the photocatalytic degradation of various dyes using metallic nanoparticles.

11.3.2 Degradation of Benzene and Its Derivatives

Aromatic compounds, including benzene, toluene, and xylene, are known to be carcinogenic and acutely toxic derivatives. These adverse properties are due to their low biodegradability and relatively high-water solubility, leading to major concerns regarding carbon fuel spills in aquatic environments [135]. The World Health Organization (WHO) states that the maximum allowed amounts of benzene, toluene, and xylenes in drinking water are, respectively, 0.01, 0.7, and 0.5 mg/L (WHO, 2004). Therefore, the degradation and removal of these persistent organic pollutants from wastewater effluents have become an urgent and much-needed environmental requirement to minimize the risk of pollution of the aquatic ecosystem. To date, many interesting methods, including adsorption, advanced oxidation processes, and

TABLE 11.1
Metal Nanoparticle-Based Photocatalysts for the Degradation of Several Dyes

Type of Dyes	Nanoparticles	Light source	Wavelength length (nm)	% of degrad ation	References
	ZnS NPs	Hg-Lamp	555	65	[109]
	CdS NPs	Xe-lamp	552	92.5	[110]
Rhodamine B	CuS NPs	Xe-lamp	554	83.22	[111]
	Ag2S NPs	Solar	554	70	[112]
	PbS/SnO_2	Xe-lamp	555	84	[113]
	ZnS/CdS	Visible	462	90	[114]
	CdS/TiO_2	Xe-lamp	464	100	[115]
Methyl orange	CuS/CdS	Xe-lamp	464	93	[116]
	Ag_2S/Ag_3PO4	Xe-lamp	464	100	[117]
	FeS_2 NCs	Xe-lamp	463	80	[118]
	ZnS QDs	Solar	625	88	[119]
Brilliant green	TiO_2	Solar	624	97	[120]
	Ag_2S/Co_3O4	Solar	625	99	[121]
Metaphenylene blue	ZnS/GO	UV	465	90.61	[122]
	CoS/AIMCM	W-lamp	664	98	[123]
Metaphenylene blue	ZnS/GO	UV	465	90.61	[122]
	CoS/AIMCM	W-lamp	664	98	[123]
Rose Bengal	ZnS NPs	UV	598	100	[119]
	CdS NSs	Xe-lamp	553	100	[124]
	ZnS NPs	Visible	617	65	[125]
	ZnS/CuS	Xe-lamp	617	78.51	[126]
Malachite green	$CdS/BiVO_4$ NCs	Hg-lamp	617	98.3	[128]
	CuS/TiO_2	Solar	621	78	[129]
	Ag_2S	Solar	516	72.51	[130]
	CuS NPs	Solar	516	91.97	[131]
Eosin Y	CdS	Visible	516	91.12	[132]
	CuS NPs	UV	516	79.49	[131]

biodegradation, have been developed to eliminate benzene and its derivatives from contaminated wastewater [136]. Besides, photocatalysis, as an efficient alternative to the previous methods, has demonstrated its capacity to mineralize any aqueous solution polluted with an organic substrate. Numerous research studies on the functionalization of the photocatalyst, commonly titanium oxide, by metallic nanoparticles have been developed to improve its photocatalytic capacity [137]. In this framework, the photocatalytic-based heterogeneous oxidation process using titanium dioxide and UV light has attracted huge interest in the mineralization of a broad spectrum of persistent organic-based contaminants by converting them into relatively more biodegradable and less toxic substances, including CO_2 and H_2O. Interestingly, N.R.C. Fernandes-Machado et al. prepared N-doped TiO_2/ZnO catalysts and applied them to the photocatalytic degradation of benzene, toluene, and xylenes [138]. The results demonstrated that the hybrid's photocatalytic activity was greater than that of unmodified TiO_2, resulting in a reduction of more than 80 weight percent after only 120 minutes of exposure to radiation. Moreover, Parida et al. have investigated the impact of ZnO nanoparticles on the photodegradation of para-nitrophenol utilizing solar irradiation [139]. In this study, several zinc oxide nanoparticle doses ranging from 0.2 to 2.0 g/L were utilized to assess the effect of concentration on the photocatalyst's photocatalytic activity. The results demonstrated that a high degradation rate was observed with the highest concentration of ZnO nanoparticles, around 0.6 g/L. The authors concluded that a greater catalyst concentration would lead to more catalyst active sites. According to a similar study reported by Selvam et al., as the dose of functionalized TiO_2 increased from 50 to 150 mg, the constant rate of the photodegradation of 4-fluorophenol increased correspondingly from 0.0152 to 0.0358 min^{-1} [140]. However, the results revealed that an additional rise in catalyst dose from 150 to 250 mg exhibited a loss in the photocatalytic rate from 0.0358 to 0.0296 min^{-1} due to the aggregation of free catalyst particles, resulting in a significant decrease in the transparency of the solution. The suggested phenomenon was also confirmed by the study of the photodegradation of the same phenol derivative by ZnO nanoparticles. Similarly, Lathasree et al. studied the effect of zinc oxide nanoparticles on the photodegradation of phenols using a concentration from 1.0 to 3.0 g/L [141]. According to the authors, the optimal photodegradation rate was obtained using a zinc oxide concentration of 2.0 g/L. However, it was determined that the extra catalyst particles in the solution were limiting light, which caused the photodegradation rate to decline above the catalyst dose of 2.0 g/L.

More interestingly, many studies have investigated the low photoactivity of TiO_2 and the reason behind this imperfection [142]. The findings revealed that the reduction in photodegradation efficiency is due mainly to the recombination of photogenerated electrons and holes. To overcome this issue, the TiO_2 photocatalyst has been doped with certain metallic nanoparticles, including silver, palladium, and gold. For instance, gold nanoparticles modified with titania hollow microspheres were applied to investigate the photodegradation of phenol under visible light irradiation [143]. Results showed that compared to unmodified titania hollow microspheres, the gold-titanium dioxide nanocomposite had the lowest band gap energy and the highest photocatalytic efficiency. According to the authors, the designed photocatalyst is capable of decomposing 97.5% of phenol after being exposed to visible light for one hour. Furthermore, V. Shetty and collaborators assessed the ability of silver-titania nanoparticles to photodegrade phenol by solar irradiation [144]. The investigations indicated that the developed nanohybrid is an efficient solar photocatalyst, making it more powerful than UV photocatalysts in the decomposition of phenol. Recently, the effect of doping titania nanoparticles with cobalt on the photocatalytic activity of the degradation of phenols was evaluated by the M. Siddiq group [145]. Compared to an un-doped and solely sulfur- or cobalt-doped titania catalyst, a co-doped titania nano-photocatalyst with 5 weight percent cobalt and 1 weight percent sulfur generated a higher photocatalytic activity. The authors suggested that cobalt and sulfur co-doped titania nanocomposites induced efficient photocatalytic activity due to their remarkable structural and optical properties. More recently, K. Ravi et al.

described the application of Cu_2O-TiO_2 nanocomposite for the photodegradation of benzene and its derivatives under light waves [146]. The Cu_2O-TiO_2 nanocomposite was prepared using the solvothermal approach. The authors assessed the Cu_2O-TiO_2 nanocomposite's photocatalytic activity for the decomposition of benzene, toluene, and chlorobenzene in polluted water and under exposure to sunlight. The outcomes revealed that, within two hours of treatment under visible irradiation, Cu_2O-TiO_2 nanoparticles obtained an overall 95% photodegradation rate of benzene and related derivatives in contaminated water.

11.3.3 Degradation of Pesticides

As described earlier (Part II.4), the use of pesticides is steadily increasing worldwide. According to FAO's statistical database, over 4.2 million tons of pesticides are utilized annually to boost crops and keep them free from pests (FAO, 2022). Based on various key criteria such as the entry path, the active group of their chemical structures, and the target organism of the pest, pesticides are categorized into four major groups: herbicides, insecticides, fungicides, and rodenticides [147]. However, excessive and uncontrollable use of pesticides provokes soil, water, and air contamination, resulting in the non-biodegradable pesticide's bioaccumulation in the human body. This causes several serious damages, including birth abnormalities, genetic modification, neurological problems, and cancer [148]. Recent statistics claim that pesticides are responsible for the deaths of 5,000 to 20,000 people and the poisoning of over 500,000 people each year [149]. Besides, the common sources of pesticide contamination are plant residues and the discharge of agricultural waste, which are highly polluted with pesticides. These pesticide wastes are widely disseminated in aquatic ecosystems and soil, thus entering the food chain and eventually threatening humans. Therefore, the destruction and degradation of pesticide residues are highly desirable. Despite the many reported methods for pesticide degradation, photocatalytic degradation is still the most cutting-edge and appropriate method for cleaning up pesticide-contaminated wastewater due to its effectiveness and sustainability [150]. Nowadays, nanomaterial-based photocatalysis is a highly demanding nanotechnology that effectively improves the degradation rate of persistent pesticides into non-toxic intermediates and end products. Hence, several types of nanoparticles have been used in the photodegradation of pesticides, including metal and metal oxide nanoparticles and their nanohybrids [151]. In this framework, both zinc oxide and titanium oxide nanoparticles are widely used as potential catalysts in a variety of catalytic processes, including the photodegradation of hazardous compounds in water. Indeed, these metallic oxide nanoparticles possess excellent physicochemical properties such as chemical stability, long-term corrosion resistance, high photostability and refractive index, exceptional ultraviolet absorption, excellent incident photoelectric conversion efficiency, and unique dielectric constants [152]. For instance, the photocatalytic degradation capacity of the pesticide acetophenone was investigated by Yesodharan et al. using ZnO nanoparticle catalysts under solar irradiation [153]. In this work, the authors studied the effect of several experimental parameters, namely the effect of oxidizing agents on the performance of photocatalytic degradation. The results showed that 1-hydroxyacetophenone had the highest photocatalytic degradation ability compared to hydrogen peroxide and dioxygen, so it became the main degradation intermediate. According to the study, photocatalysis using zinc oxide and sunlight can be an effective method for acetophenone decomposition in contaminated water. Interestingly, Baruah et al. demonstrated that the rate of photocatalysis can be enhanced by the formation of imperfections in the crystal structure of zinc oxide nanoparticles [154]. Furthermore, Malato and collaborators have assessed the photocatalytic degradation of lindane and methamidophos pesticides using TiO_2 as a catalyst under both sunlight and UV irradiation [155]. According to the report, the photodegradation of the hazardous compounds was demonstrated by the total organic carbon reduction and phosphate analyses, and the authors suggested the use of oxidizing agents like peroxide to decompose pesticides into non-toxic byproducts. Recently, the phosphamidon pesticide's decomposition using silver-doped zinc oxide nanocomposite was assessed by the Korake group [156]. The outcomes showed that visible light irradiation of zinc oxide doped with silver (0.2 mol%) was effective for phosphamidon degradation. Indeed, the modification with silver improved the photocatalytic activity of the nanocomposite photocatalyst owing to the unique optical vibration of the surface plasmon and electronic properties, thereby increasing the photocatalytic activity by creating a local electric field. As a result, changes in surface characteristics such as oxygen vacancies and crystal defects would be the cause of the increased photocatalytic activity of silver-modified zinc oxide nanocomposite. More interestingly, P. Veerakumar et al. demonstrated the efficient photodegradation activity of noble metal nanoparticles (silver and palladium) modified zinc oxide nanocomposite and their application in the decomposition of methyl parathion pesticide, pendimethalin, and trifluralin herbicides [157]. The analysis indicated that when exposed to visible light, both Ag@ZnO and Pd@ZnO nanocomposites had good photocatalytic activities toward the photodegradation of certain pesticides and herbicides. According to the authors, Pd and Ag nanoparticles function as electron sources, facilitating interfacial charge transport by reducing the rate of charge recombination. More recently, graphene oxide-modified metal ferrites nanocomposites were used to catalyze the photodegradation of acetamiprid pesticide [158]. The findings showed that under UV radiation and under optimized reaction conditions, acetamiprid loading is in the order of ~97 and ~90% of acetamiprid for GO-$CoFe_2O_4$ and GO-Fe_3O_4, respectively. The designed catalysts have demonstrated remarkable performance, with high rates of degradation, superior magnetic properties, ease of recovery, and effective reuse up to five cycles without noticeably losing catalytic activity.

11.3.4 Degradation of Antibiotics

Recently, antibiotics have been extensively applied in many fields, including healthcare, aquaculture, and livestock farming [159]. Besides, the overuse of antibiotics and their delayed metabolism caused serious aquatic pollution. Therefore, antibiotic-resistant bacteria will proliferate as a result of the presence of antibiotics in aquatic ecosystems, endangering human health and the efficacy of antibiotic medications [160]. Nevertheless, traditional methods of wastewater treatment cannot efficiently eliminate antibiotic residue and thus distribute it in ecosystems and eventually through the food industry or drinking water to the human body [161]. As a result, nanotechnology-based wastewater treatment methods, including adsorption, photocatalysis, biodegradation, electrochemical treatment, and so on, have attracted considerable interest in effectively addressing the issues mentioned [162]. On the other side, antibiotics were completely mineralized into harmless byproducts like CO_2, H_2O, and other mineral components thanks to photocatalysis degradation. In this context, metallic nanoparticles are among the most broadly used nanomaterials as photocatalysts for wastewater treatment, as most of them are nontoxic, inexpensive, stable, readily available, and highly photoactive [163]. Since their introduction in the field of photocatalysis, TiO_2 nanoparticles have attracted the interest of scientists as photocatalysts with their superior redox ability, which has made them a top candidate in many applications, including the decomposition of water pollutants such as antibiotic residues [164]. Recently, the photocatalytic oxidation activity of TiO_2 was evaluated for the photodegradation of various antibiotics, including tetracycline, β-lactam, quinolone, and sulfamethoxazole. The findings showed high photodegradation capacities, higher than 90% [165]. Besides, Elmolla and colleagues reported photodegradation of the antibiotic β-lactam, the main active ingredient in the drugs Amoxicillin, Ampicillin, and Cloxacillin, using TiO_2 under UV irradiation and in the presence of H_2O_2 [166]. The authors indicated that complete degradation can be achieved after 0.5 h of reaction. Furthermore, levofloxacin and chloramphenicol have been the subject uu+of several research reports on photocatalytic degradation as they can readily appear in discharged wastewater. For instance, Gan et al. studied the use of different crystallographic states of TiO_2, calcined, hydrated, or a mixture thereof, in the photodegradation of levofloxacin [167]. The results indicated that calcined titanium oxide shows the highest total organic carbon removal rate. Moreover, the latter showed a strong photocatalytic degradation of the antibiotic chloramphenicol, with a quantitative degradation after 4 h of irradiation. Likewise, silver nanoparticles have been broadly applied in the photocatalytic decomposition of several antibiotics, such as ampicillin and ciprofloxacin. In this framework, the S. Kumar group reported the use of silver nanoparticles, prepared from white rot fungi, for the photodegradation of ampicillin [168]. Under natural solar irradiation, the photodegradation of ampicillin in water was assessed. The results showed that the maximum degradation of ampicillin, of 96.5%, was obtained when the solution was exposed for 4 hours. Interestingly, J. Osajima et al. described the use of a silver-zinc oxide nanocomposite for the photodegradation of the antibiotic ciprofloxacin in water [169]. The results revealed that the corresponding nanocomposite induced high capacity under visible light irradiation, showing about 90% antibiotic degradation efficiency after 120 min of reaction. Therefore, they suggested that the designed photocatalytic system may be an effective method to eliminate the target antibiotic. Recently, P. Krysinsky et al. described the use of iron oxide nanoparticles for the adsorption and photodegradation of tetracycline [170]. The results showed that after being exposed to UV-Vis light for 60 minutes, roughly 40% of the tetracycline was destroyed. The authors demonstrated that hydroxyl radicals produced during irradiation caused the photodegradation of tetracycline to proceed according to a pseudo-first-order mechanism. They also demonstrated that photogenerated hydrogen peroxide can cause heterogeneous photo-Fenton processes on the surface of iron oxide nanoparticles, which further produce hydroxyl and hydroperoxyl radicals and aid in the photodegradation of tetracycline.

More interestingly, A. Dargahi et al. evaluated the ability of copper oxide nanoparticles to photodegrade metronidazole, identified as a remarkable antibiotic resistant to biological degradation [171]. According to the authors, the photocatalytic process was activated by the addition of hydrogen peroxide under UV irradiation. The outcomes indicated that the metronidazole removal efficiencies, chemical oxygen demand, and total organic carbon, respectively, were 98.36%, 73.0%, and 56.52% under the optimal experimental conditions. The system UV/H_2O_2/copper oxide nanoparticles approach is suggested in this research as a potential process to decompose the antibiotic metronidazole and increase its biodegradability. As the last example, magnesium ferrite nanoparticles have been applied to photocatalyze the degradation of tetracycline antibiotics as a model under solar light irradiation [172]. The findings demonstrated that the studied pesticide may be readily latched onto magnesium ferrite nanoparticles, serving as a potential adsorbent for antibiotic extraction from wastewater, with over 75% tetracycline decomposition achieved in just 2 hours. According to the authors, the functionalization of the surface of the unmodified magnesium ferrite nanoparticles considerably enhanced the photocatalytic activity of the system, where an exchange between capped agent and antibiotic was observed.

11.4 CONCLUSIONS

Photocatalytic degradation of water contaminants and their removal is a promising technology in wastewater remediation due to its exceptional degradation advantages over pollutants compared to traditional water treatment techniques. Indeed, these two processes are able to remove a broad range of pollutants from water, including heavy metals, pesticides, antibiotics, and dyes. This chapter summarizes advanced studies on the development of wastewater treatment techniques using adsorbents and photocatalysts based on metallic nanoparticles. Numerous approaches have been explained in depth in this chapter, such as the development of nanoadsorbents and nanophotocatalysts and the engineering of adsorption and photocatalysis. In the studies presented, several metallic nanomaterials were used, namely nanoparticles of titania, silver, gold, copper, and so on. Interestingly, the presented studies have shown that titanium oxide nanomaterials are widely used in wastewater systems. This is highly anticipated because of their unique electronic and optical properties. Despite the fact that comprehensible studies on the development of water remediation systems based on metallic nanoparticles have been carried out, many aspects of the photodegradation process have not yet been well elucidated. Besides, further research on the degradation of the real constituents of wastewater is needed to better understand the applications of the process. Therefore, efforts should be made to improve the sustainability, efficiency, and applicability of the wastewater treatment process.

ACKNOWLEDGMENTS

The authors are grateful to thank the Tunisian Ministry of High Education and Scientific Research and the Tunisian research project "Plateformes d'analyses chimiques portables à base des nanomatériaux pour la surveillance sur site de la qualité de l'eau PRECISE" VRR 38/21 for financial support of this work.

CONFLICT OF INTEREST

The authors declare that there is no conflict of interest.

REFERENCES

1. Schaider LA, Rudel RA, Ackerman JM, Dunagan SC, Brody JG. Pharmaceuticals, perfluorosurfactants, and other organic wastewater compounds in public drinking water wells in a shallow sand and gravel aquifer. Science of the Total Environment. 2014 Jan 15;468: 384–93.
2. Brunetti A, Pomilla FR, Marcì G, Garcia-Lopez EI, Fontananova E, Palmisano L, Barbieri G. CO_2 reduction by C_3N_4-TiO_2 Nafion photocatalytic membrane reactor as a promising environmental pathway to solar fuels. Applied Catalysis B: Environmental. 2019 Oct 15;255: 117779.
3. Vilların MC, Merel S. Assessment of current challenges and paradigm shifts in wastewater management. Journal of Hazardous Materials. 2020;390:122139.
4. Alventosa-deLara E, Barredo-Damas S, Alcaina-Miranda MI, Iborra-Clar MI. Ultrafiltration technology with a ceramic membrane for reactive dye removal: Optimization of membrane performance. Journal of Hazardous Materials. 2012 Mar 30;209:492–500.
5. Javed MN, Dahiya ES, Ibrahim AM, Alam M, Khan FA, Pottoo FH. Recent advancement in clinical application of nanotechnological approached targeted delivery of herbal drugs. In Nanophytomedicine 2020 (pp. 151–72). Springer, Singapore.
6. Mishra S, Sharma S, Javed MN, Pottoo FH, Barkat MA, Alam MS, Amir M, Sarafroz M. Bioinspired nanocomposites: Applications in disease diagnosis and treatment. Pharmaceutical Nanotechnology. 2019 Jun 1;7(3):206–19.
7. Javed MN, Dahiya ES, Ibrahim AM, Alam M, Khan FA, Pottoo FH. Recent advancement in clinical application of nanotechnological approached targeted delivery of herbal drugs. In Nanophytomedicine 2020 (pp. 151–72). Springer, Singapore.
8. Aslam M, Javed M, Deeb HH, Nicola MK, Mirza M, Alam M, Akhtar M, Waziri A. Lipid nanocarriers for neurotherapeutics: Introduction, challenges, blood-brain barrier, and promises of delivery approaches. CNS & Neurological Disorders-Drug Targets (Formerly Current Drug Targets-CNS & Neurological Disorders). 2022;21(10):952–65.
9. Javed MN, Pottoo FH, Shamim A, Hasnain MS, Alam MS. Design of experiments for the development of nanoparticles, nanomaterials, and nanocomposites. In Design of Experiments for Pharmaceutical Product Development 2021 (pp. 151–69). Springer, Singapore.
10. Javed MN, Alam MS, Waziri A, Pottoo FH, Yadav AK, Hasnain MS, Almalki FA. QbD applications for the development of nano pharmaceutical products. In Pharmaceutical Quality by Design 2019 Jan 1 (pp. 229–53). Academic Press. https://doi.org/10.1016/B978-0-12-815799-2.00013-7.
11. Kumar R, Dhamija G, Ansari JR, Javed MN, Alam MS. C-dot nanoparticulated devices for biomedical applications. In Nanotechnology 2022 (pp. 271–99). CRC Press. Boca Raton, FL.
12. Bharti C, Alam MS, Javed MN, Khalid M, Saifullah FA, Manchanda R. Silica based nanomaterial for drug delivery. Nanomaterials: Evolution and Advancement Towards Therapeutic Drug Delivery (Part II). 2021 Jun 2:57.
13. Sangeet Kumar Mall SK, Yadav T, Waziri A, Alam MS. Treatment opportunities with Fernandoa adenophylla and recent novel approaches for natural medicinal phytochemicals as a drug delivery system. Exploration of Medicine. 2022;3:516–39.
14. Naseh MF, Ansari JR, Alam MS, Javed MN. Sustainable nanotorus for biosensing and therapeutical applications. In Handbook of Green and Sustainable Nanotechnology: Fundamentals, Developments and Applications 2022 Aug 19 (pp. 1–21). Springer International Publishing, Cham.

15. Sunilbhai CA, Alam M, Sadasivuni KK, Ansari JR. SPR assisted diabetes detection. In Advanced Bioscience and Biosystems for Detection and Management of Diabetes 2022 (pp. 91–131). Springer, Cham.
16. Singhal S, Gupta M, Alam MS, Javed MN, Ansari JR. Carbon allotropes-based nanodevices: Graphene in biomedical applications. In Nanotechnology 2022 (pp. 241–69). CRC Press. Boca Raton.
17. Pottoo FH, Tabassum N, Javed M, Nigar S, Rasheed R, Khan A, Barkat M, Alam M, Maqbool A, Ansari MA, Barreto GE. The synergistic effect of raloxifene, fluoxetine, and bromocriptine protects against pilocarpine-induced status epilepticus and temporal lobe epilepsy. Molecular Neurobiology. 2019 Feb;56(2):1233–47.
18. Pottoo FH, Sharma S, Javed MN, Barkat MA, Harshita, Alam MS, Naim MJ, Alam O, Ansari MA, Barreto GE, Ashraf GM. Lipid-based nanoformulations in the treatment of neurological disorders. Drug Metabolism Reviews. 2020 Jan 2;52(1):185–204.
19. Pottoo FH, Javed M, Barkat M, Alam M, Nowshehri JA, Alshayban DM, Ansari MA. Estrogen and serotonin: Complexity of interactions and implications for epileptic seizures and epileptogenesis. Current Neuropharmacology. 2019 Mar 1;17(3):214–31.
20. Pottoo FH, Tabassum N, Javed MN, Nigar S, Sharma S, Barkat MA, Alam MS, Ansari MA, Barreto GE, Ashraf GM. Raloxifene potentiates the effect of fluoxetine against maximal electroshock induced seizures in mice. European Journal of Pharmaceutical Sciences. 2020 Apr 15;146:105261.
21. Waziri A, Bharti C, Aslam M, Jamil P, Mirza M, Javed MN, Pottoo U, Ahmadi A, Alam MS. Probiotics for the chemoprotective role against the toxic effect of cancer chemotherapy. Anti-Cancer Agents in Medicinal Chemistry (Formerly Current Medicinal Chemistry-Anti-Cancer Agents). 2022 Feb 1;22(4):654–67.
22. Javed MN, Akhter MH, Taleuzzaman M, Faiyazudin M, Alam MS. Cationic nanoparticles for treatment of neurological diseases. In Fundamentals of Bionanomaterials 2022 Jan 1 (pp. 273–92). Elsevier. https://doi.org/10.1016/B978-0-12-824147-9.00010-8.
23. Kumari N, Daram N, Alam MS, Verma AK. Rationalizing the use of polyphenol nano-formulations in the therapy of neurodegenerative diseases. CNS & Neurological Disorders-Drug Targets (Formerly Current Drug Targets-CNS & Neurological Disorders). 2022 Dec 1;21(10): 966–76.
24. Raj S, Manchanda R, Bhandari M, Alam M. Review on natural bioactive products as radioprotective therapeutics: Present and past perspective. Current Pharmaceutical Biotechnology. 2022;23(14):1721–38.
25. Ibrahim AM, Chauhan L, Bhardwaj A, Sharma A, Fayaz F, Kumar B, Alhashmi M, AlHajri N, Alam MS, Pottoo FH. Brain-derived neurotropic factor in neurodegenerative disorders. Biomedicines. 2022 May;10(5):1143.
26. Alam MS, Garg A, Pottoo FH, Saifullah MK, Tareq AI, Manzoor O, Mohsin M, Javed MN. Gum ghatti mediated, one pot green synthesis of optimized gold nanoparticles: Investigation of process-variables impact using Box-Behnken based statistical design. International Journal of Biological Macromolecules. 2017 Nov 1;104:758–67.
27. Alam MS, Javed MN, Pottoo FH, Waziri A, Almalki FA, Hasnain MS, Garg A, Saifullah MK. QbD approached comparison of reaction mechanism in microwave synthesized gold nanoparticles and their superior catalytic role against hazardous nirto-dye. Applied Organometallic Chemistry. 2019 Sep;33(9):e5071.
28. Pandit J, Alam MS, Ansari JR, Singhal M, Gupta N, Waziri A, Sharma K, Potto FH. Multifaced applications of nanoparticles in biological science. In Nanomaterials in the Battle Against Pathogens and Disease Vectors 2022 (pp. 17–50). CRC Press. Boca Raton, FL.
29. Alam MS, Naseh MF, Ansari JR, Waziri A, Javed MN, Ahmadi A, Saifullah MK, Garg A. Synthesis approaches for higher yields of nanoparticles. In Nanomaterials in the Battle Against Pathogens and Disease Vectors 2022 (pp. 51–82). CRC Press. Boca Raton, FL.

30. Javed MN, Pottoo FH, Alam MS. Metallic nanoparticle alone and/or in combination as novel agent for the treatment of uncontrolled electric conductance related disorders and/ or seizure, epilepsy & convulsions. Patent Acquired on October. 2016;10:40.
31. Hasnain MS, Javed MN, Alam MS, Rishishwar P, Rishishwar S, Ali S, Nayak AK, Beg S. Purple heart plant leaves extract-mediated silver nanoparticle synthesis: Optimization by Box-Behnken design. Materials Science and Engineering: C. 2019 Jun 1;99:1105–14.
32. Chai WS, Cheun JY, Kumar PS, Mubashir M, Majeed Z, Banat F, Ho SH, Show PL. A review on conventional and novel materials towards heavy metal adsorption in wastewater treatment application. Journal of Cleaner Production. 2021 May 10;296:126589.
33. Kunduru KR, Nazarkovsky M, Farah S, Pawar RP, Basu A, Domb AJ. Nanotechnology for water purification: Applications of nanotechnology methods in wastewater treatment. Water Purification. 2017 Jan 1:33–74.
34. Oskam G. Metal oxide nanoparticles: Synthesis, characterization and application. Journal of Sol-Gel Science and Technology. 2006 Mar;37(3):161–4.
35. Saleem H, Zaidi SJ. Developments in the application of nanomaterials for water treatment and their impact on the environment. Nanomaterials. 2020 Sep 7;10(9):1764.
36. Farid YH. Adsorption of Water Contaminates by $MnFe_2O_4$ Nano Particles. International Research Journal of Pure and Applied Chemistry. 2021 Jul 10;22(5):36–53.
37. Rashed MN. Adsorption technique for the removal of organic pollutants from water and wastewater. Organic Pollutants-Monitoring, Risk and Treatment. 2013 Jan 30;7:167–94.
38. Zamora-Ledezma C, Negrete-Bolagay D, Figueroa F, Zamora-Ledezma E, Ni M, Alexis F, Guerrero VH. Heavy metal water pollution: A fresh look about hazards, novel and conventional remediation methods. Environmental Technology & Innovation. 2021 May 1;22:101504.
39. Alisawi HA. Performance of wastewater treatment during variable temperature. Applied Water Science. 2020 Apr;10(4):1–6.
40. Palani G, Arputhalatha A, Kannan K, Lakkaboyana SK, Hanafiah MM, Kumar V, Marella RK. Current trends in the application of nanomaterials for the removal of pollutants from industrial wastewater treatment—a review. Molecules. 2021 May 10;26(9):2799.
41. Dugyala VR, Muthukuru JS, Mani E, Basavaraj MG. Role of electrostatic interactions in the adsorption kinetics of nanoparticles at fluid–fluid interfaces. Physical Chemistry Chemical Physics. 2016;18(7):5499–508.
42. Jamkhande PG, Ghule NW, Bamer AH, Kalaskar MG. Metal nanoparticles synthesis: An overview on methods of preparation, advantages and disadvantages, and applications. Journal of Drug Delivery Science and Technology. 2019 Oct 1;53:101174.
43. Samyn P, Barhoum A, Ohlund T, Dufresne A. Nanoparticels and Nanostructures Materials in Papermaking. Journal of Materials Science. 2018;53:146–84.
44. Lellis B, Fávaro-Polonio CZ, Pamphile JA, Polonio JC. Effects of textile dyes on health and the environment and bioremediation potential of living organisms. Biotechnology Research and Innovation. 2019;3:275–90.
45. Dinh QT, Moreau-Guigon E, Labadie P, Alliot F, Teil MJ, Blanchard M, Chevreuil M. Occurrence of antibiotics in rural catchments. Chemosphere. 2017 Feb 1;168:483–90.
46. Góralczyk-Bińkowska A, Długoński A, Bernat P, Długoński J, Jasińska A. Environmental and molecular approach to dye industry waste degradation by the ascomycete fungus Nectriella pironii. Scientific Reports. 2021 Dec 13;11(1):1–3.
47. Kerkez D, Bečelić-Tomin M, Gvoić V, Dalmacija B. Metal nanoparticles in dye wastewater treatment–smart solution for clear water. Recent Patents on Nanotechnology. 2021 Sep 1;15(3):270–94.
48. Upham P, Smith B. Using the Rapid Impact Assessment Matrix to synthesize biofuel and bioenergy impact assessment results: The example of medium scale bioenergy heat options. Journal of Cleaner Production. 2014 Feb 15;65:261–9.
49. Pan F, Yu Y, Xu A, Xia D. Application of magnetic OMS-2 in sequencing batch. Process Safety and Environmental Protection. 2015;95:255–64.

50. Khan FS, Mubarak NM, Tan YH, Karri RR, Khalid M, Walvekar R, Abdullah EC, Mazari SA, Nizamuddin S. Magnetic nanoparticles incorporation into different substrates for dyes and heavy metals removal—A review. Environmental Science and Pollution Research. 2020 Dec;27(35):43526–41.
51. Mashkoor F, Nasar A. Carbon nanotube-based adsorbents for the removal of dyes from waters: A review. Environmental Chemistry Letters. 2020 May;18(3):605–29.
52. Absalan G, Asadi M, Kamran S, Sheikhian L, Goltz DM. Removal of reactive red-120 and 4-(2-pyridylazo) resorcinol from aqueous samples by Fe_3O_4 magnetic nanoparticles using ionic liquid as modifier. Journal of Hazardous Materials. 2011 Aug 30;192(2):476–84.
53. Xu YY, Zhou M, Geng HJ, Hao JJ, Ou QQ, Qi SD, Chen HL, Chen XG. A simplified method for synthesis of Fe_3O_4@ PAA nanoparticles and its application for the removal of basic dyes. Applied Surface Science. 2012 Feb 1;258(8):3897–902.
54. Zhang J, Li B, Yang W, Liu J. Synthesis of magnetic Fe_3O_4@ hierarchical hollow silica nanospheres for efficient removal of methylene blue from aqueous solutions. Industrial & Engineering Chemistry Research. 2014 Jul 2;53(26):10629–36.
55. Becker A, Kirchberg K, Marschall R. Magnesium ferrite (MgFe2O4) nanoparticles for photocatalytic antibiotics degradation. Zeitschrift für Physikalische Chemie. 2020 Apr 1;234(4):645–54.
56. Umar A, Khan MS, Alam S, Zekker I, Burlakovs J, dC Rubin SS, Bhowmick GD, Kallistova A, Pimenov N, Zahoor M. Synthesis and characterization of Pd-Ni bimetallic nanoparticles as efficient adsorbent for the removal of acid orange 8 present in wastewater. Water. 2021 Apr 15;13(8):1095.
57. Smith SC, Rodrigues DF. Carbon-based nanomaterials for removal of chemical and biological contaminants from water: A review of mechanisms and applications. Carbon. 2015 Sep 1;91:122–43.
58. Naeem H, Ajmal M, Muntha S, Ambreen J, Siddiq M. Synthesis and characterization of graphene oxide sheets integrated with gold nanoparticles and their applications to adsorptive removal and catalytic reduction of water contaminants. RSC Advances. 2018;8(7):3599–610.
59. Vardhan KH, Kumar PS, Panda RC. A review on heavy metal pollution, toxicity and remedial measures: Current trends and future perspectives. Journal of Molecular Liquids. 2019 Sep 15;290:111197.
60. Kicińska A, Wikar J. Ecological risk associated with agricultural production in soils contaminated by the activities of the metal ore mining and processing industry-example from southern Poland. Soil and Tillage Research. 2021 Jan 1;205:104817.
61. Rottstock T, Göttert T, Zeller U. Relatively undisturbed African savannas-an important reference for assessing wildlife responses to livestock grazing systems in European rangelands. Global Ecology and Conservation. 2020 Sep 1;23: e01124.
62. García-Sánchez A, Contreras F, Adams M, Santos F. Mercury contamination of surface water and fish in a gold mining region (Cuyuni river basin, Venezuela). International Journal of Environment and Pollution. 2008 Jan 1;33(2–3):260–74.
63. Bolisetty S, Peydayesh M, Mezzenga R. Sustainable technologies for water purification from heavy metals: Review and analysis. Chemical Society Reviews. 2019;48(2):463–87.
64. Meenambigai P, Vijayaraghavan R, Gowri RS, Rajarajeswari P, Prabhavathi P. Biodegradation of heavy metals-A review. International Journal of Current Microbiology and Applied Sciences. 2016;5(4):375–83.
65. Cabral Pinto MM, Marinho-Reis P, Almeida A, Pinto E, Neves O, Inácio M, Gerardo B, Freitas S, Simões MR, Dinis PA, Diniz L. Links between cognitive status and trace element levels in hair for an environmentally exposed population: A case study in the surroundings of the estarreja industrial area. International Journal of Environmental Research and Public Health. 2019 Nov;16(22):4560.

66. Abdel-Raouf MS, Abdul-Raheim AR. Removal of heavy metals from industrial waste water by biomass-based materials: A review. Journal of Pollution Effects & Control. 2017;5(1):1–3.
67. Samuel MS, Datta S, Chandrasekar N, Balaji R, Selvarajan E, Vuppala S. Biogenic synthesis of iron oxide nanoparticles using enterococcus faecalis: Adsorption of hexavalent chromium from aqueous solution and in vitro cytotoxicity analysis. Nanomaterials. 2021 Dec 3;11(12):3290.
68. Lisha KP, Pradeep T. Towards a practical solution for removing inorganic mercury from drinking water using gold nanoparticles. Gold Bulletin. 2009 Jun;42(2):144–52.
69. Ojea-Jiménez I, López X, Arbiol J, Puntes V. Citrate-coated gold nanoparticles as smart scavengers for mercury (II) removal from polluted waters. ACS Nano. 2012 Mar 27;6(3):2253–60.
70. Sumesh E, Bootharaju MS, Pradeep T. A practical silver nanoparticle-based adsorbent for the removal of Hg^{2+} from water. Journal of Hazardous Materials. 2011 May 15;189(1–2):450–7.
71. Suhasini R, Thiagarajan V. Magnetic nanomaterials for wastewater remediation. In Nanomaterials for Water Treatment and Remediation 2021 Dec 28 (pp. 279–308). CRC Press. Boca Raton, FL.
72. Seyedi SM, Rabiee H, Shahabadi SM, Borghei SM. Synthesis of zero-valent iron nanoparticles via electrical wire explosion for efficient removal of heavy metals. CLEAN–Soil, Air, Water. 2017 Mar;45(3):1600139.
73. O'Carroll D, Sleep B, Krol M, Boparai H, Kocur C. Nanoscale zero valent iron and bimetallic particles for contaminated site remediation. Advances in Water Resources. 2013 Jan 1;51:104–22.
74. Zhang Z, Hao ZW, Liu WL, Xu XH. Synchronous treatment of heavy metal ions and nitrate by zero-valent iron. Huan jing ke xue= Huanjing kexue. 2009 Mar 1;30(3):775–9.
75. Huang DL, Chen GM, Zeng GM, Xu P, Yan M, Lai C, Zhang C, Li NJ, Cheng M, He XX, He Y. Synthesis and application of modified zero-valent iron nanoparticles for removal of hexavalent chromium from wastewater. Water, Air, & Soil Pollution. 2015 Nov;226(11):1–4.
76. Zarime NA, Yaacob WZ, Jamil H. Removal of heavy metals using bentonite supported nano-zero valent iron particles. In AIP Conference Proceedings 2018 Apr 4 (Vol. 1940, No. 1, p. 020029). AIP Publishing LLC. Selangor, Malaysia.
77. Daija L, Selberg A, Rikmann E, Zekker I, Tenno T, Tenno T. The influence of lower temperature, influent fluctuations and long retention time on the performance of an upflow mode laboratory-scale septic tank. Desalination and Water Treatment. 2016 Aug 26;57(40):18679–87.
78. Masudi A, Harimisa GE, Ghafar NA, Jusoh NW. Magnetite-based catalysts for wastewater treatment. Environmental Science and Pollution Research. 2020 Feb;27(5):4664–82.
79. Giraldo L, Erto A, Moreno-Piraján JC. Magnetite nanoparticles for removal of heavy metals from aqueous solutions: Synthesis and characterization. Adsorption. 2013 Apr;19(2):465–74.
80. Kumar KY, Muralidhara HB, Nayaka YA, Balasubramanyam J, Hanumanthappa H. Low-cost synthesis of metal oxide nanoparticles and their application in adsorption of commercial dye and heavy metal ion in aqueous solution. Powder Technology. 2013 Sep 1;246:125–36.
81. Somu P, Paul S. Casein based biogenic-synthesized zinc oxide nanoparticles simultaneously decontaminate heavy metals, dyes, and pathogenic microbes: A rational strategy for wastewater treatment. Journal of Chemical Technology & Biotechnology. 2018 Oct;93(10):2962–76.
82. Sharma A, Kumar V, Shahzad B, Tanveer M, Sidhu GP, Handa N, Kohli SK, Yadav P, Bali AS, Parihar RD, Dar OI. Worldwide pesticide usage and its impacts on ecosystem. SN Applied Sciences. 2019;1(11):1446.

83. Senthil Kumar P, Femina Carolin C, Varjani SJ. Pesticides bioremediation. In Bioremediation: Applications for Environmental Protection and Management 2018 (pp. 197–222). Springer, Singapore.
84. Rajmohan KS, Chandrasekaran R, Varjani S. A review on occurrence of pesticides in environment and current technologies for their remediation and management. Indian Journal of Microbiology. 2020 Jun;60(2):125–38.
85. Gacem MA, Telli A, Khelil AO. Nanomaterials for detection, degradation, and adsorption of pesticides from water and wastewater. In Aquananotechnology 2021 Jan 1 (pp. 325–46). Elsevier. https://doi.org/10.1016/B978-0-12-821141-0.00003-3.
86. Ghosh N, Das S, Biswas G, Haldar PK. Review on some metal oxide nanoparticles as effective adsorbent in wastewater treatment. Water Science and Technology. 2022;85 (12): 3370–95.
87. Lan J, Cheng Y, Zhao Z. Effective organochlorine pesticides removal from aqueous systems by magnetic nanospheres coated with polystyrene. Journal of Wuhan University of Technology-Mater Science Education. 2014 Feb;29(1):168–73.
88. Dehaghi SM, Rahmanifar B, Moradi AM, Azar PA. Removal of permethrin pesticide from water by chitosan–zinc oxide nanoparticles composite as an adsorbent. Journal of Saudi Chemical Society. 2014 Sep 1;18(4):348–55.
89. Mitchell MB, Sheinker VN, Cox WW, Gatimu EN, Tesfamichael AB. The room temperature decomposition mechanism of dimethyl methylphosphonate (DMMP) on alumina-supported cerium oxide– participation of nano-sized cerium oxide domains. The Journal of Physical Chemistry B. 2004 Feb 5;108 (5):1634–45.
90. Momić T, Pašti TL, Bogdanović U, Vodnik V, Mraković A, Rakočević Z, Pavlović VB, Vasić V. Adsorption of organophosphate pesticide dimethoate on gold nanospheres and nanorods. Journal of Nanomaterials. 2016 Dec 1;2016.
91. Moustafa M, Abu-Saied MA, Taha T, Elnouby M, El-Shafeey M, Alshehri AG, Alamri S, Shati A, Alrumman S, Alghamdii H, Al-Khatani M. Chitosan functionalized AgNPs for efficient removal of imidacloprid pesticide through a pressure-free design. International Journal of Biological Macromolecules. 2021 Jan 31;168:116–23.
92. Palani G, Arputhalatha A, Kannan K, Lakkaboyana SK, Hanafiah MM, Kumar V, Marella RK. Current trends in the application of nanomaterials for the removal of pollutants from industrial wastewater treatment—A review. Molecules. 2021 May 10;26(9):2799.
93. Rawat D, Mishra V, Sharma RS. Detoxification of azo dyes in the context of environmental processes. Chemosphere. 2016 Jul 1;155:591–605.
94. Maynard AD, Aitken RJ, Butz T, Colvin V, Donaldson K, Oberdörster G, Philbert MA, Ryan J, Seaton A, Stone V, Tinkle SS. Safe handling of nanotechnology. Nature. 2006 Nov;444(7117):267–9.
95. Dutta T, Kim KH, Deep A, Szulejko JE, Vellingiri K, Kumar S, Kwon EE, Yun ST. Recovery of nanomaterials from battery and electronic wastes: A new paradigm of environmental waste management. Renewable and Sustainable Energy Reviews. 2018 Feb 1;82:3694–704.
96. Nancharaiah YV, Mohan SV, Lens PN. Biological and bioelectrochemical recovery of critical and scarce metals. Trends in Biotechnology. 2016 Feb 1;34(2):137–55.
97. Ahmed J, Thakur A, Goyal A. Industrial wastewater and its toxic effects. Chemistry In the Environment: Biological Treatment Of Industrial Wastewater 2021 (pp. 1–14). https://doi.org/10.1039/9781839165399-00001.
98. Guo Y, Qi PS, Liu YZ. A review on advanced treatment of pharmaceutical wastewater. In IOP Conference Series: Earth and Environmental Science 2017 May 1 (Vol. 63, No. 1, p. 012025). IOP Publishing. Suzhou.
99. Ahmed SN, Haider W. Heterogeneous photocatalysis and its potential applications in water and wastewater treatment: A review. Nanotechnology. 2018 Jun 13;29(34):342001.

100. Stolarczyk HS. JK Photocatalytic reduction of CO 2 on TiO 2 and other semiconductors. Angewandte Chemie International Edition. 2013;52(29):7372–408.
101. Sapawe N, Jalil AA, Triwahyono S. One-pot electro-synthesis of ZrO_2–ZnO/HY nanocomposite for photocatalytic decolorization of various dye-contaminants. Chemical Engineering Journal. 2013 Jun 1;225:254–65.
102. Chincholi M, Sagwekar P, Nagaria C, Kulkarni S, Dhokpande S. International Journal of Scientific & Technology Research. 2014;3:835e840.
103. Yu J, Han Y, Li Y, Gao P, Li W. Mechanism and kinetics of the reduction of hematite to magnetite with CO–CO2 in a micro-fluidized bed. Minerals. 2017 Nov 1;7(11):209.
104. Murphy M, Ting K, Zhang X, Soo C, Zheng Z. Current development of silver nanoparticle preparation, investigation, and application in the field of medicine. Journal of Nanomaterials. 2015 Oct; 5–5. https://doi.org/10.1155/2015/696918.
105. Zhou H, Zhou J, Wang T, Zeng J, Liu L, Jian J, Zhou Z, Zeng L, Liu Q, Liu G. In-situ preparation of silver salts/collagen fiber hybrid composites and their photocatalytic and antibacterial activities. Journal of Hazardous Materials. 2018 Oct 5;359:274–80.
106. Lee SY, Kang D, Jeong S, Do HT, Kim JH. Photocatalytic degradation of rhodamine B dye by TiO_2 and gold nanoparticles supported on a floating porous polydimethylsiloxane sponge under ultraviolet and visible light irradiation. ACS Omega. 2020 Feb 18;5(8):4233–41.
107. Lam SM, Sin JC, Abdullah AZ, Mohamed AR. Degradation of wastewaters containing organic dyes photocatalysed by zinc oxide: A review. Desalination and Water Treatment. 2012 Mar 1;41(1–3):131–69.
108. Blažeka D, Car J, Klobučar N, Jurov A, Zavašnik J, Jagodar A, Kovačević E, Krstulović N. Photodegradation of methylene blue and rhodamine B using laser-synthesized ZnO nanoparticles. Materials. 2020 Sep 30;13(19):4357.
109. Shah E, Vaghasiya JV, Soni SS, Panchal CJ, Suryavanshi PS, Chavda M, Soni HP. Ni doped ZnS nanoparticles as photocatalyst: Can mixed phase be optimized for better performance. Journal of Environmental Chemical Engineering. 2016;4(4) Part A:2213–3437.
110. Longxing Hu, Guihua D, Wencong L, Siwei P, Xing H. Deposition of CdS nanoparticles on MIL-53(Fe) metal-organic framework with enhanced photocatalytic degradation of RhB under visible light irradiation. Applied Surface Science. 2017;2017(410):401–13, 0169–4332.
111. Subramanyam K, Sreelekha N, Amaranatha Reddy, D, Murali G, Rahul Varma K, Vijayalakshmi RP. Chemical synthesis, structural, optical, magnetic characteristics and enhanced visible light active photocatalysis of Ni doped CuS nanoparticles. Solid State Sciences. 2017;(65):68–78, 1293–2558.
112. Sudheer KY, Jeevanandam P. Synthesis of Ag_2S–TiO_2 nanocomposites and their catalytic activity towards rhodamine B photodegradation. Journal of Alloys and Compounds. 2015;649:483–90, ISSN 0925–8388.
113. Arik K, Sumanta S, David R, Knappet BR, Pradhan SK, Wheatley AEH. Facile synthesis of SnO_2–PbS nanocomposites with controlled structure for applications in photocatalysis. Nanoscale. 2016;(8):2727–39.
114. Ch. Venkata R, Jaesool S, Migyung C. Synthesis, structural, optical and photocatalytic properties of CdS/ZnS core/shell nanoparticles. Journal of Physics and Chemistry of Solids. 2017;2017(103): 209–17, 0022–3697.
115. Panpan Z, Yu X, Jing F, Yun L, Changling Y, Xiaoming L, Yuhua D, Yuancheng Q, Dan Z. CdS quantum dots confined in mesoporous TiO2 with exceptional photocatalytic performance for degradation of organic polutants. Chemosphere. 2017;2017(178):1–10, 0045–6535.

116. Deng X, Wang C, Yang H, et al. One-pot hydrothermal synthesis of CdS decorated CuS microflower-like structures for enhanced photocatalytic properties. Scientific Reports. 2017(1);7:3877 (1–12). DOI:10.1038/s41598-017-04270-y 1.
117. Jun T, Tingjiang Y, Zheng Q, Linlin W, Wenjuan L, Jinmao Y, Baibiao H. Anion-exchange synthesis of Ag2S/Ag3PO4 core/shell composites with enhanced visible and NIR light photocatalytic performance and the photocatalytic mechanisms. Applied Catalysis B: Environmental. 2017;2017(209):566–78, 092–3373.
118. Alec L, and Shenqiang R. Thermodynamic control of iron pyrite nanocrystal synthesis with high photoactivity and stability. Journal of Materials Chemistry A. 2013(1):49–54.
119. Kaur B, Singh K, Malik AK. Effect of ligands on crystallography, morphology and photo-catalytic ability of ZnS nanostructures. Dyes and Pigments. 2017; 142:153–160. https://doi.org/10.1016/j.dyepig.2017.03.013.
120. Prasannalakshmi P, Shanmugam N. Photocatalytic decolourization of brilliant green and methylene blue by TiO2/CdS nanorods. Journal of Solid State Electrochemistry. 2017;(21):1751–66. https://doi.org/10.1007/s10008-017-3522-6.
121. Qiu XP, Yu JS, Xu HM, Chen WX, Hu W, Bai HY, Chen GL. Interfacial effect of the nanostructured Ag_2S/Co_3O_4 and its catalytic mechanism for the dye photodegradation under visible light. Applied Surface Science. 2016;362:498–505, 0169–4332.
122. Lellala K, Namratha K, Byrappa K, Sol-gel assisted hydrothermal synthesis and characterization of hybrid ZnS-RGO nanocomposite for efficient photodegradation of dyes. Journal of Alloys and Compounds. 2017;695:799–809, 0925–8388.
123. Pourahmad A, Sohrabnezhad S, Radaee E. Degradation of basic blue 9 dye by CoS/nanoAlMCM-41 catalyst under visible light irradiation. Journal of Porous Materials. 2017;17:367–75.
124. Luo M, Liu Y, Hu J, Li J, Liu J, Richards RM. General strategy for one-pot synthesis of metal sulfide hollow spheres with enhanced photocatalytic activity. Applied Catalysis B: Environmental. 2012;125:180–88, 0926–3373.
125. Khaparde R, Acharya S. Effect of isovalent dopants on photodegradation ability of ZnS nanoparticles. Spectrochimica Acta Part A: Molecular and Biomolecular Spectroscopy. 2016;163:49–57, 1386–1425.
126. Liu Y, Xin G, Guang-Sheng W, Xiao-Hui G, Bo J. Synthesis of ZnS/CuS nanospheres loaded on reduced graphene oxide as high-performance photocatalysts under simulated sunlight irradiation. New Journal of Chemistry. 2017;41(13):5732–44. https://doi.org/10.1039/C7NJ00801E.
128. Shanshan F, Su X, Can W, Guanqiu W, Xi W, Qian L, Zhongyu L, Song X. Fabrication and characterization of $CdS/BiVO_4$ nanocomposites with efficient visible light driven photocatalytic activities. Ceramics International. 2016;3(42):4421–28, 0272–8842.
129. He H-Y. Facile synthesis of ultrafine CuS nanocrystalline/TiO_2: Fe nanotubes hybrids and their photocatalytic and Fenton-like photocatalytic activities in the dye degradation. Microporous and Mesoporous Materials. 2016;(227):31–8, 1387-1811.
130. Dasari A, Guttena V. Hydrothermally generated and highly efficient sunlight responsive SiO_2 and TiO_2 capped Ag_2S nanocomposites for photocatalytic degradation of organic dyes. Journal of Environmental Chemical Engineering. 2018;1(6):311–324, 2213–3437.
131. Dasari A, Venkatesham M, Santoshi kumari A., Bhagavanth Reddy G, Ramakrishna D, Veerabhadram G. Photocatalytic degradation of dye pollutants under solar, visible and UV lights using green synthesised CuS nanoparticles, Journal of Experimental Nanoscience. 2016;6(11):418–32.
132. Dasari A, Guttena V, One-pot green synthesis, characterization, photocatalytic, sensing and antimicrobial studies of Calotropis gigantea leaf extract capped CdS NPs. Materials Science and Engineering: B. 2017;2017(225): 33–44, ISSN 0921–5107.

133. Elmolla ES, Chaudhuri M. Photocatalytic degradation of amoxicillin, ampicillin and cloxacillin antibiotics in aqueous solution using UV/TiO_2 and UV/H_2O_2/TiO_2 photocatalysis. Desalination. 2010 Mar 1;252(1–3):46–52.
134. Said M, Rizki WT, Asri WR, Desnelli D, Rachmat A, Hariani PL. SnO_2– Fe_3O_4 nanocomposites for the photodegradation of the Congo red dye. Heliyon. 2022 Apr 1;8(4):e09204.
135. Pérez C, Velando A, Munilla I, López-Alonso M, Oro D. Monitoring polycyclic aromatic hydrocarbon pollution in the marine environment after the Prestige oil spill by means of seabird blood analysis. Environmental Science & Technology. 2008 Feb 1;42(3):707–13.
136. Aivalioti M, Vamvasakis I, Gidarakos E. BTEX and MTBE adsorption onto raw and thermally modified diatomite. Journal of Hazardous Materials. 2010 Jun 15;178(1–3):136–43.
137. Dharma HN, Jaafar J, Widiastuti N, Matsuyama H, Rajabsadeh S, Othman MH, Rahman MA, Jafri NN, Suhaimin NS, Nasir AM, Alias NH. A review of titanium dioxide (TiO_2)-based photocatalyst for oilfield-produced water treatment. Membranes. 2022 Mar 19;12(3):345.
138. Ferrari-Lima AM, De Souza RP, Mendes SS, Marques RG, Gimenes ML, Fernandes-Machado NR. Photodegradation of benzene, toluene and xylenes under visible light applying N-doped mixed TiO_2 and ZnO catalysts. Catalysis Today. 2015 Mar 1;241:40–6.
139. Parida KM, Dash SS, Das DP. Physico-chemical characterization and photocatalytic activity of zinc oxide prepared by various methods. Journal of Colloid and Interface Science. 2006 Jun 15;298(2):787–93.
140. Selvam K, Muruganandham M, Muthuvel I, Swaminathan M. The influence of inorganic oxidants and metal ions on semiconductor sensitized photodegradation of 4-fluorophenol. Chemical Engineering Journal. 2007 Mar 15;128(1):51–7.
141. Abreu FR, Lima DG, Hamú EH, Wolf C, Suarez PA. Utilization of metal complexes as catalysts in the transesterification of Brazilian vegetable oils with different alcohols. Journal of Molecular Catalysis A: Chemical. 2004 Feb 16;209(1–2):29–33.
142. Brunetti A, Pomilla FR, Marcì G, Garcia-Lopez EI, Fontananova E, Palmisano L, Barbieri G. CO_2 reduction by C_3N_4-TiO_2 Nafion photocatalytic membrane reactor as a promising environmental pathway to solar fuels. Applied Catalysis B: Environmental. 2019 Oct 15;255:117779.
143. Chowdhury IH, Roy M, Kundu S, Naskar MK. TiO_2 hollow microspheres impregnated with biogenic gold nanoparticles for the efficient visible light-induced photodegradation of phenol. Journal of Physics and Chemistry of Solids. 2019 Jun 1;129:329–39.
144. Shet A, Vidya SK. Solar light mediated photocatalytic degradation of phenol using Ag core–TiO_2 shell (Ag@ TiO_2) nanoparticles in batch and fluidized bed reactor. Solar Energy. 2016 Apr 1;127:67–78.
145. Siddiqa A, Masih D, Anjum D, Siddiq M. Cobalt and sulfur co-doped nano-size TiO_2 for photodegradation of various dyes and phenol. Journal of Environmental Sciences. 2015 Nov 1;37:100–9.
146. Nune SV, Golimidi RK. Cu_2O-TiO_2 Composite for photocatalytic degradation of benzene and its derivatives using visible light. Current Nanoscience. 2021 Nov 1;17(6):904–9.
147. Yadav IC, Devi NL. Pesticides classification and its impact on human and environment. Environmental Science & Engineering. 2017 Feb;6:140–58.
148. Nicolopoulou-Stamati P, Maipas S, Kotampasi C, Stamatis P, Hens L. Chemical pesticides and human health: The urgent need for a new concept in agriculture. Frontiers in Public Health. 2016 Jul 18;4:148.
149. Boedeker W, Watts M, Clausing P, Marquez E. The global distribution of acute unintentional pesticide poisoning: Estimations based on a systematic review. BMC Public Health. 2020 Dec;20(1):1–9.

150. Vagi MC, Petsas AS. Recent advances on the removal of priority organochlorine and organophosphorus biorecalcitrant pesticides defined by Directive 2013/39/EU from environmental matrices by using advanced oxidation processes: An overview (2007–2018). Journal of Environmental Chemical Engineering. 2020 Feb 1;8(1):102940.
151. Pirkanniemi K, Sillanpää M. Heterogeneous water phase catalysis as an environmental application: A review. Chemosphere. 2002 Sep 1;48(10):1047–60.
152. Takanabe K. Solar water splitting using semiconductor photocatalyst powders. Solar Energy for Fuels. 2015:73–103.
153. Rajeev B, Yesodharan S, Yesodharan EP. Sunlight activated ZnO mediated photocatalytic degradation of acetophenone in water. IOSR Journal of Applied Chemistry. 2016:55–70.
154. Baruah S, Mahmood MA, Myint MT, Bora T, Dutta J. Enhanced visible light photocatalysis through fast crystallization of zinc oxide nanorods. Beilstein Journal of Nanotechnology. 2010 Nov 22;1(1):14–20.
155. Malato S, Blanco J, Richter C, Milow B, Maldonado MI. Solar photocatalytic mineralization of commercial pesticides: Methamidophos. Chemosphere. 1999 Feb 1;38(5):1145–56.
156. Korake PV, Sridharkrishna R, Hankare PP, Garadkar KM. Photocatalytic degradation of phosphamidon using Ag-doped ZnO nanorods. Toxicological & Environmental Chemistry. 2012 Jul 1;94(6):1075–85.
157. Veerakumar P, Sangili A, Saranya K, Pandikumar A, Lin KC. Palladium and silver nanoparticles embedded on zinc oxide nanostars for photocatalytic degradation of pesticides and herbicides. Chemical Engineering Journal. 2021 Apr 15;410:128434.
158. Tabasum A, Alghuthaymi M, Qazi UY, Shahid I, Abbas Q, Javaid R, Nadeem N, Zahid M. UV-accelerated photocatalytic degradation of pesticide over magnetite and cobalt ferrite decorated graphene oxide composite. Plants. 2020 Dec 23;10(1):6.
159. Yang L, Guan X, Wang GS, Guan XH, Jia B. Synthesis of ZnS/CuS nanospheres loaded on reduced graphene oxide as high-performance photocatalysts under simulated sunlight irradiation. New Journal of Chemistry. 2017;41(13):5732–44.
160. Kerrigan JF, Sandberg KD, Engstrom DR, LaPara TM, Arnold WA. Small and large-scale distribution of four classes of antibiotics in sediment: Association with metals and antibiotic resistance genes. Environmental Science: Processes & Impacts. 2018;20(8):1167–79.
161. Dinh QT, Moreau-Guigon E, Labadie P, Alliot F, Teil MJ, Blanchard M, Chevreuil M. Occurrence of antibiotics in rural catchments. Chemosphere. 2017 Feb 1;168:483–90.
162. Gacem MA, Telli A, Khelil AO. Nanomaterials for detection, degradation, and adsorption of pesticides from water and wastewater. In Aquananotechnology 2021 Jan 1 (pp. 325–46). Elsevier. https://doi.org/10.1016/B978-0-12-821141-0.00003-3.
163. Kotrange H, Najda A, Bains A, Gruszecki R, Chawla P, Tosif MM. Metal and metal oxide nanoparticle as a novel antibiotic carrier for the direct delivery of antibiotics. International Journal of Molecular Sciences. 2021 Sep 4;22(17):9596.
164. Yang X, Chen Z, Zhao W, Liu C, Qian X, Zhang M, Wei G, Khan E, Ng YH, Ok YS. Recent advances in photodegradation of antibiotic residues in water. Chemical Engineering Journal. 2021 Feb 1;405:126806.
165. Beltran FJ, Aguinaco A, García-Araya JF, Oropesa A. Ozone and photocatalytic processes to remove the antibiotic sulfamethoxazole from water. Water Research. 2008 Aug 1;42(14):3799–808.
166. Elmolla ES, Chaudhuri M. Photocatalytic degradation of amoxicillin, ampicillin and cloxacillin antibiotics in aqueous solution using UV/TiO_2 and UV/H_2O_2/TiO_2 photocatalysis. Desalination. 2010 Mar 1;252(1–3):46–52.
167. Wu S, Lin Y, Hu YH. Strategies of tuning catalysts for efficient photodegradation of antibiotics in water environments: A review. Journal of Materials Chemistry A. 2021;9(5):2592–611.

168. Jassal P, Khajuria R, Sharma R, Debnath P, Verma S, Johnson A, Kumar S. Photocatalytic degradation of ampicillin using silver nanoparticles biosynthesised by Pleurotus ostreatus. BioTechnologia. 2020;101(1):5–14.
169. Damacena D, Macedo V, Silva A, Honório L, Cavalcante E, Trigueiro P, Osajima J. Ag@ ZnO-saponite nanocomposite for ciprofloxacin photodegradation. In Proceedings of the 2nd International Electronic Conference on Catalysis Sciences—A Celebration of Catalysts 10th Anniversary, 15–30 October 2021, MDPI: Basel.
170. Olusegun SJ, Larrea G, Osial M, Jackowska K, Krysinski P. Photocatalytic degradation of antibiotics by superparamagnetic iron oxide nanoparticles. Tetracycline Case Catalysts. 2021 Oct 15;11(10):1243.
171. Seid-Mohammadi A, Ghorbanian Z, Asgari G, Dargahi A. Photocatalytic degradation of metronidazole (MnZ) antibiotic in aqueous media using copper oxide nanoparticles activated by H_2O_2/UV process: Biodegradability and kinetic studies. Desalination and Water Treatment. 2020 Jul 1;193:369–80.
172. Becker A, Kirchberg K, Marschall R. Magnesium ferrite ($MgFe_2O_4$) nanoparticles for photocatalytic antibiotics degradation. Zeitschrift für Physikalische Chemie. 2020 Apr 1;234(4):645–54.

12 Toxicology and Regulatory Challenges of Metallic Nanoparticles

Shushay Hagos Gebre and Berhane Gebremedhin Gebrezgabher

12.1 INTRODUCTION

The natural environment is vital for every living thing, including human beings, and provides all services such as water, food, air, biodiversity, regulation of the climate, nutrient cycling, waste degradation, accommodation, etc. However, human beings themselves innovate new products and technologies for their needs, such as plastics, engineered nanomaterials (ENMs), cosmetic products, and more, which later become serious problems and deteriorate the environment's air, soil, and water quality [1]. The continuing advancements of nanotechnology for diverse applications like catalysts [2], nanomedicine, environmental remediation [3], energy storage and conversion [4], electronics, cosmetic additives, and agriculture are serious problems for the ecosystem and human health. Metallic nanoparticles (MNPs) with a 1–100 nm size have a significantly higher surface area and better surface reactivity than bulk materials. The nanomaterials are highly reactive and sensitive to their surroundings if discharged into the environment. They changed their physicochemical properties due to agglomeration, aggregation, dissolution, and sulfidation [5–9].

Nanotechnology is rapidly expanding worldwide in multi-billion USD in the food, agricultural, and medicine sectors. However, the impact of nanomaterials on the environment has not been well studied [10]. The effects of nanomaterials may arise due to their unique properties and interactions with the environment. Still, challenges exist in determining regulations, guidelines, and risk assessments for workers' and consumers' safety. On the other hand, the scientific community still does not understand the nature and exposure of MNPs to health problems [11, 12]. MNPs with small sizes can easily penetrate the bodies to accommodate and enter the cells and tissues of living organisms and human beings. The toxicity levels and properties of the bulk materials from which the MNPs are made are known [13–15]. Still, the toxicity level of these materials changes at the nanoscale, and it becomes difficult to determine whether the dosage and size of the MNP will have a toxicity impact [12]. Most studies confirmed that smaller-size nanoparticles (NPs) showed high toxicity to microorganisms, algae, mammalian cells, and yeasts due to the more extraordinary penetration ability of the smaller-size NPs than the large-size ones [16]. However,

DOI: 10.1201/9781003317319-12

it is impossible to generalize that "the smaller, the more toxic" because some NPs showed that the toxicity is reduced when the particle size is decreased [13, 17–20]. In another instance, increasing the size of NPs causes the toxicity to increase [21]. For instance, BALB/C mice treated with 8 mg/kg/week of citrate-capped AuNPs of small (3 and 5 nm) and large (50 and 1000 nm) sizes do not exhibit toxicity [22–25]. However, the mice were poisonous to intermediate sizes (8–37 nm). It resulted in weight loss, weariness, color changes, and loss of appetite, and most of the mice perished within 21 days. Mice exposed to intermediate-sized AuNPs develop more Kupffer cells in the liver, their lungs lose structural integrity, and white pulp diffuses in the spleen [26]. As a result, a combination of the MNPs' size, surface area, shape, surface charge, solubility, and other physicochemical features contribute to their toxicity. In general, the size of the NPs resembles the DNA helix (2 nm), the thickness of the cell membrane (10 nm), and protein globules (2–10 nm). Therefore, NPs can penetrate the nucleus of organelles and cells. Triphenylphosphine-stabilized AuNPs of different sizes were tested for their toxicity against macrophages, connective tissue fibroblasts, melanoma, and epithelial cells. The AuNPs with 0.8–15 nm were tested on the cells [25–28]. The NPs with a 1.4 nm size exhibited 30–56 mm half maximal inhibitory concentration (IC_{50}) values and caused cell death via a necrosis mechanism within 12 h. Similarly, the 1.2-nm-sized AuNPs showed a predominant cell death via apoptosis. However, the AuNPs with 15 nm and gold thiomalate were not toxic to the cells even at 60 and 100 times higher concentrations [31]. In another study, ultrasmall AuNPs (<6 nm) were easily penetrated and entered the nucleus in gene therapy; in contrast, large AuNPs (>10 or 16 nm) were joined to the membrane and deposited in the cytoplasm [32]. MNPs such as Ag, Au, nanoscale zero-valent iron (nZVI), Cu, Co, and Ni are fabricated for various applications [33, 34]. AgNPs are used as antimicrobials even at low concentrations for food packaging applications due to their broad-spectrum inhibition of microorganisms [35]. Similarly, AuNPs are used in biomedical applications, gene/drug delivery, cancer treatment, and biological imaging because of their unique optical properties, relatively low toxicity, better stability, and biocompatibility with human cells [36]. Due to its large surface area and high level of reactivity, nZVI is frequently utilized for environmental cleanup of inorganic, organic, and microbe-based pollutants from soil and wastewater [37].

12.2 PATHWAYS OF NANOPARTICLES INTO THE NATURAL ECOSYSTEMS

The MNPs are introduced into the environment through the soil, water, or air through human activities. The NPs may be released during production; NPs manufactured in huge volumes can contribute a significant byproduct to the ecosystem. Depending on their application and type, NPs can enter directly or indirectly into different pathways [38].

On the other hand, NPs may be absorbed by ingestion or inhalation into the various organs and tissues of animals and humans. Once the NPs are released into the aquatic system, the dissolved MNP can exert toxic effects on living organisms such as bacteria, algae, fish, and daphnia [40, 41]. In addition to being intentionally released into the environment, MNPs can also enter via natural and anthropogenic activities.

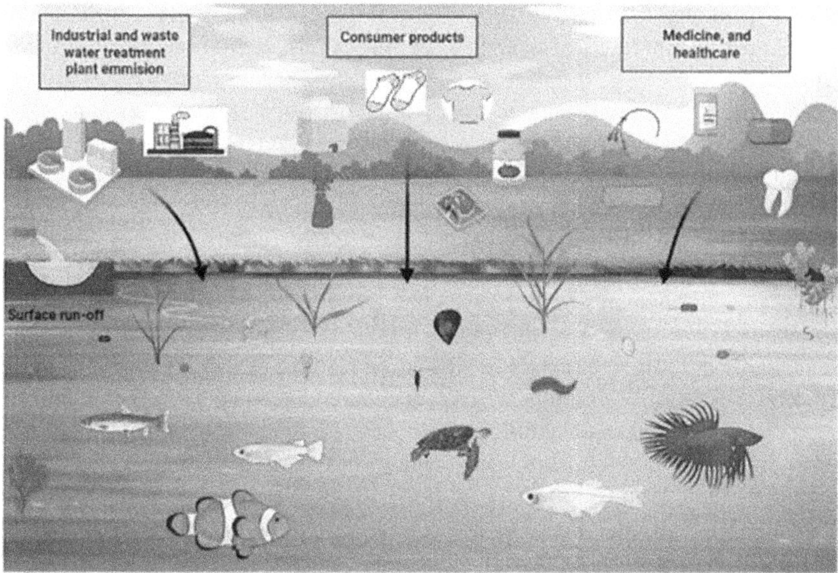

FIGURE 12.1 Routes of AgNPs from industrial, agriculture and health center into the aquatic system [39].

Source: Reproduced with permission from Springer Nature.

Incident disasters like volcanoes, erosion, dust storms, force fires, and flooding also significantly influence the discharging of the NPs into the whole ecosystem [42]. The discharged NPs from industries, wastewater treatment plants, transportation, agricultural, and healthcare sectors end up in waterways. Then water is the fastest medium to spread the discharge NPs, like AgNPs, to the whole aquatic organism (fish, algae, fungi, bacteria, plants, and large animals) and influence the food chains [39] (Figure 12.1).

12.3 FATE OF NANOPARTICLES IN THE ENVIRONMENT

The transfer of electrons between the aquatic system's chemical moieties and the NPs leads to the redox process. MNPs are transformed via physical, chemical, and biological processes. In physical transformation, MNPs can undergo aggregation or agglomeration resulting from the strong binding between the particles, decreasing the surface charge. In contrast, the size of the MPNs is increased, and the particles' mobility, reactivity, and toxicity are changed. The toxicity of the MNPs is decreased due to reducing their surface area, which influences reactive oxygen species (ROS) formation. The aggregation may be homo-aggregation with identical particles binding together or hetero-aggregation in which the opposite charge of the MNPs binds with the organic or inorganic chemicals of the environment. The chemical transformation of the MNPs depends on the type of environment, typically the oxidizing environment (an oxygen-rich environment), the reductive environment (limited or no oxygen), and the dynamic redox. Therefore, soft metals like Ag, Zn, and Cu can undergo dissolution, aggregation, and sulfidation [42, 43], as demonstrated in Figure 12.2a.

When the MNPs are discharged into the environment during their synthesis or after use, accidentally or deliberately, especially in the aquatic environment, their fate and interaction are heavily influenced by their physicochemical characteristics, such as shape, size, defects, synthetic processes, and the receiving environment. Several factors, like pH, natural organic matter (NOM), salinity, ionic strength, pH, temperature, water hardness, phosphate ion concentration, affect the fate of the NPs. The aquatic environments are highly complex due to the presence of many natural particulates that result from the degradation of leaves and animals and dissolved minerals in the rocks. The interaction of NPs is complicated [44, 45]. The MNPs change physically due to agglomeration, aggregation, and deposition, as well as chemically due to dissolution and chlorination, in the context of the complex aquatic ecosystem. Living things also play a crucial role in biological transformation through oxidation, sulfidation, and other processes. As an example, when silver nanoparticles (AgNPs) are oxidized to Ag_2O and form a core structure of (AgNPs)-Ag_2O, it releases Ag^+ ions when it dissolves in water [46]. On the other hand, AgNPs can undergo sulfurization to form silver sulfide in water, which reduces the toxicity of the NPs due to their lower water solubility [47]. In addition, the aquatic environment consists of multi-species of microbial aggregates living together (microalgae, bacteria, fungi, archaea, protozoa, etc.), which are robust to the environment. Thus, the microbial aggregate and community structure protect against prolonged NP exposure and minimize the negative impact of NPs [48] (Figure 12.2b).

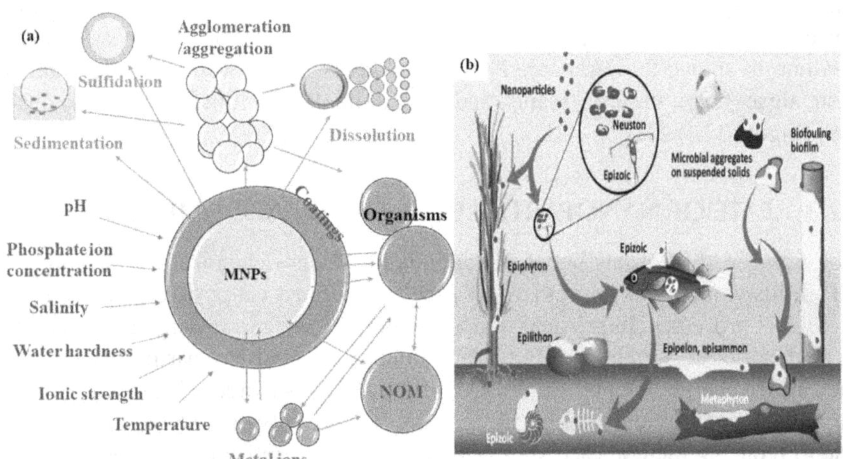

FIGURE 12.2 (a) Possible transformation and interaction processes of engineered nanoparticles (ENPs) [42]. Adapted with permission from Elsevier. (b) The fates of NPs within the sediments and the water column in an aquatic ecosystem [48].

Source: Reproduced with permission from Elsevier.

The presence of NOM in the environment played a significant role in the transformation of the MNPs. The NOM's functional groups, such as the hydroxy, ketone, amine, and thiols, directly engage with the metal ions from the MNPs to rebuild the MNPs. In order to convert the metal ions to MNPs, the functional groups of the NOM work as a reducing agent [49]. Studies on AgNPs confirmed the release of Ag^+ and the reform of AgNPs by the action of the NOM. In addition, the Au^{3+} form of AuNPs is formed in the presence of sunlight through the action of the functional groups of the NOM because of its high reduction potential. Iron oxide NPs could also be formed from the reduction of Fe^{2+} and Fe^{3+} naturally [49]. The MNPs can undergo surface modification, transformation, and reconstruction due to their reactive surface area by oxidation reaction in a NOM-rich environment. For example, Mudunkotuwa et al. [50] confirmed that both aged and fresh CuNPs were oxidized to Cu_2O by the action of citric and oxalic acids. Similarly, Dryer et al. [51] reported that Pb^{4+} present in PbO_2 NPs was transformed to Pb^{2+} by the action of the aromatic hydroxy groups of the NOM [52–56].

The oxidative dissolution and sulfidation of Ag nanowires (AgNWs) were investigated in the aquatic system using protein models (lysozyme (LYZ)). The presence of proteins influences the chemical transformation of the AgNWs, similar to NOM. A strong bond between the negatively charged AgNWs and positively charged LYZ was confirmed via fluorescence spectroscopy and isothermal titration calorimetry (ITC). The other positive part of the AgNWs was also convenient for the sulfide ions. Thus, protein-generated microbes in anaerobic zones and sulfide ions impact the nano-Ag's sulfidation and reduce its toxicity in the aquatic environment. However, the kinetics of oxidative dissolution and sulfidation of AgNWs with ovalbumin (OVA) and bovine serum albumin (BSA) were insignificant due to the weak interaction between the proteins and the AgNWs [57]. The sulfidation of AgNPs is carried out either directly or indirectly. Suppose a high concentration of sulfide ions is present in the environment. In that case, direct sulfidation is carried out (Eq. 12.1). However, indirect sulfidation is carried out at a lower concentration of sulfide ions, and dissolved oxygen (DO) is essential. Initially, the Ag^+ ion is formed from AgNPs, followed by precipitation by HS^-or H_2S [58], as shown in Eq. 12.2 given Ag_2S [59].

$$4Ag + O_2 + 2H_2S \rightarrow 2Ag_2S + 2H_2O \text{ or } 4Ag + O_2 + 2HS \rightarrow 2Ag_2S + 2HO^- \quad 12.1$$

$$4Ag + O_2 + 2H_2O \xrightarrow{slow} 4Ag^+ + 4HO^- \text{ then } 4Ag^+ + 2HS^- + 2HO^- \xrightarrow{fast} 2Ag_2S + 2H_2O \quad 12.2$$

The toxicity effect of iron-based NPs was tested on a green alga (*Chlorella pyrenoidosa*). Four different nZVI, two Fe_2O_3 NPs, and Fe_3O_4 were investigated with regard to particle size, oxidation, crystal phase, and aging. The nZVI was oxidized to $Fe^0 \rightarrow Fe(OH)_2 \rightarrow Fe_3O_4 \rightarrow \gamma-Fe_2O_3$. Small-size particles significantly inhibit the algal growth; however, oxidation of the NPs decreases their toxicity as nZVI > Fe_3O_4 > Fe_2O_3 and α-Fe_2O_3 NPs. Furthermore, aging the nZVI for three months in distilled and surface water increases its oxidation and decreases its toxicity. Higher oxidation of iron-derived MNPs was more stable, and their interaction with algae was weak [60].

12.4 TOXICITY OF METALLIC NANOPARTICLES

Biocompatibility and toxicity investigations of new material before it is employed for application, especially for biological activities, are very important. The toxicity of the MNPs depends on the test environment (DO, total suspended solids (TSS), NOM), contact time, physicochemical properties of the MNPs like size, surface area, surface charge, dose, the tendency to aggregate or agglomerate, hydrophobicity or hydrophilicity, and bioassay tests either in vitro or in vivo [61–63]. The NPs interact with and are adsorbed by organic and inorganic species that can change their transport rate. Thus, NPs can be taken by microorganisms, algae, fish, mammals, and other living things. In some cases, plant extracts and microorganisms, including algae, can be used to synthesize NPs intracellularly or extracellularly [25, 56]. The plant extract and microorganism biomolecules are used as bioreductants, stabilizers, and capping agents during the biosynthesis approach. The toxic heavy metals bioremediated in the species are converted into more amenable and less harmful forms. In another case, the NPs are synthesized using supporting materials like carbohydrates, polymers, ionic liquids, silicon, carbon materials to prevent agglomeration. Therefore, the presence of supporting materials may reduce the toxicity of the NPs [16].

Shape-dependent cellular uptake of AuNPs was reported in RAW264.7 cells. Three anisotropic AuNPs, namely star, rod, and triangle shapes, were fabricated and coated with methoxy polyethylene glycol (mPEG) to study their effect on cellular uptake. The triangle shape showed the most efficient cellular uptake, and the star shape exhibited the lowest cellular uptake. Different shapes have different endocytosis pathways. Thus, NPs are designed based on their cellular uptake for drug delivery [64]. In another study, green tea-mediated synthesized nanosphere, nanostars, and nanorod AuNPs were investigated for cellular uptake and cytotoxicity in human hepatocyte carcinoma cells (HepG2). The IC_{50} values of the HepG2 were 127.2, 81.8, and 22.7 μM, corresponding to nanospheres, nanostars, and nanorods; thus, the nanorods were the most cytotoxic, and the nanosphere was the least cytotoxic against the HepG2 cells. On the other hand, the nanosphere showed the highest cellular uptake (58.0%), followed by nanorods (52.7%) and nanostars (41.5%). This induces the most increased cellular uptake, showing the most negligible cytotoxicity, and vice versa. The highest uptake capacity does not always cause the highest cytotoxicity [65]. Depending on the occasion, the uptake capacity of the NPs by the cellular cells is related to the binding and contact areas on the cell membrane of receptors. For example, shorter nanorods have the highest uptake than longer nanorod NPs. However, the rod-shaped NPs have lower uptake than the nanosphere-shaped NPs. Because the longitudinal axis of the nanorod-shaped NPs directly interacts with the receptors, reducing the accessible active site of the receptors for binding, the rod-shaped NPs have a bigger contact area with the cell membrane receptors than the nanosphere-shaped NPs [58]. Sphere- and rod-shaped AgNPs were prepared by chemical reduction for antimicrobial activities against gram-negative and gram-positive bacteria. The killing kinetics of AgNPs found that the death rate of *Klebsiella pneumoniae* was higher for the sphere-shaped AgNPs than the rod-like shape. The sphere-shaped AgNPs showed similar activity against, *P. aeruginosa, Escherichia coli (E. coli), B. subtilis*, and *S. aureus* at a dose of 188, 190, 185 and 190 μg/mL, respectively. However, 358, 350,

340, and 348 μg/mL of the rod-shaped AgNPs were required for the aforementioned bacteria strains. *Klebsiella pneumoniae* was inhibited with a minimum concentration of both NPs [66].

Surface charge is another determinant factor in the toxicity of nanomaterials. In this regard, El Badawy et al. [67] fabricated AgNPs coated on citrate, polyvinylpyrrolidone (PVP), branched polyethyleneimine (BPEI), and uncoated H_2-AgNPs, respectively, to study their toxicity effects on *Bacillus* species. The positively charged BPEI-AgNPs were more toxic to *Bacillus* sp. while the negatively charged citrate-AgNPs were the least toxic. Due to the presence of phosphate, carboxyl, and amino functional groups, which result in a negative charge, the citrate-AgNPs have an ionization potential of –38 mV, which is similar to that of gram-positive bacteria (–37 mV). As a result, there is a strong electrostatic repulsion between the citrate-AgNPs and the *Bacillus cells*, which creates an electrostatic barrier that prevents further interaction and lowers the toxicity of the citrate-AgNPs. Similarly, the negative potentials of the PVP-AgNPs (–10 mV) and uncoated H_2-AgNPs (–22 mV) have decreased, which has increased the interaction between the NPs and *Bacillus* sp. However, the positively charged BPEI-AgNPs, which have a potential of +40 mV interacted with the bacteria more and evolved into the most lethal NPs of all. The transmission electron micrograph (TEM) image showed physical interactions that significantly influence the cellular damage of the membrane using the BPEI-AgNPs (Figure 12.3a–d). The nanoparticle movements in the fluid are measured by zeta potential. It is electrostatic, which divides the compact layer from the diffuse layer of the colloidal particles. In this regard, MNPs with a zeta potential value >+30 mV or <–30 mV are stable because the MNPs will tend to have a sizeable electrostatic repulsion among the particles, preventing them from aggregation. Conversely, MNPs with lower zeta potential values tend to stick together. Therefore, the sticking cohesion of the MNPs is called aggregation, carried out among the particles with strong bonding, and the loose connection of the individual particles results in agglomeration due to van der Waal's forces.

Therefore, microorganisms such as bacteria's membranes have negative charges that can directly bind with metal ions like Ag^+ than MNPs. Hence, the metal ion (Ag^+) can impact more toxicological responses than AgNPs [68]. In an aquatic environment, the surface of microorganisms is negatively charged. As a result, the large molecular weight of NOM sterically prevents the NPs from making direct contact with the surface of the organism, which determines the NPs' toxicity [69–71].

A recent study by Souza et al. [72] investigated the toxicity of AgNPs in the aquatic plant Lemna minor. Different concentrations of AgNPs (30, 85, and 110 nm) were treated with the culture media for 30 days. In all the sizes, 50 μgmL^{-1} of AgNPs was toxic and negatively impacted plant growth. More Ag^+ ions were released from the 30 nm AgNPs than the other sizes. After one month, 60% more mortality of the plant was observed for the 30 nm AgNPs than the different sizes. The small size of the NPs was more toxic than the 85 and 110 nm ones. For comparison, the toxicity of $AgNO_3$ was measured, and the release of Ag^+ was faster than that of AgNPs. Another study was conducted to investigate the toxicity of Au, Ag, and bimetallic Ag-AuNPs against *Daphnia magna*. It was found that the toxicity of the MNPs depended on their dose and composition. 65–75 mg/L of Au was used to kill 50% of

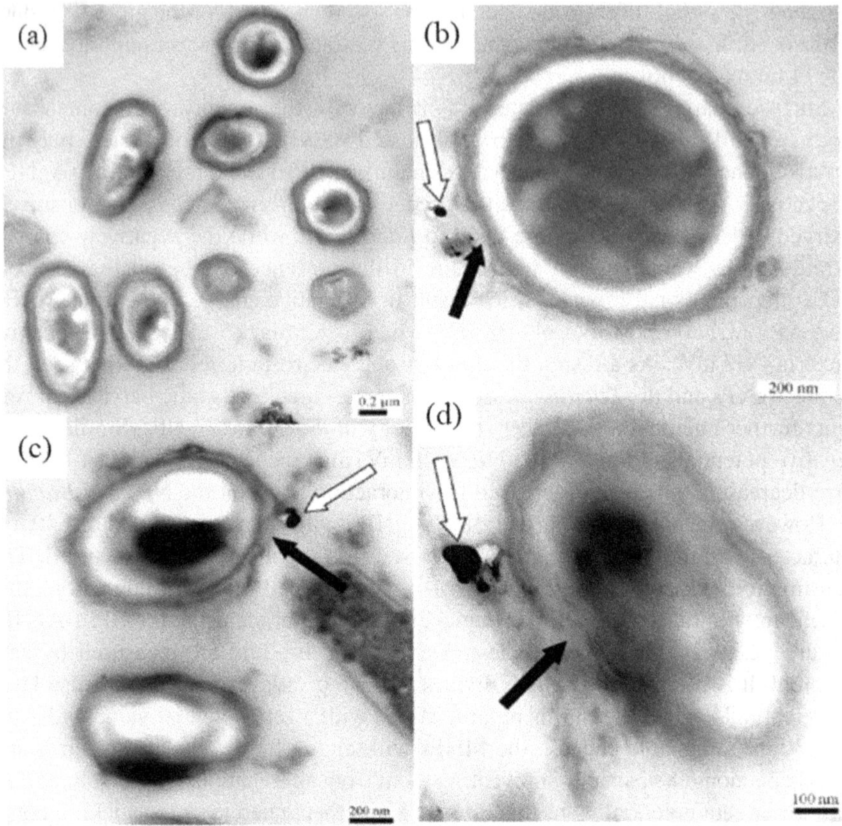

FIGURE 12.3 TEM images of (a) unexposed cells (b–d) cells exposed to BPEI-AgNPs, the white and black arrows showed the AgNPs and its impact on the cellular membrane, respectively [67].

Source: Reproduced with permission from American Chemical Society.

the test organism (LC_{50}), whereas only 3 μg/L of Ag dose was enough to obtain LC_{50}. The toxicity of the Ag-Au was between the Au and AgNPs. Ag:Au (4:1) and (1:4) were used at 15 μg/L and 12 μg/L to exhibit the LC_{50} of *Daphnia magna*. Thus, the AuNPs were 1000-fold less toxic than the AgNPs, and the toxicity of the bimetallic NP arises from AgNPs [73]. Co-exposure of AgNPs and Cd^{2+} was carried out against HepG2 cells for 24. The AgNPs + Cd^{2+} showed more toxicity than the individual AgNPs and Cd^{2+}. The lack of antioxidant defense activity results in high levels of oxidative stress and disruption of the metabolic system [74]. Manivannan et al. [75] used two different sizes of AuNPs, namely citrate-capped AuNPs (CIT30 AuNPs), CIT30 AuNPs, citrate–polyvinyl pyrrolidone-capped AuNPs (CIT30-PVP-capped AuNPs) and CIT40-PVP-capped AuNPs, to study the toxicity against algae, bacteria, cell lines (in vitro), and mice (in vivo). The CIT30 AuNPs were more toxic to bacteria, algae, and SiHa cells; however, the addition of PVP decreased the toxicity of AuNPs in the in vitro study, but the genotoxicity was increased in the mouse model [59].

Cobalt-based NPs, namely, Co, CoO, and Co_3O_4 were assessed for their possible (geno)toxicity on different cell lines. The CoNP showed higher ROS formation ability than the CoO, but no reactivity of Co_3O_4 was established due to its chemical stability in vitro. The metal (Co) NPs were highly cytotoxic due to their unstable surface with a concentration of 5.4 mg/mL, whereas the CoO was medium with a 30.8 mg/mL dose. It was found that the cellular uptake of Co and CoO in lung cells causes oxidative damage and the induction of DNA strand breaks, but the Co_3O_4 is inactive. The CoNPs were transformed into oxide surfaces, and further dissolution of the oxides released more Co metal [76]. Similar to this, 12 weeks of exposure to low-dose CoNPs (0.05 and 0.1 mg/mL) against the isogenic 8-oxoguanine glycosylase (Ogg1) knockout partner and the wild-type mouse embryonic fibroblast (MEF Ogg1$^{+/+}$) demonstrated the generation of ROS and oxidative DNA damage. The MEF Ogg1$^{-/-}$ cells responded to NP with greater sensitivity. All cell transformations and morphological changes were seen after a protracted period of treatment. Thus, from all the given information, it was confirmed that the CoNPs possess a possible carcinogenic effect on the cells [77].

NiNPs is used for various applications, such as a catalysts, sensors, fuel cells and medicine. The toxicity study of the NiNPs at a cellular and animal level confirmed that the NP causes cytotoxicity, genotoxicity, and possible carcinogenicity [78, 79]. Though there are no epidemiological studies, in vivo and in vitro studies indicate that nickel and nickel-based NPs have possible genotoxicity and carcinogenicity in different cells [80, 81]. The NPs exposure to the reproductive systems of female rats was studied. The high levels of ROS, oxidants, MDA, and NO significantly affect the activity of antioxidant enzymes such as catalase (CAT) and superoxide dismutase (SOD), which lead to cell apoptosis. The ultrastructure of the ovaries of the rat showed a loss of mitochondrial cristae, swelling, and enlargement of the endoplasmic reticulum compared to the control groups compared to the nickel microparticles (NiMPs) due to the high surface area and small size of the NiNPs [82]. Similar results have been reported about the toxicity of NiNPs on the reproduction of *Caenorhabditis elegans*. A decrease in brood size and prolongation of the generation time of *C. elegans* was observed using the NiNPs rather than the NiMPs [83]. The Iavicoli group [84] investigated the toxicity of PdNPs against Rat-1 embryo fibroblasts and the A549 lung carcinoma epithelial cell line. They have studied the cells growth inhibition, proliferation, ROS, and level of DNA damage. Employing 1 μg/mL of PdNPs on the Rate-1 cells resulted in 10% and 70% growth inhibition after 48 and 120 h of incubation. But by raising the concentration to 2 μg/mL, the cell growth inhibition was found to be 30% and 80% after 48 and 120 h. Similarly, 10% and 30% growth inhibition were observed for A549 cells after 48 and 120 h. Using 2 μg/mL of PdNPs the cell growth inhibition was 20 and 50 after 48 and 120 h of incubations, respectively.

In general, the correlation between MNP toxicity and their physical and chemical properties is attributed to MNPs with electropositive surface charge, solubility, nanospheres, and needle-like shapes being highly toxic, whereas MNPs with large size, good stability, electronegative surface charge, nanorod shapes, coated with different materials, and higher concentration doses are less toxic, or their toxicity tends to reduce [85]. The toxicity effect of selected MNPs on different organisms is summarized in Table 12.1.

TABLE 12.1
Toxicity of Selected MNPs for Different Organism or Species and Their Effect

NPs	Size (nm)	Shape	Reducing agent/ stabilizing agent	Tested organism/ species	Concentration (mg/L)	Effect	Reference
Ag	10–15	Spherical	–	*Microcystis aeruginosa*	0.1	Cell growth decline Changes in cell density and morphology	[86]
Ag	2–18	Spherical	Polyvinyl alcohol	*Pseudokirchneriella subcapitata*	0.15	Growth inhibition	[87]
Ag	6.20	Spherical	Polyvinyl pyrrolidone gum Arabic	*Spirodela polyrhiza*	5, 10	ROS formation	[88]
nZVI	–	–	Linear alkylbenzene sulfonate (LAS)	*cenedesmus obliquus*	7.885 nZVI: 0.266 mmol/L LAS	Growth inhibition (100%)	[89]
nZVI	–	–	Sulfide	*Escherichia coli*	10	ROS formation	[90]
nZVI	–	Spherical	Neem leaves extract	*Artemia salina*	100	ROS formation	[91]
Cu	40, 60	Aggregate	–	Rat	40–100 μM	Cell death, reduction in cell viability	[92]
Ag	50	Spherical	VP	*Glyptotendipes tokunagai*	297.36	Membrane deformation & DNA damage	[93]
Ag	34 ± 18	Spherical and	–	*Pseudokirchneriella subcapitata*	5.0	100% growth inhibition	[94]
Pt	51 ± 12	Irregular polyhedral	–	*Pseudokirchneriella subcapitata*	22.2	100% growth inhibition	[94]
Pb	76	–	–	*Chlorella Vulgaris*	100	39% reduction of chlorophyll *a*	[95]
Ag	30	Spherical		*Lemna minor*	50 μg mL–	60% mortality	[72]
Ni	60	Dendritic	–	Zebrafish	1000	100%mortality	[96]

12.5 MECHANISMS OF NANOMATERIALS TOXICITY

NPs disrupt microbial cells on their membrane due to ROS formation, damaging DNA, protein denaturation, inactivating enzyme activity, and releasing free metal ions. In this regard, metal and metal oxide NPs possess higher toxicity than carbon-containing nanomaterials [48].

Most studies confirm that the toxicity mechanism of the NPs is carried out by oxidative stress. Oxidative stress happens as a consequence of immune system induction, which results in cell proliferation or DNA damage followed by apoptosis or cell death [97]. When the ROS level becomes moderate in animal or plant and algal cells, the body's scavenging ability is exceeded and irreversible damage occurs. The function of mitochondria is affected by excess ROS, which finally leads to apoptosis [78]. Part of the essential biomolecules is also lost [98]. Free radicals, oxygen ions, and peroxide are just a few of the ROS species that can be created by MNPs. The cell membrane may be harmed as a result of the metal NPs' cellular absorption, leading to the leaking of the cell's contents and cell death. The release of dissolved metals is another method by which MNPs are hazardous [21]. Further, MNPs cause nonspecific damage to the cell membranes of organisms. The toxicity of the MNPs depends on their size, surface area, shape, functionalizing, and coating materials. In terms of size, small MNPs expose cellular uptake, which results in a high concentration inside the cell [99]. The uptake of NPs also depends on the organism's intracellular structures and cell membrane. For example, gram-positive bacteria resist more MNPs than gram-negative bacteria due to the thick peptidoglycan layer in their cell walls. Furthermore, it contains negatively charged lipopolysaccharide, which is susceptible to positive change MNP ions. The negative charge of both the gram-positive and gram-negative bacteria cell walls also played a key role during the interaction with the NPs and releasing the ions [100]. The electrostatic interactions between the NPs and the cells affect the MNPs' toxicity. Negatively charged bacterial membranes cause positively charged MNPs to be released, which can damage negatively charged lipids and charged phospholipids in cells [101]. For instance, gram-negative (*Proteus vulgaris and E. coli*) and gram-positive (*Streptococcus mutans*, *Staphylococcus aureus*, and *Streptococcus pyogenes*) bacteria were more effectively inhibited by positively charged AgNPs than by negatively charged AgNPs. The bactericidal activity of the neutrally charged AgNPs was also superior to that of the negatively charged AgNPs but lower than that of the positively charged AgNPs. Of all charged kinds of AgNPs, *Proteus vulgaris* was the bacterium with the highest resistance [102].

nZVI is among the noble NPs extensively used to remove toxicants and contaminants from wastewater. nZVI is highly reactive with a high surface area of decontamination of water inorganic and organic pollutants. Thus, the nZVI shows toxicity toward microorganisms [103]. The formation of ROS in the presence of DO or physical damage to the microbes, algal cells, and fungi due to the strong binding of the nZVI on the cell surface of the microorganism results in decreasing cell mobility and nutrient flow. The nZVI particles are adsorbed on the surface cells, which start to release Fe^{2+} and react with H_2O_2 produced by the mitochondria to form ROS (OH^*, O^{2-}, and FeO^{2+}) through the Fenton reaction. An excess concentration of ROS causes oxidative stress and damage to the critical biomolecules (nucleic acids, proteins, and lipids) (Eq. 12.3–12.5). In another mechanism, the interaction between the nZVI and microbial cells directly causes the outer membrane damage to inactivate the cells. Aerobic microorganisms can easily be damaged by the nZVI due to its limited concentration of DO and oxidation-reduction potential [104].

$$Fe^0 + O_2 + 2H^+ \rightarrow Fe^{2+} + H_2O_2 \qquad 12.3$$

$$Fe^0 + H_2O_2 + 2H^+ \rightarrow Fe^{2+} + 2H_2O \quad 12.4$$

$$Fe^{2+} + H_2O_2 \rightarrow Fe^{3+} + {}^*OH + OH^- \quad 12.5$$

It is also known that iron-oxidizing bacteria benefit from nZVI. Carbohydrate and carboxymethyl cellulose, chitosan, and sodium carboxymethyl cellulose-modified nZVI reduced the toxicity of the NPs, and they are suitable for the microbes as a source of nutrients [105]. The toxicity of bare nZVI, sodium alginate, and bentonite-modified nZVI (Ben-nZVI) was tested for gram-negative *E. coli.* Complete damage of the outer membrane and incomplete inhibition of the bacteria within 30 min using a 7.2-log CFU/mL initial concentration and 0.1 g/L of bare nZVI were observed. However, the toxicity of sodium alginate (SA-nZVI) and bentonite (Ben-ZVI) decreased. 6.4-log CFU/mL and 7.0-log CFU/mL viable cell densities were recorded after treatment with 0.1 g/L of 0.5SA-nZVI and 15Ben-nZVI within one hour, respectively. Both bentonite and sodium alginate are used as stabilizers for nZVI and reduce toxicity levels against *E. coli.* Neither of the stabilizers showed toxicity to the bacteria [106].

A study of NiNPs in rats confirmed that the functional groups of amino acids and proteins bind with Ni and enhance the generation of ROS [107]. The activity of antioxidant enzymes like CAT and SOD in cells becomes active to counteract ROS damage at the beginning if ROS levels become higher and the antioxidant resistance declines. The levels of malondialdehyde (MDA) and nitric oxide (NO) were also increased. Finally, excess levels of ROS and induced oxidative stress lead to the death of cells [108].

12.6 REGULATORY CHALLENGES OF METALLIC NANOPARTICLES

Setting regulatory definitions for NPs is difficult for various reasons. One of the current challenges is the lack of accurate data and the knowledge gap about the nanoparticle's interaction with cells and the fate of the nanoparticle in a complex environment. The unique physicochemical properties of the NPs and the difficulty in quantifying the particles after exposure are still challenges for regulators to establish clear-cut regulatory rules and definitions [109]. Further, the toxicity of the NPs depends on the medium. Thus, the toxicity of the medium cannot be used for other mediums. Test duration is another factor; exposure of the NPs to the target organism may be static or dynamic, pulsed or continuous [110].

To date, the United States of America (USA) and China are among the producers and consumers of ENMs. However, the regulatory frameworks and guidelines of the ENMs still get little attention from consumers and procedures [111]. Mostly, soft rules are not practical because they are not well organized. Due to a knowledge gap, a lack of awareness, and the complexity of the NPs in the aquatic system, it is difficult to set rules and regulations. Furthermore, the lack of exposure and hazard data on the MNPs is limited to dealing with the environmental and health impacts to strengthen and enforce the rules and legislation on nanomaterials. Other challenges that should be studied and considered in setting regulations are the lack of quantification of the NPs, the absence of nonspecific labeling, the lack of the exact dose of the NPs, the absence of studies on human beings, and the generalization and extrapolation of in vivo or in vitro studies to humans [112].

Nowadays, the USA and the European Union (EU) pay attention to regulations on nanomaterials. In the USA, organizations such as the Federal Insecticide, Fungicide, and Rodenticide Act (FIFRA), Food and Drug Administration (FDA), and Environmental Protection Agency (EPA) under the Toxic Substances Control Act (TSCA) are trying to set regulations for nanomaterials [113]. However, the FDA prepared guidelines for drugs, food, cosmetics, and animal foods containing nanomaterials. It does not cover nanomedicine, and it does not categorize nanomaterials [114]. Similarly, the EU prepared nano-specific provisions for cosmetics, food, and biocides, which include the term nanomaterials. They have defined nanomaterials as natural, incidental, and intentionally fabricated with one or more dimensions, higher surface area, and a 1–100 nm particle size [115]. To clearly define and establish rules and regulations for the nanomaterials, a categorical strategy is important based on their similar toxicological profiles, physicochemical properties, exposure, and applications. A medicine should be a nanomaterial, defined as a nano-drug or equipment [116].

12.7 WAYS OF MEASURING THE TOXICITY OF MNPS

The toxicity of the NPs is measured based on different frameworks, extending from in vitro cell culture assessment to basic model organisms like sea *urchin* and *daphnia* and higher vertebrate animals such as primates and rodents. Small organisms and cell lines are essential to assess the cell-level genotoxicity or nanotoxicity. However, higher vertebrates are used to investigate complex and detailed physiological interactions [117]. Studying the toxicity of the NPs using the in vitro method gives an outcome such as calculating the half maximal effective concentration (EC_{50}), IC_{50}, minimum inhibitory concentration (MIC), and minimum bactericidal concentration (MBC) results and studying cell lines, cell cycles, and death. On the other hand, the in vivo (cell-based assay) studies exhibited a deep understanding of the pharmacokinetics, pharmacodynamics, cytokine storm, absorption, distribution, metabolism and excretion (ADME), and hematological and histopathological study results [118]. Researchers, individually or in groups, frequently used the in vitro method to measure the toxicity of the NPs. Because the in vitro experiment is simple, cheap, and devoid of any ethical issues. However, studying the toxicity of the NPs using the in vitro method missed different effects such as complex cells and diversity of cell types, cell–matrix interactions, and hormonal effects. The long-term chronic effects cannot study using in vitro experiments. On the other hand, the in vivo method is time-consuming and expensive and requires ethical considerations. The exposed NPs in the in vivo study were detected in live or dead animals via radiolabels [119].

The in vitro assessment of toxic NPs is based on proliferation, DNA damage, oxidative stress, apoptosis, and necrosis assays. In the proliferation assay, the cellular metabolism of metabolically active cells is measured using 3-(4,5-dimet hylthiazol-2-yl)-2,5-diphenyltetrazolium bromide (MTT), which is the most common salt employed to measure the toxicity of the NPs. DNA synthesis assays are another effective and reliable laboratory method to detect proliferation cells. Visual inspection of proliferation cell counting via the clonogenic assay is also a critical assay [120]. In oxidative stress, ROS and reactive nitrogen species are formed and detected by the reaction of stable O^{2-} radicals with 2,2,6,6-tetramethylpiperidine (TEMP) and characterized by X-band electron paramagnetic resonance (EPR). But this is highly costly, and other approaches, such as fluorescent probe molecules, can be used as an alternative. Apoptosis, a programmed cell death, occurs due to the overproduction of free radicals. Thus, apoptosis and DNA damage are measured based on the Comet assay, Annexin-V assay, TdT-mediated dUTP-biotin nick end labeling (TUNEL) assay, and inspection of morphological changes. Changes in cell size and DNA fragmentation are also indications of apoptosis. DNA damages are detected by combined instruments such as liquid chromatography with mass spectrometry (MS), liquid chromatography with electrochemical detection (ED), and high-performance liquid chromatography (HPLC) with ultraviolet detection (UD), and electrophoresis [120, 121]. In necrosis assays, the death or injury of cells in a tissue is measured based on membrane integrity, which determines the viability of the cell by measuring the uptake of dyes. In another case, necrosis can be detected based on the appearance of the lactate dehydrogenase (LDH) enzyme [122].

Toxicity measurement of NPs using in vivo methods is detected or measured using histopathology, biodistribution, clearance, hematology, serum chemistry, etc. [120]. Organization for Economic Cooperation and Development (OECD) recommendations advise using additional methods to assess the acute in vivo toxicity of NPs, including the lethal dosage 50 (LD_{50}), oral toxicity tests, corrosion, eye irritation, and skin toxicity tests [123].

12.8 CHALLENGES IN MEASURING THE TOXICITY OF MNPS

The extraordinarily small size and low mass concentration of the NPs make it difficult for the instruments to measure and quantify water or soil contaminants. Because most of the light microscopy techniques work at 200 nm, they are not able to detect the NPs. The transformation of the NPs original structure, shape, size, reactivity, surface area, etc., in the environment by the action of natural matter and biological nature into different forms such as ions, particulates, colloids, and interfacial reactions results in an inaccurate measurement. The NPs have an unpredictable response to the biological/environmental systems. The lack of standard or universal measurements of NPs toxicity mechanisms is also another challenge [124, 125].

Most of the guidelines for toxicity measurement, like OECD and ISO 8692:2004, for algal growth inhibition by the NPs required water soluble test chemicals. However, NPs are slightly soluble or not soluble; hence, the NPs are agglomerated/aggregated instead of dissolving. Therefore, it is uncertain to measure the toxicity of the NPs using these guidelines, which raises concerns about the results [126].

12.9 CONCLUSION AND FUTURE PROSPECTS

The application of NPs is growing along with the advancement and developments of nanotechnology for drug delivery, catalysts, environmental remediation, biosensors, energy, food packaging, etc., and MNPs are the basic elements of nanotechnology.

Numerous investigations on the synthesis of NPs utilizing physical, chemical, and biological techniques have been conducted. The NPs performance is enhanced by their small size and large surface area. But nothing is known about how hazardous NPs are to the environment. For NPs, there are no precise regulatory frameworks or rules. The MNPs are spread into the aquatic system from different sources and interact with the receiving environment. The MNPs have the potential to affect microorganisms and vertebrate animals via DNA damage, releasing metal ions, inactivation, and denaturation of enzymes, proteins, and metabolic systems. The toxicity measurements of the MNPs are carried out via in vitro and in vivo experiments. The in vitro assay is simple in assessing the NPs toxicity, but limited information is obtained.

The small size of the MNPs gives them the ability to penetrate the cells of organisms. Therefore, a detailed study is desired to investigate the toxicity of the MNPs in higher animals via in vivo methods. The chemical or physical synthesis methods of the NPs may have their own environmental impact. Thus, green synthesis strategies have to be investigated as an alternative approach for large-scale production due to their environmentally friendly procedures.

The fast growth of nanotechnology realizes various functional materials in different fields. Therefore, the use of small-size NPs in food packing, cosmetics, and other applications can pose high risks to consumers, and strict regulations should be stated. Safety and ethical guideline frameworks related to the toxicity of NPs should be prepared. Government agencies should also proactively engage in developing management strategies for nanomaterials. Guidelines such as neutralization of the NPs before disposal, recycling procedures for the particles, dilution, and other methods that reduce the reactivity of the NPs should be developed [127]. An awareness campaign has to be created for the consumers of MNPs.

The NP's interaction with in vivo or in vitro studies should be investigated using artificial-based computational and theoretical approaches combined with advanced characterization techniques. Standardization of the NP concentration, pH, and solubility of each NP should be done to exploit more data, which may help to ensure the regulations. Most of the studies that confirm the toxicity of the MNPs are obtained from in vitro cell cultures of small organisms. However, toxicity measurement of NPs in higher animals is limited due to the lack of appropriate tools for in vivo assessment and characterization. A collaborative study is required to overcome such challenges. Furthermore, models and sensors that are used to assess the aged NPs and their fate in the complex environment should be designed.

The interaction, transformation mechanisms, and fate of the NPs in the aquatic system are not studied for most of the MNPs, which hinders the development of regulatory frameworks and guidelines. There are only a few preliminary reports of specific MNPs, like AgNPs, dealing with toxicity, fate, and interaction in the aquatic system at subcellular and cellular levels. Therefore, the transformation mechanism and their fate in soil and aquatic water should be studied for every MNPs, including their effect after a prolonged time. The MNPs are synthesized using different supporting, coating, and surfactant agents to prevent agglomeration and modify the surface charge. However, the surfactants and supporting materials themselves are toxic. Other alternative approaches have to be used, such as compatible molecules derived from plants or other microorganisms.

ACKNOWLEDGMENTS

The authors are also thankful to Dr. Jamilur R. Ansari, Dronacharya College of Engineering, Gurgaon, India, for the needful editing in the Chapter. SHG and BGG gratefully acknowledge the support from Jigjiga University, Ethiopia

REFERENCES

[1] M. Mortimer, P.A. Holden, Fate of engineered nanomaterials in natural environments and impacts on ecosystems, in: N. Marmiroli, J. White, J. Song (Eds.), Exposure to Engineered Nanomaterials in the Environment, Elsevier Inc., 2019: pp. 59–104. https://doi.org/10.1016/B978-0-12-814835-8.00003-0.

[2] S.H. Gebre, Recent developments in the fabrication of magnetic nanoparticles for the synthesis of trisubstituted pyridines and imidazoles: A green approach, Synthetic Communications. 51 (2021) 1669–1699. https://doi.org/10.1080/00397911.2021.1900257.

[3] S.H. Gebre, Synthesis and potential applications of trimetallic nanostructures, New Journal of Chemistry. 46 (2022) 5438–5459. https://doi.org/10.1039/D1NJ06074K.

[4] S.H. Gebre, M.G. Sendeku, Trimetallic nanostructures and their applications in electrocatalytic energy conversions, Journal of Energy Chemistry. 65 (2022) 329–352. https://doi.org/10.1016/j.jechem.2021.06.006.

[5] J. Zhang, W. Guo, Q. Li, Z. Wang, S. Liu, The effects and the potential mechanism of environmental transformation of metal nanoparticles on their toxicity in organisms, Environmental Science: Nano. 5 (2018) 2482–2499. https://doi.org/10.1039/c8en00688a.

[6] I.S. Yunus, A. Kurniawan, D. Adityawarman, A. Kurniawan, D. Adityawarman, A. Indarto, Nanotechnologies in water and air pollution treatment, Environmental Technology Reviews. 1 (2012) 136–148. https://doi.org/10.1080/21622515.2012.733966.

[7] C. Peng, C. Shen, S. Zheng, W. Yang, H. Hu, J. Liu, Transformation of CuO nanoparticles in the aquatic environment: Influence of pH, electrolytes and natural organic matter, Nanomaterials. 7 (2017) 326. https://doi.org/10.3390/nano7100326.

[8] S.H. Gebre, M.G. Sendeku, New frontiers in the biosynthesis of metal oxide nanoparticles and their environmental applications: An overview, SN Applied Sciences. 1 (2019) 928. https://doi.org/10.1007/s42452-019-0931-4.

[9] S.H. Gebre, Recent developments of supported and magnetic nanocatalysts for organic transformations: An up - to - date review, Applied Nanoscience. 13 (2023) 15–63. https://doi.org/10.1007/s13204-021-01888-3.

[10] C. Fajardo, G. Martinez-rodriguez, J. Blasco, J. Miguel, B. Thomas, M. De Donato, Nanotechnology in aquaculture: Applications, perspectives and regulatory challenges, Aquaculture and Fisheries. 7 (2023) 185–200. https://doi.org/10.1016/j.aaf.2021.12.006.

[11] N. Abbas, Y. Li, J. Ding, J.P. Liu, H. Luo, J. Du, W. Xia, A. Yan, F. Wang, J. Zhang, A facile synthesis of directly gas-phase ordered high anisotropic Sm-Co based non-segregated nanoalloys by cluster beam deposition method, Materials and Design. 181 (2019) 108052. https://doi.org/10.1016/j.matdes.2019.108052.

[12] R. Singla, A. Guliani, A. Kumari, S.K. Yadav, Metallic nanoparticles, toxicity issues and applications in medicine, in: S.K. Yadav (Ed.), Nanoscale Materials in Targeted Drug Delivery, Theragnosis and Tissue Regeneration, Springer Science, 2016: pp. 41–80. https://doi.org/10.1007/978-981-10-0818-4.

[13] M.N. Javed, E.S. Dahiya, A.M. Ibrahim, Md.S. Alam, F.A. Khan, F.H. Pottoo, Recent advancement in clinical application of nanotechnological approached targeted delivery of herbal drugs, in: S. Beg, M.A. Barkat, F.J. Ahmad (Eds.), Nanophytomedicine, Springer Singapore, Singapore, 2020: pp. 151–172. https://doi.org/10.1007/978-981-15-4909-0_9.

[14] M.S. Alam, M.N. Javed, F.H. Pottoo, A. Waziri, F.A. Almalki, M.S. Hasnain, A. Garg, M.K. Saifullah, QbD approached comparison of reaction mechanism in microwave synthesized gold nanoparticles and their superior catalytic role against hazardous nirto-dye, Applied Organometallic Chemistry. (2019). https://doi.org/10.1002/aoc.5071.

[15] M.N. Javed, F.H. Pottoo, A. Shamim, M.S. Hasnain, M.S. Alam, Design of experiments for the development of nanoparticles, nanomaterials, and nanocomposites, in: S. Beg (Ed.), Design of Experiments for Pharmaceutical Product Development, Springer Singapore, Singapore, 2021: pp. 151–169. https://doi.org/10.1007/978-981-33-4351-1_9.

[16] G. Domingo, M. Bracale, C. Vannini, Phytotoxicity of silver nanoparticles to aquatic plants, algae, and microorganisms, in: Nanomaterials in Plants, Algae and Microorganisms, Elsevier Inc., 2019: pp. 143–168. https://doi.org/10.1016/B978-0-12-811488-9.00008-1.

[17] F.H. Pottoo, N. Tabassum, Md.N. Javed, S. Nigar, R. Rasheed, A. Khan, Md.A. Barkat, Md.S. Alam, A. Maqbool, M.A. Ansari, G.E. Barreto, G.M. Ashraf, The synergistic effect of raloxifene, fluoxetine, and bromocriptine protects against pilocarpine-induced status epilepticus and temporal lobe epilepsy, Mol Neurobiol. 56 (2019) 1233–1247. https://doi.org/10.1007/s12035-018-1121-x.

[18] S. Raj, R. Manchanda, M. Bhandari, Md.S. Alam, Review on natural bioactive products as radioprotective therapeutics: Present and past perspective, CPB. 23 (2022) 1721–1738. https://doi.org/10.2174/1389201023666220110104645.

[19] N. Kumari, N. Daram, Md.S. Alam, A.K. Verma, Rationalizing the use of polyphenol nano-formulations in the therapy of neurodegenerative diseases, CNSNDDT. 21 (2022) 966–976. https://doi.org/10.2174/1871527321666220512153854.

[20] F.H. Pottoo, N. Tabassum, Md.N. Javed, S. Nigar, S. Sharma, Md.A. Barkat, Harshita, Md.S. Alam, M.A. Ansari, G.E. Barreto, G.M. Ashraf, Raloxifene potentiates the effect of fluoxetine against maximal electroshock induced seizures in mice, European Journal of Pharmaceutical Sciences. 146 (2020) 105261. https://doi.org/10.1016/j.ejps.2020.105261.

[21] S.A. Khan, Metal nanoparticles toxicity: Role of physicochemical aspects, in: Metal Nanoparticles for Drug Delivery and Diagnostic Applications, Elsevier Inc., 2020: pp. 1–12. https://doi.org/10.1016/B978-0-12-816960-5.00001-X.

[22] M. Kumar, M. Na, J.R Ansari, Analysis of the heating ability by varying the size of Fe_3O_4 magnetic nanoparticles for hyperthermia, Nanoscience and Technology: An International Journal. (2022). https://doi.org/10.1615/NanoSciTechnolIntJ.2022040075.

[23] R. Kumar, G. Dhamija, J.R. Ansari, Md.N. Javed, Md.S. Alam, C-Dot nanoparticulated devices for biomedical applications, in: Nanotechnology, 1st ed., CRC Press, Boca Raton, 2022: pp. 271–299. https://doi.org/10.1201/9781003220350-15.

[24] S. Singhal, M. Gupta, Md.S. Alam, Md.N. Javed, J.R. Ansari, Carbon allotropes-based nanodevices, in: Nanotechnology, 1st ed., CRC Press, Boca Raton, 2022: pp. 241–269. https://doi.org/10.1201/9781003220350-14.

[25] J.R. Ansari, N. Singh, R. Ahmad, D. Chattopadhyay, A. Datta, Controlling self-assembly of ultra-small silver nanoparticles: Surface enhancement of Raman and fluorescent spectra, Optical Materials. 94 (2019) 138–147. https://doi.org/10.1016/j.optmat.2019.05.023.

[26] Y.-S. Chen, Y.-C. Hung, I. Liau, G.S. Huang, Assessment of the in vivo toxicity of gold nanoparticles, Nanoscale Research Letters. 4 (2009) 858–864. https://doi.org/10.1007/s11671-009-9334-6.

[27] S. Mishra, S. Sharma, M.N. Javed, F.H. Pottoo, M.A. Barkat, Harshita, M.S. Alam, M. Amir, M. Sarafroz, Bioinspired nanocomposites: Applications in disease diagnosis and treatment, PNT. 7 (2019) 206–219. https://doi.org/10.2174/2211738507666190425121509.

[28] F.H. Pottoo, Md.N. Javed, Md.A. Barkat, Md.S. Alam, J.A. Nowshehri, D.M. Alshayban, M.A. Ansari, Estrogen and serotonin: Complexity of interactions and implications for epileptic seizures and epileptogenesis, CN. 17 (2019) 214–231. https://doi.org/10.2174/1570159X16666180628164432.

[31] Y. Pan, S. Neuss, A. Leifert, M. Fischler, F. Wen, U. Simon, G. Schmid, W. Brandau, W. Jahnen-dechent, Size-dependent cytotoxicity of gold nanoparticles, Small. 3 (2007) 1941–1949. https://doi.org/10.1002/smll.200700378.

[32] S. Huo, S. Jin, X. Ma, X. Xue, K. Yang, A. Kumar, P.C. Wang, J. Zhang, Z. Hu, X. Liang, Ultrasmall gold nanoparticles as carriers for nucleus-based gene therapy due to size-dependent, ACS Nano. 8 (2014) 5852–2862.

[33] S.B. Yaqoob, R. Adnan, R.M.R. Khan, M. Rashid, Gold, silver, and palladium nanoparticles: A chemical tool for biomedical applications, Frontiers in Chemistry. 8 (2020) 376. https://doi.org/10.3389/fchem.2020.00376.

[34] T.X.S. Liang, L.S. Wong, A.C.T.A. Dhanapal, S. Djearamane, Toxicity of metals and metallic nanoparticles on nutritional properties of microalgae, Water Air Soil Pollution. 231 (2020) 52.

[35] S.H. Gebre, Bio-inspired synthesis of metal and metal oxide nanoparticles: The key role of phytochemicals, Journal of Cluster Science. 34 (2023) 665–704. https://doi.org/10.1007/s10876-022-02276-9.

[36] S.J. Soenen, P. Rivera-gil, J. Montenegro, W.J. Parak, S.C. De Smedt, K. Braeckmans, Cellular toxicity of inorganic nanoparticles: Common aspects and guidelines for improved nanotoxicity evaluation, Nano Today. 6 (2011) 446–465. https://doi.org/10.1016/j.nantod.2011.08.001.

[37] R.A. Crane, T.B. Scott, Nanoscale zero-valent iron: Future prospects for an emerging water treatment technology, Journal of Hazardous Materials. 211–212 (2012) 112–125. https://doi.org/10.1016/j.jhazmat.2011.11.073.

[38] M. Bundschuh, J. Filser, S. Lüderwald, M.S. Mckee, G. Metreveli, G.E. Schaumann, R. Schulz, S. Wagner, Nanoparticles in the environment: Where do we come from, where do we go to? Environmental Sciences Europe. 30 (2018). https://doi.org/10.1186/s12302-018-0132-6.

[39] A.N. Banu, N. Kudesia, A.M.R.I. Pakrudheen, J. Wahengbam, Toxicity, bioaccumulation, and transformation of silver nanoparticles in aqua biota: A review, Environmental Chemistry Letters. 19 (2021) 4275–4296. https://doi.org/10.1007/s10311-021-01304-w.

[40] I. Khan, K. Saeed, I. Khan, Nanoparticles: Properties, applications and toxicities, Arabian Journal of Chemistry. 12 (2017) 908–931. https://doi.org/10.1016/j.arabjc.2017.05.011.

[41] W. Xue, D. Huang, G. Zeng, J. Wan, M. Cheng, C. Zhang, C. Hu, J. Li, Performance and toxicity assessment of nanoscale zero valent iron particles in the remediation of contaminated soil: A review, Chemosphere. (2018). https://doi.org/10.1016/j.chemosphere.2018.07.118.

[42] N.B. Turan, H.S. Erkan, G.O. Engin, M.S. Bilgili, Nanoparticles in the aquatic environment: Usage, properties, transformation and toxicity-A review, Process Safety and Environmental Protection. 130 (2019) 238–249. https://doi.org/10.1016/j.psep.2019.08.014.

[43] P.C. Ray, H. Yu, P.P. Fu, Toxicity and environmental risks of nanomaterials: Challenges and future needs, Journal of Environmental Science and Health-Part C Environmental Carcinogenesis and Ecotoxicology Reviews. 27 (2010) 1–35. https://doi.org/10.1080/10590500802708267.Toxicity.

[44] H. Selck, R.D. Handy, T.F. Fernandes, S.J. Klaine, E.J. Petersen, Nanomaterials in the aquatic environment: A European union–United States perspective on the status of ecotoxicity testing, research priorities, and challenges ahead, Environmental Toxicology and Chemistry. 35 (2016) 1055–1067. https://doi.org/10.1002/etc.3385.

[45] J. Eixenberger, C. Anders, R. Hermann, K.M. Reddy, A. Punnoose, D. Wingett, Rapid dissolution of ZNO nanoparticles induced by biological buffers significantly impacts cytotoxicity, Chemical Research in Toxicology. 30 (2017) 1641–1651. https://doi.org/10.1021/acs.chemrestox.7b00136.

[46] R. Arif, S. Jadoun, Rahisuddin, A review on recent developments in the biosynthesis of silver nanoparticles and its biomedical applications, Medical Devices and Sensors. 4 (2021) e10158. https://doi.org/10.1002/mds3.10158.

[47] J. Singh, T. Dutta, K.H. Kim, M. Rawat, P. Samddar, P. Kumar, "Green" synthesis of metals and their oxide nanoparticles: Applications for environmental remediation, Journal of Nanobiotechnology. 16 (2018) 84. https://doi.org/10.1186/s12951-018-0408-4.

[48] J. Tang, Y. Wu, S. Esquivel-elizondo, S.J. Sørensen, B.E. Rittmann, How microbial aggregates protect against nanoparticle toxicity, Trends in Biotechnology. 36 (2018) 1171–1182. https://doi.org/10.1016/j.tibtech.2018.06.009.

[49] Z. Wang, L. Zhang, J. Zhao, B. Xing, Environmental processes and toxicity of metallic nanoparticles in aquatic systems as affected by natural organic matter, Environmental Science: Nano. 3 (2016) 240–255. https://doi.org/10.1039/c5en00230c.

[50] I.A. Mudunkotuwa, J.M. Pettibone, V.H. Grassian, Environmental implications of nanoparticle aging in the processing and fate of copper-based nanomaterials, Environmental Science Technology. 46 (2012) 7001 – 7010.

[51] D.J. Dryer, G. V Korshin, Investigation of the Reduction of Lead Dioxide by Natural Organic Matter, Environmental Science Technology. 41 (2007) 5510–5514.

[52] Md.S. Alam, Md.F. Naseh, J.R. Ansari, A. Waziri, Md.N. Javed, A. Ahmadi, M.K. Saifullah, A. Garg, Synthesis Approaches for Higher Yields of Nanoparticles, in: Nanomaterials in the Battle Against Pathogens and Disease Vectors, 1st ed., CRC Press, Boca Raton, 2022: pp. 51–82. https://doi.org/10.1201/9781003126256-3.

[53] N. Singh, J.R. Ansari, M. Pal, N.T.K. Thanh, T. Le, A. Datta, Synthesis and magnetic properties of stable cobalt nanoparticles decorated reduced graphene oxide sheets in the aqueous medium, Journal of Materials Science: Materials in Electronics. 31 (2020) 15108–15117. https://doi.org/10.1007/s10854-020-04075-2.

[54] Md.F. Naseh, J.R. Ansari, Md.S. Alam, Md.N. Javed, Sustainable nanotorus for biosensing and therapeutical applications, in: U. Shanker, C.M. Hussain, M. Rani (Eds.), Handbook of Green and Sustainable Nanotechnology, Springer International Publishing, Cham, 2022: pp. 1–21. https://doi.org/10.1007/978-3-030-69023-6_47-1.

[55] M. Kumar, Madhavi, J.R. Ansari, studies on the heating ability by varying the size of fe3o4 magnetic nanoparticles for hyperthermia, Nanoscience and Technology: An International Journal. 13 (2022) 33–45. https://doi.org/10.1615/NanoSciTechnolIntJ.2022040075.

[56] C.A. Sunilbhai, Md.S. Alam, K.K. Sadasivuni, J.R. Ansari, SPR assisted diabetes detection, in: K.K. Sadasivuni, J.-J. Cabibihan, A.K. A M Al-Ali, R.A. Malik (Eds.), Advanced Bioscience and Biosystems for Detection and Management of Diabetes, Springer International Publishing, Cham, 2022: pp. 91–131. https://doi.org/10.1007/978-3-030-99728-1_6.

[57] Y. Zhang, J. Xu, Y. Yang, B. Sun, K. Wang, L. Zhu, Impacts of Proteins on Dissolution and Sulfidation of Silver Nanowires in an Aquatic Environment: Importance of Surface Charges, Environmental Science & Technology. 54 (2020) 5560–5568. https://doi.org/10.1021/acs.est.0c00461.

[58] D.B. Chithrani, Intracellular uptake, transport, and processing of gold nanostructures, Molecular Membrane Biology. 27 (2010) 299–311. https://doi.org/10.3109/09687688.2010.507787.

[59] J.R. Ansari, N. Singh, S. Mohapatra, R. Ahmad, N.R. Saha, D. Chattopadhyay, M. Mukherjee, A. Datta, Enhanced near infrared luminescence in Ag@Ag2S core-shell nanoparticles, Applied Surface Science. 463 (2019) 573–580. https://doi.org/10.1016/j.apsusc.2018.08.244.

[60] C. Lei, L. Zhang, K. Yang, L. Zhu, D. Lin, Toxicity of iron-based nanoparticles to green algae: Effects of particle size, crystal phase, oxidation state and environmental aging, Environmental Pollution. 218 (2016) 505–512. https://doi.org/10.1016/j.envpol.2016.07.030.

[61] A. Tesfaye, Y. Liu, D.N. Bekele, Z. Dong, R. Naidu, G. Neda, Sustainability and environmental ethics for the application of engineered nanoparticles, Environmental Science and Policy. 103 (2020) 85–98. https://doi.org/10.1016/j.envsci.2019.10.013.

[62] A. Manuja, B. Kumar, R. Kumar, D. Chhabra, M. Ghosh, M. Manuja, B. Brar, Y. Pal, B.N. Tripathi, M. Prasad, Metal/metal oxide nanoparticles: Toxicity concerns associated with their physical state and remediation for biomedical applications, Toxicology Reports. 8 (2021) 1970–1978. https://doi.org/10.1016/j.toxrep.2021.11.020.

[63] A. Woźniak, A. Malankowska, G. Nowaczyk, B.F. Grześkowiak, K. Tuśnio, R. Słomski, A. Zaleska-Medynska2, S. Jurga, Size and shape-dependent cytotoxicity profile of gold nanoparticles for biomedical applications, Journal of Materials Science: Materials in Medicine. 28 (2017) 92.

[64] X. Xie, J. Liao, X. Shao, Q. Li, Y. Lin, The effect of shape on cellular uptake of gold nanoparticles in the forms of stars, rods, and triangles, Scientific Reports. 7 (2017) 3827. https://doi.org/10.1038/s41598-017-04229-z.

[65] Y.J. Lee, E. Ahn, Y. Park, Shape-dependent cytotoxicity and cellular uptake of gold nanoparticles synthesized using green tea extract, Nanoscale Research Letters. 14 (2019) 129.

[66] D. Acharya, K.M. Singha, P. Pandey, B. Mohanta, J. Rajkumari, L.P. Singha, Shape dependent physical mutilation and lethal effects of silver nanoparticles on bacteria, Scientific Reports. 8 (2018) 201. https://doi.org/10.1038/s41598-017-18590-6.

[67] A.M.E.L. Badawy, R.G. Silva, B. Morris, K.G. Scheckel, M.T. Suidan, Surface Charge-Dependent Toxicity of Silver Nanoparticles, Environmental Science & Technology. 45 (2011) 283–287.

[68] W. Zhang, B. Xiao, T. Fang, Chemical transformation of silver nanoparticles in aquatic environments: Mechanism, morphology and toxicity, Chemosphere. 191 (2018) 324–334. https://doi.org/10.1016/j.chemosphere.2017.10.016.

[69] S. Ma, K. Zhou, K. Yang, D. Lin, Hetero-agglomeration of oxide nanoparticles with algal cells: Effects of particle type, ionic strength and pH, Environmental Science & Technology. (2014). https://doi.org/10.1021/es504730k.

[70] J. Zhao, Z. Wang, Y. Dai, B. Xing, Mitigation of CuO nanoparticle-induced bacterial membrane damage by dissolved organic matter, Water Research. (2013) 1–10. https://doi.org/10.1016/j.watres.2012.11.058.

[71] Z. Li, K. Greden, P.J.J. Alvarez, G.V. Lowry, Adsorbed polymer and NOM limits adhesion and toxicity of nano scale zerovalent iron to E coli, Environmental Science & Technology. 44 (2010) 3462–3467.

[72] L.R.R. Souza, T.Z. Corrêa, A.T. Bruni, M.A.M.S. da Veiga, The effects of solubility of silver nanoparticles, accumulation, and toxicity to the aquatic plant Lemna minor, Environmental Science and Pollution Research. 28 (2021) 16720–16733.

[73] T. Li, B. Albee, M. Alemayehu, R. Diaz, L. Ingham, S. Kamal, M. Rodriguez, S.W. Bishnoi, comparative toxicity study of Ag, Au, and Ag-Au bimetallic nanoparticles on Daphnia magna, Analytical and Bioanalytical Chemistry. 398 (2010) 689–700. https://doi.org/10.1007/s00216-010-3915-1.

[74] R.R. Miranda, V. Gorshkov, B. Korzeniowska, J. Stefan, F.F. Neto, F. Kjeldsen, R. Rank, V. Gorshkov, B. Korzeniowska, J. Kempf, F.F. Neto, F. Kjeldsen, Co-exposure to silver nanoparticles and cadmium induce metabolic adaptation in HepG2 cells, Nanotoxicology. 12 (2018) 781–795. https://doi.org/10.1080/17435390.2018.1489987.

[75] V. Iswarya, J. Manivannan, A. De, S. Paul, R. Roy, J.B. Johnson, R. Kundu, N. Chandrasekaran, A. Mukherjee, A. Mukherjee, Surface capping and size-dependent toxicity of gold nanoparticles on different trophic levels, Environmental Science and Pollution Research. 23 (2015) 4844–4858. https://doi.org/10.1007/s11356-015-5683-0.

[76] F. Cappellini, Y. Hedberg, S. Mccarrick, J. Hedberg, R. Derr, G. Hendriks, I.O. Wallinder, H.L. Karlsson, Mechanistic insight into reactivity and (Geno) toxicity of well-characterized nanoparticles of cobalt metal and oxides, Nanotoxicology. 12 (2018) 602–620. https://doi.org/10.1080/17435390.2018.1470694.

[77] B. Annangi, J. Bach, G. Vales, L. Rubio, R. Marcos, A. Herna´ndez, Long-term exposures to low doses of cobalt nanoparticles induce cell transformation enhanced by

oxidative damage, Nanotoxicology. 9 (2014) 1–10. https://doi.org/10.3109/17435390.2014.900582.
[78] Y. Wu, L. Kong, Advance on toxicity of metal nickel nanoparticles, Environmental Geochemistry and Health. 42 (2020) 2277–2286. https://doi.org/10.1007/s10653-019-00491-4.
[79] R. Magaye, Q. Zhou, L. Bowman, B. Zou, G. Mao, J. Xu, V. Castranova, J. Zhao, M. Ding, Metallic nickel nanoparticles may exhibit higher carcinogenic potential than fine particles in JB6 cells, PLoS One. 9 (2014) e92418. https://doi.org/10.1371/journal.pone.0092418.
[80] R. Magaye, J. Zhao, Recent progress in studies of metallic nickel and nickel-based nanoparticles' genotoxicity and carcinogenicity, Environmental Toxicology and Pharmacology. 34 (2012) 644–650. https://doi.org/10.1016/j.etap.2012.08.012.
[81] R. Magaye, Y. Gu, Y. Wang, H. Su, Q. Zhou, G. Mao, H. Shi, X. Yue, B. Zou, J. Xu, J. Zhao, In vitro and in vivo evaluation of the toxicities induced by metallic nickel nano and fine particles, Journal of Molecular Histology. 47 (2016) 273–286. https://doi.org/10.1007/s10735-016-9671-6.
[82] L. Kong, X. Gao, J. Zhu, K. Cheng, M. Tang, Mechanisms involved in reproductive toxicity caused by nickel nanoparticle in female rats, Environmental Toxicology. 31 (2016) 1674–1683. https://doi.org/10.1002/tox.
[83] L. Kong, X. Gao, J. Zhu, T. Zhang, Y. Xue, M. Tang, Reproductive toxicity induced by nickel nanoparticles in Caenorhabditis elegans, Environmental Toxicology. 32 (2016) 1530–1538. https://doi.org/10.1002/tox.
[84] I. Iavicoli, M. FarinaL, L. Fontana, D. Lucchetti, V. Leso, C. Fanali, V. Cufino, A. Boninsegna, K. Leopold, R. Schindl, D. Brucker, A. Sgambato, In Vitro Evaluation of the Potential Toxic Effects of Palladium Nanoparticles on Fibroblasts and Lung Epithelial Cells, Elsevier Ltd, 2017. https://doi.org/10.1016/j.tiv.2017.04.024.
[85] G. Crisponi, V.M. Nurchi, J.I. Lachowicz, S. Medici, M.A. Zoroddu, Toxicity of nanoparticles: Etiology and mechanisms, in: Antimicrobial Nanoarchitectonics, Elsevier Inc., 2017: pp. 511–546. https://doi.org/10.1016/B978-0-323-52733-0/00018-5.
[86] T.T. Duong, T.S. Le, T. Thu, H. Tran, T.K. Nguyen, C.T. Ho, T.H. Dao, T. Phuong, Q. Le, H.C. Nguyen, D.K. Dang, T. Thu, H. Le, P.T. Há, Inhibition effect of engineered silver nanoparticles to bloom forming cyanobacteria, Advances in Natural Sciences: Nanoscience and Nanotechnology. 7 (2016) 035018.
[87] A.A. Becaro, C.M. Jonsson, F.C. Puti, M. Célia, L.H.C. Mattoso, D.S. Correa, M.D. Ferreira, Toxicity of PVA-stabilized silver nanoparticles to algae and microcrustaceans, Environmental Nanotechnology, Monitoring & Management. 3 (2015) 22–29. https://doi.org/10.1016/j.enmm.2014.11.002.
[88] H.-S. Jiang, X.-N. Qiu, G.-B. Li, L.-Y. Yin, Silver nanoparticles induced accumulation of reactive oxygen species and alteration of antioxidant systems in the aquatic plant Spirodela Polyrhiza, Environmental Toxicology and Chemistry. 33 (2014) 1398–1405. https://doi.org/10.1002/etc.2577.
[89] R. Cheng, Y. Liu, Y. Chen, L. Shen, J. Wu, L. Shi, X. Zheng, Combined effect of nanoscale zero-valent iron and linear alkylbenzene sulfonate (LAS) to the freshwater algae Scenedesmus obliquus, Ecotoxicology. 30 (2020) 1366–1375. https://doi.org/10.1007/s10646-020-02294-1.
[90] Y. Cheng, H. Dong, Y. Lu, K. Hou, Y. Wang, Q. Ning, L. Li, B. Wang, L. Zhang, G. Zeng, Chemosphere toxicity of sulfide-modified nanoscale zero-valent iron to Escherichia coli in aqueous solutions, Chemosphere. 220 (2019) 523–530. https://doi.org/10.1016/j.chemosphere.2018.12.159.
[91] D. Kumar, R. Roy, A. Parashar, A.M. Raichur, N. Chandrasekaran, A. Mukherjee, A. Mukherjee, Toxicity assessment of zero valent iron nanoparticles on Artemia salina, Environmental Toxicology. 32 (2017) 1617–1627. https://doi.org/10.1002/tox.22389.

[92] B.M. Prabhu, S.F. Ali, R.C. Murdock, S.M. Hussain, M. Srivatsan, Copper nanoparticles exert size and concentration dependent toxicity on somatosensory neurons of rat, Nanotoxicology. 4 (2010) 150–160. https://doi.org/10.3109/17435390903337693.
[93] S. Choi, S. Kim, Y. Bae, J. Park, J. Jung, Size-dependent toxicity of silver nanoparticles to Glyptotendipes tokunagai, Environmental Health and Toxicology. 30 (2015) 1–6.
[94] K. Małgorzata, M. Asztemborska, R. Steborowski, G. Bystrzejewska-Piotrowska, Toxic effect of silver and platinum nanoparticles toward the freshwater microalga pseudokirchneriella subcapitata, Bulletin of Environmental Contamination and Toxicology. 94 (2015) 554–558. https://doi.org/10.1007/s00128-015-1505-9.
[95] G. Sibi, D.A. Kumar, T. Gopal, K. Harinath, S. Banupriya, S. Chaitra, Metal nanoparticle triggered growth and lipid production in chlorella vulgaris, International Journal of Scientific Research in Environmental Science and Toxicology. 2 (2017) 1–8.
[96] C. Ispas, D. Andreescu, A. Patel, D.V. Goia, S. Andreescu, K.N. Wallace, Toxicity and developmental defects of different sizes and shape nickel nanoparticles in zebrafish, Environmental Science and Technology. 43 (2009) 6349–6356. https://doi.org/10.1021/es9010543.
[97] A. Châtel, C. Mouneyrac, Signaling pathways involved in metal-based nanomaterial toxicity towards aquatic organisms, Comparative Biochemistry and Physiology, Part C. (2017). https://doi.org/10.1016/j.cbpc.2017.03.014.
[98] F. Chen, Z. Xiao, L. Yue, J. Wang, Y. Feng, X. Zhu, Z. Wang, B. Xing, Algae response to engineered nanoparticles: Current understanding, mechanisms and implications, Environmental Science: Nano. 6 (2019) 1026–1042. https://doi.org/10.1039/C8EN01368C.
[99] V. García-Torra, A. Cano, M. Espina, M. Ettcheto, A. Camins, E. Barroso, M. Vazquez-carrera, M.L. García, E. Sánchez-López, E.B. Souto, State of the art on toxicological mechanisms of metal and metal oxide nanoparticles and strategies to reduce toxicological risks, Toxics. 9 (2021) 195.
[100] Y.N. Slavin, J. Asnis, U.O. Häfeli, H. Bach, Metal nanoparticles: Understanding the mechanisms behind antibacterial activity, Journal of Nanobiotechnology. 15 (2017) 1–20. https://doi.org/10.1186/s12951-017-0308-z.
[101] A.B. Djuris, Y.H. Leung, A.M.C. Ng, X.Y. Xu, P.K.H. Lee, N. Degger, R.S.S. Wu, Toxicity of metal oxide nanoparticles: Mechanisms, characterization, and avoiding experimental artefacts, Small. (2014) 1–19. https://doi.org/10.1002/smll.201303947.
[102] A. Abbaszadegan, Y. Ghahramani, A. Gholami, B. Hemmateenejad, S. Dorostkar, M. Nabavizadeh, H. Sharghi, The effect of charge at the surface of silver nanoparticles on antimicrobial activity against gram-positive and gram-negative bacteria: A preliminary study, Journal of Nanomaterials. 2015 (2015) 1–8.
[103] M.T. G´omez-Sagasti, L. Epelde, M. Anza, J. Urra, I. Alkorta, C. Garbisu, The impact of nanoscale zero-valent iron particles on soil microbial communities is soil dependent, Journal of Hazardous Materials. 364 (2018) 591–599. https://doi.org/10.1016/j.jhazmat.2018.10.034.
[104] N.H.A. Nguyen, R. Špánek, V. Kasalický, D. Ribas, D. Vlková, H. Řeháková, P. Kejzlar, A. Ševců, Different effects of nano-scale and micro-scale zero-valent iron particles on planktonic microorganisms from natural reservoir water, Environmental Science: Nano. 5 (2018) 1117–1129. https://doi.org/10.1039/c7en01120b.
[105] Y. Xie, H. Dong, G. Zeng, L. Tang, Z. Jiang, C. Zhang, J. Deng, L. Zhang, Y. Zhang, The interactions between nanoscale zero-valent iron and microbes in the subsurface environment: A review, Journal of Hazardous Materials. 321 (2017) 390–407. https://doi.org/10.1016/j.jhazmat.2016.09.028.
[106] M. Zhang, K. Yi, X. Zhang, P. Han, W. Liu, M. Tong, Modification of zero valent iron nanoparticles by sodium alginate and bentonite: Enhanced transport, effective hexavalent chromium removal and reduced bacterial toxicity, Journal of Hazardous Materials. 388 (2019) 121822. https://doi.org/10.1016/j.jhazmat.2019.121822.

[107] K.S. Cameron, V. Buchner, P.B. Tchounwou, Exploring the molecular mechanisms of nickel-induced genotoxicity and carcinogenicity: A literature review, Reviews on Environmental Health. 26 (2011) 81–92.

[108] L. Kong, W. Hu, C. Lu, K. Cheng, M. Tang, Mechanisms underlying nickel nanoparticle induced reproductive toxicity and chemo- protective effects of vitamin C in male rats, Chemosphere. 218 (2018) 259–265. https://doi.org/10.1016/j.chemosphere.2018.11.128.

[109] R.W.S. Lai, K.W.Y. Yeung, M.M.N. Yung, A.B. Djurišić, J.P. Giesy, K.M.Y. Leung, Regulation of engineered nanomaterials: Current challenges, insights and future directions, Environmental Science and Pollution Research. 25 (2018) 3060–3077. https://doi.org/10.1007/s11356-017-9489-0.

[110] S. Lekamge, A.S. Ball, R. Shukla, D. Nugegoda, The Toxicity of Nanoparticles to Organisms in Freshwater, Springer Nature, 2018. https://doi.org/10.1007/398.

[111] R.W.S. Lai, K.W.Y. Yeung, M.M.N. Yung, A.B. Djurišić, J.P. Giesy, K.M.Y. Leung, Regulation of engineered nanomaterials: Current challenges, insights and future directions, Environmental Science and Pollution Research. 25 (2018) 3060–3077. https://doi.org/10.1007/s11356-017-9489-0.

[112] H.S. Jasreen Kaur, M. Khatri, Regulatory considerations for safety of nanomaterials, in: M. Rahman, S. Beg, V. Kumar, F.J. Ahmad (Eds.), Nanomedicine for Bioactives: Healthcare Applications, Springer Nature, Singapore. 2020: pp. 431–450.

[113] K. Hegde, Ks.K. Brar, M. Verma, R.Y. Surampalli, Current understandings of toxicity, risks and regulations of engineered nanoparticles with respect to environmental microorganisms, Nanotechnology for Environmental Engineering. 1 (2016) 1–12. https://doi.org/10.1007/s41204-016-0005-4.

[114] R. Foulkes, E. Man, J. Thind, S. Yeung, A. Joy, C. Hoskins, Biomaterials Science nanomedicines for clinical application: Current and future perspectives, Biomaterials Science. 44 (2020) 4653–4664. https://doi.org/10.1039/d0bm00558d.

[115] M. Miernicki, T. Hofmann, I. Eisenberger, F. Von Der Kammer, A. Praetorius, nanomaterials according to regulatory definitions, Nature Nanotechnology. 14 (2019) 208–216. https://doi.org/10.1038/s41565-019-0396-z.

[116] H. Godwin, C. Nameth, D. Avery, L.L. Bergeson, D. Bernard, E. Beryt, W. Boyes, S. Brown, A.J. Clippinger, O.Y. Cohen, M. Doa, C.O. Hendren, P. Holden, K. Houck, A.B. Kane, F. Klaessig, T. Kodas, R. Landsiedel, I. Lynch, T. Malloy, M.B. Miller, J. Muller, G. Oberdorster, E.J. Petersen, R.C. Pleus, P. Sayre, V. Stone, O.O. Kristie, M. Sullivan, J. Tentschert, P. Wallis, A.E. Nel, Nanomaterial categorization for assessing risk potential to facilitate regulatory decision-making, ACS Nano. 9 (2015) 3409–3417.

[117] E. Haque, A.C. Ward, Zebrafish as a model to evaluate nanoparticle toxicity, Nanomaterials. 8 (2018) 561. https://doi.org/10.3390/nano8070561.

[118] D. Gupta, P. Yadav, D. Garg, T.K. Gupta, Pathways of nanotoxicity: Modes of detection, impact, and challenges, Frontiers of Materials Science. 15 (2021) 512–542.

[119] A. Dhawan, V. Sharma, Toxicity assessment of nanomaterials: Methods and challenges, Analytical and Bioanalytical Chemistry. 398 (2010) 589–605. https://doi.org/10.1007/s00216-010-3996-x.

[120] V. Kumar, N. Sharma, S.S. Maitra, In vitro and in vivo toxicity assessment of nanoparticles, International Nano Letters. 7 (2017) 243–256. https://doi.org/10.1007/s40089-017-0221-3.

[121] B. Dogan-topal, B. Uslu, S.A. Ozkan, Detection of DNA damage induced by nanomaterials, in: Nanoscale Fabrication, Optimization, Scale-up and Biological Aspects of Pharmaceutical Nanotechnology, Elsevier Inc., 2018: pp. 547–578. https://doi.org/10.1016/B978-0-12-813629-4.00014-0.

[122] F.K.-M. Chan, K. Moriwaki, M.J. De Rosa, Detection of necrosis by release of lactate dehydrogenase (LDH) activity, Methods Mol Biol. 979 (2014) 65–70. https://doi.org/10.1007/978-1-62703-290-2.

[123] S. Clichici, A. Filip, In vivo assessment of nanomaterials toxicity, in: O. Soloneski, M.L. Larramendy (Eds.), Nanomaterials -Toxicity and Risk Assessment, INTECH, 2015: pp. 93–121. http://dx.doi.org/10.5772/60707.
[124] M. Zhang, J. Yang, Z. Cai, Y. Feng, Y. Wang, D. Zhang, X. Pan, Detection of engineered nanoparticles in aquatic environment: State-of-art and challenges in enrichment, separation and analysis, Environmental Science: Nano. 9 (2019) 709–735. https://doi.org/10.1039/C8EN01086B.
[125] A. Ramanathan, Toxicity of nanoparticles-challenges and opportunities, Applied Microscopy. 49 (2019) 1–11.
[126] N.I. Bloch, N.B. Hartmann, C. Engelbrekt, J. Zhang, J. Ulstrup, K.O. Kusk, A. Baun, The challenges of testing metal and metal oxide nanoparticles in algal bioassays: Titanium dioxide and gold nanoparticles as case studies, Nanotoxicology. 7 (2013) 1082–1094. https://doi.org/10.3109/17435390.2012.710657.
[127] A. Sani, C. Cao, D. Cui, Toxicity of gold nanoparticles (AuNPs): A review, Biochemistry and Biophysics Reports. 26 (2021) 100991. https://doi.org/10.1016/j.bbrep.2021.100991.

13 Toxicity of Metallic Nanoparticles

Assessment and Impacts

Kalyani Chepuri, Tulasi C D S L N, Manikantha Dunna, Shivani Munagala, and CH. Shilpa Chakra

13.1 INTRODUCTION

The field of nanotechnology is wide and crosses several disciplines, including physics, biology, engineering, and chemistry. It mostly deals with nanoparticles with dimensions between 1 and 100 [1–3]. "Why is it so interesting with these tiny materials?" is one primary point that comes to mind while dealing with nanomaterials. "What is the fundamental difference between these nanomaterials and their counterparts?". Before delving deep into the details of nanoparticles (NPs), answers to some inescapable questions are required. NPs possess unique inherent qualities like good electrical, catalytic, magnetic, and optical behavior; mechanical and chemical stability; huge surface areas for reactions; and ease of surface modification. In present times, metallic NPs have become the prime focus of research among nanomaterials [1, 4]. Nanoparticles of metals are primarily made or synthesized by using two methods: (a) "top-down" and (b) "bottom-up". A variety of chemical and physical methods have been devised using these two approaches for synthesizing metallic NPs [5–7]. Sonochemical decomposition, chemical reduction, thermal decomposition, microemulsion, polyol technique, microwave, laser cutting, and many other methods are used [4]. Physical, as well as chemical, approaches are used to create stable and monodisperse NPs. However, these approaches necessitate the use of high energy consumption, dangerous chemical substrates, damaging end products, and expensive reaction setups [8]. As a result, based on the eventual usage of products generated at the end, an alternative method that is environmentally acceptable and with less expenditure has been developed where microbial cultures or plants/plant parts have been utilized for the synthesis of NPs [9].

Characterizing the shape, morphology, and size of metal NPs is a vital step in the synthesis of metal NPs [10]. Altering the size and shape of nanoparticles is largely dependent on factors such as temperature, incubation time, ratio of metal precursor to reducing or binding agent, metal substrate concentrations, surfactant types used, quantity of microbial culture, plant extract concentration based on pH, and culture age [11, 12]. Particle properties (shape, surface functionality, and size) are strongly dependent on their characteristics, which in the long run dictate the function of NPs in biological contexts [13]. The nanoparticles are smaller in size and have a shorter wavelength than

DOI: 10.1201/9781003317319-13

light. Because of these characteristics, NPs can effectively get across the cells and tissues of the body to reach their destination. Metal NPs have the potential to be used in a wide range of biological applications, including imaging for illness detection, targeted drug administration, gene therapy, cancer treatment, bacteriocidal action, tissue engineering, and wound healing [14]. Many NPs have been synthesized to date to scan the condition of sick organs using modalities such as computed tomography (CT), magnetic resonance imaging (MRI), ultrasound imaging techniques, and positron-emission tomography (PET) [15]. Many of the metal NPs can be used in imaging modalities because they have distinct physicochemical characteristics. Despite the abundance of remarkable physicochemical features, there have been indications of toxicity also associated with metallic NPs. It is necessary to determine the dangers associated with metal NPs before using them in applications involving biological systems. As a result, metal NP surfaces must be designed in such a way that their toxicity is minimized while their potentialities are maximized. This chapter covers the physicochemical properties of metallic nanoparticles, diverse production methods (chemical, biological, and physical), factors impacting NP properties and biochemical behavior, and uses of NPs in treatment, imaging, tissue engineering, and drug administration. The toxicity concerns connected with the use of these NPs are also addressed.

13.2 TYPES OF NANOPARTICLES

Nanoparticles were classified based on their size, shape, and chemical characteristics. NPs are classified into various groups, but some of the well-known classes of NPs based on their physical and chemical features are listed in Figure 13.1.

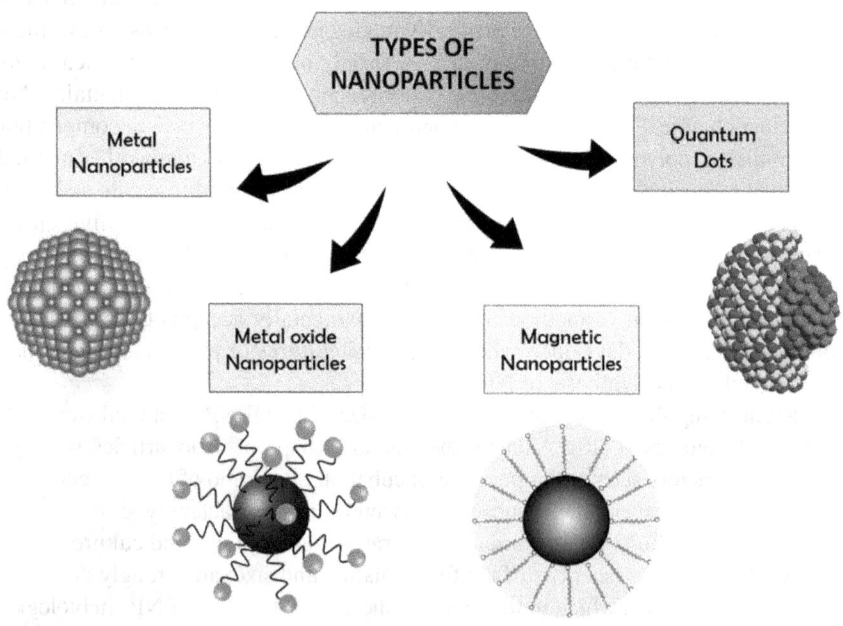

FIGURE 13.1 Types of nanoparticles.

13.2.1 Metal NPs

The major precursor material in the manufacture of metal NPs is metal. The localized surface plasmon resonance (LSPR) characteristic was considered a unique opto-electrical feature in these NPs. It is well known that NPs made of noble and alkali metals, such as Ag, Au, and Cu, have a large absorption spectrum in the visible zone of the electromagnetic spectrum. Metal NPs possessing regulated size, shape, and facet are significant in present-era pioneering particles [16, 17]. These metal NPs are synthesized from alkali metals in nanometric sizes using either constructive or destructive processes. More or less every metal [18] can be synthesized into nanoparticles. Cadmium (Cd), aluminum (Al), cobalt (Co), gold (Au), copper (Cu), iron (Fe), silver (Ag), zinc (Zn), and lead (Pb) have been the most commonly utilized metals in the process of creating nanoparticles. NPs have demarcated properties like unique surface characteristics, sizes ranging from 10 to 100 nm, high surface area to volume ratio, surface charge, density, pore size, amorphous and crystalline structures, cylindrical and spherical shapes, color, sensitivity, and reactivity to environmental factors such as heat, air, sunlight, and moisture.

13.2.2 Metal Oxide Nanoparticles

Iron nanoparticles (Fe) oxidize rapidly to iron oxide (Fe_2O_3) when oxygen acts as a catalyst at normal temperatures. The produced iron oxide nanoparticles show an increased reactivity rate compared to FeNPs. The improved reactivity and efficiency of metal oxide NPs are fundamental reasons for their creation [19]. Silicon dioxide (SiO_2), aluminum oxide (Al_2O_3), titanium oxide (TiO_2), iron oxide (Fe_2O_3), cerium oxide (CeO_2), magnetite (Fe_3O_4), and zinc oxide (ZnO) are mostly synthesized metal oxide NPs. When the metal counterparts are compared with oxide NPs, these NPs offer remarkable characteristics compared to metal NPs.

13.2.3 Magnetic Nanoparticles

Magnetic nanoparticles are manufactured from a wide range of compositions and materials, each with unique physical and magnetic properties that are important for their anticipated uses. However, in the biomedical field, potential biocompatibility/toxicity, as well as long-term in vivo destiny and removal, must be considered [20, 21]. Due to these limitations, only a subset of ferrite nanoparticles (magnetite: MFe_{3-} O_4 here M could be Ni, Mn, Fe, or Zn) is suitable for biomedical applications [22–25]. Because of how they interact with magnetic fields and their intrinsic biocompatibility, magnetic nanoparticles (MNPs) created by external sources are of interest in a variety of biological applications. MNPs can alter the magnetic field around them, allowing for improved contrast in MRI. There has been a significant amount of study in the last decade on the synthesis, characterization, and post-synthesis application like in-particular modification of magnetic Fe_2O_3 and substituted ferrite NP, resulting in a slew of new applications in a variety of sectors. There is literature regarding the same that serves as an ideal pioneer for a general assessment of MNP applications for cancer, antibacterial, neurological, and various biomedical applications [26–53].

13.2.4 Quantum Dot Nanoparticles

Quantum dots (QDs) are semiconductor nanoparticles with optical and electronic (optoelectronic) capabilities that are size- and composition-dependent. QDs are extremely tiny, measuring between 1.5 to 10.0 nm in size. QD nanotechnology has recently made inroads into several technological and biological areas. Due to their unique qualities, including excellent photostability, size-dependent optical properties, high extinction coefficient, brightness, and substantial Stokes shift, QDs have been effectively shown. Organic dyes are limited in their ability to exhibit all of these characteristics, making them unsuitable for many imaging and biosensing applications. Furthermore, due to their enormous surface area and in vitro/in vivo optical trackability, QDs may have an advantage over comparably bigger silica and polymer-based nanoparticles for constructing multimodal/multifunctional probes.

13.3 APPLICATIONS OF METALLIC NPS

13.3.1 Biomedical Imaging and Sensing

Metallic NPs have several applications in biological sectors, and there is a high potential for further expansion in this field (Figure 13.2). The antimicrobial properties of metallic NPs are frequently employed, like the use of silver NPs in implants, bone cement, and wound coverings [54]. Gold nanoparticles (AuNPs) exhibit anticancer and optical properties that are important in medicine [55]. Discussed that the surface

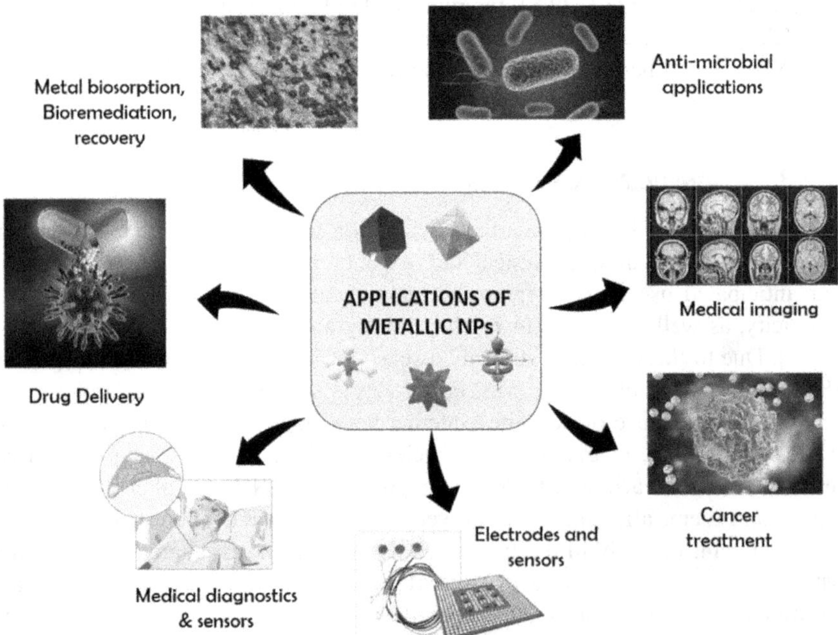

FIGURE 13.2 Applications of metallic nanoparticles.

plasmon light scattering and absorption of surface plasmon have been employed for therapy and diagnosis [56] and conversed about the manufacture as well as utilization of nanoparticles of gold AuNPs in cancer therapeutics. Magnetic NPs show potential for targeted medication administration and hyperthermia uses accordingly [57].

13.3.1.1 Antimicrobial Applications

Metallic NPs with antimicrobial, antifungal, and antiviral properties include Ti, Cu, Mg, and, mainly, Ag and Au [58]. The mentioned NPs are being studied as food processing enhancers, disinfectants, clothing additives, and medical device additives. AgNP toxicity in bacteria is caused mostly by ions liberated by Ag into the aqueous solution after its partial oxidation. Small NPs and the generated Ag ions attach tightly to the cell membrane, preventing functions like respiration or permeability and eventually causing lysis. By denaturing ribosomes or binding to DNA, AgNP can impede DNA replication and protein synthesis [59].

Biosynthesized NPs have been shown to have antifungal action in several investigations. Gajbhyie et al. reported antifungal effects of biosynthesized NPs in conjunction with fluconazole (a triazole antifungal medication) against *Phomaglomerata*, *Trichoderma* sp., *Phoma herbarum*, *Candida albicans*, and *Fusarium semitectum* [60]. Fluconazole's antifungal activity was increased by AgNPs produced by another fungus, *Alternaria alternate*, against all studied strains except *F. semitectum* and *P. herbarum*. Although viruses are a severe concern in medicine and agriculture, there have been few reports of biosynthesized NPs exhibiting antiviral activity. AgNPs antiviral activity generated intracellularly in *Aspergillus ochraceus* was described by Vijayakumar and Prasad (2009). Heating treatment of the cells yielded AgNPs encased in a carbonaceous matrix, which was then tested for efficacy against the M13 phage using plaque assays.

Biosynthesized NPs are effective against a variety of disease-causing parasites and insects. Santhoshkumar et al. examined the larvicidal efficacy of gold NPs produced from various samples of lotus leaves (*Nelumbonucifera*) [61]. Larvicidal activity has been assessed in the larvae (fourth instar) of *Culex quinquefasciatus* and *Anopheles subpictus*, two lymphatic filariases as well as malaria vectors, respectively.

13.3.1.2 Treatment of Cancers and Drug Delivery

Nanoparticles have been shown to react with mammalian tissues and change their functionality. Metallic NPs, for example, can disrupt the antioxidant defense system, resulting in the buildup of reactive oxygen species, mitochondrial damage, and apoptosis [62]. The metal NP exposures are significantly dependent on the capping agent, based on in vivo studies. The identical AgNPs are both non-cytotoxic and cytotoxic [63]. The use of stable novel metal nanoparticles acting like carriers might reduce some adverse effects of conventional chemotherapeutics by selectively delivering anticancer drugs to malignant cells without damaging normal cells. This could be due to its strong binding capacity to amines ($-NH_2$) and thiols (–SH). Similarly, the surface coating of the NP with biomolecules that act like chemotherapeutics and as targeted agents is comparatively simple [64].

The biocompatible nature of normal and healthy tissues is a crucial factor when NPs are employed for in vivo experiments. In one of the studies by Moulton et al.

(2010), the leaf extract of tea proved to be a good source of capping as well as a reducing agent while synthesizing AgNPs. These researchers examined mitochondrial function in human keratinocytes to determine cell survival and the integrity of the membrane. The findings revealed that the biosynthesized AgNPs using leaf extracts of tea are shown to be nontoxic, implying that the approach could be useful for the next level of in vivo applications. Biosynthesized NPs have been subjected to cytotoxicology studies against a variety of cancer cell types [65]. Reported on trials employing AuNPs produced from *Vitis vinifera* phytochemicals (grape). These AuNPs have a strong affinity for HBL-100 (human breast cancer cells), and when exposed to them, cells die. Porphyran from *Porphyravietnamensis* (marine algae) was used as a capping agent and reducing agent to produce AuNPs in a drug delivery study [66]. These NPs were employed to transport DOX (doxorubicin). The findings showed that AuNPs bound with DOX had a stronger cytotoxic effect on human glioma cells (LN-229) compared to unbound DOX. The interaction between chemotherapeutic drugs and NP is due to hydrogen bonds.

13.3.1.3 Medical Diagnostics and Sensors

In the domains of diagnosis and sensing, biosynthesized NPs are shown to be significant handouts in the medical field. Chemical sensors that can identify medically relevant substances such as peroxides and glucose were successfully developed using these materials. To produce AuNPs, the membrane of an egg was employed as one of the reduction agents, along with a scaffold. With a 17 μM detection limit, the sensor displayed a linear response to various glucose levels. A similar substance was used in assessing sugar concentration in human blood serum, and the results were found to be consistent with standard medical spectrophotometric testing [67]. The study presented and evaluated another prospective medical application of biogenic AuNPs: DNA conjugation with biosynthesized nanoparticles can be employed for genetic disease diagnosis.

13.3.1.4 Medical Imaging Applications

For ages, scientists have been fascinated by the optical features of metallic nanocrystals. Metal NPs of various sizes, shapes, and dielectric characteristics may now be manufactured thanks to the use of biosynthetic processes. Low or high refractive index, unique photoluminescence capabilities, photonic crystal, high transparency, and plasmon resonance are all optical properties linked with metallic NPs [68]. In the nanophotonics field, at some places where light interacts with any particles smaller than its wavelength, unexpected occurrences called LSPR (localized surface plasmon resonance) or a semiconductor band gap that varies with size have been discovered [69]. Biosynthesized AuNPs through the silica-encapsulated microalgae, namely *Klebsormidium flaccidum* result in a "living" bio-hybrid material [70]. Raman spectroscopy revealed the images of imprisoned cells (in situ) and the effect of gold NPs on the algae's photosynthetic system has been evaluated by researchers. Photosynthesis-based biosensors could be developed using a combination of sol-gel encapsulation and Raman imaging.

13.3.2 Environmental Cleanup

13.3.2.1 Bioremediation, Biosorption of Metals, and Biological Recovery

Since cellular metabolism is based on oxidation reduction processes, biomolecules that may oxidize or reduce any other chemical components are numerous in any live cell. As an active strategy to cleanse their surroundings, several bacteria can bind and accumulate dissolved metals. The renowned biological process, including dead biomass and metal reduction with an electron source by way of oxidation. The bioremediation process of metals in various solutions and synthesis of the metallic NPs by the biological method are in reality very closely connected processes [71] because the outcome of the cellular metal reduction process is frequently nanostructured. The evaluated revisions demonstrate metal elimination in a biological way and the production of NPs through biological-mediated methods. Some biopolymers and bio-waste materials, as well as the biomass of yeast, algae, bacteria, and fungi, have been shown to get bound and precipitate some precious metals. This biosorption approach could be a more cost-effective alternative to traditional methods (chemical) for recovering dissolved or soluble metals in aqueous solutions. In natural environments, the probable approaches/mechanisms involved in the binding of microorganisms and concentrating the dissolved metals have been thoroughly researched [72].

13.3.2.2 Contamination Degradation Coupled With Biosorption and Catalytic Nature

In many applications, metallic NPs are used for catalysis, especially heterogeneous catalysis, because of their enormous surface area per weight. Metallic NPs have a high level of stability, activity, and selectivity. The main disadvantage of metallic NP catalysis is the high price of the constituent metals, such as platinum, gold, and palladium, as well as the time and energy required to fabricate them into nanoparticles. With the biosorption of wastewater containing precious metals, combined biosynthetic methods can produce catalytically active NPs. The latest studies describe the connection between biological metal elimination and catalyst production. These biosynthesized metallic nanoparticles could subsequently be employed for organic oxidation, dehalogenation, or metal reduction. Palladium (Pd) NPs are used in the majority of biosynthesized catalyst applications for environmental cleanup [73, 74].

13.3.2.3 Degrading Pollutants (Organic) by a Catalytic Process

In the middle of the last century, the rapid invention and usage of explosives, synthetic insecticides, pharmaceuticals, and dyes were said to be responsible for dreadful causative nitro-aromatic contamination in water and soil. PNP (4-nitrophenol) or p-nitrophenol cleanup technologies may aid in the restoration of damaged ecosystems. Furthermore, using extensively available low-cost optical technologies like spectrophotometry especially, UV-Visible, the degradation of PNP can be easily evaluated. As a result, the degradation (catalytic) capacity corresponding to PNP is one of the standard test techniques for evaluating the catalytic reducing properties of biosynthesized NPs [75].

13.3.2.4 Catalytic Cr (VI) Reduction

Majorly, biosynthesized NPs have been shown to have a role in the environment by reducing (catalytically) Cr (VI), which is a potent oxidant, to its relatively non-toxic form of Cr (III). PdNPs biosynthesized by bacteria such as *E. coli*, *Clostridium*, *Esulfovibrio*, and *Serratia* can successfully catalyze the process of reduction by converting Cr(VI) into the harmless form of Cr(III) through both batch as well as continuous fluid flow, according to several recent studies [76]. For example, one study examined whether three kinds of enzymes usually hydrogenases expressed by the organism *E. coli* affect the performance of a biosynthesized NP as a Cr (VI)/Cr (III) reduction catalyst. The role of hydrogenase in Pd(II) reduction was also discussed in this study.

13.3.2.5 Membrane Treatment Processes—Biosynthesized NPs

Membrane-based water treatment procedures can benefit considerably from biofouling reduction technologies. *Lactobacillus fermentum* has been used in the generation of various levels of AgNPs (Biogenic), which play a role in inhibiting biofouling on membranes of polyethersulfone [77]. *E. coli* cells and organophosphate insecticides were removed from model wastewater using the membranes. The scientists used *P. aeruginosa* bacteria and *E. coli* cells separately and in mixed culture to examine membrane structure and performance as well as biofouling. During the week's test, the nano-functionalized membrane coating had strong inhibitory activity against bacteria as well as biofilm formation on the surface of the membrane.

13.3.3 Applications of NPs—Industrial Level

13.3.3.1 Catalytic Organic Synthesis

In industry, inexpensive and selective catalysts for commercially relevant chemical synthesis reactions such as epoxidation, cross-coupling reactions, and hydrogenation are critical. Platinum, gold, and palladium catalysts made from biosynthesis were successfully used in a variety of organic synthesis pathways. For some of the reactions involved in the hydrogenation of itaconic (or methylenesuccinic) acid catalytically [78], employed cells (palladized) of a few strains of bacteria, *Bacillus sphaericus* (G^+) and *D. desulfuricans* (G^-). Biosynthesized PdNPs, as well as commercial graphite-supported catalysts, showed equal activity when compared to a stirring autoclave [79]. In his research, described a useful technique for the synthesis of PdNPs biologically, which does not need the presence of an H_2 donor, unlike the previous studies. Palladium chloride was reduced with the help of Gardenia jasminoides crude extract to produce biosynthetic PdNPs. For the release of the product p-methyl-cyclohexylamine, the researchers compared p-nitrotoluene's catalytic hydrogenation and, after 2 hours, found a huge transition of 100% at 5 MPa pressure at 150 °C, with a selective percentage of 26.3. The catalyst kept its activity after five cycles, according to reusability tests.

13.3.3.2 Applications of Nanoparticles—Generation of Energy

Many elements of fuel cell performance have been improved by using organisms' ability to decrease and consume precious metal salts and produce NPs. NPs synthesized biologically have been employed in making fuel from H_2, catalyzing oxidation chemically, and increasing power recovery. For H_2 generation, several researchers have employed biosynthesized PdNPs. Biosynthesized PdNPs for the generation of H_2 catalytically from hypophosphite chemical using different bacteria species, *Paracoccus denitrificans*, *Pseudomonas putida*, and *Cupriavidus necator* [80]. Wu et al. used gardenia extract in biosynthesizing PdNPs along TiO_2 to improve pure water through the evolution of H_2 photocatalytically.

13.3.3.3 Nanoparticles in Performance of Electrodes and Sensors

Nanoscale research and nanotechnology could help in understanding fundamental electrochemical phenomena, and biosynthesized nanoparticles can be used to change or enhance the performance of electrodes and sensors. The electrical conductivity, biological compatibility, and surface chemistry of biosynthesized metallic nanoparticles were all thoroughly investigated. AuNPs have acknowledged a lot of consideration for nanoelectrochemical advantages, as they are capable of being a good platform for biomolecule immobilization while at the same time not affecting its biological activity and showing outstanding electron transmission between the biomolecule and electrode surface [81]. When compared to conventional materials, the addition of surfactants and capping agents of AuNPs, which are biologically synthesized, could adjust the functionality of electrodes and increase the selectivity and range of operation. For the production of AuNPs, a dried entire *Scutellaria barbata* plant extract was utilized in one study [82]. The scientists observed that adding AuNPs to a GCE (glassy carbon electrode) improved electrical transmission between the PNP and the modified electrode. Torres-Chavolla et al. have reported that three actinomycetes have been used in the biosynthesis of AuNPs: *T. fusca*, *T. chromogena*, and *Thermomonospora curvata* [83]. They were stabilized with the crosslinker glutaraldehyde for improved biosensing applications.

13.4 TOXICITY OF NANOPARTICLES

13.4.1 Nanoparticles With *In Vitro* Toxicity

13.4.1.1 Cell Viability and Lethality

The two general criteria used to assess the induction of cytotoxicity by the nanoparticles are lethality and cell viability. Carbon nanoparticles (CNTs) are one of the most utilized nanoparticles for determining the viability and lethality of cells. Because of their distinctive features, they are extensively used in industrial, biomedical, and pharmaceutical processes [84, 85]. Carbon nanotubes, either single-walled or multi-walled SWCNTs/MWCNTs, are used to manufacture them [86]. Because of the damage induced by the nanotubes mechanically, research on different bacteria has shown that CNTs exhibit antimicrobial capabilities [87–89]. Functionalized carbon nanotubes seemed to alter soil bacterial diversity in a recent study [90]. Toxicological studies on *Daphnia magna* (crustacean), *Oryziaslatipes* (fish), *Chlorella vulgaris*, and *Raphidocelis subcapitata* (freshwater microalgae) revealed that the *R. subcapitata* and C. *vulgaris* algae were inhibited by SWCNT concentrations of 29.99 mg/mL and 30.96 mg/mL, respectively [91]. A study by Naqvi et al. showed that iron oxide NPs are shown to have significant toxicity toward human liver cancer cells, murine macrophage cells, mesenchymal stem cells from a rat, and human macrophages. Of all these cells, murine macrophage cells are shown to have much lethality against iron oxide nanoparticles at concentrations of 25–200 g/mL after a 2-hour exposure [92].

13.4.1.2 Toxicity Toward Cell Lines

Nanoparticle toxicity has been tested on a variety of cell lines. SWCNTs are shown to have potent toxicity against many cancer cell lines, such as human cell, A549 cells, human epithelial-like cervical cancer cells, and HeLa cells, and have been disclosed in many studies [93–95]. In a study [96], CNTs were used to test the toxicity of lung fibroblast cells. After treating A549 cells with SWCNTs at 250–500 g/mL for 72 hours, they developed an oxidative response and damage to the membrane as a result of the inflammatory response [97]. In vitro reduction of inflammatory molecules like MCP-1, IL-8, and IL-6 was found in another study [98]. The impacts of multi-wall carbon nanotubes on human keratinocytes were also studied [99]. It's been proposed that MWCNT toxicity is regulated by pro-inflammatory effects aided by ROS and NF-kB [100]. *In vitro* investigations in mammalian cell lines revealed apoptosis, oxidative stress, and DNA damage, among the additional toxic effects of MWCNTs. In human aortic endothelial cells, further impacts include the integrity of actin filaments and VE-cadherin distribution [101–104].

In the MRC-5 cell line (human lung fibroblasts), gold nanoparticles (AuNPs) promoted autophagy in the presence of oxidative stress [105]. On animal cells, the cytotoxicity of semiconductor and metal nanoparticles was demonstrated using cellular motility in a study. The study makes use of electric cell-substrate impedance measurement as a very suitable method for quantifying in vitro quantum dots and gold nanorod cytotoxicity. Fluorescent and dark-field microscopy were used to validate the procedure [106]. The silver nanoparticles' toxicity was further assessed using the DNA microarray analysis, viability assay, and micronucleus test in the HepG2 (human hematoma cell line) by Kawata et al. (2009). In HeLa cells, silver nanoparticles were found to be hazardous. Mt-2A, oxidative stress genes, and Ho-1 were all upregulated after exposure [107]. Three distinct forms of silver nanoparticle treatments on *E. coli* indicated that the risk gene's replication fidelity was degraded [108].

When MCF-7 cells are treated with zinc sulfide/cadmium selenide (ZnS/CdSe), cadmium telluride (CdTe), nanoparticles capped with N-acetyl cysteine conjugated to cysteamine, mercaptopropionic acid, and cysteamine, high amount of intracellular Cd^{2+} levels have been observed. CdTe quantum dots were found to be cytotoxic, causing lysosomal damage and generation of harmful reactive oxygen species [109]. The neurotoxicity of Cd Se quantum dots was investigated using a hippocampus neuronal culture model. The study concentrated on the voltage-gated sodium ion channels and the calcium levels in the cytoplasm. Induced neuronal damage and increased calcium levels in the cytoplasm were reported. On HBMVECs (human brain microvascular endothelial cells), aluminum nanoparticles in the range of sizes 1–10 μM were utilized for 24 hours. Following treatment, mitochondrial functionality is reduced, cell viability differs, and an elevated level of oxidative stress has been observed in the reported study [110]. Aluminum nanoparticles at concentrations of 0–5000 g/mL were found to cause damage to DNA in mammalian cell lines following 2 hours of exposure [111].

13.4.1.3 Mechanistic Studies

The effects of nanoparticles in vitro were studied using mechanistic investigations. CNTs have been shown to disrupt metabolic activity, membrane integrity, cellular reproduction, and membrane potential in vitro [112, 113]. AgNPs cause damage to the mitochondria and have an impact on oxidative stress, autophagy, and cellular micro-mobility [114]. Interference with the instability of chromosomes, cytotoxicity, intracellular calcium levels, fidelity, DNA replication, oxidative stress, cell cycle arrest, and apoptosis has been observed in most of the studies done in vitro [115]. In mammalian cell lines, fullerenes cause oxidative stress and DNA damage [116]. Epithelial cells were studied in an FE1-Muta™ mouse lung study to see how SWCNTs and C60 fullerenes affected ROS generation, genotoxicity, and cytotoxicity. The toxicity effects of metallic NPs in vitro and in vivo models are depicted in Table 13.1 [117–139].

13.4.2 Nanoparticles With *In Vivo* Toxicity

13.4.2.1 Dose Exposure

The nanoparticle toxicity is in general estimated by parameters like conditions, period, and dose of exposure. After exposure with 1–10 mm SWCNTs for about 1 h duration, 3T3 cells, the treatment resulted in viability of 20% [140]. In another study, 80% of confluent epidermal keratinocytes were exposed to SWCNTs at different concentrations. After 4 hours of treatment, viability was reduced by 65% when exposed to 0.24 mg/mL nanoparticles [141]. The treatment of 3T3/NIH cells and HeLa to gold nanoparticles for 3 hours of size 150 pm showed reduced viability of cells from 100 to 20% and 5%, respectively [142]. Au Nanoparticles with particle sizes ranging from 10 to 100 μm were treated in macrophage cells for 24 to 72 hours. When human fibroblast cells were treated with Fe_3O_4 NPs at concentrations of 0–1000 μg/mL for 24 hours, cell viability was reduced by 25–50% [143]. Fe_3O_4 nanoparticles were treated in mouse macrophages for 1 to 4 days at 0.2 mg/mL in another experiment [144]. The study found that cytotoxicity was dosage-dependent. 90% of Hela cells survived after being treated for 2 hours with a mixture of cadmium and selenium quantum dots with a diameter of 1 to 100 nm in diameter range [145]. The metabolic activity of rat pheochromocytoma cells was reduced by 50% after 24 hours of treatment using 0.01–100 g/mL CdTe quantum dots [146].

13.4.2.2 Effects on Organ Systems

The carbon nanotubes effects were found to play a vital role in the administration strategy in in vivo models. According to several studies, exposure to nanoparticles in the respiratory system is causative and leads to lung cancer, emphysema, and asthma. Human colon cancer and Crohn's disease could be caused by nanoparticles invading the digestive tract. Moreover, contact with nanoparticles through the vascular system has been linked to clot formation in the blood and heart disease [147]. Several investigations on the in vivo toxicity of carbon nanotubes in animals, including guinea pigs and rodents, were conducted. On guinea pigs, the action of soot-containing carbon

TABLE 13.1
Toxicity Effects of Metallic NPs In Vitro and In Vivo Models

S. No.	Type of Nanoparticles	Experimental model	Mode of toxicity	Reference
1.	AgNPs	i) Human germ cell tumor cells (embryonic) ii) Testicular cells from mouse	Cytotoxicity	[117]
2.	ZnO NPs	i) Human umbilical vein endothelial cells ii) H9c2 cardiomyoblast Cells iii) Aortic VSMC cells	Cardiovascular pathogenesis	[118]
3.	AuNPs	Mouse fibroblast cells	Cytotoxicity	[119]
4.	AuNPs	i) Cl7.2 cells ii) PC12 iii) HUVEs	Cytotoxicity	[120]
5.	AgNPs	T-cells	Cytotoxicity	[121]
6.	CuO NPs	i) HBECs ii) A549 cells	Cytotoxicity	[122]
7.	Superparamagnetic iron oxide nanoparticles (SPION)s	Rats	Toxic effects on the organs like lungs, liver, and kidney	[20]
8.	ZnO NPs	Human bronchial epithelial cells	Cytotoxicity	[123]
9.	TiO_2 NPs	Rats	Lethargy, tremors, and loss of appetite	[124]
10.	CuO, ZnO, and MgO NPs	Human cardiac microvascular endothelial cells	Cytotoxicity, genotoxicity, endothelial activation: and impaired NO signaling	[125]
11.	ZnO, Al_2O_3, and TiO_2 NPs	*Coenorhabditis elegans*	Inhibition of growth and reproductive capability	[126]
12.	ZnO NPs	i) Human hepatocyte (L02) ii) Human embryonic kidney (HEK293) cells	Morphological alterations, reduction of SOD, mitochondrial dysfunction, depletion of GSH, and oxidative	[127]
13.	AgNPs	*Seeds of Allium cepa*	Cytotoxicity and genotoxicity	[128]
14.	ZnO NPs	Human epidermal cells-431	Cytotoxicity and genotoxicity	[129]
15.	ZnO NPs	i) Raw 264.7 cell line ii) C57BL/6 mice	Immunotoxicity and immunosuppression	[130]
16.	ZnO NPs	C57BL/6J mice	Allergic airway inflammation	[131]
17.	Fe_2O_3 NPs	HepG2 cell line	DNA damage and apoptosis	[132]
18.	Oxidized graphene nanoribbons (O- GNR)s	NIH-3T3, SKBR3	Decrease in cell viability	[133]
19.	ZnO NPs	Human-induced pluripotent stem cell-derived cardiomyocytes	Cytotoxicity, mitochondrial dysfunction, and trigger cardiac electro-physiological alterations	[134]

TABLE 13.1 *(Continued)*
Toxicity Effects of Metallic NPs In Vitro and In Vivo Models

S. No.	Type of Nanoparticles	Experimental model	Mode of toxicity	Reference
20.	TiO_2 NPs	Mice	Liver architecture damage, Genetic disturbance	[135]
21.	AuNPs	i) Human osteoblast cell line ii) Human pancreatic duct cell line	Cytotoxicity and Ultrastructural changes	[136]
22.	Si NPs	Human umbilical vein endothelial cells	Apoptosis	[137
23.	ZnO NPs	HUVEC	DNA damage and functional impairment	[138]
24.	ZnO NPs	BALB/c mice	Airway inflammation	[139]

nanotubes comprising Ni/Co was seen. Various researchers studied the effects of SWCNTs containing metals on mice and rats [148]. The acute lung toxicity of intratracheally administered SWCNTs in rats was investigated in one of the experiments mentioned earlier [149]. Cell damage, transitory inflammatory reactions, and multifocal granulomas were all detected in this investigation. In another study, early neutrophil accumulation, lymphocyte influx, pro-inflammatory cytokine increase, macrophage influx, and fibrogenic transforming growth factor were all reported [150].

The effects of MWCNTs were also examined in a study on a rat. In the experiment, rats were given MWCNTs intravenously. Fibrotic reactions, lung persistence, and inflammation were studied biochemically and histologically. The bronchial lumen was found to have pulmonary lesions characterized by granulomas rich in collagen and the generation of TNF-α due to stimulation. Carbon nanotube cytotoxicity has also been discovered in aquatic organisms like rainbow trout [151]. Rainbow trout were administered SWCNTs at 0.1 to 0.5 mg/L for 10 days in the study. SWCNTs were identified as a respiratory toxicant that caused cell cycle abnormalities and neurotoxicity in the study. Chromosomal aberrations and enhanced micronuclei intensity have also been observed in MWCNTs, as well as cyclooxygenase enzyme activation via downregulation of immune system functionality in organs like the spleen, promotion of allergic responses in mice, and alterations in the level of gene expression in the liver [152–154]. Phenotypic abnormalities, apoptosis, bacterial toxicity, and the production of aberrant spinal cords in zebrafish embryos are among the other consequences of MWCNTs [155].

In vivo gold nanoparticle exposure caused bioaccumulation in organs, apoptosis, sperm head and tail region penetration, and acute inflammation in the liver [156]. The toxicity of 13-nm-sized gold nanoparticles coated with PEG in the liver was studied in a study. The particles aggregated in the spleen and liver after about 7 days following the administration. In the liver, the mentioned nanoparticles caused inflammation as well as apoptosis. The existence of gold nanoparticles coated with PEG in liver lysosomes and macrophages in the spleen was also discovered [157]. Some of the in vivo studies revealed that the AgNPs cause oxidative stress, blood-brain barrier damage, and astrocyte swelling caused by free radicals, neuronal degeneration, and gene expression alterations. Subcutaneous injections of AgNPs were used to study their dispersion and accumulation in rats. Particles were found in the brain, kidney, spleen, lung, and liver according to the studies [158].

13.4.2.3 Mechanistic Studies

To determine the toxicity of nanoparticles, mechanistic experiments were conducted in vivo. Silver nanoparticles have also been linked to cytotoxicity and genotoxicity in fish, including gill tissue deposition and lysosomal instability in adult oysters. They've also been linked to problems with oxidative stress, oyster embryonic development, and expression of the guardian gene of a cell, p53 in zebrafish [159–163]. The toxicology studies performed on zebrafish livers revealed that silver nanoparticles produced oxidative stress and apoptosis progressively. The p53-related pro-apoptotic genes p21, Bax, and Noxa, Bax, were upregulated after nanoparticle therapy [163]. The harmful effects of AuNPs on embryogenesis were studied using an oyster embryo. Metallothionein genetic expression was shown to be significantly higher in embryos [160]. They are recognized as oxidative stress, overexpression of p53 proteins, and heat shock stress in *D. melanogaster.* Oxidative stress and a reduction in reproductive capacity are recognized in *Caenorhabditis elegans* [162].

Another study that produced cadmium sulfate (CdS) quantum dots investigated the cytotoxicity of quantum dots. CdS quantum dots are much more harmful than microsized CdS, according to the study, and they increase ROS generation by 20–30%. According to the study [164], the cytotoxicity of CdS quantum dots is thought to be mediated by GSH depletion, cadmium ions (Cd^{2+}) release, and intracellular ROS generation. In *Daphnia magna*, quantum dots were essential for phototoxicity under UV-B radiation [165]. In vivo research has linked fullerenes to increased MHC-II class gene expression, elevated levels of cytokines related to inflammation, and enhanced distribution of the T cell population in the lungs [166].

Human bronchial epithelial cells, human hepatocytes, and cervical cancer cells, namely HEK 293, and HEp-2, have all been shown to be hazardous to zinc nanoparticles. On HEp-2 cells, zinc nanoparticles in the range of 10 to 100 μg/mL were utilized for 24 to 48 hours. The study's findings revealed the ZnO nanoparticles are harmful to healthy cells, showing a decrease in cell viability and DNA damage [167]. Apoptosis, differentiation, cell proliferation, and migration are all affected by titanium nanoparticle toxicity [168]. When HaCaT (keratinocyte cell line), SZ95 (human immortalized sebaceous gland cell lines), and human dermal fibroblasts were employed, titanium oxide nanoparticle toxicity was detected. Apoptosis was caused by cytotoxicity, which disrupted cellular processes such as differentiation, migration, and cell proliferation [169].

13.5 SAFETY MEASURES

In a real-time scenario, worldwide, around 20% of NPs that are synthesized either chemically or biologically are being rejected in clinical trials because of safety issues. The fate and toxicity of NPs must be assessed as part of the approval procedure for human consumption. This is normally performed through various pre-clinical and clinical stages with permission from international regulatory authorities like the EMA (European Medicines Agency) or FDA (Food and Drug Administration). One difficulty with this procedure is that these agencies' judgments of NP efficacy and safety may differ. This necessitates the establishment of safety assessment standards [170].

The major limitation of the generation of safer NPs from manufacturers is the continuity of cancer after the administration or dosage of NPs, which ultimately leads to failure in pre-clinical trials. Polyethylene glycol (PEG) surface coating is a commonly utilized technique to minimize NP toxicity in the clinic [171]. Even NPs that have been licensed for therapeutic use might cause harm. Doxil®, a liposomal formulation of doxorubicin, has shown success in treating severe human breast cancer abnormalities and Kaposi's sarcoma patients. It's unclear whether these side effects are caused by the medication doxorubicin, liposomal carriers, or both. While PEG is usually deemed non-toxic, it has raised certain concerns. Clinically, antibodies and an immunological reaction toward PEG is being detected, although more investigations are required [172]. Moreover, the data underscore the critical necessity for more research into the extensive effects and toxicity of NP therapies.

It is always impossible to attain safe nanoparticles in hand and too difficult to maintain the comprehensive safety profile of all designed or synthesized NPs. Further, it is impossible to predict the unanticipated toxicities that are lethal until the usage of designed or marketed nanoparticle products in inpatient therapy. In the field of medicine, the situation is the same for any sort of drug, irrespective of type. We can overcome this by practicing or motivating more collaborations among academia, researchers, and pharmaceutical industries, which may surely help to speed up progress and improve safety concerns related to dosage and administration of designed NPs.

13.6 FUTURE ASPECTS

As the effects of metallic NPs are so unpredictable, it's critical to handle them carefully. Overall studies on nanoparticle designs indicate toxicity, at least to an extent benign, irrespective of their magnitude and mechanism of toxicity. Furthermore, many chemicals used in the design or synthesis of NPs were found to be harmful. Overall, it is very necessary at this point to understand the properties of NPs at the level of pharmacokinetics, and a detailed analysis of health concerns needs to be documented. Present nanotoxicity research has centered on developing libraries and databases as well as computational ways of forecasting the toxicity levels associated with metallic nano formulations. This integrated data documentation could pave the way for more research on in vitro and in vivo NP testing. The processes of NP transmission, accumulation, extensive safety/toxicity, when in contact with the cells, their receptors, respective signaling pathways, and ultimately phagocytosis should all be studied. Keen observation and analysis of the relationships between new building materials in the form of newly designed NPs and biological systems might help preserve these objects in medical sectors like diagnostics and therapy.

ACKNOWLEDGMENTS

The authors are thankful to Dr. Jamilur R. Ansari, Dronacharya College of Engineering, Gurgaon, India, for the needful editing in the Chapter. We also thank Dr. M. Sabir Alam (SGT University), and Dr. M. Noushad (University of Texas Rio-Grande Valley, Edinburg, Texas TX 78539, USA), for giving us the opportunity and assistance to write this chapter.

REFERENCES

1. Rosarin FS, Mirunalini S. Nobel metallic nanoparticles with novel biomedical properties. Journal of Bioanalysis & Biomedicine. 2011;3(4):85–91.
2. Zamborini FP, Bao L, Dasari R. Nanoparticles in measurement science. Analytical Chemistry. 2012 Jan 17;84(2):541–76.
3. Mishra S, Sharma S, Javed MN, Pottoo FH, Barkat MA, Alam MS, Amir M, Sarafroz M. Bioinspired nanocomposites: Applications in disease diagnosis and treatment. Pharmaceutical Nanotechnology. 2019 Jun 1;7(3):206–19.
4. Kumari A, Singla R, Guliani A, Yadav SK. Nanoencapsulation for Drug Delivery. EXCLI Journal. 2014;13:265.
5. Tripathy A, Raichur AM, Chandrasekaran N, Prathna TC, Mukherjee A. Process variables in biomimetic synthesis of silver nanoparticles by aqueous extract of Azadirachta indica (Neem) leaves. Journal of Nanoparticle Research. 2010 Jan;12(1):237–46.
6. Pandit J, Alam MS, Ansari JR, Singhal M, Gupta N, Waziri A, Sharma K, Potto FH. Multifaced applications of nanoparticles in biological science. In Nanomaterials in the Battle Against Pathogens and Disease Vectors 2022 (pp. 17–50). CRC Press, London.
7. Alam MS, Naseh MF, Ansari JR, Waziri A, Javed MN, Ahmadi A, Saifullah MK, Garg A. Synthesis approaches for higher yields of nanoparticles. In Nanomaterials in the Battle Against Pathogens and Disease Vectors 2022 (pp. 51–82). CRC Press, London.
8. Khodashenas B, Ghorbani HR. Synthesis of silver nanoparticles with different shapes. Arabian Journal of Chemistry. 2019 Dec 1;12(8):1823–38.
9. Iravani S, Korbekandi H, Mirmohammadi SV, Zolfaghari B. Synthesis of silver nanoparticles: Chemical, physical and biological methods. Research in Pharmaceutical Sciences. 2014 Nov;9(6):385.
10. Korbekandi H, Ashari Z, Iravani S, Abbasi S. Optimization of biological synthesis of silver nanoparticles using Fusarium oxysporum. Iranian Journal of Pharmaceutical Research: IJPR. 2013;12(3):289.
11. Dang TM, Le TT, Fribourg-Blanc E, Dang MC. Synthesis and optical properties of copper nanoparticles prepared by a chemical reduction method. Advances in Natural Sciences: Nanoscience and Nanotechnology. 2011 Mar 7;2(1):015009.
12. Hyeon T. Chemical synthesis of magnetic nanoparticles. Chem Comm. 2003;8:927–34.
13. Makarov VV, Love AJ, Sinitsyna OV, Makarova SS, Yaminsky IV, Taliansky ME, Kalinina NO. "Green" nanotechnologies: Synthesis of metal nanoparticles using plants. Acta Naturae (англоязычная версия). 2014;6(1 (20)):35–44.
14. Mody VV, Siwale R, Singh A, Mody HR. Introduction to metallic nanoparticles. Journal of Pharmacy and Bioallied Sciences. 2010 Oct;2(4):282.
15. Thomas R, Park IK, Jeong YY. Magnetic iron oxide nanoparticles for multimodal imaging and therapy of cancer. International Journal of Molecular Sciences. 2013 Jul 31;14(8):15910–30.
16. Dreaden EC, Alkilany AM, Huang X, Murphy CJ, El-Sayed MA. The golden age: Gold nanoparticles for biomedicine. Chemical Society Reviews. 2012;41(7):2740–79.

17. Javed MN, Pottoo FH, Alam MS. Metallic nanoparticle alone and/or in combination as novel agent for the treatment of uncontrolled electric conductance related disorders and/or seizure, epilepsy & convulsions. Patent Acquired on October. 2016;10:40.
18. Salavati-Niasari M, Davar F, Mir N. Synthesis and characterization of metallic copper nanoparticles via thermal decomposition. Polyhedron. 2008 Nov 25;27(17):3514–8.
19. Tai CY, Tai CT, Chang MH, Liu HS. Synthesis of magnesium hydroxide and oxide nanoparticles using a spinning disk reactor. Industrial & Engineering Chemistry Research. 2007 Aug 15;46(17):5536–41.
20. Lewinski N, Colvin V, Drezek R. Cytotoxicity of nanoparticles. Small. 2008 Jan 18;4(1):26–49.
21. Kong B, Seog JH, Graham LM, Lee SB. Experimental considerations on the cytotoxicity of nanoparticles. Nanomedicine. 2011 Jul;6(5):929–41.
22. Umut E. Surface modification of nanoparticles used in biomedical applications. Modern Surface Engineering Treatments. 2013 May 22;20:185–208.
23. Srinivasan SY, Paknikar KM, Bodas D, Gajbhiye V. Applications of cobalt ferrite nanoparticles in biomedical nanotechnology. Nanomedicine. 2018 May;13(10):1221–38.
24. Sharifi I, Shokrollahi H, Amiri S. Ferrite-based magnetic nanofluids used in hyperthermia applications. Journal of Magnetism and Magnetic Materials. 2012 Mar 1;324(6):903–15.
25. Xu C, Sun S. New forms of superparamagnetic nanoparticles for biomedical applications. Advanced Drug Delivery Reviews. 2013 May 1;65(5):732–43.
26. Torres-Díaz I, Rinaldi C. Recent progress in ferrofluids research: Novel applications of magnetically controllable and tunable fluids. Soft Matter. 2014;10(43):8584–602.
27. Wu W, Wu Z, Yu T, Jiang C, Kim WS. Recent progress on magnetic iron oxide nanoparticles: Synthesis, surface functional strategies and biomedical applications. Science and Technology of Advanced Materials. 2015 Apr 28;16(2):023501.
28. Weissleder R, Nahrendorf M, Pittet MJ. Imaging macrophages with nanoparticles. Nature Materials. 2014 Feb;13(2):125–38.
29. Wu L, Mendoza-Garcia A, Li Q, Sun S. Organic phase syntheses of magnetic nanoparticles and their applications. Chemical Reviews. 2016 Sep 28;116(18):10473–512.
30. Kozissnik B, Dobson J. Biomedical applications of mesoscale magnetic particles. Mrs Bulletin. 2013 Nov;38(11):927–32.
31. Ling D, Lee N, Hyeon T. Chemical synthesis and assembly of uniformly sized iron oxide nanoparticles for medical applications. Accounts of Chemical Research. 2015 May 19;48(5):1276–85.
32. Javed MN, Dahiya ES, Ibrahim AM, Alam M, Khan FA, Pottoo FH. Recent advancement in clinical application of nanotechnological approached targeted delivery of herbal drugs. In Nanophytomedicine 2020 (pp. 151–72). Springer, Singapore.
33. Javed MN, Pottoo FH, Shamim A, Hasnain MS, Alam MS. Design of experiments for the development of nanoparticles, nanomaterials, and nanocomposites. In Design of Experiments for Pharmaceutical Product Development 2021 (pp. 151–69). Springer, Singapore.
34. Javed MN, Pottoo FH, Alam MS. Metallic nanoparticle alone and/or in combination as novel agent for the treatment of uncontrolled electric conductance related disorders and/or seizure, epilepsy & convulsions. Patent Acquired on October. 2016;10:40.
35. Javed MN, Alam MS, Waziri A, Pottoo FH, Yadav AK, Hasnain MS, Almalki FA. QbD applications for the development of nanopharmaceutical products. In Pharmaceutical Quality by Design 2019 Jan 1 (pp. 229–53). Academic Press. https://doi.org/10.1016/B978-0-12-815799-2.00013-7.
36. Kumar R, Dhamija G, Ansari JR, Javed MN, Alam MS. C-Dot nanoparticulated devices for biomedical applications. In Nanotechnology 2022 (pp. 271–99). CRC Press, London.
37. Bharti C, Alam MS, Javed MN, Khalid M, Saifullah FA, Manchanda R. Silica based nanomaterial for drug delivery. Nanomaterials: Evolution and Advancement Towards Therapeutic Drug Delivery (Part II). 2021 Jun 2:57.

38. Sangeet Kumar Mall SK, Yadav T, Waziri A, Alam MS. Treatment opportunities with *Fernandoa adenophylla* and recent novel approaches for natural medicinal phytochemicals as a drug delivery system. Exploration of Medicine. 2022;3:516–39.
39. Naseh MF, Ansari JR, Alam MS, Javed MN. Sustainable nanotorus for biosensing and therapeutical applications. In Handbook of Green and Sustainable Nanotechnology: Fundamentals, Developments and Applications 2022 Aug 19 (pp. 1–21). Springer International Publishing, Cham.
40. Sunilbhai CA, Alam M, Sadasivuni KK, Ansari JR. SPR Assisted diabetes detection. In Advanced Bioscience and Biosystems for Detection and Management of Diabetes 2022 (pp. 91–131). Springer, Cham.
41. Singhal S, Gupta M, Alam MS, Javed MN, Ansari JR. Carbon allotropes-based nanodevices: Graphene in biomedical applications. In Nanotechnology 2022 (pp. 241–69). CRC Press, London.
42. Alam MS, Garg A, Pottoo FH, Saifullah MK, Tareq AI, Manzoor O, Mohsin M, Javed MN. Gum ghatti mediated, one pot green synthesis of optimized gold nanoparticles: Investigation of process-variables impact using Box-Behnken based statistical design. International Journal of Biological Macromolecules. 2017 Nov 1;104:758–67.
43. Alam MS, Javed MN, Pottoo FH, Waziri A, Almalki FA, Hasnain MS, Garg A, Saifullah MK. QbD approached comparison of reaction mechanism in microwave synthesized gold nanoparticles and their superior catalytic role against hazardous Nirto-dye. Applied Organometallic Chemistry. 2019 Sep;33(9):e5071.
44. Pottoo FH, Tabassum N, Javed M, Nigar S, Rasheed R, Khan A, Barkat M, Alam M, Maqbool A, Ansari MA, Barreto GE. The synergistic effect of raloxifene, fluoxetine, and bromocriptine protects against pilocarpine-induced status epilepticus and temporal lobe epilepsy. Molecular Neurobiology. 2019 Feb;56(2):1233–47.
45. Pottoo FH, Sharma S, Javed MN, Barkat MA, Harshita, Alam MS, Naim MJ, Alam O, Ansari MA, Barreto GE, Ashraf GM. Lipid-based nanoformulations in the treatment of neurological disorders. Drug Metabolism Reviews. 2020 Jan 2;52(1):185–204.
46. Pottoo FH, Javed M, Barkat M, Alam M, Nowshehri JA, Alshayban DM, Ansari MA. Estrogen and serotonin: Complexity of interactions and implications for epileptic seizures and epileptogenesis. Current Neuropharmacology. 2019 Mar 1;17(3):214–31.
47. Pottoo FH, Tabassum N, Javed MN, Nigar S, Sharma S, Barkat MA, Alam MS, Ansari MA, Barreto GE, Ashraf GM. Raloxifene potentiates the effect of fluoxetine against maximal electroshock induced seizures in mice. European Journal of Pharmaceutical Sciences. 2020 Apr 15;146:105261.
48. Aslam M, Javed MN, Deeb HH, Nicola MK, Mirza M, Alam MS, Akhtar MH, Waziri A. Lipid nanocarriers for neurotherapeutics: Introduction, challenges, blood-brain barrier, and promises of delivery approaches. CNS & Neurological Disorders-Drug Targets (Formerly Current Drug Targets-CNS & Neurological Disorders). 2022 Dec 1;21(10):952–965.
49. Waziri A, Bharti C, Aslam M, Jamil P, Mirza M, Javed MN, Pottoo U, Ahmadi A, Alam MS. Probiotics for the chemoprotective role against the toxic effect of cancer chemotherapy. Anti-Cancer Agents in Medicinal Chemistry (Formerly Current Medicinal Chemistry-Anti-Cancer Agents). 2022 Feb 1;22(4):654–67.
50. Javed MN, Akhter MH, Taleuzzaman M, Faiyazudin M, Alam MS. Cationic nanoparticles for treatment of neurological diseases. In Fundamentals of Bionanomaterials 2022 Jan 1 (pp. 273–92). Elsevier, United States.
51. Kumari N, Daram N, Alam MS, Verma AK. Rationalizing the use of polyphenol nano-formulations in the therapy of neurodegenerative diseases. CNS & Neurological Disorders-Drug Targets (Formerly Current Drug Targets-CNS & Neurological Disorders). 2022 Dec 1;21(10):966–76.

52. Raj S, Manchanda R, Bhandari M, Alam MS. Review on natural bioactive products as radioprotective therapeutics: Present and past perspective. Current Pharmaceutical Biotechnology. 2022;23(14), 1721–38.
53. Ibrahim AM, Chauhan L, Bhardwaj A, Sharma A, Fayaz F, Kumar B, Alhashmi M, AlHajri N, Alam MS, Pottoo FH. Brain-derived neurotropic factor in neurodegenerative disorders. Biomedicines. 2022 May;10(5):1143.
54. Chaloupka K, Malam Y, Seifalian AM. Nanosilver as a new generation of nanoproduct in biomedical applications. Trends in Biotechnology. 2010 Nov 1;28(11):580–8.
55. Alanazi FK, Radwan AA, Alsarra IA. Biopharmaceutical applications of nanogold. Saudi Pharmaceutical Journal. 2010 Oct 1;18(4):179–93.
56. Patra CR, Bhattacharya R, Mukhopadhyay D, Mukherjee P. Fabrication of gold nanoparticles for targeted therapy in pancreatic cancer. Advanced Drug Delivery Reviews. 2010 Mar 8;62(3):346–61.
57. Pankhurst QA, Thanh NT, Jones SK, Dobson J. Progress in applications of magnetic nanoparticles in biomedicine. Journal of Physics D: Applied Physics. 2009 Nov 6;42(22):224001.
58. Rai M, Yadav A, Gade A. Silver nanoparticles as a new generation of antimicrobials. Biotechnology Advances. 2009 Jan 1;27(1):76–83.
59. Marambio-Jones C, Hoek E. A review of the antibacterial effects of silver nanomaterials and potential implications for human health and the environment. Journal of Nanoparticle Research. 2010 Jun;12(5):1531–51.
60. Gajbhiye M, Kesharwani J, Ingle A, Gade A, Rai M. Fungus-mediated synthesis of silver nanoparticles and their activity against pathogenic fungi in combination with fluconazole. Nanomedicine: Nanotechnology, Biology and Medicine. 2009 Dec 1;5(4):382–6.
61. Santhoshkumar T, Rahuman AA, Rajakumar G, Marimuthu S, Bagavan A, Jayaseelan C, Zahir AA, Elango G, Kamaraj C. Synthesis of silver nanoparticles using Nelumbo nucifera leaf extract and its larvicidal activity against malaria and filariasis vectors. Parasitology Research. 2011 Mar;108(3):693–702.
62. Valodkar M, Jadeja RN, Thounaojam MC, Devkar RV, Thakore S. Biocompatible synthesis of peptide capped copper nanoparticles and their biological effect on tumor cells. Materials Chemistry and Physics. 2011 Jul 15;128(1–2):83–9.
63. Moulton MC, Braydich-Stolle LK, Nadagouda MN, Kunzelman S, Hussain SM, Varma RS. Synthesis, characterization and biocompatibility of "green" synthesized silver nanoparticles using tea polyphenols. Nanoscale. 2010;2(5):763–70.
64. Patra CR, Bhattacharya R, Mukhopadhyay D, Mukherjee P. Fabrication of gold nanoparticles for targeted therapy in pancreatic cancer. Advanced Drug Delivery Reviews. 2010 Mar 8;62(3):346–61.
65. Amarnath K, Mathew NL, Nellore J, Siddarth CR, Kumar J. Facile synthesis of biocompatible gold nanoparticles from Vites vinefera and its cellular internalization against HBL-100 cells. Cancer Nanotechnology. 2011 Dec;2(1):121–32.
66. Venkatpurwar V, Shiras A, Pokharkar V. Porphyran capped gold nanoparticles as a novel carrier for delivery of anticancer drug: In vitro cytotoxicity study. International Journal of Pharmaceutics. 2011 May 16;409(1–2):314–20.
67. Zheng B, Qian L, Yuan H, Xiao D, Yang X, Paau MC, Choi MM. Preparation of gold nanoparticles on eggshell membrane and their biosensing application. Talanta. 2010 Jun 30;82(1):177–83.
68. Iskandar F. Nanoparticle processing for optical applications–A review. Advanced Powder Technology. 2009 Jul 1;20(4):283–92.
69. Talapin DV, Lee JS, Kovalenko MV, Shevchenko EV. Prospects of colloidal nanocrystals for electronic and optoelectronic applications. Chemical Reviews. 2010 Jan 13;110(1):389–458.

70. Sicard C, Brayner R, Margueritat J, Hémadi M, Couté A, Yéprémian C, Djediat C, Aubard J, Fiévet F, Livage J, Coradin T. Nano-gold biosynthesis by silica-encapsulated micro-algae: A "living" bio-hybrid material. Journal of Materials Chemistry. 2010;20(42):9342–7.
71. Hennebel T, De Gusseme B, Boon N, Verstraete W. Biogenic metals in advanced water treatment. Trends in Biotechnology. 2009 Feb 1;27(2):90–8.
72. Das N. Recovery of precious metals through biosorption—a review. Hydrometallurgy. 2010 Jun 1;103(1–4):180–9.
73. Hennebel T, De Corte S, Verstraete W, Boon N. Microbial production and environmental applications of Pd nanoparticles for treatment of halogenated compounds. Current Opinion in Biotechnology. 2012 Aug 1;23(4):555–61.
74. De Corte S, Hennebel T, De Gusseme B, Verstraete W, Boon N. Bio-palladium: From metal recovery to catalytic applications. Microbial Biotechnology. 2012 Jan;5(1):5–17.
75. Sharma NC, Sahi SV, Nath S, Parsons JG, Gardea-Torresde JL, Pal T. Synthesis of plant-mediated gold nanoparticles and catalytic role of biomatrix-embedded nanomaterials. Environmental Science & Technology. 2007 Jul 15;41(14):5137–42.
76. Deplanche K, Caldelari I, Mikheenko IP, Sargent F, Macaskie LE. Involvement of hydrogenases in the formation of highly catalytic Pd (0) nanoparticles by bioreduction of Pd (II) using Escherichia coli mutant strains. Microbiology. 2010 Sep 1;156(9):2630–40.
77. Zhang M, Zhang K, De Gusseme B, Verstraete W. Biogenic silver nanoparticles (bio-Ag0) decrease biofouling of bio-Ag0/PES nanocomposite membranes. Water Research. 2012 May 1;46(7):2077–87.
78. Creamer NJ, Mikheenko IP, Yong P, Deplanche K, Sanyahumbi D, Wood J, Pollmann K, Merroun M, Selenska-Pobell S, Macaskie LE. Novel supported Pd hydrogenation bionanocatalyst for hybrid homogeneous/heterogeneous catalysis. Catalysis Today. 2007 Oct 15;128(1–2):80–7.
79. Jia L, Zhang Q, Li Q, Song H. The biosynthesis of palladium nanoparticles by antioxidants in Gardenia jasminoides Ellis: Long lifetime nanocatalysts for p-nitrotoluene hydrogenation. Nanotechnology. 2009 Aug 28;20(38):385601.
80. Bunge M, Søbjerg LS, Rotaru AE, Gauthier D, Lindhardt AT, Hause G, Finster K, Kingshott P, Skrydstrup T, Meyer RL. Formation of palladium (0) nanoparticles at microbial surfaces. Biotechnology and Bioengineering. 2010 Oct 1;107(2):206–15.
81. Pingarrón JM, Yañez-Sedeño P, González-Cortés A. Gold nanoparticle-based electrochemical biosensors. Electrochimica Acta. 2008 Aug 1;53(19):5848–66.
82. Wang H, Wick RL, Xing B. Toxicity of nanoparticulate and bulk ZnO, Al_2O_3 and TiO_2 to the nematode Caenorhabditis elegans. Environmental Pollution. 2009 Apr 1;157(4):1171–7.
83. Torres-Chavolla E, Ranasinghe RJ, Alocilja EC. Characterization and functionalization of biogenic gold nanoparticles for biosensing enhancement. IEEE Transactions on Nanotechnology. 2010 Jun 14;9(5):533–8.
84. Guo NL, Wan YW, Denvir J, Porter DW, Pacurari M, Wolfarth MG, Castranova V, Qian Y. Multiwalled carbon nanotube-induced gene signatures in the mouse lung: Potential predictive value for human lung cancer risk and prognosis. Journal of Toxicology and Environmental Health, Part A. 2012 Sep 15;75(18):1129–53.
85. Sathyanarayana S, Hübner C. Thermoplastic nanocomposites with carbon nanotubes. In Structural Nanocomposites 2013 (pp. 19–60). Springer, Berlin, Heidelberg.
86. Madani SY, Mandel A, Seifalian AM. A concise review of carbon nanotube's toxicology. Nano Reviews. 2013 Jan 1;4(1):21521.
87. Amarnath S, Hussain MA, Nanjundiah V, Sood AK. β-Galactosidase leakage from Escherichia coli points to mechanical damageas likely cause of carbon nanotube toxicity. Soft Nanoscience Letters. 2012 Jun 18;2(3):41–5.

88. Kumar V, Sharma N, Maitra SS. In vitro and in vivo toxicity assessment of nanoparticles. International Nano Letters. 2017 Dec;7(4):243–56.
89. Pasquini LM, Hashmi SM, Sommer TJ, Elimelech M, Zimmerman JB. Impact of surface functionalization on bacterial cytotoxicity of single-walled carbon nanotubes. Environmental Science & Technology. 2012 Jun 5;46(11):6297–305.
90. Kerfahi D, Tripathi BM, Singh D, Kim H, Lee S, Lee J, Adams JM. Effects of functionalized and raw multi-walled carbon nanotubes on soil bacterial community composition. PLoS One. 2015 Mar 31;10(3):e0123042.
91. Sohn EK, Chung YS, Johari SA, Kim TG, Kim JK, Lee JH, Lee YH, Kang SW, Yu IJ. Acute toxicity comparison of single-walled carbon nanotubes in various freshwater organisms. BioMed Research International. 2015 Jan 14;2015.
92. Naqvi S, Samim M, Abdin MZ, Ahmed FJ, Maitra AN, Prashant CK, Dinda AK. Concentration-dependent toxicity of iron oxide nanoparticles mediated by increased oxidative stress. International Journal of Nanomedicine. 2010;5:983.
93. Yehia HN, Draper RK, Mikoryak C, Walker EK, Bajaj P, Musselman IH, Daigrepont MC, Dieckmann GR, Pantano P. Single-walled carbon nanotube interactions with HeLa cells. Journal of Nanobiotechnology. 2007 Dec;5(1):1–7.
94. Fiorito S, Serafino A, Andreola F, Bernier P. Effects of fullerenes and single-wall carbon nanotubes on murine and human macrophages. Carbon. 2006 May 1;44(6):1100–5.
95. Davoren M, Herzog E, Casey A, Cottineau B, Chambers G, Byrne HJ, Lyng FM. In vitro toxicity evaluation of single walled carbon nanotubes on human A549 lung cells. Toxicology in Vitro. 2007 Apr 1;21(3):438–48.
96. Kisin ER, Murray AR, Keane MJ, Shi XC, Schwegler-Berry D, Gorelik O, Arepalli S, Castranova V, Wallace WE, Kagan VE, Shvedova AA. Single-walled carbon nanotubes: Geno-and cytotoxic effects in lung fibroblast V79 cells. Journal of Toxicology and Environmental Health, Part A. 2007 Nov 13;70(24):2071–9.
97. Choi SJ, Oh JM, Choy JH. Toxicological effects of inorganic nanoparticles on human lung cancer A549 cells. Journal of Inorganic Biochemistry. 2009 Mar 1;103(3):463–71.
98. Herzog E, Byrne HJ, Casey A, Davoren M, Lenz AG, Maier KL, Duschl A, Oostingh GJ. SWCNT suppress inflammatory mediator responses in human lung epithelium in vitro. Toxicology and Applied Pharmacology. 2009 Feb 1;234(3):378–90.
99. Monteiro-Riviere NA, Nemanich RJ, Inman AO, Wang YY, Riviere JE. Multi-walled carbon nanotube interactions with human epidermal keratinocytes. Toxicology Letters. 2005 Mar 15;155(3):377–84.
100. Ye SF, Wu YH, Hou ZQ, Zhang QQ. ROS and NF-κB are involved in upregulation of IL-8 in A549 cells exposed to multi-walled carbon nanotubes. Biochemical and Biophysical Research Communications. 2009 Feb 6;379(2):643–8.
101. Cveticanin J, Joksic G, Leskovac A, Petrovic S, Sobot AV, Neskovic O. Using carbon nanotubes to induce micronuclei and double strand breaks of the DNA in human cells. Nanotechnology. 2009 Nov 30;21(1):015102.
102. Patlolla A, Patlolla B, Tchounwou P. Evaluation of cell viability, DNA damage, and cell death in normal human dermal fibroblast cells induced by functionalized multiwalled carbon nanotube. Molecular and Cellular Biochemistry. 2010 May;338(1):225–32.
103. Ravichandran P, Periyakaruppan A, Sadanandan B, Ramesh V, Hall JC, Jejelowo O, Ramesh GT. Induction of apoptosis in rat lung epithelial cells by multiwalled carbon nanotubes. Journal of Biochemical and Molecular Toxicology. 2009 Sep;23(5):333–44.
104. Reddy AR, Reddy YN, Krishna DR, Himabindu V. Multi wall carbon nanotubes induce oxidative stress and cytotoxicity in human embryonic kidney (HEK293) cells. Toxicology. 2010 Jun 4;272(1–3):11–6.
105. Li JJ, Hartono D, Ong CN, Bay BH, Yung LY. Autophagy and oxidative stress associated with gold nanoparticles. Biomaterials. 2010 Aug 1;31(23):5996–6003.

106. Tarantola M, Schneider D, Sunnick E, Adam H, Pierrat S, Rosman C, Breus V, Sonnichsen C, Basche T, Wegener J, Janshoff A. Cytotoxicity of metal and semiconductor nanoparticles indicated by cellular micromotility. ACS Nano. 2009 Jan 27;3(1):213–22.
107. Miura N, Shinohara Y. Cytotoxic effect and apoptosis induction by silver nanoparticles in HeLa cells. Biochemical and Biophysical Research Communications. 2009 Dec 18;390(3):733–7.
108. Yang W, Shen C, Ji Q, An H, Wang J, Liu Q, Zhang Z. Food storage material silver nanoparticles interfere with DNA replication fidelity and bind with DNA. Nanotechnology. 2009 Feb 2;20(8):085102.
109. Cho SJ, Maysinger D, Jain M, Röder B, Hackbarth S, Winnik FM. Long-term exposure to CdTe quantum dots causes functional impairments in live cells. Langmuir. 2007 Feb 13;23(4):1974–80.
110. Chen L, Yokel RA, Hennig B, Toborek M. Manufactured aluminum oxide nanoparticles decrease expression of tight junction proteins in brain vasculature. Journal of Neuroimmune Pharmacology. 2008 Dec;3(4):286–95.
111. Kim YJ, Choi HS, Song MK, Youk DY, Kim JH, Ryu JC. Genotoxicity of aluminum oxide () nanoparticle in mammalian cell lines. Molecular & Cellular Toxicology. 2009;5(2):172–8.
112. Clark KA, O'Driscoll C, Cooke CA, Smith BA, Wepasnick K, Fairbrother DH, Lees PS, Bressler JP. Evaluation of the interactions between multiwalled carbon nanotubes and Caco-2 cells. Journal of Toxicology and Environmental Health, Part A. 2012 Jan 1;75(1):25–35.
113. Kim J, Park Y, Yoon TH, Yoon CS, Choi K. Phototoxicity of CdSe/ZnSe quantum dots with surface coatings of 3-mercaptopropionic acid or tri-n-octylphosphine oxide/gum arabic in Daphnia magna under environmentally relevant UV-B light. Aquatic Toxicology. 2010 Apr 15;97(2):116–24.
114. Pan Z, Lee W, Slutsky L, Clark RA, Pernodet N, Rafailovich MH. Adverse effects of titanium dioxide nanoparticles on human dermal fibroblasts and how to protect cells. Small. 2009 Feb 20;5(4):511–20.
115. Foldbjerg R, Dang DA, Autrup H. Cytotoxicity and genotoxicity of silver nanoparticles in the human lung cancer cell line, A549. Archives of Toxicology. 2011 Jul;85(7):743–50.
116. Zhang LW, Yang J, Barron AR, Monteiro-Riviere NA. Endocytic mechanisms and toxicity of a functionalized fullerene in human cells. Toxicology Letters. 2009 Dec 15;191(2–3):149–57.
117. Asare N, Instanes C, Sandberg WJ, Refsnes M, Schwarze P, Kruszewski M, Brunborg G. Cytotoxic and genotoxic effects of silver nanoparticles in testicular cells. Toxicology. 2012 Jan 27;291(1–3):65–72.
118. Nagarajan M, Maadurshni GB, Tharani GK, Udhayakumar I, Kumar G, Mani KP, Sivasubramanian J, Manivannan J. Exposure to zinc oxide nanoparticles (ZnO-NPs) induces cardiovascular toxicity and exacerbates pathogenesis–Role of oxidative stress and MAPK signaling. Chemico-Biological Interactions. 2022 Jan 5;351:109719.
119. Coradeghini R, Gioria S, García CP, Nativo P, Franchini F, Gilliland D, Ponti J, Rossi F. Size-dependent toxicity and cell interaction mechanisms of gold nanoparticles on mouse fibroblasts. Toxicology Letters. 2013 Mar 13;217(3):205–16.
120. Soenen SJ, Manshian B, Montenegro JM, Amin F, Meermann B, Thiron T, Cornelissen M, Vanhaecke F, Doak S, Parak WJ, De Smedt S. Cytotoxic effects of gold nanoparticles: A multiparametric study. ACS Nano. 2012 Jul 24;6(7):5767–83.
121. Greulich C, Braun D, Peetsch A, Diendorf J, Siebers B, Epple M, Köller M. The toxic effect of silver ions and silver nanoparticles towards bacteria and human cells occurs in the same concentration range. RSC Advances. 2012;2(17):6981–7.

122. Jing X, Park JH, Peters TM, Thorne PS. Toxicity of copper oxide nanoparticles in lung epithelial cells exposed at the air–liquid interface compared with in vivo assessment. Toxicology in Vitro. 2015 Apr 1;29(3):502–11.
123. Heng BC, Zhao X, Xiong S, Ng KW, Boey FY, Loo JS. Toxicity of zinc oxide (ZnO) nanoparticles on human bronchial epithelial cells (BEAS-2B) is accentuated by oxidative stress. Food and Chemical Toxicology. 2010 Jun 1;48(6):1762–6.
124. Chen J, Dong X, Zhao J, Tang G. In vivo acute toxicity of titanium dioxide nanoparticles to mice after intraperitioneal injection. Journal of Applied Toxicology. 2009 May;29(4):330–7.
125. Sun J, Wang S, Zhao D, Hun FH, Weng L, Liu H. Cytotoxicity, permeability, and inflammation of metal oxide nanoparticles in human cardiac microvascular endothelial cells. Cell Biology and Toxicology. 2011 Oct;27(5):333–42.
126. Wang H, Wick RL, Xing B. Toxicity of nanoparticulate and bulk ZnO, Al_2O_3 and TiO_2 to the nematode Caenorhabditis elegans. Environmental Pollution. 2009 Apr 1;157(4):1171–7.
127. Guan R, Kang T, Lu F, Zhang Z, Shen H, Liu M. Cytotoxicity, oxidative stress, and genotoxicity in human hepatocyte and embryonic kidney cells exposed to ZnO nanoparticles. Nanoscale Research Letters. 2012 Dec;7(1):1–7.
128. Scherer MD, Sposito JC, Falco WF, Grisolia AB, Andrade LH, Lima SM, Machado G, Nascimento VA, Gonçalves DA, Wender H, Oliveira SL. Cytotoxic and genotoxic effects of silver nanoparticles on meristematic cells of Allium cepa roots: A close analysis of particle size dependence. Science of the Total Environment. 2019 Apr 10;660:459–67.
129. Senapati VA, Kumar A. ZnO nanoparticles dissolution, penetration and toxicity in human epidermal cells. Influence of pH. Environmental Chemistry Letters. 2018 Sep;16(3):1129–35.
130. Huang KL, Chang HL, Tsai FM, Lee YH, Wang CH, Cheng TJ. The effect of the inhalation of and topical exposure to zinc oxide nanoparticles on airway inflammation in mice. Toxicology and Applied Pharmacology. 2019 Dec 1;384:114787.
131. Kim CS, Nguyen HD, Ignacio RM, Kim JH, Cho HC, Maeng EH, Kim YR, Kim MK, Park BK, Kim SK. Immunotoxicity of zinc oxide nanoparticles with different size and electrostatic charge. International Journal of Nanomedicine. 2014;9(Suppl 2):195.
132. Sadeghi L, Tanwir F, Babadi VY. In vitro toxicity of iron oxide nanoparticle: Oxidative damages on Hep G2 cells. Experimental and Toxicologic Pathology. 2015 Feb 1;67(2):197–203.
133. Li Y, Li F, Zhang L, Zhang C, Peng H, Lan F, Peng S, Liu C, Guo J. Zinc oxide nanoparticles induce mitochondrial biogenesis impairment and cardiac dysfunction in human iPSC-derived cardiomyocytes. International Journal of Nanomedicine. 2020;15:2669.
134. Chowdhury SM, Lalwani G, Zhang K, Yang JY, Neville K, Sitharaman B. Cell specific cytotoxicity and uptake of graphene nanoribbons. Biomaterials. 2013 Jan 1; 34(1):283–93.
135. Rizk MZ, Ali SA, Hamed MA, El-Rigal NS, Aly HF, Salah HH. Toxicity of titanium dioxide nanoparticles: Effect of dose and time on biochemical disturbance, oxidative stress and genotoxicity in mice. Biomedicine & Pharmacotherapy. 2017 Jun 1;90:466–72.
136. Steckiewicz KP, Barcinska E, Malankowska A, Zauszkiewicz–Pawlak A, Nowaczyk G, Zaleska-Medynska A, Inkielewicz-Stepniak I. Impact of gold nanoparticles shape on their cytotoxicity against human osteoblast and osteosarcoma in in vitro model. Evaluation of the safety of use and anti-cancer potential. Journal of Materials Science: Materials in Medicine. 2019 Feb;30(2):1–5.
137. Wang W, Zeng C, Feng Y, Zhou F, Liao F, Liu Y, Feng S, Wang X. The size-dependent effects of silica nanoparticles on endothelial cell apoptosis through activating the p53-caspase pathway. Environmental Pollution. 2018 Feb 1;233:218–25.

138. Poier N, Hochstöger J, Hackenberg S, Scherzad A, Bregenzer M, Schopper D, Kleinsasser N. Effects of zinc oxide nanoparticles in HUVEC: Cyto-and genotoxicity and functional impairment after long-term and repetitive exposure in vitro. International Journal of Nanomedicine. 2020;15:4441.
139. Chuang HC, Chuang KJ, Chen JK, Hua HE, Shen YL, Liao WN, Lee CH, Pan CH, Chen KY, Lee KY, Hsiao TC. Pulmonary pathobiology induced by zinc oxide nanoparticles in mice: A 24-hour and 28-day follow-up study. Toxicology and Applied Pharmacology. 2017 Jul 15;327:13–22.
140. Pantarotto D, Briand JP, Prato M, Bianco A. Translocation of bioactive peptides across cell membranes by carbon nanotubes. Chemical Communications. 2004(1):16–7.
141. Shvedova A, Castranova V, Kisin E, Schwegler-Berry D, Murray A, Gandelsman V, Maynard A, Baron P. Exposure to carbon nanotube material: Assessment of nanotube cytotoxicity using human keratinocyte cells. Journal of Toxicology and Environmental Health Part A. 2003 Jun 1;66(20):1909–26.
142. Tkachenko AG, Xie H, Liu Y, Coleman D, Ryan J, Glomm WR, Shipton MK, Franzen S, Feldheim DL. Cellular trajectories of peptide-modified gold particle complexes: Comparison of nuclear localization signals and peptide transduction domains. Bioconjugate Chemistry. 2004 May 19;15(3):482–90.
143. Gupta AK, Wells S. Surface-modified superparamagnetic nanoparticles for drug delivery: Preparation, characterization, and cytotoxicity studies. IEEE Transactions on Nanobioscience. 2004 Mar 15;3(1):66–73.
144. Hu F, Neoh KG, Cen L, Kang ET. Cellular response to magnetic nanoparticles "PEGylated" via surface-initiated atom transfer radical polymerization. Biomacromolecules. 2006 Mar 13;7(3):809–16.
145. Chen F, Gerion D. Fluorescent CdSe/ZnS nanocrystal– peptide conjugates for long-term, nontoxic imaging and nuclear targeting in living cells. Nano Letters. 2004 Oct 13;4(10):1827–32.
146. Lovrić J, Cho SJ, Winnik FM, Maysinger D. Unmodified cadmium telluride quantum dots induce reactive oxygen species formation leading to multiple organelle damage and cell death. Chemistry & Biology. 2005 Nov 1;12(11):1227–34.
147. Madani SY, Mandel A, Seifalian AM. A concise review of carbon nanotube's toxicology. Nano Reviews. 2013 Jan 1;4(1):21521.
148. Lam CW, James JT, McCluskey R, Arepalli S, Hunter RL. A review of carbon nanotube toxicity and assessment of potential occupational and environmental health risks. Critical Reviews in Toxicology. 2006 Jan 1;36(3):189–217.
149. Warheit DB, Laurence BR, Reed KL, Roach DH, Reynolds GA, Webb TR. Comparative pulmonary toxicity assessment of single-wall carbon nanotubes in rats. Toxicological Sciences. 2004 Jan 1;77(1):117–25.
150. Shvedova AA, Kisin ER, Mercer R, Murray AR, Johnson VJ, Potapovich AI, Tyurina YY, Gorelik O, Arepalli S, Schwegler-Berry D, Hubbs AF. Unusual inflammatory and fibrogenic pulmonary responses to single-walled carbon nanotubes in mice. American Journal of Physiology-Lung Cellular and Molecular Physiology. 2005 Nov;289(5):L698–708.
151. Smith CJ, Shaw BJ, Handy RD. Toxicity of single walled carbon nanotubes to rainbow trout,(Oncorhynchus mykiss): Respiratory toxicity, organ pathologies, and other physiological effects. Aquatic Toxicology. 2007 May 1;82(2):94–109.
152. Mitchell LA, Lauer FT, Burchiel SW, McDonald JD. Mechanisms for how inhaled multiwalled carbon nanotubes suppress systemic immune function in mice. Nature Nanotechnology. 2009 Jul;4(7):451–6.
153. Nygaard UC, Hansen JS, Samuelsen M, Alberg T, Marioara CD, Løvik M. Single-walled and multi-walled carbon nanotubes promote allergic immune responses in mice. Toxicological Sciences. 2009 May 1;109(1):113–23.
154. Park EJ, Cho WS, Jeong J, Yi J, Choi K, Park K. Pro-inflammatory and potential allergic responses resulting from B cell activation in mice treated with multi-walled carbon nanotubes by intratracheal instillation. Toxicology. 2009 May 17;259(3):113–21.

155. Cheng J, Chan CM, Veca LM, Poon WL, Chan PK, Qu L, Sun YP, Cheng SH. Acute and long-term effects after single loading of functionalized multi-walled carbon nanotubes into zebrafish (Danio rerio). Toxicology and Applied Pharmacology. 2009 Mar 1;235(2):216–25.
156. Lasagna-Reeves C, Gonzalez-Romero D, Barria MA, Olmedo I, Clos A, Ramanujam VS, Urayama A, Vergara L, Kogan MJ, Soto C. Bioaccumulation and toxicity of gold nanoparticles after repeated administration in mice. Biochemical and Biophysical Research Communications. 2010 Mar 19;393(4):649–55.
157. Cho WS, Cho M, Jeong J, Choi M, Cho HY, Han BS, Kim SH, Kim HO, Lim YT, Chung BH, Jeong J. Acute toxicity and pharmacokinetics of 13 nm-sized PEG-coated gold nanoparticles. Toxicology and Applied Pharmacology. 2009 Apr 1;236(1):16–24.
158. Sharma HS, Hussain S, Schlager J, Ali SF, Sharma A. Influence of nanoparticles on blood–brain barrier permeability and brain edema formation in rats. In Brain Edema XIV 2010 (pp. 359–64). Springer, Vienna.
159. Scown TM, Santos EM, Johnston BD, Gaiser B, Baalousha M, Mitov S, Lead JR, Stone V, Fernandes TF, Jepson M, van Aerle R. Effects of aqueous exposure to silver nanoparticles of different sizes in rainbow trout. Toxicological Sciences. 2010 Jun 1;115(2):521–34.
160. Choi JE, Kim S, Ahn JH, Youn P, Kang JS, Park K, Yi J, Ryu DY. Induction of oxidative stress and apoptosis by silver nanoparticles in the liver of adult zebrafish. Aquatic Toxicology. 2010 Oct 15;100(2):151–9.
161. Ringwood AH, McCarthy M, Bates TC, Carroll DL. The effects of silver nanoparticles on oyster embryos. Marine Environmental Research. 2010 Jan 1;69:S49–51.
162. Wise Sr JP, Goodale BC, Wise SS, Craig GA, Pongan AF, Walter RB, Thompson WD, Ng AK, Aboueissa AM, Mitani H, Spalding MJ. Silver nanospheres are cytotoxic and genotoxic to fish cells. Aquatic Toxicology. 2010 Apr 1;97(1):34–41.
163. Choi SJ, Oh JM, Choy JH. Toxicological effects of inorganic nanoparticles on human lung cancer A549 cells. Journal of Inorganic Biochemistry. 2009 Mar 1;103(3):463–71.
164. Li KG, Chen JT, Bai SS, Wen X, Song SY, Yu Q, Li J, Wang YQ. Intracellular oxidative stress and cadmium ions release induce cytotoxicity of unmodified cadmium sulfide quantum dots. Toxicology in Vitro. 2009 Sep 1;23(6):1007–13.
165. Kim JS, Song KS, Joo HJ, Lee JH, Yu IJ. Determination of cytotoxicity attributed to multiwall carbon nanotubes (MWCNT) in normal human embryonic lung cell (WI-38) line. Journal of Toxicology and Environmental Health, Part A. 2010 Nov 1;73(21–22):1521–9.
166. Folkmann JK, Risom L, Jacobsen NR, Wallin H, Loft S, Møller P. Oxidatively damaged DNA in rats exposed by oral gavage to C60 fullerenes and single-walled carbon nanotubes. Environmental Health Perspectives. 2009 May;117(5):703–8.
167. Osman IF, Baumgartner A, Cemeli E, Fletcher JN, Anderson D. Genotoxicity and cytotoxicity of zinc oxide and titanium dioxide in HEp-2 cells. Nanomedicine. 2010 Oct;5(8):1193–203.
168. Pan Y, Leifert A, Ruau D, Neuss S, Bornemann J, Schmid G, Brandau W, Simon U, Jahnen-Dechent W. Gold nanoparticles of diameter 1.4 nm trigger necrosis by oxidative stress and mitochondrial damage. Small. 2009 Sep 18;5(18):2067–76.
169. Kiss B, Bíró T, Czifra G, Tóth BI, Kertész Z, Szikszai Z, Kiss ÁZ, Juhász I, Zouboulis CC, Hunyadi J. Investigation of micronized titanium dioxide penetration in human skin xenografts and its effect on cellular functions of human skin-derived cells. Experimental Dermatology. 2008 Aug;17(8):659–67.
170. Schütz CA, Juillerat-Jeanneret L, Mueller H, Lynch I, Riediker M. Therapeutic nanoparticles in clinics and under clinical evaluation. Nanomedicine. 2013 Mar;8(3):449–67.
171. Anselmo AC, Mitragotri S. Nanoparticles in the clinic: An update. Bioengineering & Translational Medicine. 2019 Sep;4(3):e10143.
172. Kozma GT, Shimizu T, Ishida T, Szebeni J. Anti-PEG antibodies: Properties, formation, testing and role in adverse immune reactions to PEGylated nano-biopharmaceuticals. Advanced Drug Delivery Reviews. 2020 Jan 1;154:163–75.

Index

Note: Page numbers in *italics* indicate a figure and page numbers in **bold** indicate a table on the corresponding page.

D

E

F

Q

R

S

T

U

V

W

X

Z

For Product Safety Concerns and Information please contact our EU representative GPSR@taylorandfrancis.com
Taylor & Francis Verlag GmbH, Kaufingerstraße 24, 80331 München, Germany

www.ingramcontent.com/pod-product-compliance
Lightning Source LLC
Chambersburg PA
CBHW060759220326
41598CB00022B/2486

* 9 7 8 1 0 3 2 3 2 9 1 3 0 *